Math Study Skills

Your overall success in mastering the material in your course depends on you. You must be ***committed*** to doing your best in this course. This commitment means dedicating the time needed to study math and to do your homework.

In order to succeed in math, you must know how to study it. The goal is to study math so that you understand and not just memorize it. The following tips and strategies will help you develop good study habits.

General Tips

ATTEND EVERY CLASS Be on time. If you must miss class, be sure to talk with your instructor or a classmate about what was covered.

MANAGE YOUR TIME School, work, family, and other commitments place a lot of demand on your time. To be successful, you must be able to devote time to study math every day. Writing out a weekly schedule that lists your class schedule, work schedule, and all other commitments with times that are not flexible will help you to determine when you can study. Use the companion resources that accompany the book, such as MyMathLab, so you can study online with tutoring help whenever you have the time.

Instructor Contact Information

Name:

Office Hours:

Office Location:

Phone Number:

E-mail Address:

Campus Tutoring Center

Location:

Hours:

DO NOT WAIT TO GET HELP If you are having difficulty, get help immediately. Since the material presented in class usually builds on previous material, it is very easy to fall behind. Ask your instructor if he or she is available during office hours or get help at the tutoring center on campus.

FORM A STUDY GROUP A study group provides an opportunity to discuss class material and homework problems. Find at least two other people in your class who are committed to being successful. Exchange contact information and plan to meet or work together regularly throughout the semester either in person or via e-mail, MyMathLab, or phone.

USE YOUR BOOK'S STUDY RESOURCES There are additional resources and support materials available with this book to help you succeed. See the list below and in the preface.

Notebook and Note Taking

Taking good notes and keeping a neat, well-organized notebook are important factors in being successful. If you do your homework online through MyMathLab, you should still keep a notebook to stay organized.

YOUR NOTEBOOK Use a loose-leaf binder divided into four sections: (1) notes, (2) homework, (3) graded tests (and quizzes), and (4) handouts. Or combine the resources in MyMathLab with the MyWorkBook with Chapter Summaries.

TAKING NOTES

- ❒ Copy all important information. Also, write all points that are not clear to you so that you can discuss them with your instructor, a tutor, or your study group.
- ❒ Write explanations of what you are doing in your own words next to each step of a practice problem.
- ❒ Listen carefully to what your instructor emphasizes and make note of it.

The following resources are available in MyMathLab, through your college bookstore, and at **www.pearsonhighered.com:**

- Student's Solutions Manual
- Video Resources with Chapter Test Prep Videos
- MyMathLab
- MyWorkBook with Chapter Summaries

Full descriptions are available in the preface.

Basic College Mathematics

through Applications

Fifth Edition

GEOFFREY AKST • SADIE BRAGG

Borough of Manhattan Community College, The City University of New York

PEARSON

Boston Columbus Indianapolis New York San Francisco Upper Saddle River

Amsterdam Cape Town Dubai London Madrid Milan Munich Paris Montréal Toronto

Delhi Mexico City São Paulo Sydney Hong Kong Seoul Singapore Taipei Tokyo

Editorial Director: Christine Hoag
Editor in Chief: Maureen O'Connor
Executive Content Editor: Kari Heen
Content Editor: Katie DePasquale
Editorial Assistant: Rachel Haskell
Senior Managing Editor: Karen Wernholm
Senior Production Supervisor: Ron Hampton
Senior Cover Designer: Barbara T. Atkinson
Text Design: Leslie Haimes
Composition: PreMediaGlobal
Media Producer: Aimee Thorne
Software Development: TestGen: Mary Durnwald; MathXL: Jozef Kubit
Executive Marketing Manager: Michelle Renda
Marketing Manager: Rachel Ross
Marketing Assistant: Ashley Bryan
Procurement Manager/Boston: Evelyn Beaton
Procurement Specialist: Debbie Rossi
Media Procurement Specialist: Ginny Michaud
Cover Photo: Bamboo on white © Subotina Anna/Shutterstock

Library of Congress Cataloging-in-Publication Data
Akst, Geoffrey.
Basic college mathematics through applications / Geoffrey Akst, Sadie Bragg.—5th ed.
p. cm
Includes index.
ISBN-13: 978-0-321-73339-9 ISBN-10: 0-321-73339-8 (student ed. : alk. paper)
1. Mathematics—Textbooks. I. Bragg, Sadie II. Title.
QA39.3.A47 2013
510—dc22 2011005564

2 3 4 5 6 7 8 9 10—CRK—15 14 13 12

pearsonhighered.com

For our students at BMCC

Contents

Preface

FROM THE AUTHORS

Our goal in writing *Basic College Mathematics through Applications* was to create a text that would help students progress and succeed in their college developmental math course. Throughout, we emphasize an applied approach, which has two advantages. First of all, it can help students prepare to meet their future mathematical demands—across disciplines, in subsequent coursework, in everyday life, and on the job. Secondly, this approach can be motivating, convincing students that mathematics is worth learning and more than just a school subject.

We have attempted to make the text readable, with understandable explanations and exercises for honing skills. We have also put together a set of easy-to-grasp features, consistent across sections and chapters.

In an effort to address many of the issues raised by national professional organizations, including AMATYC, NCTM, and NADE, we have been careful to stress connections to other disciplines; to incorporate the appropriate use of technology; to integrate quantitative reasoning skills; to include problem sets that facilitate student writing, critical thinking, and collaborative activities; and to emphasize real world data in examples and exercises. We have also introduced algebra early in the text to show an algebraic solution to a broad range of problems in successive chapters.

Above all, we have tried to develop a flexible text that can meet the needs of students in both traditional and redesigned developmental courses.

This text is part of the *through Applications* series that includes the following:

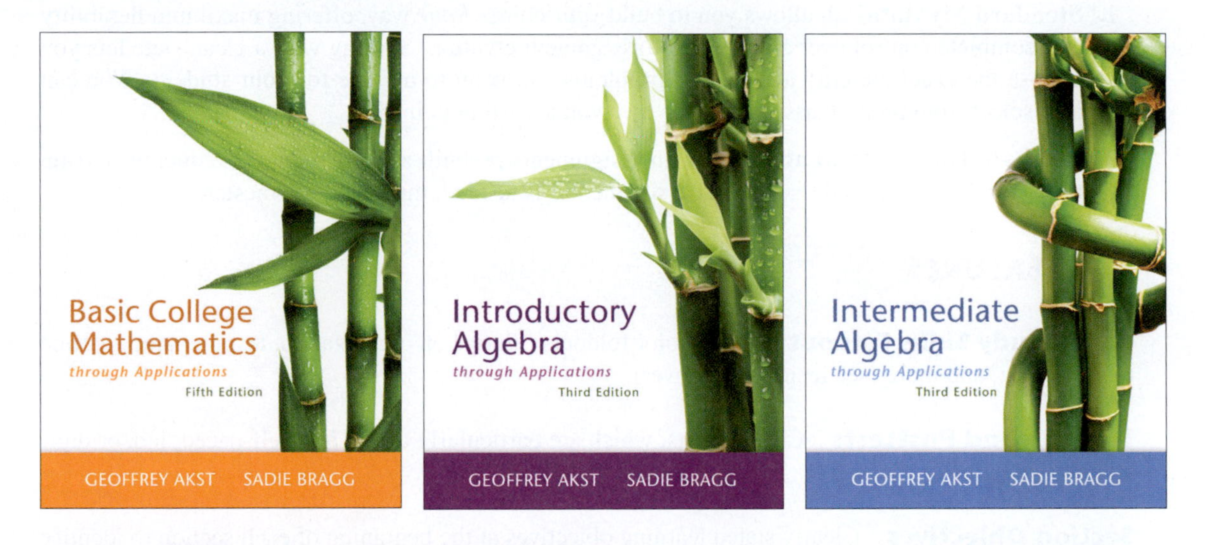

WHAT'S NEW IN THE FIFTH EDITION?

Say Why Exercises New fill-in-the-blank problems, located at the beginning of each chapter review, providing practice in reasoning and communicating mathematical ideas (see page 243).

Updated Content Adjusted content reflecting changing real-world needs. For instance, updates extend the place value concept from billions to trillions and the discussion of units to include the prefixes "mega-," "giga-," and "micro-," which are increasingly common in technology and medicine (see page 446).

Updated and Expanded Section Exercise Sets Additional practice in mastering skills (see pages 359–361).

Chapter Openers Extended real-world applications at the beginning of each chapter to motivate student interest and demonstrate how mathematics is used (see page 275).

Lengthening of Cumulative Review Exercise Sets Twice as many review exercises in response to user demand (see page 247).

Greater Emphasis on Learning Objectives End-of-section exercises closely aligned with the learning objectives in order to encourage and facilitate review (see pages 3 and 9–11).

More Examples and Exercises Based on Real Data Additional and more varied applied problems that are useful, realistic, and authentic (see page 8).

Parallel Paired Exercises Odd/even pairs of problems that more closely reflect the same learning objective (see page 298).

Easy-to-Locate Features Color borders added for back-of-book answer, glossary, and index pages.

Highlighting of Quantitative Literacy Skills Additional exercises that provide practice in number sense, proportional reasoning, and the interpretation of tables and graphs (see pages 28–30).

Increased Attention to Photos and Graphics Carefully selected photos to make problems seem more realistic, and relevant graphics to better meet the needs of visual learners (see pages 203 and 283).

Newly Expanded and Robust MyMathLab Coverage! One of *every* problem type is now assignable in MyMathLab.

Now Two MyMathLab Course Options

1. **Standard MyMathLab** allows you to build *your* course *your* way, offering maximum flexibility and complete control over all aspects of assignment creation. Starting with a clean slate lets you choose the exact quantity and type of problems you want to include for your students. You can also select from prebuilt assignments to give you a starting point.
2. **Ready-to-Go MyMathLab** comes with assignments prebuilt and preassigned, reducing start-up time. You can always edit individual assignments, as needed, through the semester.

KEY FEATURES

Math Study Skills Foldout A full-color foldout with tips on organization, test preparation, time management, and more (see inside front cover).

Pretests and Posttests Chapter tests, which are particularly useful in a self-paced, lab, or digital environment (see page 214).

Section Objectives Clearly stated learning objectives at the beginning of each section to identify topics to be covered (see page 3).

Side-by-Side Example/Practice Format Distinctive side-by-side format that pairs each example with a corresponding practice exercise and gets students actively involved from the start (see page 175).

Tips Helpful suggestions and cautions for avoiding mistakes (see page 83).

Journal Entries Writing assignments in response to probing questions interspersed throughout the text (see page 233).

Calculator Inserts Optional calculator and computer software instruction to solve section problems (see page 23).

Cultural Notes Glimpses of how mathematics has evolved across cultures and throughout history (see page 332).

For Extra Help Boxes at the beginning of every section's exercise set that direct students to helpful resources that will aid in their study of the material (see page 80).

Mathematically Speaking Exercises Vocabulary exercises in each section to help students understand and use standard mathematical terminology (see page 90).

Mixed Practice Exercises Problems in synthesizing section material (see page 63).

Application Exercises End-of-section problems to apply the topic at hand in a wide range of contexts (see pages 106–107).

Mindstretcher Exercises Nonstandard section problems in critical thinking, mathematical reasoning, pattern recognition, historical connections, writing, and group work to deepen understanding and provide enrichment (see page 108).

Key Concepts and Skills Summary With a focus on descriptions and examples, the main points of the chapter organized into a practical and comprehensive chart (see pages 145–147).

Chapter Review Exercises Problems for reviewing chapter content, arranged by section (see pages 243–244).

Chapter Mixed Application Exercises Practice in applying topics across the chapter (see page 245).

Cumulative Review Exercises Problems to maintain and build on the mathematical content covered in previous chapters (see pages 271–272).

Scientific Notation Appendix A brief appendix of particular value to students in the sciences.

U.S. and Metric Unit Tables Located opposite the inside back cover for quick reference.

Geometric Formulas A reference on the inside back cover of the text displaying standard formulas for perimeter, circumference, area, and volume.

Coherent Development Texts with consistent content and style across the developmental math curriculum.

WHAT SUPPLEMENTS ARE AVAILABLE?

For a complete list of the supplements and study aids that accompany *Basic College Mathematics through Applications*, Fifth Edition, see pp. xi.

ACKNOWLEDGMENTS

We are grateful to everyone who has helped to shape this textbook by responding to questionnaires, participating in telephone surveys and focus groups, reviewing the manuscript, and using the text in their classes. We wish to thank Michele Bach, *Kansas City Kansas Community College;* Irma Bakenhus, *San Antonio College;* Mary Lou Baker, *Columbia State Community College;* Palma Benko, *Passaic County Community College;* Tim Bremer, *Prestonburg Community College;* Sylvia Brown, *Mountain Empire Community College;* Jennifer Caldwell, *Mesa Community College;* Edythe Carter, *Amarillo College;* James Cochran, *Kirkwood Community College;* Robert Denitti, *Westmoreland County Community College;* Eunice Everett, *Seminole Community College;* Alan Greenhalgh, *Borough of Manhattan Community College;* Barbara Gardner, *Carroll Community College;* Janet C. Guynn, *Blue Ridge Community College/CUNY;* Kate Horton, *Portland Community College;* Matthew Hudock, *St. Philip's College;* Judith M. Jones, *Valencia Community College–East Campus;* Joanne Kendall, *College of the Mainland;* Yon Kim, *Passaic County Community College;* Dan Kleinfelter, *College of the Desert;* Roberta Lacefield, *Waycross College;* Lider-Manuel Lamar, *Seminole Community College;* Lee H. LaRue, *Paris Junior College;* Theodore Lai, *Hudson County Community College;*

LeAnn L. Lotz-Todd, *Metropolitan Community College–Longview;* Carol Marinas, *Barry College;* Christopher McNally, *Tallahassee Community College;* Dena S. Messer-Herrera, *Rio Salado College;* James Morgan, *Holyoke Community College;* Kathleen Offenholley; *Borough of Manhattan Community College/CUNY;* Ferdinand O. Orock, *Hudson County Community College;* Margaret Patin, *Vernon College;* Barbara Pearl, *Bucks County Community College;* Pat Roux, *Delgado Community College;* Susan Santolucito, *Delgado Community College;* Sara R. Pries, *Sierra Community College;* Andrew Russell, *Queensborough Community College/CUNY;* Joyce Saxon, *Morehead State University;* Radha Shrinivas, *Forest Park Community College;* Larry Smyrski, *Henry Ford Community College;* Marcia Swope, *Santa Fe Community College;* Sharon A. Testone, *Onondaga Community College;* James Van Ark, *University of Detroit Mercy;* Betty Vix Weinberger, *Delgado Community College;* Harvey S. Weiner, *Marymount Manhattan College;* Lisa Winch, *Kalamazoo Valley Community College;* J. W. Wing, *Angelina College;* James C. Woodall, *Salt Lake Community College;* and Michael D. Yarborough, *Cosumnes River College*. In addition, we would like to extend our gratitude to our accuracy checkers and to those who helped us perfect the content in many ways: Janis Cimperman, St. Cloud University; Beverly Fusfield; Denise Heban; Perian Herring, Okaloosa-Walton College; Sharon O'Donnell, Chicago State University; Ann Ostberg; Lenore Parens; and Deana Richmond.

Writing a textbook requires the contributions of many individuals. Special thanks go to Greg Tobin, President, Mathematics and Statistics, Pearson Arts and Sciences, for encouraging and supporting us throughout the entire process. We thank Kari Heen and Katie DePasquale for their patience and tact, Michelle Renda, Rachel Ross, and Maureen O'Connor for keeping us abreast of market trends, Rachel Haskell for attending to the endless details connected with the project, Ron Hampton, Elka Block, Laura Hakala, Trish O'Kane, Tracy Duff, Marta Johnson, and Rachel Youdelman for their support throughout the production process, Barbara Atkinson for the cover design, and the entire Pearson developmental mathematics team for helping to make this text one of which we are very proud.

Geoffrey Akst   Sadie Bragg

Student Supplements

Student's Solutions Manual

By Beverly Fusfield

- Provides detailed solutions to the odd-numbered exercises in each exercise set and solutions to all chapter pretests and post-tests, practice exercises, review exercises, and cumulative review exercises

ISBN-10: 0-321-75712-2 ISBN-13: 978-0-321-75712-8

New Video Resources on DVD with Chapter Test Prep Videos

- Complete set of digitized videos on DVD for students to use at home or on campus
- Includes a full lecture for each section of the text
- Covers examples, practice problems, and exercises from the textbook that are marked with the icon
- Optional captioning in English is available
- Step-by-step video solutions for each chapter test
- Chapter Test Prep Videos are also available on YouTube (search by using author name and book title) and in MyMathLab

ISBN-10: 0-321-78632-7 ISBN-13: 978-0-321-78632-6

MyWorkBook with Chapter Summaries

By Denise Heban

- Provides one worksheet for each section of the text, organized by section objective, along with the end-of-chapter summaries from the textbook
- Each worksheet lists the associated objectives from the text, provides fill-in-the-blank vocabulary practice, and exercises for each objective

ISBN 10: 0-321-75977-X ISBN-13: 978-0-321-75977-1

MathXL Online Course (access code required)

InterAct Math Tutorial Website www.interactmath.com

- Get practice and tutorial help online
- Provides algorithmically generated practice exercises that correlate directly to the textbook exercises
- Retry an exercise as many times as desired with new values each time for unlimited practice and mastery
- Every exercise is accompanied by an interactive guided solution that gives the student helpful feedback when an incorrect answer is entered
- View the steps of a worked-out sample problem similar to the one that has been worked on

Instructor Supplements

Annotated Instructor's Edition

- Provides answers to all text exercises in color next to the corresponding problems
- Includes teaching tips

ISBN-10: 0-321-63935-9 ISBN-13: 978-0-321-63935-6

Instructor's Solutions Manual (download only)

By Beverly Fusfield

- Provides complete solutions to even-numbered section exercises
- Contains answers to all Mindstretcher problems

ISBN-10: 0-321-75713-0 ISBN-13: 978-0-321-75713-5

Instructor's Resource Manual with Tests and Mini-Lectures (download only)

By Deana Richmond

- Contains three free-response and one multiple-choice test form per chapter, and two final exams
- Includes resources designed to help both new and adjunct faculty with course preparation and classroom management, including sample syllabi, tips for using supplements and technology, and useful external resources
- Offers helpful teaching tips correlated to the sections of the text

ISBN-10: 0-321-63937-5 ISBN-13: 978-0-321-63937-0

PowerPoint Lecture Slides (available online)

- Present key concepts and definitions from the text

TestGen® (available for download from the Instructor's Resource Center)

AVAILABLE FOR STUDENTS AND INSTRUCTORS

MyMathLab® Ready-to-Go Course (access code required)

These new Ready-to-Go courses provide students with all the same great MyMathLab features that you're used to, but make it easier for instructors to get started. Each course includes preassigned homework and quizzes to make creating your course even simpler. Ask your Pearson representative about the details for this particular course or to see a copy of this course.

MathXL—Instant Access—for *Basic College Mathematics through Applications*

MathXL® is the homework and assessment engine that runs MyMathLab. (MyMathLab is MathXL plus a learning management system.) With MathXL, instructors can

- Create, edit, and assign online homework and tests using algorithmically generated exercises correlated at the objective level to the textbook.
- Create and assign their own online exercises and import TestGen tests for added flexibility.
- Maintain records of all student work tracked in MathXL's online gradebook.

With MathXL, students can

- Take chapter tests in MathXL and receive personalized study plans and/or personalized homework assignments based on their test results.
- Use the study plan and/or the homework to link directly to tutorial exercises for the objectives they need to study.
- Access supplemental animations and video clips directly from selected exercises.

MathXL is available to qualified adopters. For more information, visit www.mathxl.com or contact your Pearson representative.

Index of Applications

Construction

Consumer

Economics

Education

Energy

Entertainment

Environment

Finance

Geology

Geometry

Government

Health/Life Sciences

Labor

Miscellaneous

INDEX OF APPLICATIONS

Physics

Sports

Statistics/Demographics

Technology

Transportation

Photo Credits

p. 1: Yakov Lapinsky; p. 11(tl): Gerry Ellis/Digital Vision/Getty Images; p. 11(tr): Gary Buss/Taxi/Getty Images; p. 11(b): Al Behrman/AP Images; p. 12(tl): GWImages/Shutterstock; p. 12(bl): Steve Herrmann/Shutterstock; p. 12(br): Karin Hildebrand Lau/Shutterstock; p. 20: Keren Su/Photodisc/Getty Images; p. 29(r): Jack Smith/AP Images; p. 29(l): Angus Oborn/Dorling Kindersley, Ltd.; p. 41: Jochen Sand/Digital Vision/Getty Images; p. 63, 77, 212, 487: NASA; p. 70(l): Stephen Frink/Digital Vision/Getty Images; p. 70(r): Library of Congress Prints and Photographs Division [LC-USZC4-13287]; p. 71: AP Images; p. 78: Richard Nowitz/National Geographic/Getty Images; p. 81: Archives du 7e Art/Columbia Pictures/DR/The Image Works; p. 107(t): Mark J. Terrill/AP Images; p. 107(b): Claudiofichera/Shutterstock; p. 152(t), 274(l), 288: PhotoLink/Photodisc/Getty Images; p. 152(bl): Album/Newscom; p. 152(br): Tim Lama/National Geographic/Getty Images; p. 153: Ramon Berk/Shutterstock; p. 157: Laurence Gough/Shutterstock; p. 167: Ryan McVay/Photodisc/Getty Images; p. 169(l): InterNetwork Media/Digital Vision/Getty Images; p. 169(r): iDesign/Shutterstock; p. 170: Library of Congress Prints and Photographs Division [DIG-ggbain-11279]; p. 171: InterNetwork Media/Photodisc/Getty Images; p. 186: Tom Brakefield/Photodisc/Getty Images; p. 203(l): Jeffrey M. Frank/Shutterstock; p. 203(r): Petr Mašek/Shutterstock; p. 209: Rolf Nussbaumer Photography/Alamy; p. 213: Tatiana Morozova/Shutterstock; p. 222: Marek Szumlas/Shutterstock; p. 230(l): Darwinsm81/Dreamstime; p. 230(r): Stacy Gold/National Geographic/Getty Images; p. 236: Frontpage/Shutterstock; p. 240(l), 268, 451: Stockbyte/Getty Images; p. 240(r): Jake Schoellkopf/AP Images; p. 245(l): Jack Hollingsworth/Photodisc/Getty Images; p. 245(r): NOAA; p. 247: Sipa/AP Images; p. 248: Purestock/Getty Images; p. 249: Robert Harding Picture Library, Ltd./Alamy; p. 255: Mark McClare/Shutterstock; p. 267(l): Don Farrall/Photodisc/Getty Images; p. 267(r): Jose Luis Pelaez, Inc./Blend Images/Getty Images; p. 272(l): Andersen Ross/Getty Images; p. 272(r): AF Archive/Alamy; p. 274(r): J. Helgason/Shutterstock; p. 275: Tom Pennington/MCT/Newscom; p. 279: Jason Straziuso/AP Images; p. 283: Ocean/Corbis; p. 289: Dorling Kindersley, Ltd.; p. 295: Dewitt/Shutterstock; p. 299(l): Dallas Events Inc/Shutterstock; p. 299(r): Photodisc/Getty Images; p. 303: Steve Reed/Shutterstock; p. 311: Yuri Arcurs/Shutterstock; p. 312: Mypokcik/Shutterstock; p. 316(l): Jose Gil/Shutterstock; p. 316(r): Tom Reed/AP Images; p. 318: DAJ/Getty Images; p. 321: Reflekta/Shutterstock; p. 327: Bellafotos/Dreamstime; p. 332: DDCoral/Shutterstock; p. 348: Rod Catanach/Newscom; p. 350: Alvaro Pantoja/Shutterstock; p. 356: Charles Smith/Corbis; p. 360(l): ABC/Everett Collection; p. 360(r): ChinaFotoPress/Newscom; p. 364: RubberBall/Alamy; p. 367: The Print Collector/Alamy; p. 400: Auremar/Shutterstock; p. 416: Michael Stokes/Shutterstock; p. 419: Kerry Vanessa McQuaid/Shutterstock; p. 420: Martin Fischer/Shutterstock; p. 422: *Madame X* (*Madame Pierre Gautreau*) (1883–1884), John Singer Sargent. Oil on canvas, 82 1/8 × 43 1/4 in. (208.6 × 109.9 cm). Arthur Hoppock Hearn Fund, 1916 (16.53). The Metropolitan Museum of Art/Art Resource, New York; p. 429: Maurie Hill/Dreamstime; p. 430: Detlev Van Ravenswaay/Science Photo Library/Photo Researchers, Inc.; p. 437(l): *Portrait of Lisa Gherardini* ("*Mona Lisa*") (1503–1506), Leonardo da Vinci. Oil on poplar wood, 77 × 53 cm. Collection of the Louvre Museum; Purchased by François I in 1518 [Inv. 779]. Photograph by Stuart Dee/Photographer's Choice/Getty Images; p. 437(r): Horst Schafer/AP Images; p. 439: Library of Congress Prints and Photographs Division [LC-DIG-ppmsca-19926]; p. 440: Interfoto/Alamy; p. 456: Dane Penland, Smithsonian Institution/AP Images; p. 457: Testing/Shutterstock; p. 461: Juice Images/Alamy; p. 463: Ragne Kabanova/Shutterstock; p. 506: Jim DeLillo/iStockphoto; p. 513: Harris Shiffman/Shutterstock

CHAPTER 1

Whole Numbers

Whole Numbers and YouTube

YouTube is a website where users can upload and view videos. These include movie clips, TV clips, music videos, and amateur content. This site made it feasible for anyone with an Internet connection to publish a video that could be seen by a worldwide audience within a few minutes.

In February 2005, the company was set up in a garage by several work colleagues. The first video posted on YouTube was *Me at the Zoo*, in which founder Jawed Karim is seen at the San Diego Zoo.

The usage of the site grew at an astonishing rate. By July 2006, more than 65,000 new videos were being uploaded every day, with about 10,000,000 visitors and 100,000,000 video views per day. Barely a year after its founding, the company was bought by Google for approximately $1,650,000,000.

YouTube has made sharing online video such an important part of Internet culture that it's been said "if it's not on YouTube, it's like it never happened."

(*Sources:* telegraph.co.uk, comscore.com, wikipedia.org, and cleancutmedia.com)

CHAPTER 1 PRETEST

To see if you have already mastered the topics in this chapter, take this test.

1. Insert commas as needed in the number 2 0 5 0 0 7. Then write the number in words.

2. Write the number one million, two hundred thirty-five thousand in standard form.

3. What place does the digit 8 occupy in 805,674?

4. Round 8,143 to the nearest hundred.

5. Add: $38 + 903 + 7{,}285$

6. Subtract 286 from 5,000.

7. Subtract: $734 - 549$

8. Find the product of 809 and 36.

9. Find the quotient: $27\overline{)7{,}020}$

10. Divide: $13{,}558 \div 44$

11. Write $2 \cdot 2 \cdot 2$, using exponents.

12. Evaluate: 6^2

Simplify.

13. $26 - 7 \cdot 3$

14. $3 + 2^3 \cdot (8 - 3)$

Solve and check.

15. The mathematician Benjamin Banneker was born in 1731 and died in 1806. About how old was he when he died? (*Source: The New Encyclopedia Britannica*)

16. At a certain college, students pay $105 for each college credit. If a student takes 9 credits and pays with a $1,000 voucher, how much change will he receive?

17. Phil Mickelson had scores of 67, 71, 67, and 67 for his four rounds at the 2010 Masters Tournament. What was his average score for a round of golf?

18. The Epson PictureMate Show Compact Photo Printer can print a 4-inch by 6-inch photo in 37 seconds, and the Epson Artisan 810 All-in-One Printer can print the same size photo in 10 seconds. How much longer would it take the Epson PictureMate Show to print twelve 4-inch by 6-inch photos? (*Source:* epson.com)

19. An insurance company offers an installment plan for paying auto insurance premiums. For a $540 policy, the plan requires a down payment of $81. The balance is paid in nine equal installments of $55, which includes a service charge. How much money would be saved by paying for this policy without using the installment plan?

20. Which of the rooms pictured has the largest area? (feet = ft)

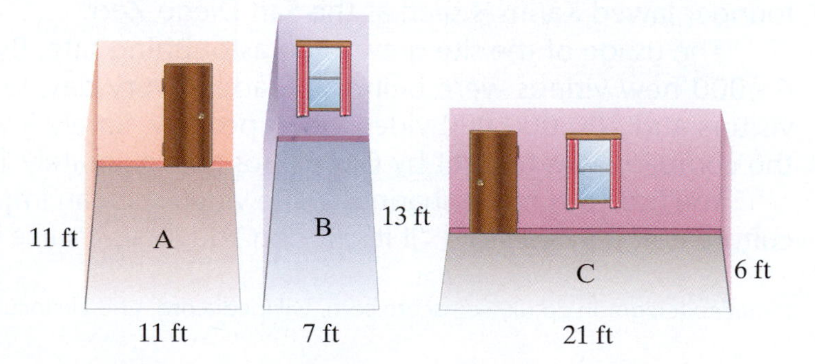

• Check your answers on page A-1.

1.1 Introduction to Whole Numbers

OBJECTIVES

- **A** To read or write whole numbers
- **B** To write whole numbers in expanded form
- **C** To round whole numbers
- **D** To solve applied problems involving reading, writing, or rounding whole numbers

What the Whole Numbers Are and Why They Are Important

We use whole numbers for counting, whether it is the number of *e*'s on this page, the number of stars in the sky, or the number of runs, hits, and errors in a baseball game.

The whole numbers are 0, 1, 2, 3, 4, 5, 6, 7, 8, 9, 10, 11, 12, 13, ... An important property of whole numbers is that there is always a next whole number. This property means that they go on without end, as the three dots above indicate.

Every whole number is either *even* or *odd*. The even whole numbers are 0, 2, 4, 6, 8, 10, 12, ... The odd whole numbers are 1, 3, 5, 7, 9, 11, 13, ...

We can represent the whole numbers on a number line. Similar to a ruler, the number line starts with 0 and extends without end to the right, as the arrow indicates.

0 1 2 3 4 5 6 7 8 ...

Reading and Writing Whole Numbers

Generally speaking, we *read* whole numbers in words, but we use the **digits** 0, 1, 2, 3, 4, 5, 6, 7, 8, and 9 to *write* them. For instance, we read the whole number *fifty-one* but write it *51*, which we call **standard form**.

Each of the digits in a whole number in standard form has a **place value**. Our place value system is very important because it underlies both the way we write and the way we compute with numbers.

The following chart shows the place values in whole numbers up to 15 digits long. For instance, in the number 1,234,056 the digit 2 occupies the hundred thousands place. Study the place values in the chart now.

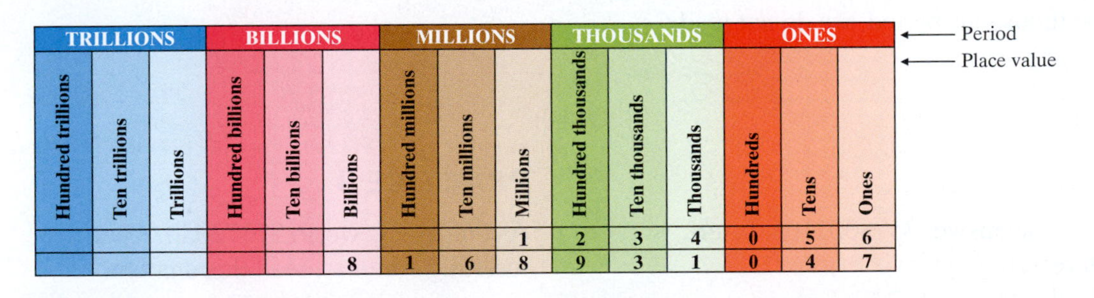

TRILLIONS			BILLIONS			MILLIONS			THOUSANDS			ONES			← Period
Hundred trillions	Ten trillions	Trillions	Hundred billions	Ten billions	Billions	Hundred millions	Ten millions	Millions	Hundred thousands	Ten thousands	Thousands	Hundreds	Tens	Ones	← Place value
								1	2	3	4	0	5	6	
					8	1	6	8	9	3	1	0	4	7	

TIP We read whole numbers from left to right, but it is easier in the place value chart to learn the names of the places *from right to left*.

When we write a large whole number in standard form, we insert *commas* to separate its digits into groups of three, called **periods**. For instance, the number 8,168,931,047 has four periods: *ones, thousands, millions*, and *billions*.

EXAMPLE 1

In each number, identify the place that the digit 7 occupies.

a. 207 **b.** 7,654,000 **c.** 5,700,000,001

Solution

a. The ones place

b. The millions place

c. The hundred millions place

PRACTICE 1

What place does the digit 8 occupy in each number?

a. 278,056

b. 803,746

c. 3,080,700,059

The following rule provides a shortcut for *reading a whole number*:

To Read a Whole Number

Working from left to right,

- read the number in each period and then
- name the period in place of the comma.

For instance, 1,234,056 is read "one million, two hundred thirty-four thousand, fifty-six." Note that the ones period is not read.

EXAMPLE 2

How do you read the number 422,000,085?

Solution Beginning at the left in the millions period, we read this number as "four hundred twenty-two million, eighty-five." Note that because there are all zeros in the thousands period, we do not read "thousands."

PRACTICE 2

Write 8,000,376,052 in words.

EXAMPLE 3

The display on a calculator shows the answer 3578002105. Insert commas in this answer and then read it.

Solution The number with commas is 3,578,002,105. It is read "three billion, five hundred seventy-eight million, two thousand, one hundred five."

PRACTICE 3

A company is worth $7372050. After inserting commas, read this amount.

Until now, we have discussed how to *read* whole numbers in standard form. Now, let's turn to the question of how they are *written* in standard form. We simply reverse the process just described. For instance, the number eight billion, one hundred sixty-eight million, nine hundred thirty-one thousand, forty-seven in standard form is 8,168,931,047. Here, we use the 0 as a **placeholder** in the hundreds place because there are no hundreds.

To Write a Whole Number

Working from left to right,

- write the number named in each period and
- replace each period name with a comma.

When writing large whole numbers in standard form, we must remember that the number of commas is always one less than the number of periods. For instance, the number one million, two hundred thirty-four thousand, fifty-six—1,234,056—has three periods and two commas. Similarly, the number 8,168,931,047 has four periods and three commas.

EXAMPLE 4

Write the number eight billion, seven in standard form.

Solution This number involves billions, so there are four periods—billions, millions, thousands, and ones—and three commas. Writing the number named in each period and replacing each period name with a comma, we get 8,000,000,007. Note that we write three 0's when no number is named in a period.

PRACTICE 4

Use digits and commas to write the amount ninety-five million, three dollars.

EXAMPLE 5

The treasurer of a company writes a check in the amount of four hundred thousand seven hundred dollars. Using digits, how would she write this amount on the check?

Solution This quantity is written with one comma, because its largest period is thousands. So the treasurer writes $400,700, as shown on the check below.

UNITED INDUSTRIES
Atlanta, Georgia
2066
DATE January 26, 2011
PAY TO THE ORDER OF American Vendors, Inc. $ 400,700—
Four Hundred Thousand Seven Hundred and 00/100 DOLLARS
SB Southern Bank
MEMO
Maxine Jefferson
⑆721107560⑆ 0220006587⑈ 2066

PRACTICE 5

A rich alumna donates three hundred seventy-five thousand dollars to her college's scholarship fund.

HARRIET YOUNG
3560 Ramstead St.
Reston, VA 22090
1434
DATE March 17, 2011
PAY TO THE ORDER OF Borough of Manhattan Community College $
Three Hundred Seventy-Five Thousand and 00/100 DOLLARS
MDBC
Bank USA Reston, VA 22090
MEMO
Harriet Young
⑆034005089⑆ 56024036⑈ 1434

Using digits, how would she write this amount on the check?

When writing checks, we write the amount in both digits and words. Why do we do this?

Writing Whole Numbers in Expanded Form

We have just described how to write whole numbers in standard form. Now, let's turn to how we write these numbers in **expanded form**.

Let's consider the whole number 4,025 and examine the place value of its digits.

$$4{,}025 = 4 \text{ thousands} + 0 \text{ hundreds} + 2 \text{ tens} + 5 \text{ ones}$$

This last expression is called the expanded form of the number, and it can be written as follows

$$4{,}000 + 0 + 20 + 5, \quad \text{or} \quad 4{,}000 + 20 + 5$$

The expanded form of a number spells out its value in terms of place value, helping us understand what the number really means. For instance, think of the numbers 92 and 29. By representing them in *expanded* form, can you explain why they differ in value even though their *standard* form consists of the same digits?

EXAMPLE 6

Write in expanded form:

a. 906 **b.** 3,203,000

Solution

a. The 6 is in the ones place, the 0 is in the tens place, and the 9 is in the hundreds place.

ONES		
Hundreds	**Tens**	**Ones**
9	0	6

So 906 is 9 hundreds + 0 tens + 6 ones = 900 + 0 + 6, or 900 + 6 in expanded form.

b. Using the place value chart, we see that

$$\begin{aligned} 3{,}203{,}000 &= 3 \text{ millions} + 2 \text{ hundred thousands} + 3 \text{ thousands} \\ &= 3{,}000{,}000 + 200{,}000 + 3{,}000. \end{aligned}$$

PRACTICE 6

Express in expanded form:

a. 27,013

b. 1,270,093

Rounding Whole Numbers

Most people equate mathematics with precision, but some problems require sacrificing precision for simplicity. In this case, we use the technique called **rounding** to approximate the exact answer with a number that ends in a given number of zeros. Rounded numbers have special advantages: They seem clearer to us than other numbers, and they make computation easier—especially when we are trying to compute in our heads.

Of these two headlines, which do you prefer? Why?

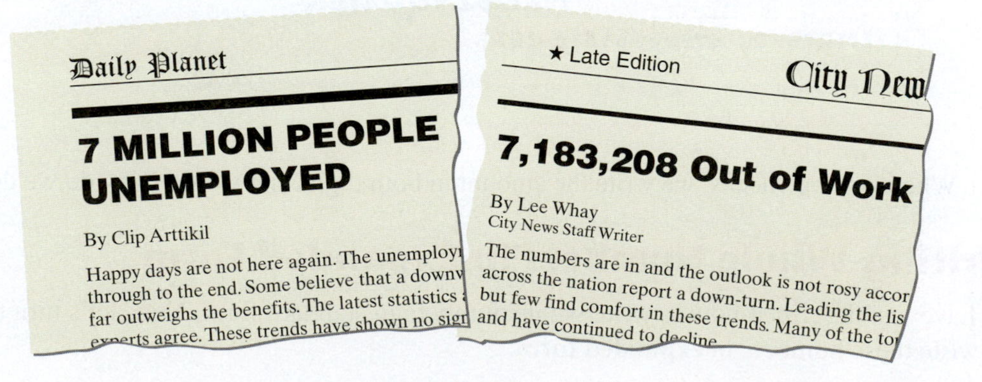

Study the following chart to see the connection between place value and rounding.

Rounding to the nearest	Means that the rounded number ends in at least
10	One 0
100	Two 0's
1,000	Three 0's
10,000	Four 0's
100,000	Five 0's
1,000,000	Six 0's

Note in the chart that the place value tells us how many 0's the rounded number must have at the end. Having more 0's than indicated is possible. Can you think of an example?

When rounding, we use an underlined digit to indicate the place to which we are rounding.

Now, let's consider the following rule for rounding whole numbers:

To Round a Whole Number

- Underline the place to which you are rounding.
- The digit to the right of the underlined digit is called the *critical digit.* Look at the critical digit—if it is 5 or more, add 1 to the underlined digit; if it is less than 5, leave the underlined digit unchanged.
- Replace all the digits to the right of the underlined digit with zeros.

EXAMPLE 7

Round 79,630 to

a. the nearest thousand

b. the nearest hundred

Solution

a. $79{,}630 = 7\underline{9}{,}630$ ← **Underline the digit in the thousands place.**

$= 7\underline{9}{,}630$ ← **The critical digit 6 is greater than 5; add 1 to the underlined digit.**

$\approx 80{,}000$ ← **Change the digits to the right of the underlined digit to 0's.**

↑ **This symbol means "is approximately equal to."**

Note that adding 1 to the underlined digit gave us 10. As a result we regroup, that is, write 0, carry 1 to the next column, and change the 7 to 8.

b. First, we underline the 6 because that digit occupies the hundreds place: $79{,}\underline{6}30$. The critical digit is **3**: $79{,}\underline{6}\mathbf{3}0$. Since 3 is less than 5, we leave the underlined digit unchanged. Then, we replace all digits to the right with 0's, getting 79,600. We write $79{,}630 \approx 79{,}600$, meaning that 79,630 when rounded to the nearest hundred is 79,600.

PRACTICE 7

Round 51,760 to

a. the nearest thousand

b. the nearest ten thousand

For Example 7a, consider this number line.

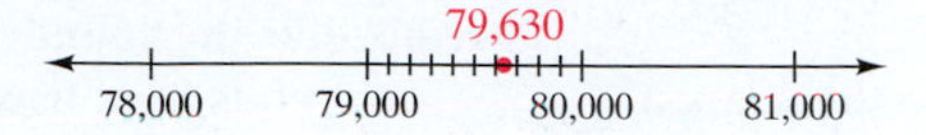

The number line shows that 79,630 lies between 79,000 and 80,000 and that it is closer to 80,000, as the rule indicates.

EXAMPLE 8

In an anatomy and physiology class, a student learned that the adult human skeleton contains 206 bones. How many bones is this to the nearest hundred bones?

Solution We first write 2̲06. The critical digit 0 is less than 5, so we do *not* add 1 to the underlined digit. However, we do change both the digits to the right of the 2 to 0's. So 2̲06 ≈ 200, and there are approximately 200 bones in the human body.

PRACTICE 8

Based on current population data, the U.S. Bureau of the Census projects that the U.S. resident population will be 419,845,000 in the year 2050. What is the projected population to the nearest million?

EXAMPLE 9

The following table lists five of the highest-grossing films of all time and the amount of money they took in.

Film	Year	World Total (in U.S. dollars)
Titanic	1997	$1,835,300,000
The Lord of the Rings: The Return of the King	2003	$1,129,219,252
Pirates of the Caribbean: Dead Man's Chest	2006	$1,060,332,628
The Dark Knight	2008	$1,001,921,825
Avatar	2009	$2,690,408,054

(*Source:* imdb.com)

a. Write in words the amount of money taken in by the film with the largest world total.

b. Round to the nearest ten million dollars the world total for *Titanic.*

Solution

a. *Avatar* has the largest world total. This total is read "two billion, six hundred ninety million, four hundred eight thousand, fifty-four dollars."

b. The world total for *Titanic* is $1,835,300,000. To round, we underline the digit in the ten millions place: 1,83̲5,300,000. Since the critical digit is 5, we add 1 to the underlined digit, and change the digits to the right to 0's. So the rounded total is $1,840,000,000.

PRACTICE 9

This chart gives the number of U.S. postsecondary teachers in the year 2008 as well as the projected number of postsecondary teachers for the year 2018.

Year	Number of Postsecondary Teachers
2008	1,699,200
2018	1,956,100

(*Source:* bls.gov)

a. Write in words the number of postsecondary teachers in the year 2008.

b. What is the number of projected postsecondary teachers in the year 2018 rounded to the nearest ten thousand?

1.1 Exercises

FOR EXTRA HELP MyMathLab MathXL PRACTICE WATCH READ REVIEW

Mathematically Speaking

Fill in each blank with the most appropriate term or phrase from the given list.

calculated	rounded	periods	odd
even	digits	whole numbers	standard form
placeholder	place value	expanded form	

1. The ____________ are 0, 1, 2, 3, 4, 5, ...

2. The numbers 0, 2, 4, 6, 8, 10, . . . are ____________.

3. The numbers 1, 3, 5, 7, 9, . . . are ____________.

4. The whole numbers are written with the ____________ 0, 1, 2, 3, 4, 5, 6, 7, 8, and 9.

5. The number thirty-seven, when written as 37, is said to be in ____________.

6. In the number 528, the ____________ of the 5 is hundreds.

7. In the number 206, the 0 is used as a ____________ in the tens place.

8. Commas separate the digits in a large whole number into groups of three called ____________.

9. When the number 973 is written as 9 hundreds + 7 tens + 3 ones, it is said to be in ____________.

10. The number 545 ____________ to the nearest hundred is 500.

A *Underline the digit that occupies the given place.*

11. 4,867 Thousands place

12. 9,752 Thousands place

13. 316 Tens place

14. 728 Tens place

15. 28,461,013 Millions place

16. 73,762,800 Millions place

Identify the place occupied by the underlined digit.

17. $\underline{6}91{,}400$

18. $7\underline{2}{,}109$

19. $7{,}\underline{3}80$

20. $3\underline{5}1$

21. $\underline{8}{,}450{,}000{,}000$

22. $\underline{3}5{,}832{,}775$

Insert commas as needed, and then write the number in words.

23. 4 8 7 5 0 0

24. 5 2 8 0 5 0

25. 2 3 5 0 0 0 0

26. 1 3 5 0 1 3 2

27. 9 7 5 1 3 5 0 0 0

28. 4 2 1 0 0 0 1 3 2

29. 2 0 0 0 0 0 0 3 5 2

30. 4 1 0 0 0 0 0 0 0 7

31. 1 0 0 0 0 0 0 0 0 0

32. 3 7 9 0 5 2 0 0 0

77. In 1990, the U.S. public debt (in dollars) was three trillion, two hundred thirty-three billion, three hundred million. (*Source: The World Almanac 2010*)

78. The light-year is a unit of length used to measure distances to stars and other distances on an astronomical scale. A light-year is equal to about five trillion, eight hundred seventy-eight billion, six hundred thirty million miles. (*Source:* wikipedia.org)

Round to the indicated place.

79. The Statue of Liberty is 152 feet high. What is its height to the nearest 10 feet?

80. The Nile, with a length of 4,180 miles, is the longest river in the world. Find this length to the nearest thousand miles.

81. In 1949, Air Force Captain James Gallagher led the first team to make an around-the-world flight. The team flew 23,452 miles. What is this distance to the nearest ten thousand miles? (*Source:* Taylor and Mondey, *Milestones of Flight*)

82. The element copper changes from a liquid to a gas at the temperature 2,567 degrees Celsius (°C). Find this temperature to the nearest hundred degrees Celsius.

83. The South American country of Colombia is home to 1,897 bird species—more than any other country. How many species is this to the nearest hundred? (*Source: Avibase—the World Bird Database,* avibase.bsc-eoc.org)

84. The Rose Bowl stadium has a seating capacity of 92,542. Round this number to the nearest ten thousand. (*Source:* Rose Bowl Operating Company)

85. This chart displays the area of seven most heavily populated countries in the world.

Country	Area (in square miles)
China	3,600,930
India	1,147,950
United States	3,537,421
Indonesia	705,189
Brazil	3,265,061
Pakistan	300,664
Bangladesh	51,703

(*Source:* census.gov)

a. Write in words the area of China.

b. Round, to the nearest thousand, the area of Pakistan.

86. This chart displays the number of degrees awarded in the United States during a recent year.

Degree	Number Awarded
Associate	750,164
Bachelor's	1,563,069
Master's	625,023
Doctorate	63,712
Professional	91,309

(*Source:* nces.ed.gov)

a. Write in words the number of bachelor's degrees awarded.

b. Round, to the nearest hundred thousand, the number of associate degrees awarded.

• Check your answers on page A-1.

MINDStretchers

Mathematical Reasoning

1. I am thinking of a certain whole number. My number, rounded to the nearest hundred, is 700. When it is rounded to the nearest ten, it is 750. What numbers could I be thinking of?

Writing

2. How does the number 10 play a special role in the way that we write whole numbers? Would it be possible to have the number 2 play this role? Explain.

Groupwork

3. Here are three ways of writing the number seven: 7 VII 𝍸 ||

Working with a partner, express each of the numbers 1, 2, . . . , 9 in these three ways.

1.2 Adding and Subtracting Whole Numbers

OBJECTIVES

- **A** To add or subtract whole numbers
- **B** To solve applied problems involving the addition or subtraction of whole numbers

The Meaning and Properties of Addition and Subtraction

Addition is perhaps the most fundamental of all operations. One way to think about this operation is as *combining sets*. For example, suppose that we have two distinct sets of pens, with 5 pens in one set and 3 in the other. If we put the two sets together, we get a single set that has 8 pens.

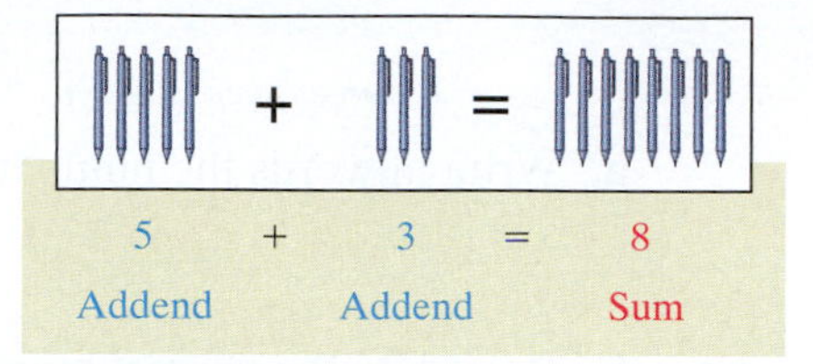

So we can say that 3 added to 5 is 8, or here, 5 pens plus 3 pens equals 8 pens. Numbers being added are called *addends*. The result is called the *sum*, or *total*.

In the above example, note that we are adding quantities of the same thing, or *like quantities*.

Another good way to think about the addition of whole numbers is as *moving to the right on a number line*. In this way, we start at the point on the line corresponding to the first number, 5. Then to add 3, we move 3 units to the right, ending on the point that corresponds to the answer, 8.

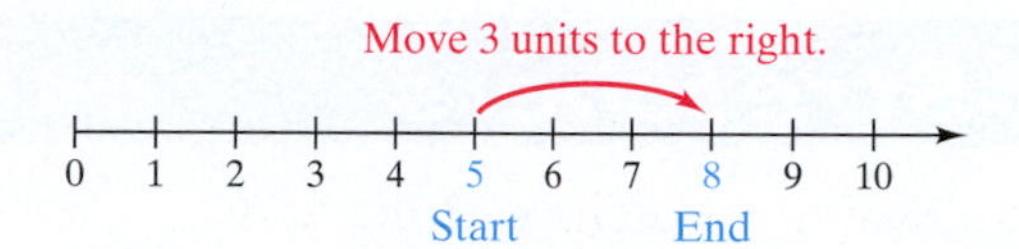

Now, let's look at subtraction. One way to look at this operation is as *taking away*. For instance, when we subtract 5 pens from 8 pens, we take 5 pens away from 8 pens, leaving 3 pens.

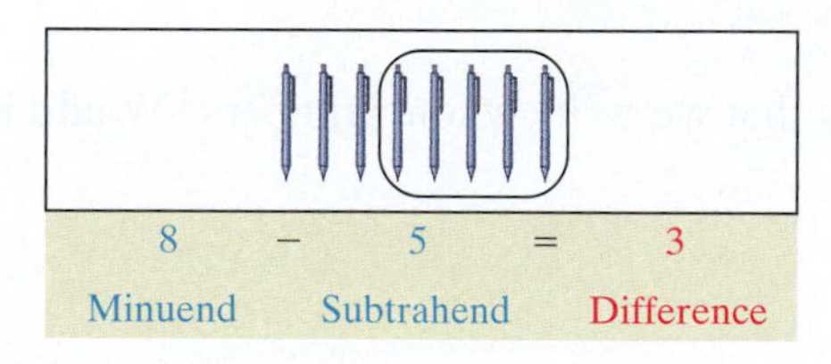

In a subtraction problem, the number from which we subtract is called the *minuend*, the number being subtracted is called the *subtrahend*, and the result is called the *difference*. In other words, the difference between two numbers is the first number take away the second number.

As in the preceding example, we can only subtract *like quantities*: we cannot subtract 5 pens from 8 scissors.

We can also think of subtraction as the *opposite of addition*.

$8 - 5 = 3$ because $5 + 3 = 8$

Subtraction **Related addition**

Note in this example that, if we add the 5 pens to the 3 pens, we get 8 pens.

Addition and subtraction problems can be written either horizontally or vertically.

$5 + 3 = 8 \qquad 8 - 5 = 3$

Horizontal

$$\begin{array}{r} 5 \\ +3 \\ \hline 8 \end{array} \qquad \begin{array}{r} 8 \\ -5 \\ \hline 3 \end{array}$$

Vertical

Either format gives the correct answer. But it is generally easier to figure out the sum and difference of large numbers if the problems are written vertically.

Now, let's briefly consider several special properties of addition that we use frequently. Examples appear to the right of each property.

The Identity Property of Addition

The sum of a number and zero is the original number.

$3 + 0 = 3$
$0 + 5 = 5$

The Commutative Property of Addition

Changing the order in which two numbers are added does not affect their sum.

$$\underbrace{3 + 2}_{5} = \underbrace{2 + 3}_{5}$$

The Associative Property of Addition

When adding three numbers, regrouping addends gives the same sum. Note that the parentheses tell us which numbers to add first.

We add inside the parentheses first

$$(4 + 7) + 2 = 4 + (7 + 2)$$
$$11 + 2 = 4 + 9$$
$$13 \quad 13$$

Adding Whole Numbers

We add whole numbers by arranging the numbers vertically, keeping the digits with the same place value in the same column. Then, we add the digits in each column.

Consider the sum $32 + 65$. In the vertical format at the right, the sum of the digits in each column is 9 or less. The sum is 97. When the sum of the digits in a column is greater than 9, we must **regroup (carry)** because only a single digit can occupy a single place. Example 1 illustrates this process.

$$\begin{array}{r} 32 \\ +65 \\ \hline 97 \end{array}$$

EXAMPLE 1

Add 47 and 28.

Solution First, we write the addends in expanded form. Then, we add down the ones column.

$$\begin{array}{rcccc} 47 & = & 4 \text{ tens} + 7 \text{ ones} & = & \overset{1 \text{ ten}}{4} \text{ tens} + 7 \text{ ones} \\ +28 & = & 2 \text{ tens} + 8 \text{ ones} & = & 2 \text{ tens} + 8 \text{ ones} \\ \hline & & 15 \text{ ones} & & 5 \text{ ones} \end{array}$$

By regrouping, we express 15 ones as 1 ten + 5 ones. Then we carry the 1 ten to the tens place.

Next, we add down the tens column.

$$\begin{array}{l} \text{1 ten} \\ 4 \text{ tens} + 7 \text{ ones} \\ \underline{2 \text{ tens} + 8 \text{ ones}} \\ 7 \text{ tens} + 5 \text{ ones} = 75 \end{array}$$

PRACTICE 1

Add: $178 + 207$

The following rule tells how to add whole numbers without using expanded form:

To Add Whole Numbers

- Write the addends vertically, lining up the place values.
- Add the digits in the ones column, writing the rightmost digit of the sum on the bottom. If the sum has two digits, carry the left digit to the top of the next column on the left.
- Add the digits in the tens column, as in the preceding step.
- Repeat this process until you reach the last column on the left, writing the entire sum of that column on the bottom.

EXAMPLE 2

Add: 9,824 + 356 + 2,976

Solution We write the problem vertically, with the addends lined up on the right.

$$\begin{array}{r} {\scriptstyle 1}\;\; \\ 9,824 \\ 356 \\ +2,976 \\ \hline 6 \end{array}$$

← **The sum of the ones digits is 16 ones. We write the 6 and carry the 1 to the tens column.**

$$\begin{array}{r} {\scriptstyle 1\,1}\;\; \\ 9,824 \\ 356 \\ +2,976 \\ \hline 56 \end{array}$$

The sum of the tens digits is 15 tens. We write the 5 and carry the 1 to the hundreds column.

$$\begin{array}{r} {\scriptstyle 2\,1\,1}\;\; \\ 9,824 \\ 356 \\ +2,976 \\ \hline 156 \end{array}$$

The sum of the hundreds digits is 21 hundreds. We write the 1 and carry the 2 to the thousands column.

$$\begin{array}{r} {\scriptstyle 2\,1\,1}\;\; \\ 9,824 \\ 356 \\ +2,976 \\ \hline 13,156 \end{array}$$

The sum of the digits in the thousands column is 13, which we write completely—no need to regroup here.

The sum is 13,156.

PRACTICE 2

Find the total: 838 + 96 + 9,502

In Example 3, let's apply the operation of addition to finding the geometric perimeter of a figure. The **perimeter** is the distance around a figure, which we can find by adding the lengths of its sides.

EXAMPLE 3

What is the perimeter of the region marked off for the construction of a swimming pool and an adjacent pool cabana?

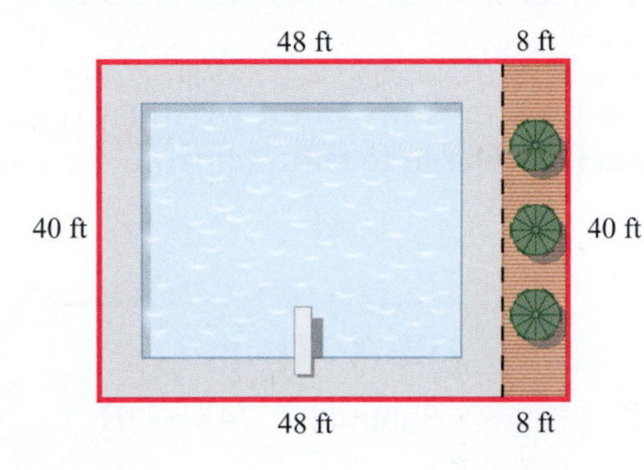

Solution This figure consists of two rectangles placed side by side. We note that the opposite sides of each rectangle are equal in length.

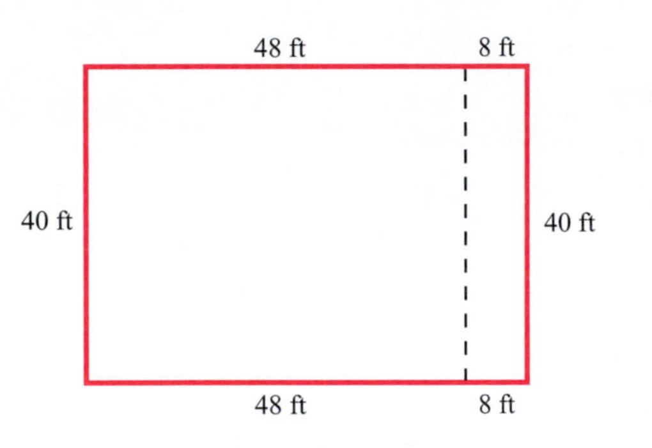

To compute the figure's perimeter, we need to add the lengths of all its sides.

$$\begin{array}{r} 40 \\ 48 \\ 8 \\ 40 \\ 8 \\ +\ 48 \\ \hline 192 \end{array}$$

The figure's perimeter is 192 feet.

PRACTICE 3

How long a fence is needed to enclose the piece of land sketched?

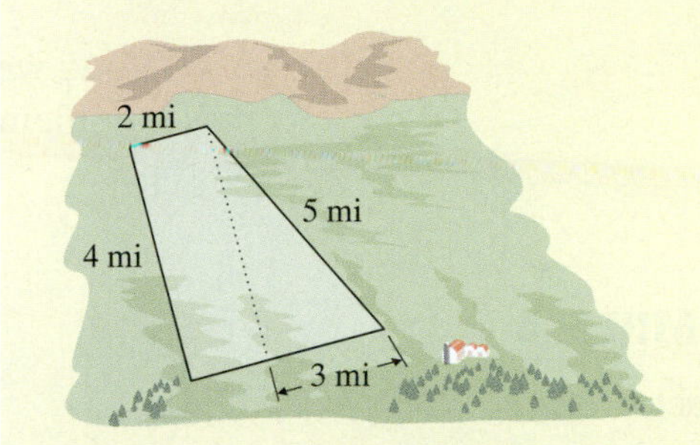

Subtracting Whole Numbers

Consider the subtraction problem 59 − 36, written vertically at the right. We write the whole numbers underneath one another, lined up on the right, so each column contains digits with the same place value. Subtracting the digits within each column, the bottom digit from the top, the result is a difference of 23.

$$\begin{array}{r} 59 \\ -36 \\ \hline 23 \end{array}$$

Keep in mind two useful properties of subtraction.

- When we subtract a number from itself, the result is 0: $6 - 6 = 0$
- When we subtract 0 from a number, the result is the original number: $25 - 0 = 25$

TIP When writing a subtraction problem vertically, be sure that
- the minuend—the number from which we are subtracting—goes on the top and that
- the subtrahend—the number being taken away—goes on the bottom.

Now, we consider subtraction problems in which we must regroup (*borrow*). In these problems a digit on the bottom is too large to subtract from the corresponding digit on top.

EXAMPLE 4

Subtract: $329 - 87$

Solution We first write these numbers vertically in expanded form.

$$\begin{array}{rcr} 329 & = & 3 \text{ hundreds} + 2 \text{ tens} + 9 \text{ ones} \\ -\ 87 & = & -\qquad\qquad 8 \text{ tens} + 7 \text{ ones} \end{array}$$

We then subtract the digits in the ones column: 7 ones from 9 ones gives 2 ones.

$$\begin{array}{r} 3 \text{ hundreds} + 2 \text{ tens} + 9 \text{ ones} \\ -\qquad\qquad 8 \text{ tens} + 7 \text{ ones} \\ \hline 2 \text{ ones} \end{array}$$

10 tens + 2 tens = 12 tens

We next go to the tens column. We cannot take 8 tens from 2 tens. But we can *borrow* 1 hundred from the 3 hundreds, leaving 2 in the hundreds place. We *exchange* this hundred for 10 tens (1 hundred = 10 tens). Then combining the 10 tens with the 2 tens gives 12 tens.

$$\begin{array}{r} \overset{2}{\cancel{3}} \text{ hundreds} + \overset{1}{2} \text{ tens} + 9 \text{ ones} \\ -\qquad\qquad 8 \text{ tens} + 7 \text{ ones} \\ \hline 2 \text{ ones} \end{array}$$

We next take 8 from 12 in the tens column, giving 4 tens. Finally, we bring down the 2 hundreds. The difference is 242 in standard form.

$$\begin{array}{r} \overset{2}{\cancel{3}} \text{ hundreds} + \overset{1}{2} \text{ tens} + 9 \text{ ones} \\ -\qquad\qquad 8 \text{ tens} + 7 \text{ ones} \\ \hline 2 \text{ hundreds} + 4 \text{ tens} + 2 \text{ ones} = 242 \end{array}$$

PRACTICE 4

Subtract: $748 - 97$

Although we can always rewrite whole numbers in expanded form so as to subtract them, the following rule provides a shortcut:

To Subtract Whole Numbers

- On top, write the number *from which* we are subtracting. On the bottom, write the number that is being taken *away*, lining up the place values. Subtract in each column separately.
- Start with the ones column.
  a. If the digit on top is *larger* than or *equal* to the digit on the bottom, subtract and write the difference below the bottom digit.
  b. If the digit on top is *smaller* than the digit on the bottom, borrow from the digit to the left on top. Then subtract and write the difference below the bottom digit.
- Repeat this process until the last column on the left is finished.

Recall that for every subtraction problem there is a related addition problem. So we can use addition to check subtraction, as in the following example.

EXAMPLE 5

Find the difference between 500 and 293.

Solution We rewrite the problem vertically.

$$\begin{array}{r} 500 \\ -293 \\ \hline \end{array}$$

We cannot subtract 3 ones from 0 ones, and we cannot borrow from 0 tens. So we borrow from the 5 hundreds.

$$\begin{array}{r} \overset{4\ 1}{\not{5}}00 \\ -293 \\ \hline \end{array} \quad \leftarrow \textbf{5 hundreds = 4 hundreds + 10 tens}$$

We now borrow from the tens column.

$$\begin{array}{r} \overset{4}{\not{5}}\ \overset{\overset{9}{\not{10}}}{\not{0}}\ \overset{1}{0} \\ -2\ 9\ 3 \\ \hline 2\ 0\ 7 \end{array} \quad \leftarrow \textbf{10 tens = 9 tens + 10 ones}$$

Check We check the difference by adding it to the subtrahend. The sum turns out to be the original minuend, so our answer is correct.

$$\begin{array}{r} 207 \\ +293 \\ \hline 500 \end{array}$$

PRACTICE 5

Subtract 3,253 from 8,000.

113. The following graph gives estimates of the number of species for various kinds of insects.

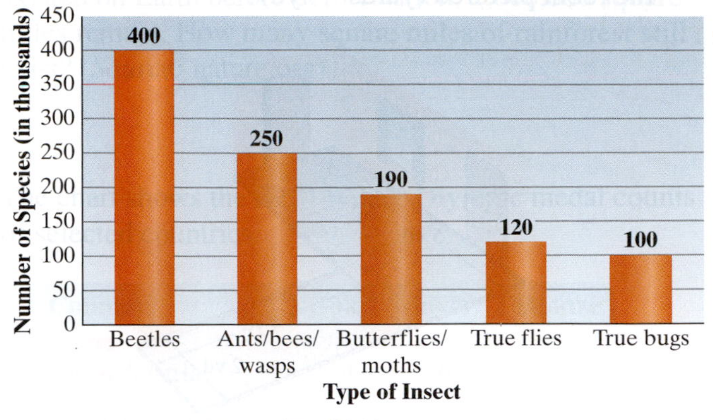

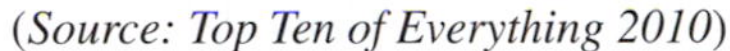
(*Source: Top Ten of Everything 2010*)

a. How many more beetle species are there than true-fly species?

b. Find the total number of species for all the insect types shown.

c. How many fewer beetle species are there than species of the other insect types shown?

114. The following graph shows the number of cases of measles for selected years in the United States as reported by the Centers for Disease Control (CDC).

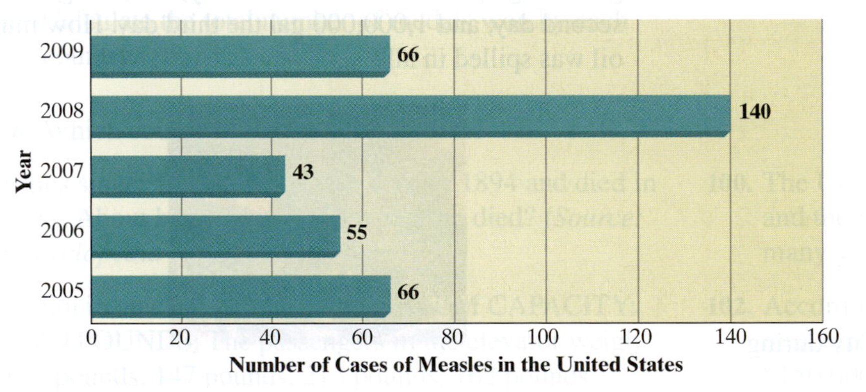

(*Source:* wonder.cdc.gov)

a. How many more cases of measles were reported in 2009 than in 2007?

b. What is the total number of cases reported for the years 2005–2009?

c. How many fewer cases were reported in 2008 than the other years combined?

115. A salesman for a car dealership works on commission. The following table gives his sales commission for each of the first six months of the year.

Month	Sales Commission
January	$5,416
February	$3,791
March	$5,072
April	$3,959
May	$6,283
June	$4,055

What is his total commission for the first half of the year?

116. At its first eight games this season, a professional baseball team had the following paid attendance:

Game	Attendance
1	11,862
2	18,722
3	14,072
4	9,713
5	25,913
6	28,699
7	19,302
8	18,780

What was the combined attendance for these games?

• Check your answers on page A-1.

MINDStretchers

Writing

1. There are many different ways of putting numerical expressions into words.

   a. For example, $3 + 2$ can be expressed as

   the sum of 3 and 2, 2 more than 3, or 3 increased by 2

   What are some other ways of reading this expression?

   b. For example, $5 - 2$ can be expressed as

   the difference between 5 and 2, 5 take away 2, or 5 decreased by 2

   Write two other ways.

Critical Thinking

2. In a **magic square**, the sum of every row, column, and diagonal is the same number. Using the given information, complete the square at the right, which contains the whole numbers from 1 to 16. (*Hint:* The sum of every row, column, and diagonal is 34.)

16	3	2	
	10	11	
	6	7	

Groupwork

3. Two methods for borrowing in a subtraction problem are illustrated as follows. In method (a)—the method that we have already discussed—we borrow by taking 1 from the top, and in method (b) by adding 1 to the bottom.

a.
$$\begin{array}{r} \overset{7}{\cancel{8}}\ \ {}^{1}5\ \ 9 \\ -3\ \ 7\ \ 6 \\ \hline 4\ \ 8\ \ 3 \end{array}$$

b.
$$\begin{array}{r} 8\ \ {}^{1}5\ \ 9 \\ -\overset{4}{\cancel{3}}\ \ 7\ \ 6 \\ \hline 4\ \ 8\ \ 3 \end{array}$$

Note that we get the same answer with both methods. Working with a partner, discuss the advantages of each method.

1.3 Multiplying Whole Numbers

OBJECTIVES

- **A** To multiply whole numbers
- **B** To solve applied problems involving the multiplication of whole numbers

The Meaning and Properties of Multiplication

What does it mean to multiply whole numbers? A good answer to this question is *repeated addition.*

For instance, suppose that you buy 4 packages of pens and each package contains 3 pens. How many pens are there altogether?

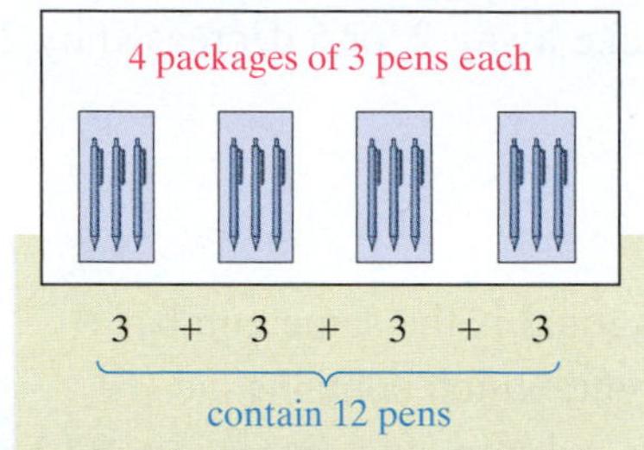
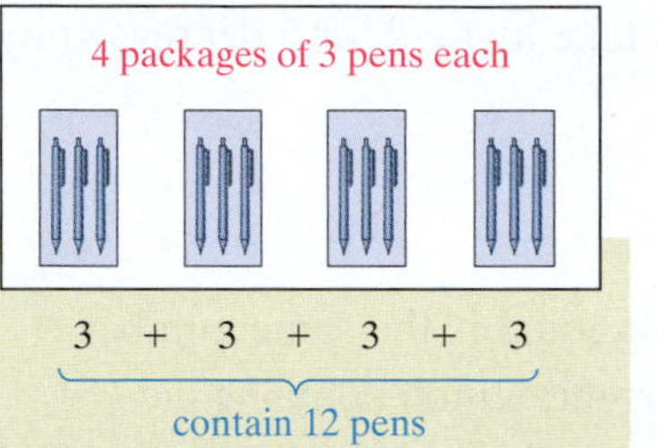

That is, $4 \times 3 = 3 + 3 + 3 + 3 = 12$. Generally, *multiplication means adding the same number repeatedly.*

We can also picture multiplication in terms of a rectangular figure, like this one, that represents 4×3.

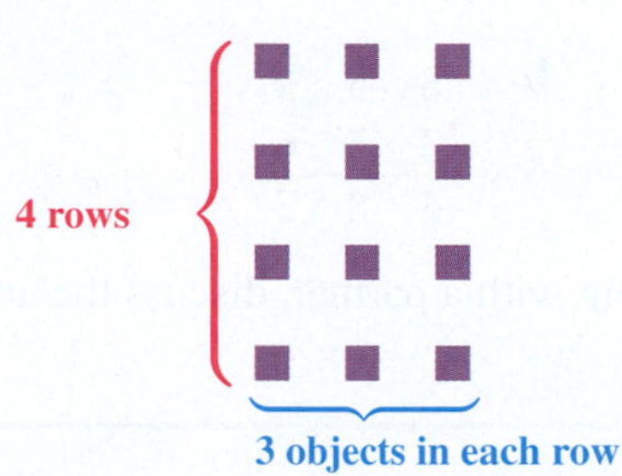

In a multiplication problem, the numbers being multiplied are called *factors*. The result is the *product*.

There are several ways to write a multiplication problem.

		Factor		Factor		Product
$\times$	the times sign	4	$\times$	3	=	12
$\cdot$	a multiplication dot	4	$\cdot$	3	=	12
( )( )	parentheses	(4)(3)			=	12

Like addition and subtraction, multiplication problems can be written either horizontally or vertically.

$$8 \times 5 = 40$$
Horizontal

$$\begin{array}{r} 8 \\ \times\ 5 \\ \hline 40 \end{array}$$
Vertical

The operation of multiplication has several important properties that we use frequently.

The Identity Property of Multiplication

The product of any number and 1 is that number.

$$1 \times 12 = 12$$
$$5 \times 1 = 5$$

The Multiplication Property of 0

The product of any number and 0 is 0.

$$49 \times 0 = 0$$
$$0 \times 8 = 0$$

The Commutative Property of Multiplication

Changing the order in which two numbers are multiplied does not affect their product.

$$2 \times 9 = 9 \times 2$$
$$18 = 18$$

The Associative Property of Multiplication

When multiplying three numbers, regrouping the factors gives the same product.

We multiply inside the parentheses first.

$$(3 \times 4) \times 5 = 3 \times (4 \times 5)$$
$$12 \times 5 = 3 \times 20$$
$$60 = 60$$

The next—and last—property of multiplication also involves addition.

The Distributive Property

Multiplying a factor by the sum of two numbers gives the same result as multiplying the factor by each of the two numbers and then adding.

$$2 \times (5 + 3) = (2 \times 5) + (2 \times 3)$$
$$2 \times 8 = 10 + 6$$
$$16 = 16$$

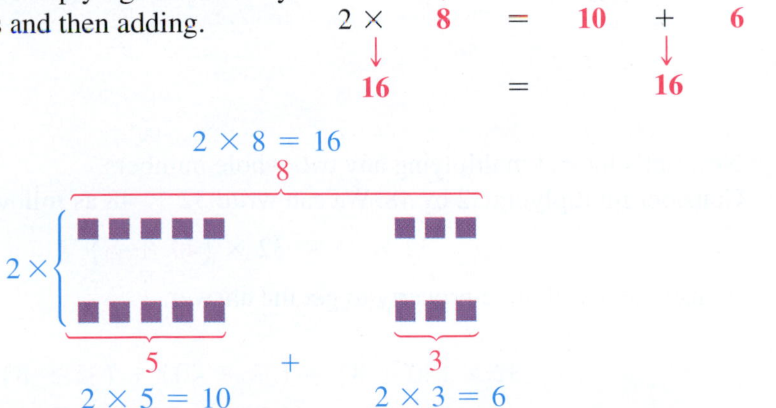

Before going on to the next section, study these properties of multiplication.

Multiplying Whole Numbers

Now, let's consider problems in which we multiply any whole number by a single-digit whole number.

Note that, to multiply whole numbers with reasonable speed, you must commit to memory the products of all single-digit whole numbers.

EXAMPLE 1

Multiply: $98 \cdot 4$

Solution We recall that the dot means multiplication. We first write the problem vertically.

We recall that the 9 in 98 means 9 tens.

So the product of 98 and 4 is 392.

$$\begin{array}{r} \overset{3}{9}\,8 \\ \times\ 4 \\ \hline 2 \end{array}$$

← The product of 4 and 8 ones is 32 ones. We write the 2 and carry the 3 to the tens column.

$$\begin{array}{r} \overset{3}{9}\,8 \\ \times\ \ 4 \\ \hline 3\,9\,2 \end{array}$$

← The product of 4 and 9 tens is 36 tens. We add the 3 tens to the 36 tens to get 39 tens.

PRACTICE 1

Find the product of 76 and 8.

EXAMPLE 2

Calculate: (806) (7)

Solution We recall that parentheses side-by-side mean to multiply. We write this problem vertically.

$$\begin{array}{r} 8\,\overset{4}{0}\,6 \\ \times\ \ \ 7 \\ \hline 5,6\,4\,2 \end{array}$$

Here, 7×0 tens $= 0$ tens. Add the 4 tens to the 0 tens to get 4 tens.

The product of 806 and 7 is 5,642.

PRACTICE 2

Find the product: (705)(6)

Now, let's look at multiplying any two whole numbers.
Consider multiplying 32 by 48. We can write 32×48 as follows.

$$32 \times 48 = 32 \times (40 + 8)$$

We then use the distributive property to get the answer.

$$\begin{aligned} 32 \times (40 + 8) &= (32 \times 40) + (32 \times 8) \\ &= 1{,}280 + 256 \\ &= 1{,}536 \end{aligned}$$

Generally, we solve this problem vertically.

$$\begin{array}{r} \overset{1}{3}\,2 \\ \times\ \ 4\,8 \\ \hline 2\,5\,6 \\ 1\,2\,8\,0 \\ \hline 1,5\,3\,6 \end{array}$$

- 256 ← Partial product (8×32)
- 1280 ← Partial product (40×32)
- 1,536 ← Add the partial products.

Shortcut

$$\begin{array}{r} 32 \\ \times\ 48 \\ \hline 256 \\ 1\,28 \\ \hline 1{,}536 \end{array}$$

- 256 ← (8×32)
- 128 ← (4×32)

If we use just the tens digit 4, we must write the product 128 leftward, starting at the tens column.

Example 2 suggests the following rule for multiplying whole numbers:

To Multiply Whole Numbers

- Multiply the top factor by the ones digit in the bottom factor, and write down this product.
- Multiply the top factor by the tens digit in the bottom factor, and write this product leftward, beginning with the tens column.
- Repeat this process until all the digits in the bottom factor are used.
- Add the partial products, writing down this sum.

EXAMPLE 3

Multiply: 300 × 50

Solution

```
  300
×  50
  000  ← 0 × 300 = 0
15 00  ← 5 × 300 = 1,500
15,000
```

In Example 3, note that the number of zeros in the product equals the total number of zeros in the factors. This result suggests a shortcut for multiplying factors that end in zeros.

```
   300 ← 2 zeros
×   50 ← 1 zero
15,000 ← 2 + 1 = 3 zeros
```

PRACTICE 3

Find the product of 1,200 and 400.

TIP When multiplying two whole numbers that end in zeros, multiply the nonzero parts of the factors and then attach the total number of zeros to the product.

EXAMPLE 4

Simplify: $739 \cdot 305$

Solution

```
    739
×   305
  3 695  ← 5 × 739
  0 00   ← 0 × 739 = 0
221 7    ← 3 × 739
225,395
```

We don't have to write the row 000. Here is a shortcut.

```
    739
×   305
  3 695
221 70   ← This one 0 represents the product of the tens digit 0
225,395    and 739. This 0 lines up the products correctly.
```

PRACTICE 4

Find the product of 987 and 208.

Now, let's apply the operation of multiplication to geometric area. Area means the number of square units that a figure contains.

In the rectangle at the right, each small square represents 1 square inch (sq in.). Finding the rectangle's area means finding the number of sq-in. units that it contains. A good strategy here is to find the number of units in each row and then multiply that number by the number of rows.

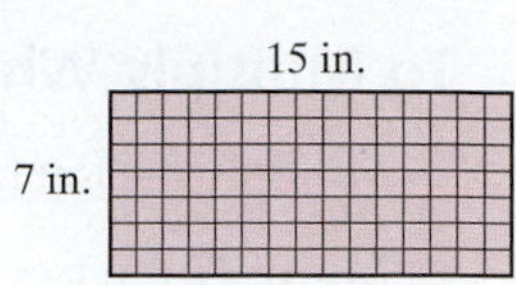

There are two ways to find that there are 15 squares in a row—either by directly counting the squares or by noting that the length of the figure is 15 in. Similarly, we find that the figure contains 7 rows. Therefore the area of the figure is 15×7, or 105 sq in.

In general, we can compute the *area of a rectangle* by finding the product of its length and its width.

EXAMPLE 5

Calculate the area of the home office shown in the diagram.

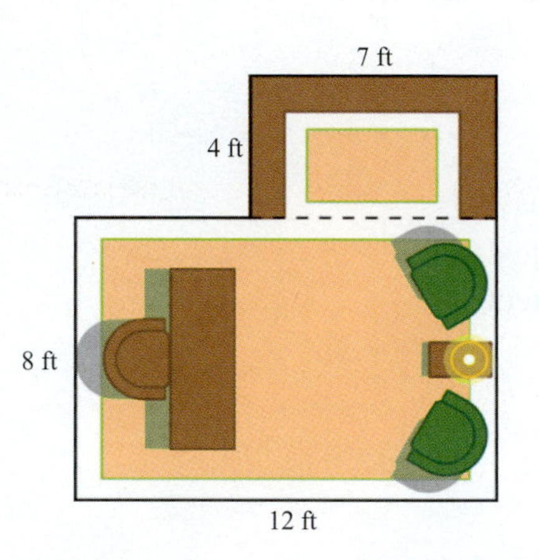

Solution The dashed line separates the office into two connected rectangles. The top rectangle measures 7 feet by 4 feet, and so its area is 7×4, or 28 square feet. The bottom rectangle measures 12 feet by 8 feet, and its area is 12×8, or 96 square feet. The entire area of the office is the sum of two smaller areas: $28 + 96$, or 124 square feet. So the area of the home office is 124 square feet.

PRACTICE 5

Find the area of the room pictured.

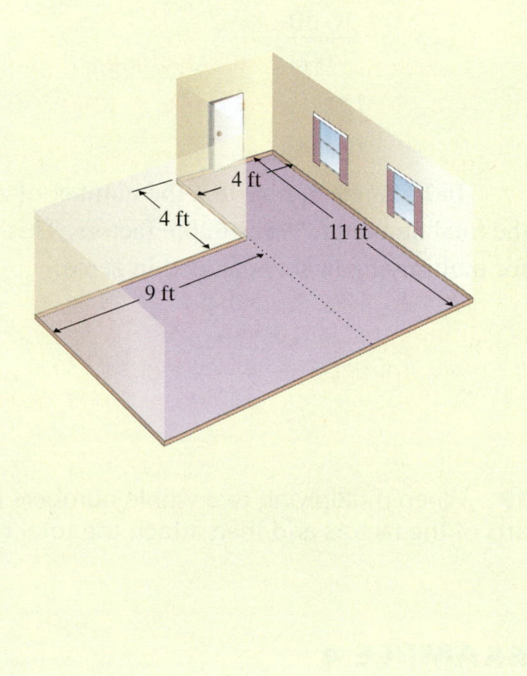

Estimating Products

As mentioned before, estimation is a valuable technique for checking an exact answer. When checking a product by estimation, round each factor to its largest place.

EXAMPLE 6

Multiply 328 by 179. Check the answer by estimation.

Solution

```
   328
 × 179
 2 952
22 96
32 8
58,712 ← Exact product
```

PRACTICE 6

Find the product of 455 and 248. Use estimation to check your answer.

Check

$$\begin{array}{rcr} 328 & \approx & 300 \leftarrow \text{The largest place is hundreds.} \\ \times\ 179 & \approx & \times\ 200 \leftarrow \text{The largest place is hundreds.} \\ \hline 58{,}712 & & 60{,}000 \leftarrow \text{Estimated product} \end{array}$$

Our exact product (58,712) and the estimated product (60,000) are fairly close.

When solving some multiplication problems, we are willing to settle for—or even prefer—an approximate answer.

EXAMPLE 7

The director of a preschool budgeted $900 to purchase supplies for an upcoming art project. She found handprint keepsake craft kits online for $12 each for the 56 children in the preschool. By estimating, decide if the director set aside enough money to purchase craft kits for all the children.

Solution The total cost of the craft kits is the product of $12 and 56. To estimate this product, we first round each factor to its largest place value so that every digit after the first digit is 0.

$$\begin{array}{rcr} 12 & \approx & 10 \leftarrow \text{The largest place is tens.} \\ \times\ 56 & \approx & 60 \leftarrow \text{The largest place is tens.} \end{array}$$

Then, we multiply the rounded factors.

$$10 \times 60 = 600$$

Since the craft kits will cost about $600 and $900 is greater than $600, we conclude that the director set aside enough money for the craft kits.

PRACTICE 7

Producing flyers for your college's registration requires 25,000 sheets of paper. If the college buys 38 reams of paper and there are 500 sheets in a ream, estimate to decide if there is enough paper to produce the flyers.

Multiplying Whole Numbers on a Calculator

Now, let's use a calculator to find a product. When you are using a calculator to multiply large whole numbers, the answer may be too big to fit in the display. When this occurs the answer may be displayed in scientific notation (see the Appendix).

EXAMPLE 8

Use a calculator to multiply: 3,192 × 41

Solution

Press	Display
3192 [×] 41 [ENTER =]	3192*41 130872.

A reasonable estimate for this product is 3,000 × 40, or 120,000, which supports our answer, 130,872.

PRACTICE 8

Find the product: 2,811 × 365

EXAMPLE 9

Calculate: 61 · 24 · 19

Solution

Press	Display
61 [×] 24 [×] 19 [ENTER =]	61*24*19 27816.

A good estimate is 60 · 20 · 20, or 24,000—in the ballpark of 27,816.

PRACTICE 9

Multiply: 2,133 · 18 · 9

1.3 Exercises

FOR EXTRA HELP MyMathLab MathXL PRACTICE WATCH READ REVIEW

Mathematically Speaking

Fill in each blank with the most appropriate term or phrase from the given list.

associative property of multiplication	identity property of multiplication	multiplication property of 0	sum product	subtraction addition
distributive property	perimeter	area		

1. The result of multiplying two factors is called their ___________.
2. The ________________ is illustrated by $3 \times (7 + 2) = (3 \times 7) + (3 \times 2)$.
3. The ____________________________ states that the product of any number and 1 is that number.
4. The ______________________ states that the product of any number and 0 is 0.
5. The multiplication of whole numbers can be thought of as repeated __________.
6. The __________ of a figure is the number of square units that it contains.

A *Compute.*

7. 4×100
8. 5×100
9. 710×200
10. 270×500
11. $8{,}500 \times 20$
12. $6{,}800 \times 30$
13. $10{,}000 \times 700$
14. $10{,}000 \times 800$

Multiply and check by estimation.

15. $6{,}350 \times 2$
16. $8{,}864 \times 7$
17. 209×2
18. 703×9
19. $812{,}000 \times 4$
20. $19{,}250 \times 8$
21. 882×74
22. 881×28
23. $43 \cdot 19$
24. $85 \cdot 72$
25. $709 \cdot 48$
26. $602 \cdot 34$
27. 273×11
28. 607×65
29. 301×12
30. 513×34
31. $3{,}001 \times 19$
32. $4{,}005 \times 72$
33. $5{,}072 \times 48$
34. $8{,}801 \times 25$
35. $5{,}003 \times 40$
36. $2{,}881 \times 70$

Find the product and check by estimation.

37. (372)(403)
38. (699)(101)
39. $8{,}500 \times 17$
40. $7{,}200 \times 27$
41. 406×305
42. 702×509
43. $46 \cdot 8 \cdot 9$
44. $13 \cdot 11 \cdot 5$

81. The state of Colorado is approximately rectangular in shape, as shown. If the area of Kansas is about 82,000 sq mi, which state is larger? (*Source: The Columbia Gazeteer of the World*)

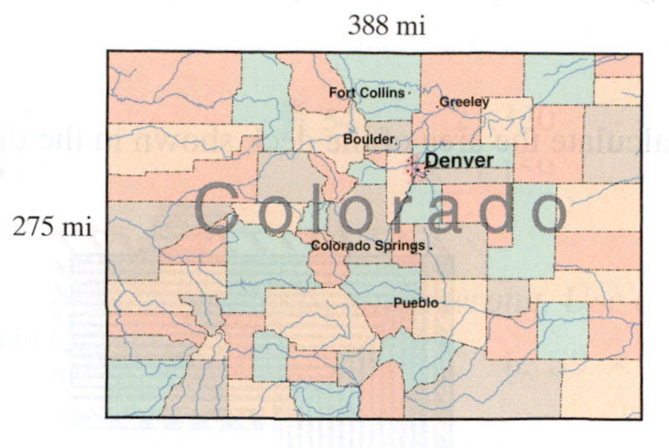

82. Tuition at a state college for full-time in-state residents is approximately $1,750 per semester. In a fall semester, there were 30,963 of these students. How much revenue did the full-time in-state resident students generate?

• Check your answers on page A-1.

MINDStretchers

Writing

1. Study the following diagram. Explain how it justifies the Distributive Property.

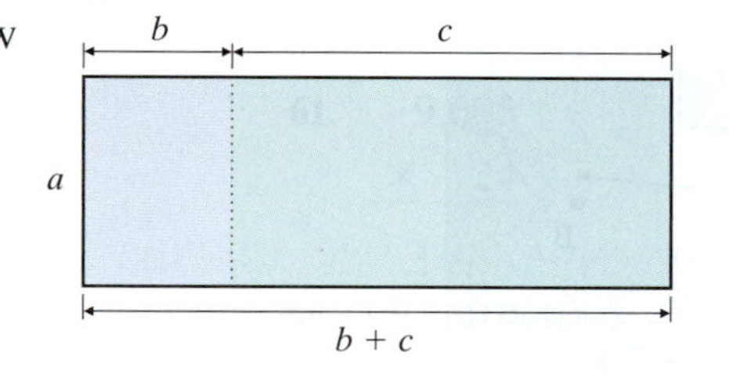

Mathematical Reasoning

2. Consider the six digits 1, 3, 5, 7, 8, and 9. Fill in the blanks with these digits, using each digit only once, so as to form the largest possible product.

__ __ __ × __ __ __

Historical

3. Centuries ago in India and Persia, the **lattice method** of multiplication was popular. The following example, in which we multiply 57 by 43, illustrates this method. Explain how it works.

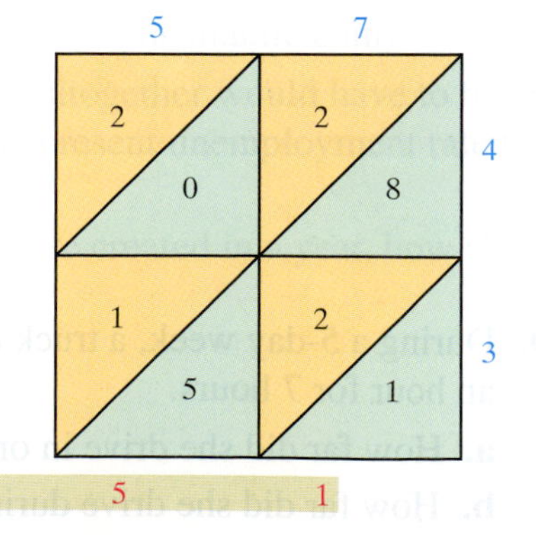

Cultural Note

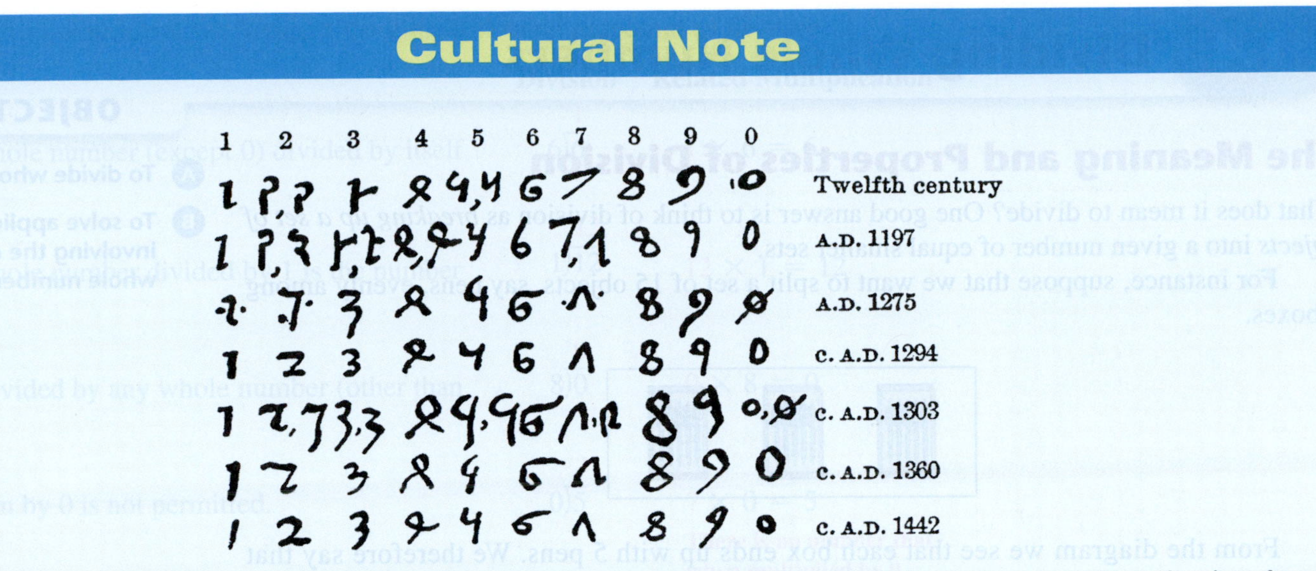

The way the ten digits are written has evolved over time. Early Hindu symbols found in a cave in India date from more than two thousand years ago. About twelve hundred years ago, an Indian manuscript on arithmetic, which had been translated into Arabic, was carried by merchants to Europe where it was later translated into Latin.

This table shows European examples of digit notation from the twelfth to the fifteenth century, when the printing press led to today's standardized notation. Through international trade, these symbols became known throughout the world.

Source: David Eugene Smith and Jekuthiel Ginsburg, *Numbers and Numerals, A Story Book for Young and Old* (New York: Bureau of Publications, Teachers College, Columbia University, 1937)

Instead of writing 0's as placeholders, we can use the following shortcut.

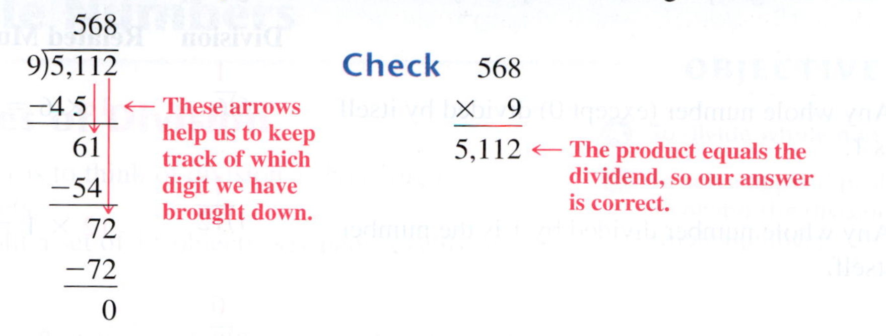

568
9)5,112
−4 5
61
−54
72
−72
0
These arrows help us to keep track of which digit we have brought down.
Check 568
× 9
5,112
The product equals the dividend, so our answer is correct.

Note that each time we subtract in a division problem, the difference is less than the divisor. Why must that be true?

EXAMPLE 1

Divide and check: 4,263 ÷ 7

Solution

Think 7)42.

```
     609
  7)4,263
   −4 2     ← 6 × 7 = 42. Subtract.
      06    ← Think 7)6. There are zero 7's in 6.
     −0     ← 0 × 7 = 0. Subtract.
      63    ← Think 7)63.
     −63    ← 9 × 7 = 63. Subtract.
       0
```

Check

```
   609
 ×   7
 4,263
```

The product agrees with our dividend. Note the 0 in the quotient. Can you explain why the 0 is needed?

PRACTICE 1

Compute 9)7,263 and then check your answer.

TIP In writing your answer to a division problem, position the first digit of the quotient over the *rightmost digit* of the number into which you are dividing (the 6 over the 2 in Example 1).

```
    609
    ↓↓↓
 7)4,263
```

EXAMPLE 2

Compute $\frac{2,709}{9}$. Then check your answer.

Solution

```
     301
  9)2,709
   −2 7
      00
     −0
      09
     −9
       0
```

Check

```
   301
 ×   9
 2,709
```

PRACTICE 2

Carry out the following division and check your answer.

8)56,016

In Examples 1 and 2, note that the remainder is 0; that is, the divisor goes evenly into the dividend. However, in some division problems, that is not the case. Consider, for instance, the problem of dividing 16 pens *equally* among 3 boxes.

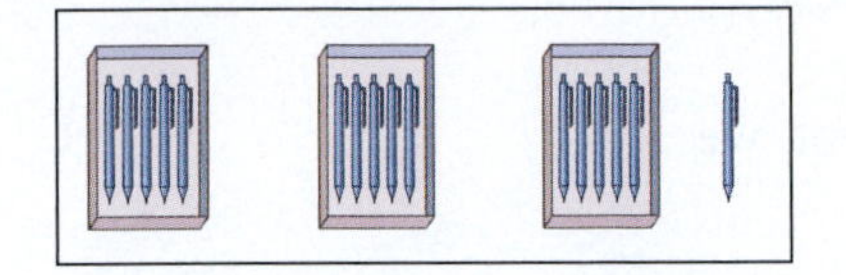

From the diagram, we see that each box contains 5 pens *but* that 1 pen—the *remainder*—is left over.

$$\begin{array}{rl} 5 & \leftarrow \text{Number of pens in each box} \\ \text{Number of boxes} \longrightarrow 3\overline{)16} & \leftarrow \text{Total number of pens} \\ \underline{-15} & \leftarrow \text{Total number of pens in the boxes} \\ 1 & \leftarrow \text{Number of pens remaining} \end{array}$$

We write the answer to this problem as 5 R1 (read "5 Remainder 1"). Note that $(3 \times 5) + 1 = 16$. The following relationship is always true.

$$(\text{Quotient} \times \text{Divisor}) + \text{Remainder} = \text{Dividend}$$

When a division problem results in a remainder as well as a quotient, we use this relationship for checking.

EXAMPLE 3

Find the quotient of 55,811 and 6. Then check.

Solution

$$\begin{array}{r} 9{,}301\text{ R5} \\ 6\overline{)55{,}811} \\ \underline{-54} \\ 1\,8 \\ \underline{-1\,8} \\ 01 \\ \underline{-0} \\ 11 \\ \underline{-6} \\ 5 \end{array}$$

Our answer is therefore 9,301 R5.

Check

(Quotient × Divisor) + Remainder = Dividend

$$(9{,}301 \times 6) + 5 =$$
$$55{,}806 + 5 = 55{,}811$$

Since this matches the dividend, our answer checks.

PRACTICE 3

Compute $8\overline{)42{,}329}$ and check.

Now, let's consider division problems in which a divisor has more than one digit. Notice that such problems involve rounding.

EXAMPLE 4

Compute $\dfrac{2,574}{34}$ and check.

Solution In order to estimate the first digit of the quotient, we round 34 to 30 and 257 to 260.

$$\begin{array}{r} 8 \\ 34\overline{)2,574} \\ \underline{-2\ 72} \end{array}$$

← Think 260 ÷ 30, or 26 ÷ 3. The quotient 8 goes over the 7 because we are dividing 34 into 257.

← 8 × 34 = 272. Try to subtract.

Because 272 is too large, we reduce our estimate in the quotient by 1 and try 7.

$$\begin{array}{r} 76 \\ 34\overline{)2,574} \\ \underline{-2\ 38} \\ 194 \\ \underline{-204} \end{array}$$

← 7 × 34 = 238. Subtract.

← Think 190 ÷ 30 or 19 ÷ 3.

← 6 × 34 = 204. Try to subtract.

Because 204 is too large, we reduce our estimate in the quotient by 1 and try 5.

$$\begin{array}{r} 75\ \text{R}24 \\ 34\overline{)2,574} \\ \underline{-2\ 38} \\ 194 \\ \underline{-170} \\ 24 \end{array}$$

So our answer is 75 R24.

Check $(75 \times 34) + 24 = 2,574$

Since 2,574 is the dividend, our answer checks.

PRACTICE 4

Divide 23 into 1,818. Then check.

EXAMPLE 5

Divide $26\overline{)1,849}$ and then check.

Solution First, we round 26 to 30 and 184 to 180. Think $180 \div 30 = 6$.

$$\begin{array}{r} 6 \\ 26\overline{)1,849} \\ \underline{-1\ 56} \\ 28 \end{array}$$

← This difference is larger than the divisor, so we increase the 6 in the quotient by 1.

$$\begin{array}{r} 71 \\ 26\overline{)1,849} \\ \underline{-182} \\ 29 \\ \underline{-26} \\ 3 \end{array}$$

Our answer is therefore 71 R3.

Check $(71 \times 26) + 3 = 1,849$

PRACTICE 5

Compute and check: $15\overline{)1,420}$

TIP If the divisor has more than one digit, estimate each digit in the quotient by rounding and then dividing. If the product is too large or too small, adjust it up or down by 1 and then try again.

EXAMPLE 6

Find the quotient of 13,559 and 44. Then check.

Solution

```
     308 R7
44)13,559
  -13 2
      35   ← This number is smaller than the divisor, so
     -0       the next digit in the quotient is 0.
     359
    -352
       7
```

Check $(308 \times 44) + 7 = 13{,}559$

PRACTICE 6

Divide 16,999 by 28. Then check your answer.

EXAMPLE 7

Divide and check: $6{,}000 \div 20$

Solution We set up the problem as before.

```
     300
20)6,000
  -60
    00
   -00
     00
    -00
      0
```

Check

```
  300
×  20
6,000
```

Because the divisor and dividend both end in zero, a quicker way to do Example 7 is by dropping zeros.

2~~0~~)6,00~~0~~ ← Drop one 0 from both the divisor and the dividend.

```
   300
2)600  ← Then divide.
```

PRACTICE 7

Compute 40)8,000 and then check.

TIP Dropping the same number of zeros at the right end of both the divisor and the dividend does not change the quotient.

Estimating Quotients

As for other operations, estimating is an important skill for division. Checking a quotient by estimation is faster than checking it by multiplication, although less exact. And in some division problems, we need only an approximate answer.

How do we estimate a quotient? A good way is to round the divisor to its greatest place. The new divisor then contains only one nonzero digit and so is relatively easy to divide by mentally. Then, we round the dividend to the place of our choice. Finally, we compute the estimated quotient by calculating its first digit and then attaching the appropriate number of zeros.

1.4 Exercises

FOR EXTRA HELP MyMathLab MathXL PRACTICE WATCH READ REVIEW

Mathematically Speaking

Fill in each blank with the most appropriate term or phrase from the given list.

subtraction	divided	divisor	multiplication
quotient	product	increased	

1. When dividing, the dividend is divided by the ________.

2. The result of dividing is called the ________.

3. The opposite operation of division is ________.

4. Any whole number ________ by 1 is equal to the number itself.

A *Divide and check.*

5. $5\overline{)2,000}$ **6.** $5\overline{)6,000}$ **7.** $5\overline{)12,800}$ **8.** $8\overline{)12,504}$

9. $9\overline{)2,709}$ **10.** $2\overline{)5,780}$ **11.** $7\overline{)21,021}$ **12.** $5\overline{)27,450}$

13. $3\overline{)24,132}$ **14.** $2\overline{)30,534}$ **15.** $9\overline{)4,500}$ **16.** $3\overline{)4,512}$

Find the quotient and check.

17. $300 \div 10$ **18.** $400 \div 20$ **19.** $700 \div 50$ **20.** $6,000 \div 20$

21. $\frac{8,400}{200}$ **22.** $\frac{7,500}{300}$ **23.** $\frac{16,000}{40}$ **24.** $\frac{48,000}{20}$

25. $6,996 \div 44$ **26.** $9,660 \div 92$ **27.** $80,295 \div 15$ **28.** $31,031 \div 13$

29. $39,078 \div 39$ **30.** $49,497 \div 21$ **31.** $249,984 \div 36$ **32.** $499,992 \div 24$

33. $52\overline{)52,052}$ **34.** $24\overline{)48,072}$ **35.** $12\overline{)36,600}$ **36.** $36\overline{)25,560}$

37. $25\overline{)22,675}$ **38.** $15\overline{)30,480}$ **39.** $49\overline{)58,849}$ **40.** $19\overline{)38,570}$

41. $6,512 \div 10$ **42.** $8,922 \div 25$ **43.** $304 \div 27$ **44.** $206 \div 45$

45. $\frac{10,175}{87}$ **46.** $\frac{21,109}{25}$ **47.** $\frac{63,002}{90}$ **48.** $\frac{12,509}{61}$

49. $47\overline{)34,000}$ **50.** $66\overline{)99,980}$ **51.** $14\overline{)6,000}$ **52.** $32\overline{)3,007}$

53. $65\overline{)65,660}$ **54.** $39\overline{)30,009}$ **55.** $42\overline{)39,000}$ **56.** $97\overline{)13,502}$

57. $537\overline{)387,177}$ **58.** $265\overline{)197,160}$ **59.** $638\overline{)98,890}$ **60.** $152\overline{)34,048}$

In each group of three quotients, one is wrong. Use estimation to identify which quotient is incorrect.

61. **a.** $18,473 \div 91 = 203$ **b.** $43,364 \div 74 = 586$ **c.** $14,562 \div 18 = 8,009$

62. **a.** $43,710 \div 93 = 47$ **b.** $71,048 \div 107 = 664$ **c.** $11,501 \div 31 = 371$

63. **a.** $455,260 \div 65 = 704$ **b.** $11,457 \div 57 = 201$ **c.** $10,044 \div 93 = 108$

64. **a.** $178,267 \div 89 = 2,003$ **b.** $350,007 \div 21 = 1,667$ **c.** $37,185 \div 37 = 1,005$

Mixed Practice

Divide and check.

65. 38,095 ÷ 42

66. $\dfrac{63{,}147}{21}$

67. $6\overline{)12{,}000}$

68. 4,907 ÷ 7

69. $\dfrac{48{,}000}{20}$

70. $36\overline{)249{,}986}$

71. $\dfrac{3{,}330}{9}$

72. 4,090 ÷ 91

Applications

B *Solve and check.*

73. A part-time student is taking 9 credit-hours this semester at a local community college. If her tuition bill is $1,215, how much does each credit-hour cost?

74. A car used 15 gallons of gas on a 300-mile trip. How many miles per gallon (mpg) of gas did the car get?

75. The area of the Pacific Ocean is about 64 million square miles, and the area of the Atlantic Ocean is approximately 32 million square miles. The Pacific is how many times as large as the Atlantic? (*Source: The New Encyclopedia Britannica*)

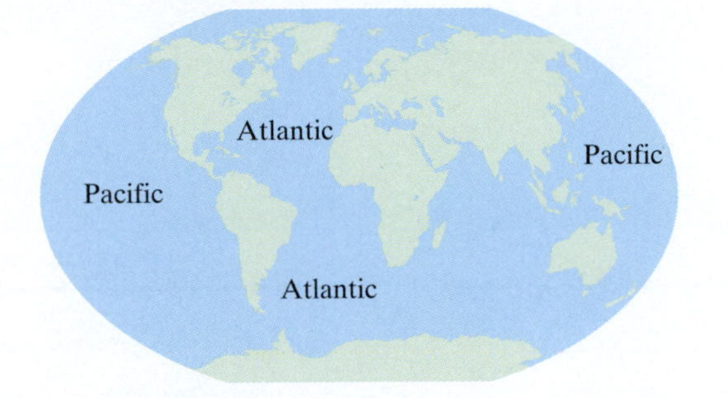

76. The diameter of Earth is about 8,000 miles, whereas the diameter of the Moon is about 2,000 miles. How many times the Moon's diameter is Earth's? (*Source: The New Encyclopedia Britannica*)

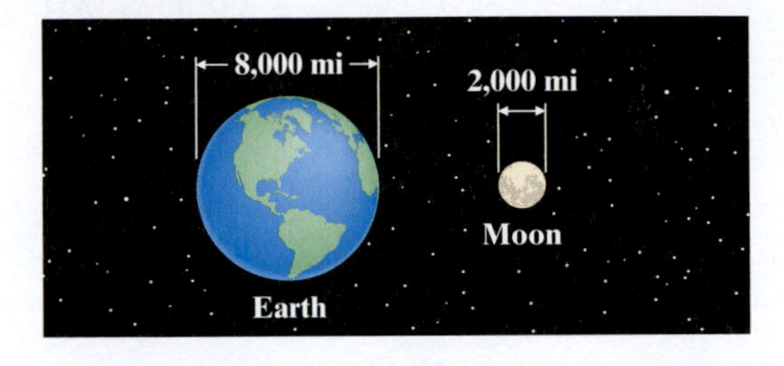

77. In the year 2030, Ohio is projected to have a population of about 12,300,000 people. If Ohio has a total land area of about 41,000 square miles, how many people per square mile will there be in 2030? (*Source:* Ohio Department of Development)

78. A certified medical assistant has an annual salary of $26,472. What is her gross monthly income?

79. A 150-pound person can burn about 360 calories in 1 hour doing yoga. How many calories are burned in 1minute? (*Source:* American Cancer Society)

80. Ryan Howard signed a 5-year, $125,000,000 contract extension with the Philadelphia Phillies in 2010. On average, what is his pay per year from the contract extension? (*Source:* sportsillustrated.com)

81. A homeowner is remodeling a bathroom with dimensions 96 inches and 114 inches. For the floor, she has selected tiles that measure 6 inches by 6 inches.

a. How many tiles must she purchase?

b. The tiles come in boxes of 12. How many boxes of tiles must she purchase?

c. If each box of tiles costs $18, how much will she spend on the tiles for the floor?

82. The group admission rate for 15 or more people at Six Flags Great Adventure amusement park is $30 per person. A student group hosted a field trip to the park and charged $47 per ticket, covering both the cost of admission to the park and the bus transportation. (*Source:* Six Flags Great Adventure, 2010)

a. If the total amount the group collected for tickets was $1,739, how many students went on the field trip?

b. Calculate the total cost of admissions for the students on the field trip.

c. What was the cost of the bus transportation?

• Check your answers on page A-2.

MINDStretchers

Writing

1. Use the problem $10 \div 2 = 5$ to help explain why division can be thought of as repeated subtraction.

Mathematical Reasoning

2. Consider the following pair of problems.

   a. $2\overline{)7}$ **b.** $4\overline{)13}$

   Are the answers the same? Explain.

Groupwork

3. In the following division problem, A, B, and C each stand for a different digit. Working with a partner, identify all the digits. (*Hint:* There are two answers.)

```
      ABA
AB ) CACAB
    -CAB
      CA
     -B
      CAB
     -CAB
```

1.5 Exponents, Order of Operations, and Averages

OBJECTIVES

- **A** To evaluate expressions involving exponents
- **B** To evaluate expressions using the rule for order of operations
- **C** To compute averages
- **D** To solve applied problems involving exponents, order of operations, or averages

Exponents

There are many mathematical situations in which we multiply a number by itself repeatedly. Writing such expressions in **exponential form** provides a shorthand method for representing this repeated multiplication of the same factor.

For instance, we can write $5 \cdot 5 \cdot 5 \cdot 5$ in exponential form as

$$5^4$$

(4 ← Exponent; 5 ← Base)

This expression is read "5 to the fourth *power*" or simply "5 to the fourth."

DEFINITION

An **exponent** (or **power**) is a number that indicates how many times another number (called the **base**) is used as a factor.

We read the power 2 or the power 3 in a special way. For instance, 5^2 is usually read "5 *squared*" rather than "5 to the second power." Similarly, we usually read 5^3 as "5 *cubed*" instead of "5 to the third power."

Let's look at a number written in exponential form—namely, 2^4. To evaluate this expression, we multiply 4 factors of 2.

$$\begin{aligned} 2^4 &= \underbrace{2 \cdot 2} \cdot 2 \cdot 2 \\ &= \underbrace{4 \cdot 2} \cdot 2 \\ &= \underbrace{8 \cdot 2} \\ &= 16 \end{aligned}$$

In short, $2^4 = 16$. Do you see the difference between 2^4 and $2 \cdot 4$?

Sometimes we prefer to shorten expressions by writing them in exponential form. For instance, we can write $3 \cdot 3 \cdot 4 \cdot 4 \cdot 4$ in terms of powers of 3 and 4.

$$\underbrace{3 \cdot 3}_{\text{2 factors of 3}} \cdot \underbrace{4 \cdot 4 \cdot 4}_{\text{3 factors of 4}} = 3^2 \cdot 4^3$$

EXAMPLE 1

Rewrite

$$6 \cdot 6 \cdot 6 \cdot 10 \cdot 10 \cdot 10 \cdot 10$$

in exponential form.

Solution

$$\underbrace{6 \cdot 6 \cdot 6}_{\text{3 factors of 6}} \cdot \underbrace{10 \cdot 10 \cdot 10 \cdot 10}_{\text{4 factors of 10}} = 6^3 \cdot 10^4$$

PRACTICE 1

Write

$$5 \cdot 5 \cdot 5 \cdot 5 \cdot 5 \cdot 2 \cdot 2$$

in terms of powers.

EXAMPLE 2

Compute:

a. 1^5

b. 22^2

Solution

a. $1^5 = \underbrace{1 \cdot 1 \cdot 1 \cdot 1 \cdot 1}$
$= 1$

Note that 1 raised to any power is 1.

b. $22^2 = 22 \cdot 22$
$= 484$

After considering this example, can you explain the difference between squaring and doubling a number?

PRACTICE 2

Calculate:

a. 1^8

b. 11^3

EXAMPLE 3

Write $4^3 \cdot 5^3$ in standard form.

Solution

$$4^3 \cdot 5^3 = (4 \cdot 4 \cdot 4) \cdot (5 \cdot 5 \cdot 5)$$
$$= 64 \cdot 125$$
$$= 8{,}000$$

From this example, do you see the difference between cubing and tripling a number?

PRACTICE 3

Express $7^2 \cdot 2^4$ in standard form.

It is especially easy to compute *powers of 10.*

$$10^2 = 10 \cdot 10 = 1\underbrace{00}_{\text{2 zeros}}, \qquad 10^3 = 10 \cdot 10 \cdot 10 = 1{,}\underbrace{000}_{\text{3 zeros}}$$

$$10^4 = 10 \cdot 10 \cdot 10 \cdot 10 = 1\underbrace{0{,}000}_{\text{4 zeros}}$$

and so on.

Note the pattern.

EXAMPLE 4

The Milky Way, the galaxy to which the Sun and Earth belong, contains about 100 billion stars. Express this number in terms of a power of 10. (*Source: The New York Times Almanac 2010*)

Solution

$$1\underbrace{00{,}000{,}000{,}000}_{\text{11 zeros}} = 10^{11}$$

So the Milky Way contains about 10^{11} stars.

PRACTICE 4

In 1804, the world population reached the milestone of one billion. Represent this number as a power of 10. (*Source:* U.S. Census Bureau)

Order of Operations

Some mathematical expressions involve more than one mathematical operation. For instance, consider $5 + 3 \cdot 2$. This expression seems to have two different values, depending on the order in which we perform the given operations.

Adding first	Multiplying first
$5 + 3 \cdot 2$	$5 + 3 \cdot 2$
$= 8 \cdot 2$	$= 5 + 6$
$= 16$	$= 11$

How are we to know which operation to carry out first? By consensus we agree to follow the rule called the **order of operations** so that everyone always gets the same value for an answer.

Order of Operations Rule

To evaluate mathematical expressions, carry out the operations *in the following order*.

1. First, perform the operations within any grouping symbols, such as parentheses ( ) or brackets [ ].
2. Then, raise any number to its power ■▪.
3. Next, perform all multiplications and divisions as they appear from left to right.
4. Finally, do all additions and subtractions as they appear from left to right.

Start ↓
( )
■▪
× ÷
+ −

Applying this rule to the preceding example gives us the following result.

$5 + 3 \cdot 2$ **Multiply first.**

$= 5 + 6$ **Then add.**

$= 11$

So 11 is the correct answer.

Let's consider more examples that depend on the order of operations rule.

EXAMPLE 5

Simplify: $18 - 7 \cdot 2$

Solution Applying the rule, we multiply first, and then subtract.

$18 - 7 \cdot 2 =$

$18 - 14 = 4$

PRACTICE 5

Evaluate: $2 \cdot 8 + 4 \cdot 3$

EXAMPLE 6

Find the value of $3 + 2 \cdot (8 + 3^2)$.

Solution

$3 + 2 \cdot (8 + 3^2) = 3 + 2 \cdot (8 + 9)$ First, perform the operations in parentheses: square the 3.

$= 3 + 2 \cdot 17$ Then, add 8 and 9.

$= 3 + 34$ Next, multiply 2 by 17.

$= 37$ Finally, add 3 and 34.

PRACTICE 6

Simplify: $(4 + 1)^2 \times 6 - 4$

TIP When a division problem is written in the format $\frac{\square}{\square}$, parentheses are understood to be around both the dividend and the divisor. For instance,

$$\frac{10 - 2}{3 + 1} \text{ means } \frac{(10 - 2)}{(3 + 1)}.$$

EXAMPLE 7

Evaluate: $6 \cdot 2^3 - \frac{21 - 11}{2}$

Solution $6 \cdot 2^3 - \frac{21 - 11}{2} = 6 \cdot 2^3 - \frac{10}{2}$ First, simplify the dividend by subtracting.

$= 6 \cdot 8 - \frac{10}{2}$ Then, cube.

$= 48 - 5$ Next, multiply and divide.

$= 43$ Finally, subtract.

PRACTICE 7

Simplify: $10 + \frac{24}{12 - 8} - 3 \times 4$

Some arithmetic expressions contain not only parentheses but also brackets. When simplifying expressions containing these grouping symbols, first perform the operations within the innermost grouping symbols and then continue to work outward.

EXAMPLE 8

Simplify: $5 \cdot [4(10 - 3^2) - 2]$

Solution

$5 \cdot [4(10 - 3^2) - 2] = 5 \cdot [4(10 - 9) - 2]$ Perform the operation in parentheses: square the 3.

$= 5 \cdot [4 \cdot 1 - 2]$ Subtract 9 from 10.

$= 5 \cdot [4 - 2]$ Multiply.

$= 5 \cdot 2$ Subtract.

$= 10$ Multiply.

PRACTICE 8

Evaluate: $[4 + 3(2^3 - 5)] \cdot 10$

Order of Operations

Some mathematical expressions involve more than one mathematical operation. For instance, consider $5 + 3 \cdot 2$. This expression seems to have two different values, depending on the order in which we perform the given operations.

Adding first	**Multiplying first**
$\underbrace{5 + 3} \cdot 2$	$5 + \underbrace{3 \cdot 2}$
$= \underbrace{8 \cdot 2}$	$= \underbrace{5 + 6}$
$= 16$	$= 11$

How are we to know which operation to carry out first? By consensus we agree to follow the rule called the **order of operations** so that everyone always gets the same value for an answer.

Order of Operations Rule

To evaluate mathematical expressions, carry out the operations *in the following order*.

1. First, perform the operations within any grouping symbols, such as parentheses ( ) or brackets [ ].
2. Then, raise any number to its power $\blacksquare^{\blacksquare}$.
3. Next, perform all multiplications and divisions as they appear from left to right.
4. Finally, do all additions and subtractions as they appear from left to right.

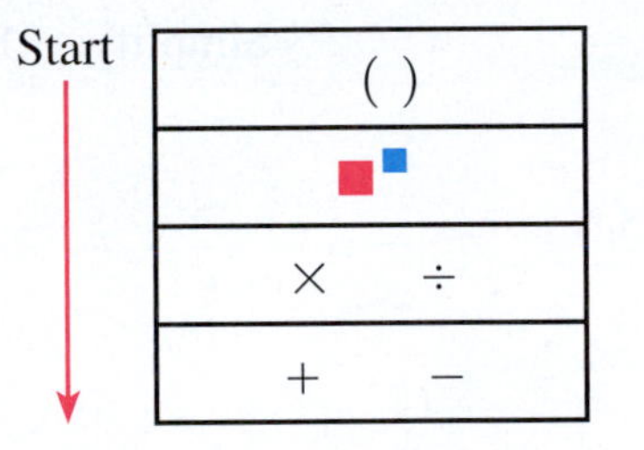

Applying this rule to the preceding example gives us the following result.

$$5 + \underbrace{3 \cdot 2} \quad \text{Multiply first.}$$
$$= \underbrace{5 + 6} \quad \text{Then add.}$$
$$= 11$$

So 11 is the correct answer.

Let's consider more examples that depend on the order of operations rule.

EXAMPLE 5

Simplify: $18 - 7 \cdot 2$

Solution Applying the rule, we multiply first, and then subtract.

$$18 - \underbrace{7 \cdot 2} =$$
$$18 - 14 = 4$$

PRACTICE 5

Evaluate: $2 \cdot 8 + 4 \cdot 3$

EXAMPLE 6

Find the value of $3 + 2 \cdot (8 + 3^2)$.

Solution

$3 + 2 \cdot (8 + 3^2) = 3 + 2 \cdot (8 + 9)$ — First, perform the operations in parentheses: square the 3.

$= 3 + 2 \cdot 17$ — Then, add 8 and 9.

$= 3 + 34$ — Next, multiply 2 by 17.

$= 37$ — Finally, add 3 and 34.

PRACTICE 6

Simplify: $(4 + 1)^2 \times 6 - 4$

TIP When a division problem is written in the format $\frac{\square}{\square}$, parentheses are understood to be around both the dividend and the divisor. For instance,

$$\frac{10 - 2}{3 + 1} \text{ means } \frac{(10 - 2)}{(3 + 1)}.$$

EXAMPLE 7

Evaluate: $6 \cdot 2^3 - \frac{21 - 11}{2}$

Solution

$6 \cdot 2^3 - \frac{21 - 11}{2} = 6 \cdot 2^3 - \frac{10}{2}$ — First, simplify the dividend by subtracting.

$= 6 \cdot 8 - \frac{10}{2}$ — Then, cube.

$= 48 - 5$ — Next, multiply and divide.

$= 43$ — Finally, subtract.

PRACTICE 7

Simplify: $10 + \frac{24}{12 - 8} - 3 \times 4$

Some arithmetic expressions contain not only parentheses but also brackets. When simplifying expressions containing these grouping symbols, first perform the operations within the innermost grouping symbols and then continue to work outward.

EXAMPLE 8

Simplify: $5 \cdot [4(10 - 3^2) - 2]$

Solution

$5 \cdot [4(10 - 3^2) - 2] = 5 \cdot [4(10 - 9) - 2]$ — Perform the operation in parentheses: square the 3.

$= 5 \cdot [4 \cdot 1 - 2]$ — Subtract 9 from 10.

$= 5 \cdot [4 - 2]$ — Multiply.

$= 5 \cdot 2$ — Subtract.

$= 10$ — Multiply.

PRACTICE 8

Evaluate: $[4 + 3(2^3 - 5)] \cdot 10$

EXAMPLE 9

Young's Rule is a rule of thumb for calculating the dose of medicine recommended for a child of a given age. According to this rule, the dose of acetaminophen in milligrams (mg) for a child who is eight years old can be calculated using the expression.

$$\frac{8 \times 500}{8 + 12}$$

What is the recommended dose?

Solution

$$\frac{8 \times 500}{8 + 12} = \frac{4{,}000}{20}$$ First, simplify the dividend and the divisor.

$$= 200$$ Then, divide.

So the recommended dose is 200 milligrams.

PRACTICE 9

The minimum distance (in feet) that it takes a car to stop if it is traveling on a particular road surface at a speed of 30 miles per hour is given by the expression.

$$\frac{10 \times 30^2}{30 \times 5}$$

What is this minimum stopping distance?

Averages

We use an **average** to represent a set of numbers. Averages allow us to compare two or more sets. (For example, do the men or the women in your class spend more time studying?) Averages also allow us to compare an individual with a set. (For example, is the amount of time you spend studying above or below the class average?) The following definition shows how to compute an average.

DEFINITION

The **average** (or **mean**) of a set of numbers is the sum of those numbers divided by however many numbers are in the set.

EXAMPLE 10

What is the average of 100, 94, and 100?

Solution

The average equals the sum of these three numbers divided by 3.

$$\frac{100 + 94 + 100}{3} = \frac{294}{3} = 98$$

PRACTICE 10

Find the average of \$30, \$0, and \$90.

EXAMPLE 11

The following map shows the five Great Lakes. The maximum depth of each of these lakes is given in the table. (*Source:* U.S. Environmental Protection Agency)

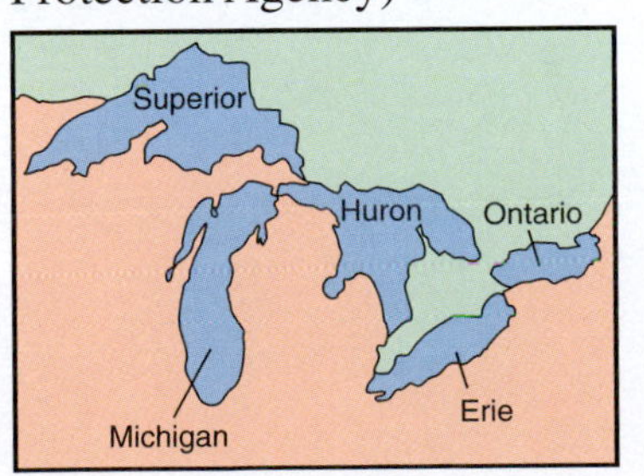

Lake	Maximum Depth (in meters)
Erie	64
Huron	229
Michigan	282
Ontario	244
Superior	406

PRACTICE 11

The table shown gives the number of at-home fatalities due to tornadoes in the United States each year from 2006 through 2009. (*Source:* spc.noaa.gov)

Year	Number of At-Home Fatalities
2006	58
2007	68
2008	99
2009	19

EXAMPLE 11 (continued)

a. What is the average maximum depth of the Great Lakes?

b. Which of the Great Lakes is deeper than the average?

Solution

a. $\frac{\text{The sum of the depths}}{\text{The number of lakes}} = \frac{64 + 229 + 282 + 244 + 406}{5}$

$= \frac{1{,}225}{5}$

$= 245$

So the average maximum depth is 245 meters.

b. Lake Michigan and Lake Superior are deeper than the average.

a. What was the average annual number of at-home fatalities for these years?

b. In which years was the number of fatalities below the average?

Powers and Order of Operations on a Calculator

Let's use a calculator to carry out computations that involve either powers or the order of operations rule.

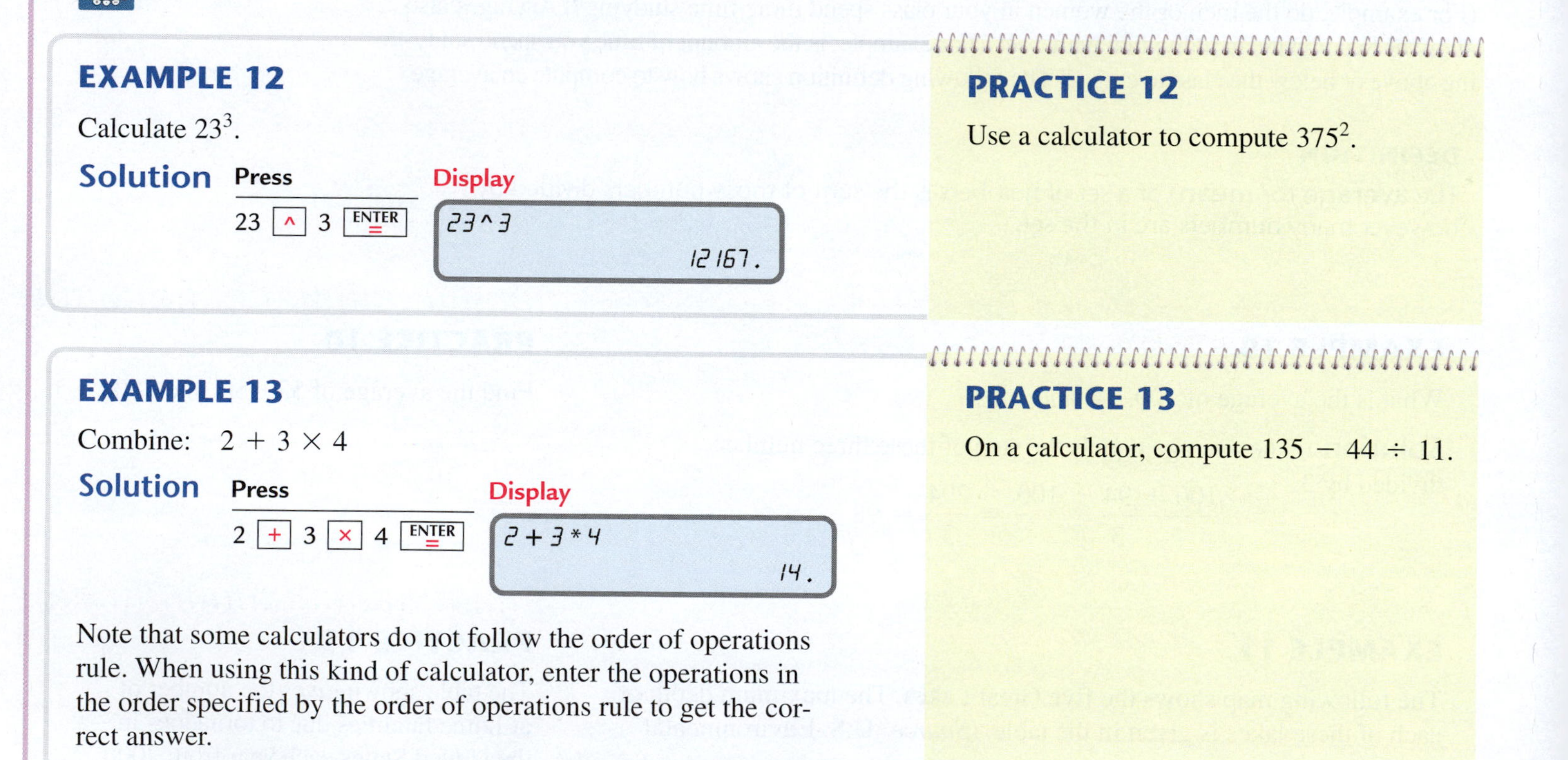

EXAMPLE 12

Calculate 23^3.

Solution

Press	Display
23 [^] 3 [ENTER]	23^3 12167.

PRACTICE 12

Use a calculator to compute 375^2.

EXAMPLE 13

Combine: $2 + 3 \times 4$

Solution

Press	Display
2 [+] 3 [×] 4 [ENTER]	2+3*4 14.

Note that some calculators do not follow the order of operations rule. When using this kind of calculator, enter the operations in the order specified by the order of operations rule to get the correct answer.

PRACTICE 13

On a calculator, compute $135 - 44 \div 11$.

1.5 Exercises

Mathematically Speaking

Fill in each blank with the most appropriate term or phrase from the given list.

product	sum	adding	listing
subtracting	grouping	power	base

1. An exponent indicates how many times the ________ is used as a factor.

2. Parentheses and brackets are examples of ___________ symbols.

3. An average of numbers on a list is found by ________ the numbers and then dividing by how many numbers there are on the list.

4. In evaluating an expression involving both a sum and a product, the ________ is evaluated first.

Complete each table by squaring the numbers given.

5.

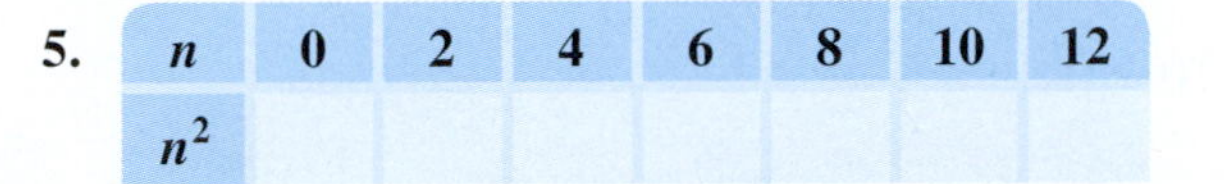

n	0	2	4	6	8	10	12
n^2							

6.

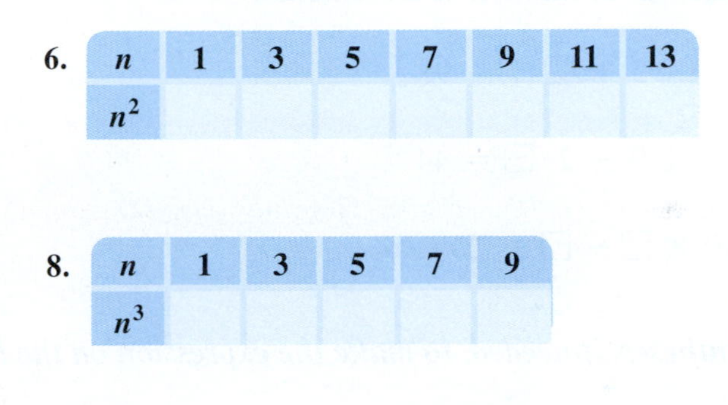

n	1	3	5	7	9	11	13
n^2							

A *Complete each table by cubing the numbers given.*

7.

n	0	2	4	6	8
n^3					

8.

n	1	3	5	7	9
n^3					

Express each number as a power of 10.

9. $100 = 10^{\square}$

10. $1{,}000 = 10^{\square}$

11. $10{,}000 = 10^{\square}$

12. $100{,}000 = 10^{\square}$

13. $1{,}000{,}000 = 10^{\square}$

14. $10{,}000{,}000 = 10^{\square}$

Write each number in terms of powers.

15. $2 \cdot 2 \cdot 3 \cdot 3 = 2^{\square} \cdot 3^{\square}$

16. $2 \cdot 2 \cdot 5 \cdot 2 \cdot 5 = 2^{\square} \cdot 5^{\square}$

17. $5 \cdot 4 \cdot 4 \cdot 4 = 4^{\square} \cdot 5^{\square}$

18. $6 \cdot 7 \cdot 6 \cdot 7 \cdot 6 \cdot 7 = 6^{\square} \cdot 7^{\square}$

Write each number in standard form.

19. $6^2 \cdot 5^2$

20. $10^3 \cdot 9^2$

21. $2^5 \cdot 7^2$

22. $3^4 \cdot 4^3$

B *Evaluate.*

23. $8 + 5 \cdot 2$

24. $9 + 10 \cdot 2$

25. $8 - 12 \div 3$

26. $12 - 6 \div 2$

27. $18 \div 2 + 4$

28. $30 \div 3 + 6$

29. $6 \cdot 3 - 16 \div 4$

30. $7 \cdot 3 - 12 \div 4$

31. $10 + 5^2$

32. $9 + 2^3$

33. $(10 + 5)^2$

34. $(9 + 2)^3$

35. 10×5^2

36. 12×2^2

37. $(12 \div 2)^2$

38. $(10 \div 5)^2$

39. $15 \div (6 - 3)$

40. $24 \div (4 + 2)$

41. $2^6 - 6^2$

42. $3^5 - 5^3$

43. $8 + 5 - 3 - 2 \times 2$

44. $7 - 1 + 2 + 3 \cdot 2$

45. $(10 - 1)(10 + 1)$

46. $(8 - 1)(8 + 1)$

47. $10^2 - 1$

48. $8^2 - 1$

49. $\left(\dfrac{8 + 2}{7 - 2}\right)^2$

50. $\left(\dfrac{9 - 1}{3 + 5}\right)^3$

51. $\dfrac{5^3 - 2^3}{3}$

52. $\dfrac{3^2 + 5^2}{2}$

53. $4 + 12(10 - 3^2)$

54. $3 + 10(20 - 2^3)$

55. $10 \cdot 3^2 + \dfrac{10 - 4}{2}$

56. $\dfrac{3^3 + 1^3 + 2^3}{4}$

57. $(21 \div 7) + [(9 - 5) \cdot 2]^2$

58. $(30 \div 3) + [(7 - 4) \cdot 3]^2$

59. $[9 + 2(3^2 - 8)] + 7$

60. $15 + [3(8 - 2^2) - 6]$

61. $[2 \cdot (3 + 4)^2 - 3 \cdot (7 - 2)^2]^2$

62. $[3 \cdot (2 + 1)^2 - 2 \cdot (5 - 4)^2]^2$

63. $32 + 9 \cdot 215 \div 5$

64. $84 \cdot 27 + 32 \cdot 27^2 \div 2$

65. $48(48 - 31)(48 - 24)(48 - 41)$

66. $137^2 - 4(36)(22)$

In each exercise, the three squares stand for the numbers 4, 6, and 8 in some order. Fill in the squares to make true statements.

67. $\square \cdot 3 + \square \cdot 5 + \square \cdot 7 = 98$

68. $\square + 10 \times \square - \dfrac{\square}{2} = 42$

69. $(\square)(3 + \square) - 2 \cdot \square = 44$

70. $\square \cdot 3 + \square \cdot 5 + \square \cdot 7 = 82$

71. $\square + 10 \times \square - \square \div 2 = 45$

72. $\dfrac{48}{\square} - \dfrac{\square}{2} + (3 + \square)^2 = 127$

Insert parentheses, if needed, to make the expression on the left equal to the number on the right.

73. $5 + 2 \cdot 4^2 = 112$

74. $5 + 2 \cdot 4^2 = 69$

75. $5 + 2 \cdot 4^2 = 169$

76. $5 + 2 \cdot 4^2 = 37$

77. $8 - 4 \div 2^2 = 1$

78. $8 - 4 \div 2^2 = 7$

Find the area of each shaded region.

79.

80.

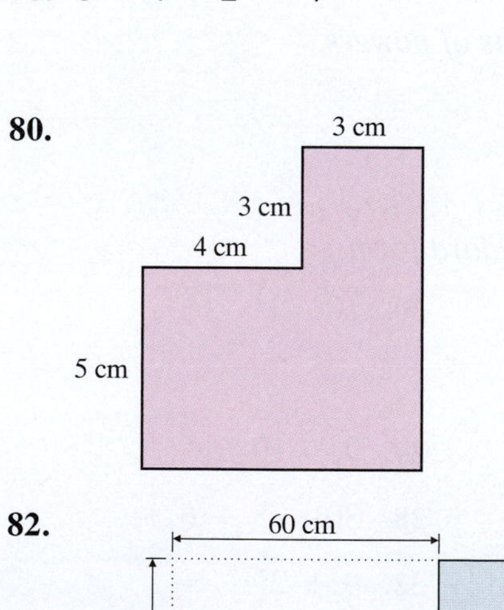

81.

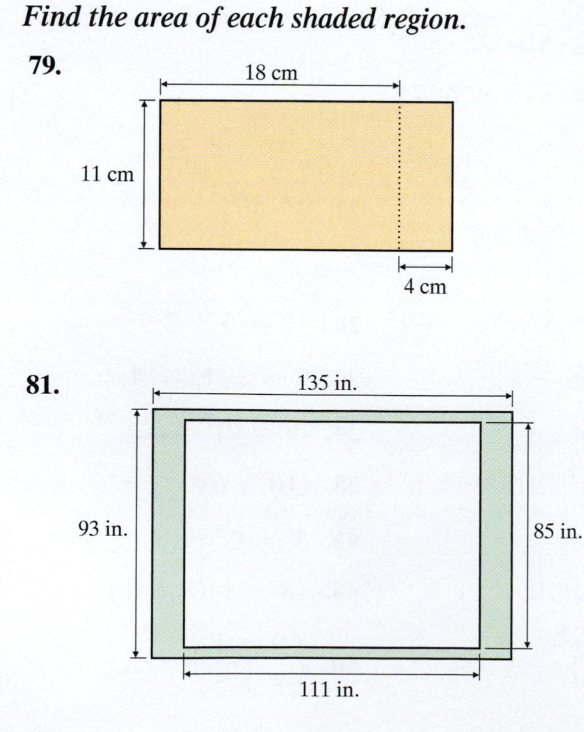

82.

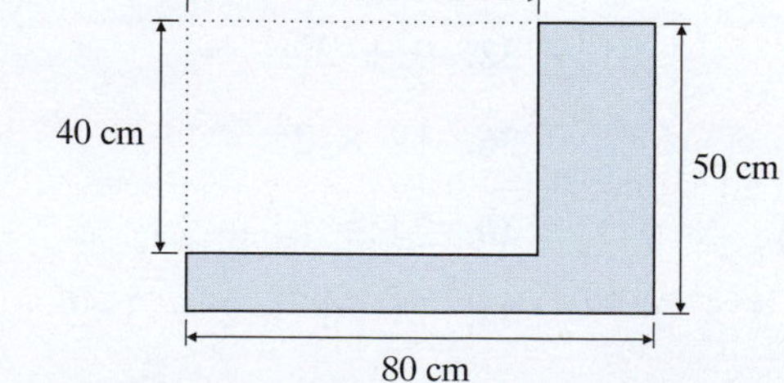

Complete each table.

83.

Input	Output
0	$21 + 3 \times \mathbf{0} =$
1	$21 + 3 \times \mathbf{1} =$
2	$21 + 3 \times \mathbf{2} =$

84.

Input	Output
0	$14 - 5 \times \mathbf{0} =$
1	$14 - 5 \times \mathbf{1} =$
2	$14 - 5 \times \mathbf{2} =$

C *Find the average of each set of numbers.*

85. 20 and 30

86. 10 and 50

87. 30, 60, and 30

88. 17, 17, and 26

89. 10, 0, 3, and 3

90. 5, 7, 7, and 17

91. 3,527 miles, 1,788 miles, and 1,921 miles

92. 3,432 miles, 1,822 miles, and 1,436 miles

93. Six 10's and four 5's

94. Sixteen 5's and four 0's

Mixed Practice

Solve.

95. Express 100,000,000 as a power of 10.

96. Find the area of the shaded region.

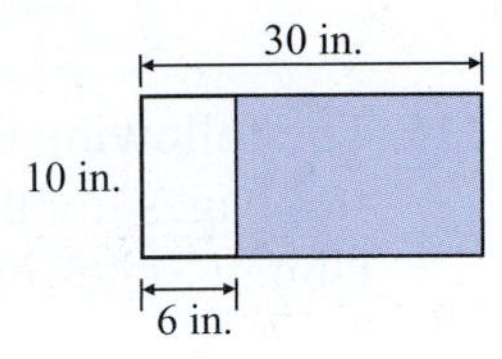

97. Square 17.

98. Rewrite in terms of powers of 2 and 7:
$2 \cdot 2 \cdot 2 \cdot 7 \cdot 7 = 2 \cdot 7$

99. Simplify: $50 - 2(10 - 3^2)$.

100. Cube 10.

101. Find the average of 10, 10, and 4.

102. Evaluate: 6×4^2

Applications

D *Solve and check.*

103. A 40-story office building has 25,000 square feet of space to rent. What is the average rental space on a floor?

104. The total area of the 50 states in the United States is about 3,700,000 square miles. If the state of Georgia's area is about 60,000 square miles, is its size above the average of all the states? Explain. (*Source: The New Encyclopedia Britannica*)

105. In a branch of mathematics called number theory, the numbers 3, 4, and 5 are called a *Pythagorean triple* because $3^2 + 4^2 = 5^2$ (that is, $9 + 16 = 25$). Show that 5, 12, and 13 are a Pythagorean triple.

106. If an object is dropped off a cliff, after 10 seconds it will have fallen $\frac{32 \cdot 10^2}{2}$ feet, ignoring air resistance. Express this distance in standard form, without exponents.

107. The solar wind streams off the Sun at speeds of about 1,000,000 miles per hour. Express this number as a power of 10. (*Source:* NASA)

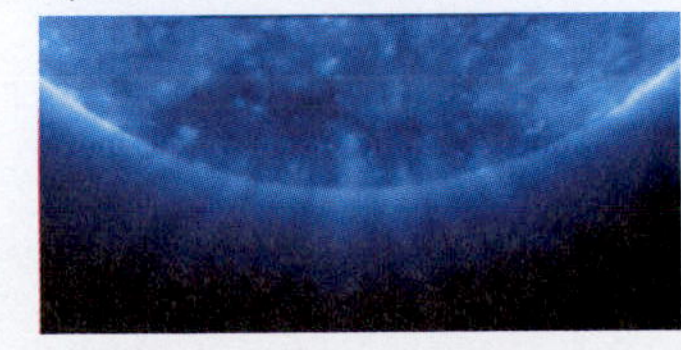

108. It has been estimated that there are 100,000,000,000,000 cells in the adult human body. Represent this number as a power of 10. (*Source:* ehd.org)

109. The following table shows a lab assistant's salary in various years.

Year	1	2	3
Salary	$19,400	$21,400	$23,700

a. Find the average salary for the three years.

b. How much greater was her average salary for the last two years than for all three years?

110. The following grade book shows a student's math test scores.

Test 1	Test 2	Test 3	Test 4
85	63	98	82

a. What is the average of his math scores?

b. If he were to get a 92 on the next math test, by how much would his average score increase?

111. In the last four home games, a college basketball team had scores of 68, 79, 57, and 72.

a. What was the average score for these games?

b. The average score for the last four away games was 64. On average, did the team score more at home or away? Explain.

112. A small theater company's production of *Romeo and Juliet* had 10 performances over two weekends. The attendance for each performance during the second weekend was 171, 297, 183, 347, and 232.

a. What was the average attendance for the performances during the second weekend?

b. If the average attendance at a performance during the first weekend was 272, was the average greater in the first or second weekend? Explain.

113. The following table gives the work stoppages for stoppages involving 1,000 or more workers in the United States in selected years.

Year	Total Number of Workers Involved	Total Number of Days Idle
2005	100,000	1,736,000
2006	70,000	2,688,000
2007	189,000	1,265,000
2008	72,000	1,954,000

(*Source:* U.S. Bureau of Labor Statistics)

a. Find the average number of workers per year involved in the work stoppages for the given four years, rounded to the nearest thousand.

b. Find the average annual number of days idle for the years 2006 through 2008. Was the number of days idle in 2008 above or below this average?

114. The chart below shows the countries that, in a recent year, had the largest military expenditures.

Country	Spending Level (in billions of U.S. dollars)	Per Capita Spending Level (in U.S. dollars)
United States	529	1,756
United Kingdom	59	990
France	53	875
China	50	37
Japan	44	341
Germany	37	447
Russia	35	244

(*Source:* infoplease.com)

a. Find the average military spending level for the five countries with the largest military spending levels.

b. What is the average per capita spending level for the seven countries shown?

115. Owners of a restaurant agreed to invest in redecorating the restaurant if the average number of customers per month for the coming 12 months is more than 500. The monthly tallies of customers for the restaurant during this period turned out to be: 372, 618, 502, 411, 638, 465, 572, 377, 521, 488, 458, and 602. Will the restaurant be redecorated?

116. The hospital chart shown is a record of a patient's temperature for two days.

Time	Temp. (°F)	Time	Temp. (°F)
6 A.M.	98	6 A.M.	101
10 A.M.	100	10 A.M.	102
2 P.M.	98	2 P.M.	101
6 P.M.	100	6 P.M.	102
10 P.M.	98	10 P.M.	100
2 A.M.	100	2 A.M.	100

What was her average temperature for this period of time?

• Check your answers on page A-2.

MINDStretchers

Writing

1. Evaluate the expressions in parts (a) and (b).

a. $7^2 + 4^2$ ___

b. $(7 + 4)^2$ ___

c. Are the answers to parts (a) and (b) the same? ___ If not, explain why not.

Mathematical Reasoning

2. The square of any whole number (called a **perfect square**) can be represented as a geometric square, as follows:

4

9

Try to represent the numbers 16, 25, 5, and 8 the same way.

16 25 5 8

Critical Thinking

3. Find the average of the whole numbers from 1 through 999.

1.6 More on Solving Word Problems

OBJECTIVE

A To solve applied problems involving the addition, subtraction, multiplication, or division of whole numbers using various problem-solving strategies

What Word Problems Are and Why They Are Important

In this section, we consider some general tips to help solve word problems.

Word problems can deal with any subject—from shopping to physics and geography to business. Each problem is a brief story that describes a particular situation and ends with a question. Our job, after reading and thinking about the problem, is to answer that question by using the given information.

Although there is no magic formula for solving word problems, you should keep the following problem-solving steps in mind.

To Solve Word Problems

- Read the problem carefully.
- Choose a strategy (such as drawing a picture, breaking up the question, substituting simpler numbers, or making a table).
- Decide which basic operation(s) are relevant and then translate the words into mathematical symbols.
- Perform the operations.
- Check the solution to see if the answer is reasonable. If it is not, start again by rereading the problem.

Reading the Problem

In a math problem, each word counts. So it is important to read the problem slowly and carefully, and not to scan it as if it were a magazine or newspaper article.

When reading a problem, we need to understand the problem's key points: *What information is given* and *what question is posed*. Once these points are clear, jot them down so as to help keep them in mind.

After taking notes, decide on a plan of action that will lead to the answer. For many problems, just thinking back to the meaning of the four basic operations will be helpful.

Operation	Meaning
$+$	Combining
$-$	Taking away
$\times$	Adding repeatedly
$\div$	Splitting up

Many word problems contain *clue words* that suggest performing particular operations. If we spot a clue word in a problem, we should consider whether the operation indicated in the table on the opposite page will lead us to a solution.

+	−	×	÷
• add	• subtract	• multiply	• divide
• sum	• difference	• product	• quotient
• total	• take away	• times	• over
• plus	• minus	• double	• split up
• more	• less	• twice	• fit into
• increase	• decrease	• triple	• per
• gain	• loss	• of	• goes into

However, be on guard—a clue word can be misleading. For instance, in the problem *What number increased by 2 is 6?*, we solve by subtracting, not adding.

Consider the following "translations" of these clues.

The patient's fever **increased by 5°**. $+ 5$

The number of unemployed people **tripled**. $\times 3$

The length of the bedroom is **8 feet less** than the kitchen's. $- 8$

The company's earnings were **split** among the **four** partners. $\div 4$

Choosing a Strategy

If no method of solution comes to mind after reading a problem, there are a number of problem-solving strategies that may help. Here we discuss four of these strategies: drawing a picture, breaking up the question, substituting simpler numbers, and making a table.

Drawing a Picture

Sketching even a rough representation of a problem—say, a diagram or a map—can provide insight into its solution, if the sketch accurately reflects the given information.

EXAMPLE 1

In an election, everyone voted for one of three candidates. The winner received 188,000 votes, and the second-place candidate got 177,000 votes. If 380,000 people voted in the election, how many people voted for the third candidate?

Solution To help us understand the given information, let's draw a diagram to represent the situation.

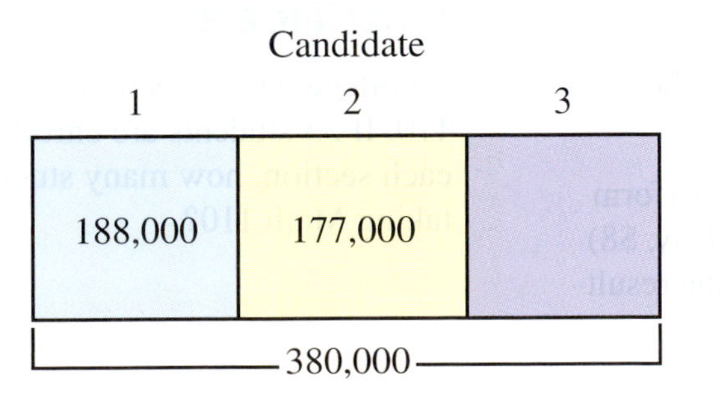

We see from this diagram that to find the answer we need to do two things.

PRACTICE 1

A company slashed its workforce by laying off 1,150 employees during one month and laying off 2,235 employees during another month. Afterward, 7,285 employees remained. How many employees worked for the company before the layoffs began?

Key Concepts and Skills

CONCEPT SKILL

Concept/Skill	Description	Example
[1.1] Place value	Thousands: Hundreds Tens Ones; Ones: Hundreds Tens Ones	846,120 (↑ under 4) 4 is in the ten thousands place.
[1.1] To read a whole number	Working from left to right, • read the number in each period, and then • name the period in place of the comma.	71,400 is read "seventy-one thousand, four hundred".
[1.1] To write a whole number	Working from left to right, • write the number named in each period, and • replace each period name with a comma.	"Five thousand, twelve" is written 5,012.
[1.1] To round a whole number	• Underline the place to which you are rounding. • The digit to the right of the underlined digit is called the *critical digit*. Look at the critical digit—if it is 5 or more, add 1 to the underlined digit; if it is less than 5, leave the underlined digit unchanged. • Replace all the digits to the right of the underlined digit with zeros.	$3\underline{8}6 \approx 390$ $4,\underline{8}17 \approx 4,800$
[1.2] Addend, sum	In an addition problem, the numbers being added are called *addends*. The result is their *sum*.	$6 + 4 = 10$ (6: Addend, 4: Addend, 10: Sum)
[1.2] The identity property of addition	The sum of a number and zero is the original number.	$4 + 0 = 4$ $0 + 7 = 7$
[1.2] The commutative property of addition	Changing the order in which two numbers are added does not affect their sum.	$7 + 8 = 8 + 7$
[1.2] The associative property of addition	When adding three numbers, regrouping addends gives the same sum.	$(5 + 4) + 1 = 5 + (4 + 1)$
[1.2] To add whole numbers	• Write the addends vertically, lining up the place values. • Add the digits in the ones column, writing the right-most digit of the sum on the bottom. If the sum has two digits, carry the left digit to the top of the next column on the left. • Add the digits in the tens column as in the preceding step. • Repeat this process until you reach the last column on the left, writing the entire sum of that column on the bottom.	carries: 1 1 1 7,385 92,551 + 2,007 101,943
[1.2] Minuend, subtrahend, difference	In a subtraction problem, the number that is being subtracted from is called the *minuend*. The number that is being subtracted is called the *subtrahend*. The answer is called the *difference*.	$10 - 6 = 4$ (10: Minuend, 6: Subtrahend, 4: Difference)

Concept/Skill	Description	Example
[1.2] **To subtract whole numbers**	• On top, write the number *from which* we are subtracting. On the bottom, write the number that is being *taken away,* lining up the place values. Subtract in each column separately. • Start with the ones column. **a.** If the digit on top is *larger* than or *equal* to the digit on the bottom, subtract and write the difference below. **b.** If the digit on top is *smaller* than the digit on the bottom, borrow from the digit to the left on top. Then subtract and write the difference below the bottom digit. • Repeat this process until the last column on the left is finished, subtracting and writing its difference below.	8 14 1 7,9 5 2 −1, 8 8 3 6, 0 6 9
[1.3] **Factor, product**	In a multiplication problem, the numbers being multiplied are called *factors*. The result is called their *product.*	Factor Product $4 \times 5 = 20$
[1.3] **The identity property of multiplication**	The product of any number and 1 is that number.	$1 \times 6 = 6$ $7 \times 1 = 7$
[1.3] **The multiplication property of 0**	The product of any number and 0 is 0.	$51 \times 0 = 0$
[1.3] **The commutative property of multiplication**	Changing the order in which two numbers are multiplied does not affect their product.	$3 \times 2 = 2 \times 3$
[1.3] **The associative property of multiplication**	When multiplying three numbers, regrouping the factors gives the same product.	$(4 \times 5) \times 6 = 4 \times (5 \times 6)$
[1.3] **The distributive property**	Multiplying a factor by the sum of two numbers gives the same result as multiplying the factor by each of the two numbers and then adding.	$2 \times (4 + 3)$ $= (2 \times 4) + (2 \times 3)$
[1.3] **To multiply whole numbers**	• Multiply the top factor by the ones digit in the bottom factor and write this product. • Multiply the top factor by the tens digit in the bottom factor and write this product leftward, beginning with the tens column. • Repeat this process until all the digits in the bottom factor are used. • Add the partial products, writing this sum.	693 × 71 693 48 51 49,203
[1.4] **Divisor, dividend, quotient**	In a division problem, the number that is being used to divide another number is called the *divisor*. The number into which it is being divided is called the *dividend*. The result is called the *quotient.*	Quotient 3 4)12 Divisor Dividend

continued

CONCEPT SKILL

Concept/Skill	Description	Example
[1.4] **To divide whole numbers**	• Divide 17 into 39, which gives 2. Multiply the 17 by 2 and subtract the result (34) from 39. Beside the difference (5), bring down the next digit (3) of the dividend. • Repeat this process, dividing the divisor (17) into 53. • At the end, there is a remainder of 2. Write it beside the quotient on top.	$\begin{array}{r} 23\text{ R2} \\ 17\overline{)393} \\ \underline{34} \\ 53 \\ \underline{51} \\ 2 \end{array}$
[1.5] **Exponent (or power), base**	An *exponent* (or *power*) is a number that indicates how many times another number (called the *base*) is used as a factor.	Exponent $5^3 = 5 \times 5 \times 5$ Base
[1.5] **Order of operations rule**	To evaluate mathematical expressions, carry out the operations *in the following order.* **1.** First, perform the operations within any grouping symbols, such as parentheses ( ) or brackets [ ]. **2.** Then, raise any number to its power. **3.** Next, perform all multiplications and divisions as they appear from left to right. **4.** Finally, do all additions and subtractions as they appear from left to right. ( ) power × ÷ + −	$8 + 5 \cdot (3 + 1)^2 = 8 + 5 \cdot 4^2$ $= 8 + 5 \cdot 16$ $= 8 + 80$ $= 88$
[1.5] **Average (or mean)**	The *average* (or *mean*) of a set of numbers is the sum of those numbers divided by however many numbers are in the set.	The average of 3, 4, 10, and 3 is 5 because $\frac{3 + 4 + 10 + 3}{4} = \frac{20}{4} = 5$
[1.6] **To solve word problems**	• Read the problem carefully. • Choose a strategy (such as drawing a picture, breaking up the question, substituting simpler numbers, or making a table). • Decide which basic operation(s) are relevant and then translate the words into mathematical symbols. • Perform the operations. • Check the solution to see if the answer is reasonable. If it is not, start again by rereading the problem.	

CHAPTER 1 Review Exercises

Say Why

Fill in each blank.

1. The place values of 4 in 410 and of 6 in 7,699 ________ (are/are not) the same because ____.

2. 5,605 rounded to the nearest thousand ________ (is/is not) 5,000 because ____.

3. The perimeter of a four-sided figure for which all sides have length 5 ________ (is/is not) 25 because ____.

4. The product of 8 and 7 ________ (is/is not) 15 because ____.

5. The area of a rectangle with length 10 and width 7 ________ (is/is not) 70 because ____.

6. In the expression 9^2, 9 ________ (is/is not) the exponent because ____.

7. $2(3 + 5)$ ________ (is/is not) equal to $2 \cdot 3 + 2 \cdot 5$ because ____.

8. The quotient of 10 and 5 is ________ (is/is not) 2 because ____.

9. The average of 7, 11, and 0 ________ (is/is not) $\frac{7 + 11 + 0}{2}$ because ____.

10. In evaluating the expression $9 + 3 \cdot 4$, we multiply ________ (before/after) adding because ____.

[1.1] *In each whole number, identify the place that the digit 3 occupies.*

11. 23

12. 30,802

13. 385,000,000

14. 30,000,000,000

Write each number in words.

15. 497

16. 2,050

17. 3,000,007

18. 85,000,000,000

Write each number in standard form.

19. Two hundred fifty-one

20. Nine thousand, two

21. Fourteen million, twenty-five

22. Three billion, three thousand

Express each number in expanded form.

23. 2,500,000

24. 42,707

Round each number to the place indicated.

25. 571 to the nearest hundred

26. 938 to the nearest thousand

27. 384,056 to the nearest ten thousand

28. 68,332 to its largest place

[1.2] *Find the sum and check.*

29. $\begin{array}{r} 102 \\ 4{,}251 \\ +\ 5{,}133 \\ \hline \end{array}$

30. $\begin{array}{r} 53{,}569 \\ 10{,}000 \\ +\ 2{,}123 \\ \hline \end{array}$

31. $\begin{array}{r} 48{,}758 \\ 37{,}226 \\ +\ 87{,}559 \\ \hline \end{array}$

32. $\begin{array}{r} 95{,}000 \\ 25{,}895 \\ +\ 30{,}000 \\ \hline \end{array}$

Add and check.

33. 972,558 + 87,055 + 36,488 + 861,724

34. \$138,865 + \$729 + \$8,002 + \$75,471

Find the difference and check.

35. $\begin{array}{r} 876 \\ -\ 431 \\ \hline \end{array}$

36. $\begin{array}{r} 56{,}000 \\ -\ 45{,}984 \\ \hline \end{array}$

37. $\begin{array}{r} 98{,}118 \\ -\ 87{,}009 \\ \hline \end{array}$

38. $\begin{array}{r} 7{,}100 \\ -\ 1{,}590 \\ \hline \end{array}$

39. 60,000,000 − 48,957,777

40. \$5,000,000 − \$2,937,148

41. From 67,502 subtract 56,496.

42. Subtract 89,724 from 92,713.

[1.3] *Find the product and check.*

43. $\begin{array}{r} 72 \\ \times\ \ 6 \\ \hline \end{array}$

44. $\begin{array}{r} 400 \\ \times\ \ 3 \\ \hline \end{array}$

45. $\begin{array}{r} 2{,}923 \\ \times\ \ 51 \\ \hline \end{array}$

46. $\begin{array}{r} 6{,}000 \\ \times\ 2{,}000 \\ \hline \end{array}$

47. $\begin{array}{r} 14{,}921 \\ \times\ \ 32 \\ \hline \end{array}$

48. $\begin{array}{r} 8{,}152 \\ \times\ \ 125 \\ \hline \end{array}$

Multiply and check.

49. $2{,}751 \cdot 508$

50. (681)(498)(555)

[1.4] *Divide and check.*

51. $\dfrac{975}{25}$

52. $21\overline{)6{,}450}$

53. $13\overline{)491}$

54. $7{,}488 \div 11$

Find the quotient and check.

55. $8\overline{)205{,}000}$

56. $347\overline{)332{,}079}$

[1.5] *Compute.*

57. 7^3

58. 1^{10}

59. $2^3 \cdot 3^2$

60. $3 \cdot 10^5$

61. $20 - 3 \times 5$

62. $(9 + 4)^2$

63. $10 - \dfrac{6 + 4}{2}$

64. $3 + (5 - 1)^2$

65. $5 + [4^2 - 3(2 + 1)]$

66. $17 + [2(3^2 - 6) - 5]$

67. $98(50 - 1)(50 - 2)(50 - 3)$

68. $\dfrac{28^3 + 29^3 + 37^3 - 10}{(7 - 1)^2}$

Rewrite each expression, using exponents.

69. $7 \cdot 7 \cdot 5 \cdot 5 = 7^{■} \cdot 5^{■}$

70. $5 \cdot 2 \cdot 5 \cdot 2 \cdot 5 = 2^{■} \cdot 5^{■}$

Find the average.

71. 34 and 44

72. 20, 0, and 1

73. 5, 8, and 5

74. 4, 6, 3, and 7

Mixed Applications

Solve and check.

75. Beetles about the size of a pinhead destroyed 2,400,000 acres of forest. Express this number in words.

76. Scientists in Utah found a dinosaur egg one hundred fifty million years old. Write this number in standard form.

77. The following graph shows the consolidated assets of the six largest banks in the United States.

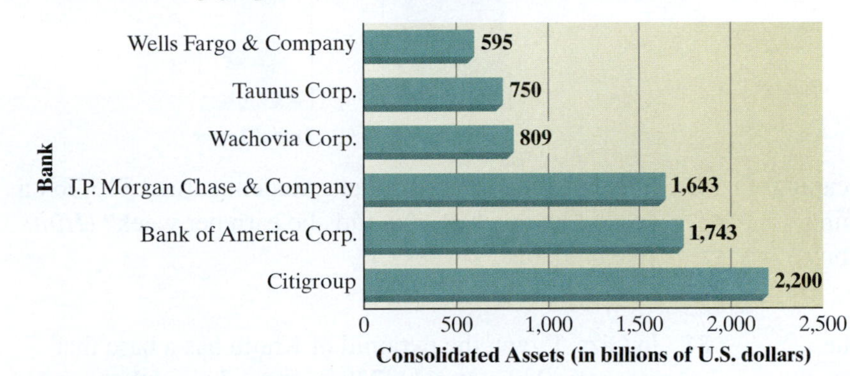

(*Source:* infoplease.com)

Find combined assets of Citigroup and Wachovia.

78. Halley's Comet is expected to visit Earth next in 2061. If the comet has a 76-year orbit, in what year did it last visit Earth? (*Source:* science.nasa.gov)

79. What is the land area of Texas to the nearest hundred thousand square miles? (*Source: Time Almanac 2010*)

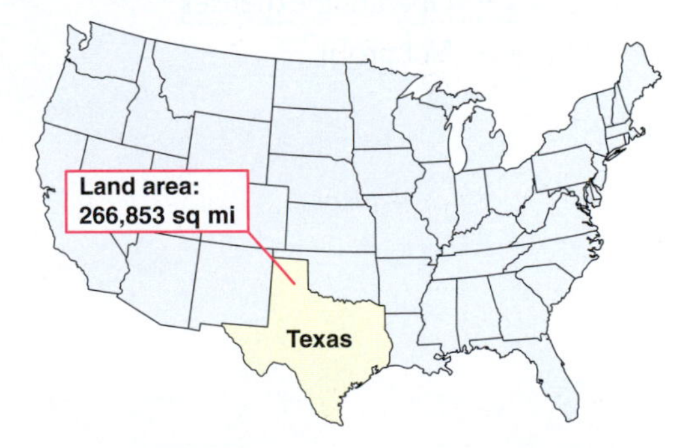

80. Apple Computer sold 31,855,000 iPods in the first two quarters of its 2010 fiscal year. How many iPods is this to the nearest million? (*Source:* wikipedia.org)

81. The Empire State Building is 1,250 feet high, and the Statue of Liberty is 152 feet in height. What is the minimum number of Statues of Liberty that would have to be stacked to be taller than the Empire State Building?

82. Both a singles tennis court and a football field are rectangular in shape. A tennis court measures 78 feet by 27 feet, whereas a football field measures 360 feet by 160 feet. About how many times the area of a tennis court is that of a football field?

83. The tallest building in the United States is Chicago's Willis Tower, which is 1,450 ft high. By contrast, the Dubai Tower, the tallest building in the world, is 1,267 ft higher. Find the height of the Dubai Tower.

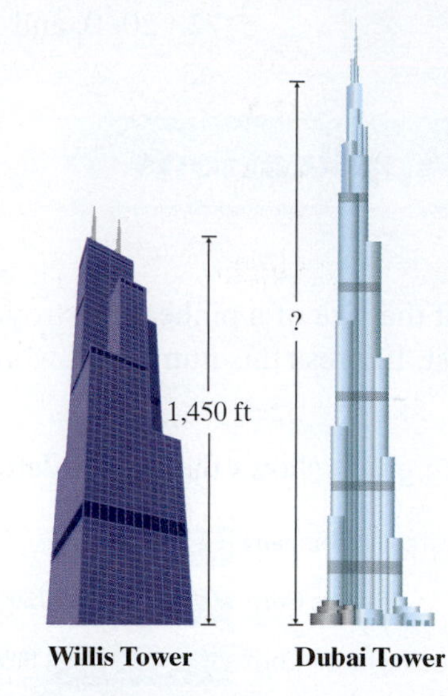

84. A landscaper needs 550 flower plants for a landscaping project. If a local garden center sells flats containing 24 plants, how many flats should the landscaper buy?

85. In a part-time job, a graduate student earned $15,964 a year. How much money did she earn per week? (*Hint:* 1 year equals 52 weeks.)

86. Compute a company's net profit by completing the following business *skeletal profit and loss statement*.

Net sales	$430,000
− Cost of merchandise sold	− 175,000
Gross margin	$
− Operating expenses	− 135,000
Net profit	$

87. In Giza, Egypt, the pyramid of Khufu has a base that measures 230 meters by 230 meters, whereas the pyramid of Khafre has a base that measures 215 meters by 215 meters. In area, how much larger than the base of the pyramid of Khafre is that of Khufu? (*Source:* pbs.org)

88. A millipede—a small insect with 68 body segments—has 4 legs per segment. How many legs does a millipede have?

89. Richard Nixon ran for the U.S. presidency three times. According to the table below, which was greater —the increase from 1960 to 1972 in the number of votes he got or the increase from 1968 to 1972? (*Source: The New York Times Almanac 2010*)

Year	Number of Votes for Nixon
1960	34,106,671
1968	31,785,148
1972	47,170,179

90. On a business trip, a sales representative flew from Chicago to Los Angeles to Boston. The chart below shows the air distances in miles between these cities.

Air Distance	Chicago	Los Angeles	Boston
Chicago	—	1,745	1,042
Los Angeles	1,745	—	2,596
Boston	1,042	2,596	—

If the sales rep earned a frequent flier point for each mile flown, how many points did he earn?

91. The Tour de France is a 20-stage bicycle race held in France annually. The chart shows the distances for the first 10 stages of the 2010 Tour de France. (*Source:* Le Tour de France)

Stage	Distance (in kilometers)
1	223
2	201
3	216
4	154
5	188
6	228
7	166
8	189
9	205
10	179

a. What was the total distance covered in the first 10 stages?

b. The entire race covered a distance of 3,632 kilometers. How many kilometers were covered in the last 10 stages?

92. The projected enrollment for public colleges in the United States from 2014 through 2018 is given in the table.

Year	Projected Enrollment
2014	14,758,000
2015	14,874,000
2016	14,981,000
2017	15,116,000
2018	15,241,000

(*Source:* census.gov)

a. What is the projected average enrollment in the years 2014 through 2018?

b. If a projected enrollment of 15,366,000 for the year 2019 was included, how would the average annual enrollment change?

93. Find the area of the figure.

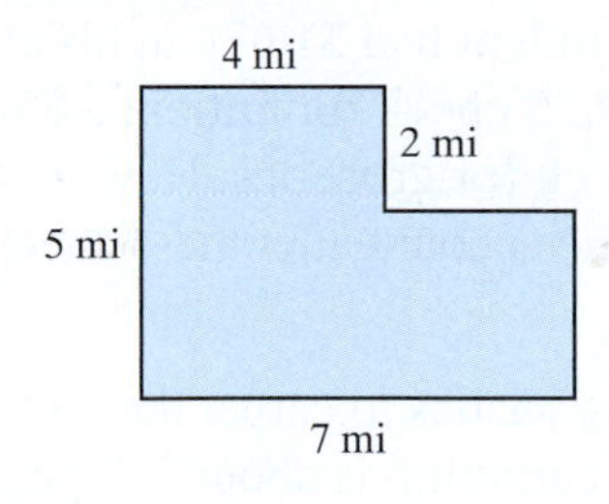

94. Find the perimeter of the figure.

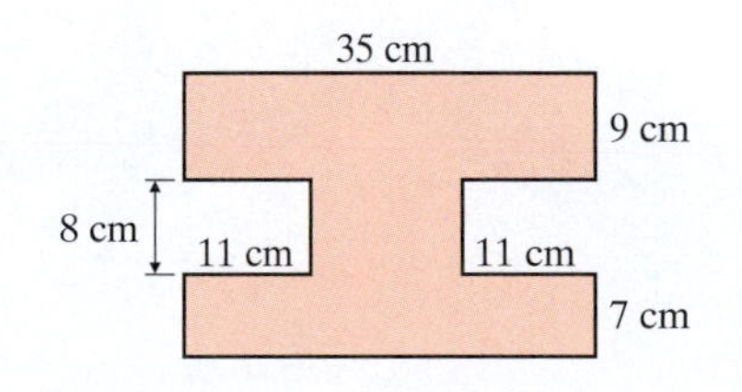

• Check your answers on page A-2.

CHAPTER 1 Posttest

To see if you have already mastered the topics in this chapter, take this test.

1. Write two hundred twenty-five thousand, sixty-seven in standard form.

2. Underline the digit that occupies the ten thousands place in 1,768,405.

3. Write 1,205,007 in words.

4. Round 196,593 to the nearest hundred thousand.

5. Find the sum of 398 and 1,496.

6. Subtract 398 from 1,005.

7. Subtract: $2{,}000 - 1{,}853$

8. Multiply: 328×907

9. Compute: $\dfrac{23{,}923}{47}$

10. Find the quotient: $59\overline{)36{,}717}$

11. Evaluate: 5^4

12. Write $5 \cdot 5 \cdot 4 \cdot 4 \cdot 4$ using exponents.

Simplify.

13. $4 \cdot 9 + 3 \cdot 4^2$

14. $29 - 3^3 \cdot (10 - 9)$

Solve and check.

15. The two largest continents in the world are Asia and Africa. To the nearest hundred thousand square miles, Asia's area is 17,200,000 and Africa's is 11,600,000. How much larger is Asia than Africa? (*Source: National Geographic Atlas of the World*)

16. In the year 2009, the state of Kansas had about 65,500 farms with an average size of 705 acres. How many acres of land in Kansas were devoted to farming? (*Source:* nossausda.gov)

17. A part-time student had $1,679 in his checking account. He wrote a $625 check for tuition, a $546 check for rent, and a $39 check for groceries. How much money remained in the account after these checks cleared?

18. A part-time taxi driver worked 4 days last week and made the following amounts of money: Monday $95, Tuesday $110, Wednesday $132, and Friday $155. How much money on the average did he make per work day?

19. A homeowner wishes to carpet the hallway shown below. If the cost of carpeting is about $10 per square foot, approximately how much will the carpeting cost?

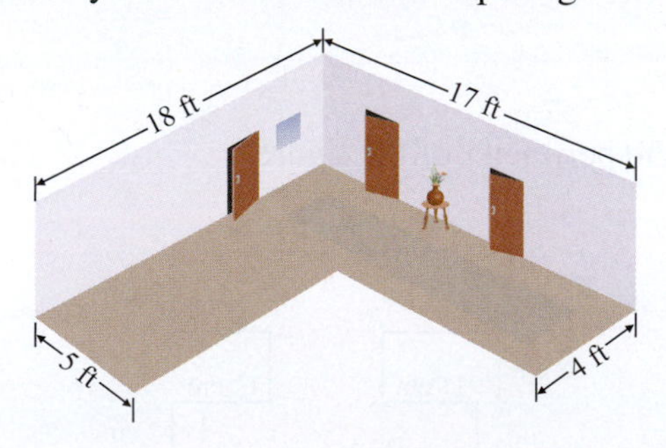

20. Many fast food chains display the caloric intake and the number of grams of fat associated with the chain's dishes.

Food	Calories	Grams of Fat
Original-recipe chicken whole wing	150	9
Original-recipe chicken breast	380	19
Original-recipe drumstick	140	8
Original-recipe thigh	360	25

How many more grams of fat are there in 2 original-recipe thighs than in 3 original-recipe drumsticks? (*Source: Washington Post*)

• Check your answers on page A-2.

Jimmy Stewart made the practice of filibustering famous in the Hollywood film *Mr. Smith Goes to Washington.*

CHAPTER 2

Fractions

Fractions and Filibustering

The rules of the United States Senate allow one or more members to *filibuster*, that is, to talk for as long as they wish and on any topic they choose. So a single member can in effect block the vote on a proposal.

The use of the filibuster to obstruct legislative action in the Senate has a long history going back to the early years of Congress. Unlimited debate was condoned on the grounds that any senator should have the right to address any issue for as long as necessary. However, in the course of time, the *cloture* procedure was developed to end a filibuster. In 1917, the Senate adopted the rule that invoking cloture requires a $\frac{2}{3}$ vote of the current 100 senators. This new rule was first put to the test in 1919, when a motion of cloture was passed ending a filibuster against the Treaty of Versailles. As a result, the state of war between Germany and the Allied Powers was over. In 1975, the Senate changed the rule, reducing the number of votes needed to stop a filibuster from $\frac{2}{3}$ to $\frac{3}{5}$ of the senators.

(*Sources:* senate.gov and wikipedia.org)

When checking to see if one number is a factor of another, it is generally faster to use the following **divisibility tests** than to divide.

The number is divisible by	if
2	the ones digit is 0, 2, 4, 6, or 8, that is, if the number is even.
3	the sum of the digits is divisible by 3.
4	the number named by the last two digits is divisible by 4.
5	the ones digit is either 0 or 5.
6	the number is even and the sum of the digits is divisible by 3.
9	the sum of the digits is divisible by 9.
10	the ones digit is 0.

EXAMPLE 2

What are the factors of 45?

Solution Let's see if 45 is divisible by 1, 2, 3, and so on, using the divisibility tests wherever they apply.

Is 45 divisible by	Answer
1?	Yes, because 1 is a factor of any number; $\frac{45}{1} = 45$, so 45 is also a factor.
2?	No, because the ones digit is not even.
3?	Yes, because the sum of the digits, $4 + 5 = 9$, is divisible by 3; $\frac{45}{3} = 15$, so 15 is also a factor.
4?	No, because 4 will not divide into 45 evenly.
5?	Yes, because the ones digit is 5; $\frac{45}{5} = 9$, so 9 is also a factor.
6?	No, because 45 is not even.
7?	No, because $45 \div 7$ has remainder 3.
8?	No, because $45 \div 8$ has remainder 5.
9?	We already know that 9 is a factor.
10?	No, because the ones digit is not 0.

The factors of 45 are, therefore, 1, 3, 5, 9, 15, and 45.

Note that we really didn't have to check to see if 9 was a factor—we learned that it was when we checked for divisibility by 5. Also, because the factors were beginning to repeat with 9, there was no need to check numbers greater than 9.

PRACTICE 2

Find all the factors of 75.

EXAMPLE 3

Identify all the factors of 60.

Solution Let's check to see if 60 is divisible by 1, 2, 3, 4, and so on.

Is 60 divisible by	Answer
1?	Yes, because 1 is a factor of all numbers; $\frac{60}{1} = 60$, so 60 is also a factor.
2?	Yes, because the ones digit is even; $\frac{60}{2} = 30$, so 30 is also a factor.
3?	Yes, because the sum of the digits, $6 + 0 = 6$, is divisible by 3; $\frac{60}{3} = 20$, so 20 is also a factor.
4?	Yes, because 4 will divide into 60 evenly; $\frac{60}{4} = 15$, so 15 is also a factor.
5?	Yes, because the ones digit is 0; $\frac{60}{5} = 12$, so 12 is also a factor.
6?	Yes, because the number is even, and the sum of the digits is divisible by 3; $\frac{60}{6} = 10$, so 10 is also a factor.
7?	No, because $60 \div 7$ has remainder 4.
8?	No, because $60 \div 8$ has remainder 4.
9?	No, because the sum of the digits, $6 + 0 = 6$, is not divisible by 9.
10?	We already know that 10 is a factor.

The factors of 60 are, therefore, 1, 2, 3, 4, 5, 6, 10, 12, 15, 20, 30, and 60.

PRACTICE 3

What are the factors of 90?

EXAMPLE 4

A presidential election takes place in the United States every year that is a multiple of 4. Was there a presidential election in 1866? Explain.

Solution The question is: Does 4 divide into 1866 evenly? Using the divisibility test for 4, we check whether 66 is a multiple of 4.

$$\frac{66}{4} = 16 \text{ R2}$$

Because $\frac{66}{4}$ has remainder 2, 4 is not a factor of 1866. So there was no presidential election in 1866.

PRACTICE 4

The doctor instructs a patient to take a pill every 3 hours. If the patient took a pill at 8:00 this morning, should she take one tomorrow at the same time? Explain.

Identifying Prime and Composite Numbers

Now, let's discuss the difference between prime numbers and composite numbers.

DEFINITIONS

A **prime number** is a whole number that has exactly two different factors: itself and 1.

A **composite number** is a whole number that has more than two factors.

Note that the numbers 0 and 1 are neither prime nor composite. But every whole number greater than 1 is either prime or composite, depending on its factors.

For instance, 5 is prime because its only factors are 1 and 5. But 8 is composite because it has more than two factors (it has four factors: 1, 2, 4, and 8).

Let's practice distinguishing between primes and composites.

EXAMPLE 5

Indicate whether each number is prime or composite.

a. 2 **b.** 78 **c.** 51 **d.** 19 **e.** 31

Solution

a. The only factors of 2 are 1 and 2. Therefore, 2 is prime.

b. Because 78 is even, it is divisible by 2. Having 2 as an "extra" factor—in addition to 1 and 78—means that 78 is composite. Do you see why all even numbers, except for 2, are composite?

c. Using the divisibility test for 3, we see that 51 is divisible by 3, because the sum of the digits 5 and 1, or 6, is divisible by 3. Because 51 has more than two factors, it is composite.

d. The only factors of 19 are itself and 1. Therefore, 19 is prime.

e. Because 31 has no factors other than itself and 1, it is prime.

PRACTICE 5

Decide whether each number is prime or composite.

a. 3 **b.** 57 **c.** 29

d. 34 **e.** 17

Finding the Prime Factorization of a Number

Every composite number can be written as the product of prime factors. This product is called its **prime factorization.** For instance, the prime factorization of 12 is $2 \cdot 2 \cdot 3$.

DEFINITION

The **prime factorization** of a whole number is the number written as the product of its prime factors.

Being able to find the prime factorization of a number is an important skill to have for working with fractions, as we show later in this chapter. A good way to find the prime factorization of a number is by making a **factor tree**, as illustrated in Example 6.

EXAMPLE 6

Write the prime factorization of 72.

Solution We start building a factor tree for 72 by dividing 72 by the smallest prime, 2. Because 72 is $2 \cdot 36$, we write both 2 and 36 underneath the 72. Then, we circle the 2 because it is prime.

Next, we divide 36 by 2, writing both 2 and 18, and circling 2 because it is prime. Below the 18, we write 2 and 9, again circling the 2. Because 9 is not divisible by 2, we divide it by the next smallest prime, 3. We stop this process because all the factors in the bottom row are prime. The prime factorization of 72 is the product of the circled factors.

72
(2) 36 ← Continue to divide by 2.
(2) 18 ← Continue to divide by 2.
(2) 9 ← 9 cannot be divided by 2, so we try 3.
(3) (3) ← 3 is prime, so we stop dividing.

$$72 = 2 \times 2 \times 2 \times 3 \times 3$$

We can also write this prime factorization as $2^3 \times 3^2$.

PRACTICE 6

Write the prime factorization of 56, using exponents.

EXAMPLE 7

Express 80 as the product of prime factors.

Solution The factor tree method for 80 is as shown.

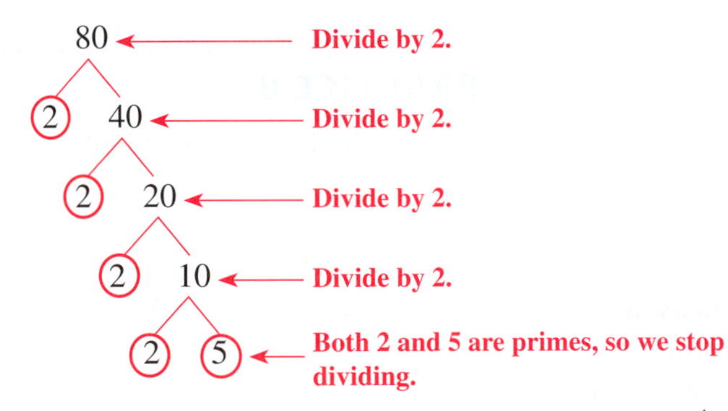

The prime factorization of 80 is $2 \times 2 \times 2 \times 2 \times 5$, or $2^4 \times 5$.

PRACTICE 7

What is the prime factorization of 75?

Finding the Least Common Multiple

The *multiples* of a number are the products of that number and the whole numbers. For instance, some multiples of 5 are the following:

$$\underbrace{0}_{0 \times 5} \qquad \underbrace{5}_{1 \times 5} \qquad \underbrace{10}_{2 \times 5} \qquad \underbrace{15}_{3 \times 5}$$

2.1 Exercises

FOR EXTRA HELP MyMathLab

Mathematically Speaking

Fill in each blank with the most appropriate term or phrase from the given list.

division	least common multiple	composite	factor tree
divisibility	prime	remainders	common multiple
prime factorization	factors	multiples	

1. 1, 2, 3, 5, 6, 10, 15, and 30 are _______ of 30.

2. A(n) _________ number is a whole number that has more than two factors.

3. A(n) _______ number has exactly two different factors: itself and 1.

4. The _____________________ of two or more numbers is the smallest nonzero number that is a multiple of each number.

5. A number written as the product of its prime factors is called its ___________________.

6. The _____________ test for 10 is to check if the ones digit is 0.

A *List all the factors of each number.*

7. 21
8. 10
9. 17
10. 9
11. 12
12. 15
13. 31
14. 47
15. 36
16. 35
17. 29
18. 73
19. 100
20. 98
21. 28
22. 48

B *Indicate whether each number is prime or composite. If it is composite, identify a factor other than the number itself and 1.*

23. 13
24. 7
25. 16
26. 24
27. 49
28. 75
29. 11
30. 31
31. 81
32. 45

C *Write the prime factorization of each number.*

33. 8
34. 10
35. 49
36. 14
37. 24
38. 18
39. 50
40. 40
41. 77
42. 63
43. 51
44. 57
45. 25
46. 49
47. 32
48. 64
49. 21
50. 22
51. 104
52. 105
53. 121
54. 169
55. 142
56. 62
57. 100
58. 200
59. 125
60. 90
61. 135
62. 400

D *Find the LCM in each case.*

63. 3 and 15
64. 9 and 12
65. 8 and 10
66. 4 and 6
67. 9 and 30
68. 20 and 21
69. 10 and 11
70. 15 and 60
71. 18 and 24
72. 30 and 150
73. 40 and 180
74. 100 and 90

75. 12, 5, and 50

76. 2, 8, and 10

77. 4, 7, and 12

78. 2, 3, and 5

79. 3, 5, and 7

80. 6, 8, and 12

81. 5, 15, and 20

82. 8, 24, and 56

Mixed Practice

Solve.

83. Write the prime factorization of 75.

84. Is 63 prime or composite? If it is composite, identify a factor other than the number itself and 1.

85. List all the factors of 72.

86. Find the LCM of 5, 10, and 12.

Applications

E *Solve.*

87. The federal government conducts a census every year that is a multiple of 10. Explain whether there will be a census in

a. 2015.
b. 2020.

88. Because of production considerations, the number of pages in a book that you are writing must be a multiple of 4. Can the book be

a. 196 pages long?
b. 198 pages long?

89. In 2006, the men's World Cup soccer tournament was held in Munich, Germany. If the tournament is held every 4 years, will there be a tournament in 2036? (*Source:* FIFA World Cup; Soccer Hall of Fame)

90. A car manufacturer recommends changing the oil every 3,000 miles. Would an oil change be recommended at 21,000 miles? Explain.

91. There are 15 players on a rugby team and 10 players on a men's lacrosse team. What is the smallest number of male students in a college that can be split evenly into either rugby or lacrosse teams? (*Source:* wikipedia.org)

92. The Fields Medal, the highest scientific award for mathematicians, is awarded every 4 years. The Dantzig Prize, an achievement award in the field of mathematical programming, is awarded every 3 years. If both were given in 2006, in what year will both be given again? (*Sources:* mathunion.org and siam.org.)

93. Two friends work in a hospital. One gets a day off every 5 days, and the other every 6 days. If they were both off today, in how many days will they again both be off?

94. A family must budget for life insurance premiums every 6 months, car insurance premiums every 3 months, and payments for a home security system every 4 months. If all these bills were due this month, in how many months will they again all fall due?

• Check your answers on page A-3.

EXAMPLE 1

In the diagram, what does the shaded portion represent?

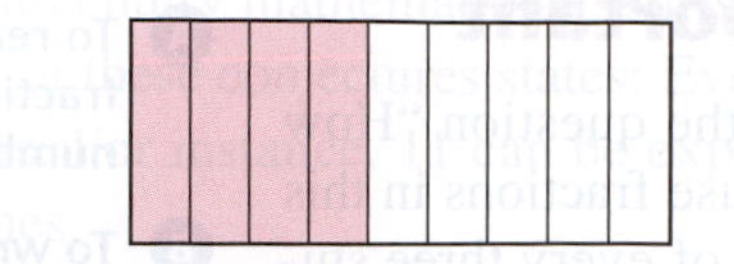

Solution In this diagram, the whole is divided into nine equal parts, so the denominator of the fraction shown is 9. Four of these parts are shaded, so the numerator is 4. The diagram represents the fraction $\frac{4}{9}$.

PRACTICE 1

The diagram illustrates what fraction?

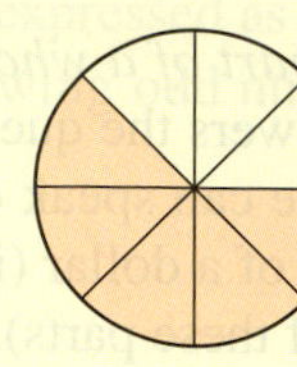

EXAMPLE 2

A college accepted 147 out of 341 applicants for admission into the nursing program. What fraction of the applicants were accepted into this program?

Solution Since there was a total of 341 applicants, the denominator of our fraction is 341. Because 147 of the applicants were accepted, 147 is the numerator. So the college accepted $\frac{147}{341}$ of the applicants into the nursing program.

PRACTICE 2

During a 30-minute television program, 7 minutes were devoted to commercials. What fraction of the time was for commercials?

EXAMPLE 3

The U.S. Senate approved a foreign-aid spending bill by a vote of 83 to 17. What fraction of the senators voted against the bill?

Solution First, we find the total number of senators. Because 83 senators voted for the bill and 17 voted against it, the total number of senators is 83 + 17, or 100.

So $\frac{17}{100}$ of the senators voted against the bill.

PRACTICE 3

A college conducted an online survey of how many students send text messages. Of the 800 students who were surveyed, about 200 did *not* respond. Express the response rate as a fraction.

Mixed Numbers and Improper Fractions

On many jobs, if you work overtime, the rate of pay increases to one-and-a-half times the regular rate. A number such as $1\frac{1}{2}$, with a whole number part and a proper fraction part, is called a mixed number. A mixed number can also be expressed as an improper fraction, that is, a fraction whose numerator is greater than or equal to its denominator. The number $\frac{3}{2}$ is an example of an improper fraction.

Diagrams help us understand that mixed numbers and improper fractions are different forms of the same numbers, as Example 4 illustrates.

EXAMPLE 4

Draw diagrams to show that $2\frac{1}{3} = \frac{7}{3}$.

Solution First, represent the mixed number and the improper fraction in diagrams.

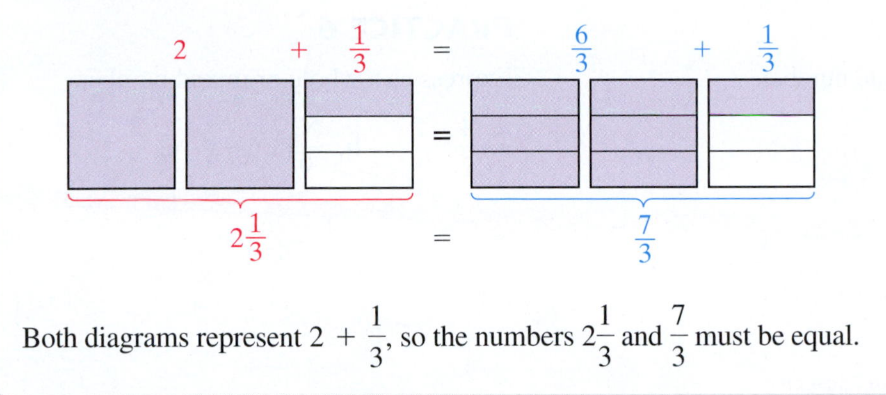

Both diagrams represent $2 + \frac{1}{3}$, so the numbers $2\frac{1}{3}$ and $\frac{7}{3}$ must be equal.

PRACTICE 4

By means of diagrams, explain why $1\frac{2}{3} = \frac{5}{3}$.

In Example 4, each unit (or square) corresponds to 1 whole, which is also three-thirds. That is why the total number of *thirds* in $2\frac{1}{3}$ is $(2 \times 3) + 1$, or 7. The number of *wholes* in $\frac{7}{3}$ is 2 wholes, with $\frac{1}{3}$ of a whole left over. We can generalize these observations into two rules.

To Change a Mixed Number to an Improper Fraction

- Multiply the denominator of the fraction by the whole-number part of the mixed number.
- Add the numerator of the fraction to this product.
- Write this sum over the denominator to form the improper fraction.

EXAMPLE 5

Write each of the following mixed numbers as an improper fraction.

a. $3\frac{2}{9}$ **b.** $12\frac{1}{4}$

Solution

a. $3\frac{2}{9} = \frac{(9 \times 3) + 2}{9}$ **Multiply the denominator 9 by the whole number 3, adding the numerator 2. Place over the denominator.**

$= \frac{27 + 2}{9} = \frac{29}{9}$ **Simplify the numerator.**

b. $12\frac{1}{4} = \frac{(4 \times 12) + 1}{4}$

$= \frac{48 + 1}{4} = \frac{49}{4}$

PRACTICE 5

Express each mixed number as an improper fraction.

a. $5\frac{1}{3}$ **b.** $20\frac{2}{5}$

To Change an Improper Fraction to a Mixed Number

- Divide the numerator by the denominator.
- If there is a remainder, write it over the denominator.

EXAMPLE 6

Write each improper fraction as a mixed or whole number.

a. $\frac{11}{2}$ **b.** $\frac{20}{20}$ **c.** $\frac{42}{5}$

Solution

a. $\frac{11}{2} = \begin{array}{r} 5\text{ R1} \\ 2\overline{)11} \end{array}$ Divide the numerator by the denominator.

$\frac{11}{2} = 5\frac{1}{2}$ Write the remainder over the denominator.

In other words, 5 R1 means that in $\frac{11}{2}$ there are 5 wholes with $\frac{1}{2}$ of a whole left over.

b. $\frac{20}{20} = 1$

c. $\frac{42}{5} = 8\frac{2}{5}$

PRACTICE 6

Express as a whole or mixed number.

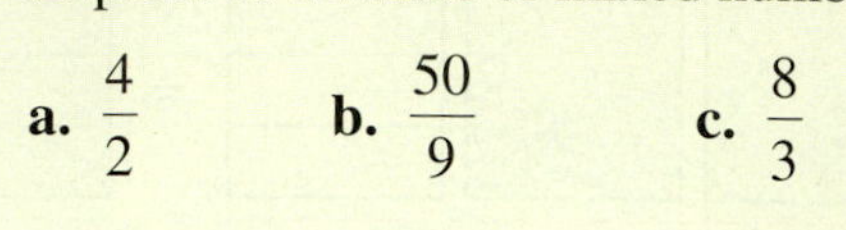

a. $\frac{4}{2}$ **b.** $\frac{50}{9}$ **c.** $\frac{8}{3}$

Changing an improper fraction to a mixed number is important when we are dividing whole numbers: It allows us to express any remainder as a fraction. Previously, we would have said that the problems $2\overline{)7}$ and $4\overline{)13}$ both have the answer 3 R1. But by interpreting these problems as improper fractions, we see that their answers are different.

$$\frac{7}{2} = 3\frac{1}{2} \quad \text{but} \quad \frac{13}{4} = 3\frac{1}{4}$$

When a number is expressed as a mixed number, we know its size more readily than when it is expressed as an improper fraction. For instance, consider the mixed number $11\frac{7}{8}$. We immediately see that it is larger than 11 and smaller than 12 (that is, between 11 and 12). We could not reach this conclusion so easily if we were to examine only $\frac{95}{8}$, its improper form. However, there are situations—when we multiply or divide fractions—in which the use of improper fractions is preferable.

Equivalent Fractions

Some fractions that at first glance appear to be different from one another are really the same.

For instance, suppose that we cut a pizza into 8 equal slices, and then eat 4 of the slices. The shaded portion of the diagram at the right represents the amount eaten. Can you explain why in this diagram the fractions $\frac{4}{8}$ and $\frac{1}{2}$ describe the same part of the whole pizza? We say that these fractions are **equivalent**.

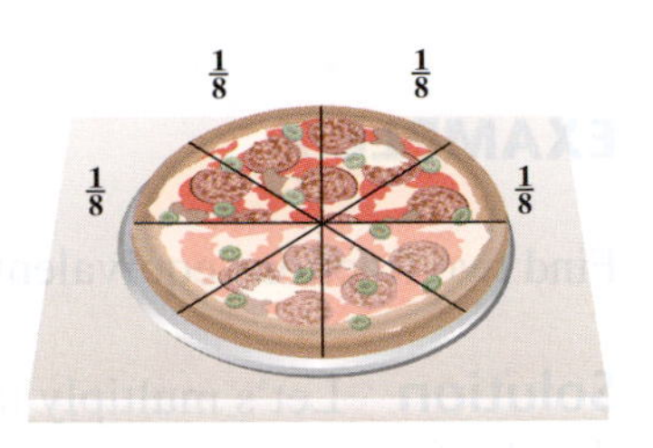

Any fraction has infinitely many equivalent fractions. To see why, let's consider the fraction $\frac{1}{3}$. We can draw different diagrams representing one-third of a whole.

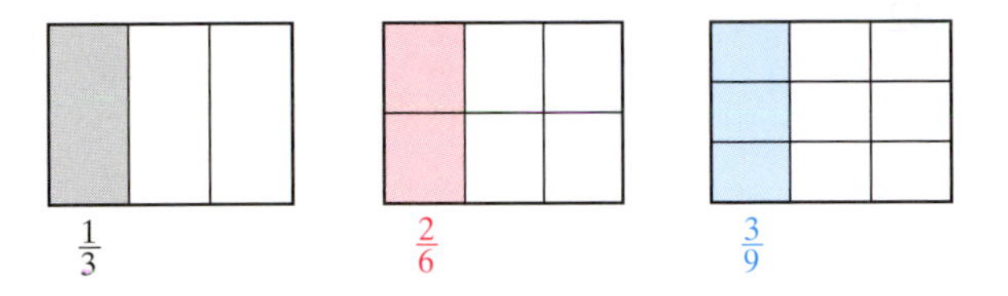

All the shaded portions of the diagrams are identical, so $\frac{1}{3} = \frac{2}{6} = \frac{3}{9}$.

A faster way to generate fractions equivalent to $\frac{1}{3}$ is to multiply both its numerator and denominator by the *same* whole number. Any whole number except 0 will do.

$$\frac{1}{3} = \frac{1 \cdot 2}{3 \cdot 2} = \frac{2}{6}$$

$$\frac{1}{3} = \frac{1 \cdot 3}{3 \cdot 3} = \frac{3}{9}$$

$$\frac{1}{3} = \frac{1 \cdot 4}{3 \cdot 4} = \frac{4}{12}$$

$$\frac{1}{3} = \frac{1 \cdot 5}{3 \cdot 5} = \frac{5}{15}$$

So $\frac{1}{3} = \frac{2}{6} = \frac{3}{9} = \frac{4}{12} = \frac{5}{15} = \ldots.$

Can you explain how you would generate fractions equivalent to $\frac{3}{5}$?

To Find an Equivalent Fraction

Multiply the numerator and denominator of $\frac{a}{b}$ by the same whole number n,

$$\frac{a}{b} = \frac{a \cdot n}{b \cdot n},$$

where both b and n are nonzero.

An important property of equivalent fractions is that their **cross products** are always equal.

$$\frac{1}{3} = \frac{2}{6}$$

In this case, $1 \cdot 6 = 3 \cdot 2 = 6$

EXAMPLE 11 (continued)

The part is \$12,000 and the whole is \$75,000, so the fraction is $\frac{12{,}000}{75{,}000}$. We can simplify this fraction by cancelling common factors:

$$\frac{12{,}000}{75{,}000} = \frac{12,\cancel{000}}{75,\cancel{000}} = \frac{12}{75}$$

Note that canceling a 0 is the same as dividing by 10.

$$= \frac{3 \cdot 4}{3 \cdot 25} = \frac{\overset{1}{\cancel{3}} \cdot 4}{\underset{1}{\cancel{3}} \cdot 25} = \frac{4}{25}$$

Therefore, $\frac{4}{25}$ of the couple's income goes for rent and food.

Comparing Fractions

Some situations require us to *compare* fractions, that is, to rank them in order of size.

For instance, suppose that $\frac{5}{8}$ of one airline's flights arrive on time, in contrast to $\frac{3}{5}$ of another airline's flights. To decide which airline has a better record for on-time arrivals, we need to compare the fractions.

Or to take another example, suppose that the drinking water in your home, according to a lab report, has 2 parts per million (ppm) of lead. Is the water safe to drink if the federal limit on lead in drinking water is 15 parts per billion (ppb)? Again, we need to compare fractions.

One way to handle such problems is to draw diagrams corresponding to the fractions in question. The larger fraction corresponds to the larger shaded region.

For instance, the diagrams to the right show that $\frac{3}{4}$ is greater than $\frac{1}{4}$. The symbol $>$ stands for "greater than."

$\frac{3}{4} > \frac{1}{4}$

Both $\frac{3}{4}$ and $\frac{1}{4}$ have the same denominator, so we can rank them simply by comparing their numerators.

$$\frac{3}{4} > \frac{1}{4} \quad \text{because} \quad 3 > 1$$

For **like fractions**, the fraction with the larger numerator is the larger fraction.

DEFINITIONS

Like fractions are fractions with the same denominator.

Unlike fractions are fractions with different denominators.

To Compare Fractions

- If the fractions are like, compare their numerators.
- If the fractions are unlike, write them as equivalent fractions with the same denominator, and then compare their numerators.

EXAMPLE 12

Compare $\frac{7}{15}$ and $\frac{4}{9}$.

Solution These fractions are unlike because they have different denominators. Therefore, we need to express them as equivalent fractions having the same denominator. But what should that denominator be?

One common denominator that we can use is the *product of the denominators*: $15 \cdot 9 = 135$.

$\frac{7}{15} = \frac{7 \cdot 9}{15 \cdot 9} = \frac{63}{135}$ 135 = 15 • 9, so the new numerator is 7 • 9 or 63.

$\frac{4}{9} = \frac{4 \cdot 15}{9 \cdot 15} = \frac{60}{135}$ 135 = 9 • 15, so the new numerator is 4 • 15 or 60.

Next, we compare the numerators of the like fractions that we just found.

Because $63 > 60$, $\frac{63}{135} > \frac{60}{135}$. Therefore, $\frac{7}{15} > \frac{4}{9}$.

Another common denominator that we can use is the least common multiple of the denominators.

$$15 = 3 \times 5 \qquad 9 = 3 \times 3 = 3^2$$

The LCM is $3^2 \times 5 = 9 \times 5 = 45$. We then compute the equivalent fractions.

$\frac{7}{15} = \frac{7 \cdot 3}{15 \cdot 3} = \frac{21}{45}$ Multiply the numerator and denominator by 3.

$\frac{4}{9} = \frac{4 \cdot 5}{9 \cdot 5} = \frac{20}{45}$ Multiply the numerator and denominator by 5.

Because $\frac{21}{45} > \frac{20}{45}$, we know that $\frac{7}{15} > \frac{4}{9}$.

PRACTICE 12

Which is larger, $\frac{13}{24}$ or $\frac{11}{16}$?

Note that in Example 12 we computed the LCM of the two denominators. This type of computation is used frequently in working with fractions.

DEFINITION

For two or more fractions, their **least common denominator** (LCD) is the least common multiple of their denominators.

In Example 13, pay particular attention to how we use the LCD.

EXAMPLE 13

Order from smallest to largest: $\frac{3}{4}$, $\frac{7}{10}$, and $\frac{29}{40}$

Solution Because these fractions are unlike, we need to find equivalent fractions with a common denominator. Let's use their LCD as that denominator.

$$4 = 2 \times 2 = 2^2$$
$$10 = 2 \times 5$$
$$40 = 2 \times 2 \times 2 \times 5 = 2^3 \times 5$$

The LCD $= 2^3 \times 5 = 8 \times 5 = 40$. Check: 4 and 10 are both factors of 40.

We write each fraction with a denominator of 40.

$$\frac{3}{4} = \frac{3 \cdot 10}{4 \cdot 10} = \frac{30}{40} \qquad \frac{7}{10} = \frac{7 \cdot 4}{10 \cdot 4} = \frac{28}{40} \qquad \frac{29}{40} = \frac{29}{40}$$

Then, we order the fractions from smallest to largest. (The symbol $<$ stands for "less than.")

$$\frac{28}{40} < \frac{29}{40} < \frac{30}{40} \quad \text{or} \quad \frac{7}{10} < \frac{29}{40} < \frac{3}{4}$$

PRACTICE 13

Arrange $\frac{9}{10}$, $\frac{23}{30}$, and $\frac{8}{15}$ from smallest to largest.

EXAMPLE 14

About $\frac{39}{50}$ of Earth's atmosphere is made up of nitrogen, and $\frac{21}{100}$ is made up of oxygen. Does nitrogen or oxygen make up more of Earth's atmosphere? (*Source: The New York Times Almanac, 2010*)

Solution We need to compare $\frac{39}{50}$ with $\frac{21}{100}$. The LCD is 100.

$$\frac{39}{50} = \frac{78}{100}$$
$$\frac{21}{100} = \frac{21}{100}$$

Since $\frac{78}{100} > \frac{21}{100}, \frac{39}{50} > \frac{21}{100}$. Therefore, nitrogen makes up more of Earth's atmosphere than oxygen does.

PRACTICE 14

In a recent year, about $\frac{3}{20}$ of the commercial radio stations in the United States had an adult contemporary format and $\frac{1}{5}$ had a country format. In that year, were there more adult contemporary stations or country stations? (*Source:* musicbizacademy.com)

2.2 Exercises

FOR EXTRA HELP MyMathLab MathXL PRACTICE WATCH READ REVIEW

Mathematically Speaking

Fill in each blank with the most appropriate term or phrase from the given list.

improper fraction	proper fraction	like fractions
greatest common factor	simplify	mixed
convert	least common denominator	composite
equivalent		

1. A fraction whose numerator is smaller than its denominator is called a(n) ______________.

2. The improper fraction $\frac{5}{2}$ can be expressed as a(n) ________ number.

3. The fractions $\frac{6}{8}$ and $\frac{3}{4}$ are ____________.

4. Divide the numerator and denominator of a fraction by the same whole number in order to ____________ it.

5. Fractions with the same denominator are said to be ______________.

6. The ____________________ of two or more fractions is the least common multiple of their denominators.

A *Identify a fraction or mixed number represented by the shaded portion of each figure.*

7.

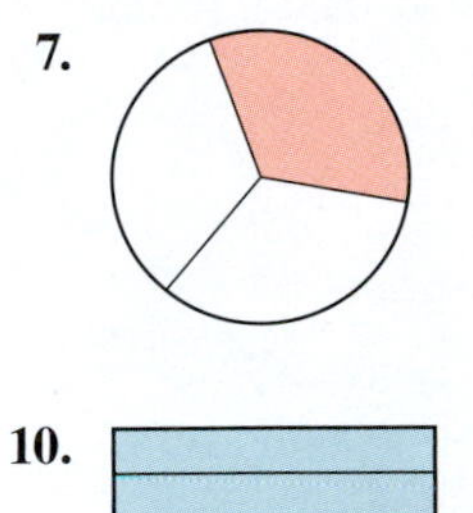

8.

9.

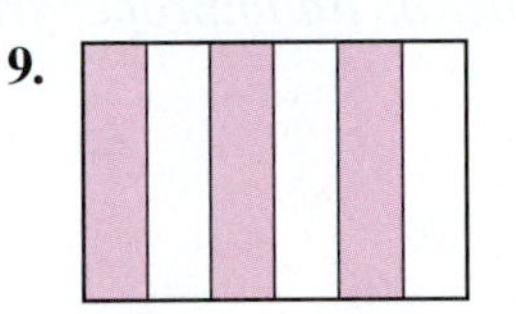

10.

11.

12.

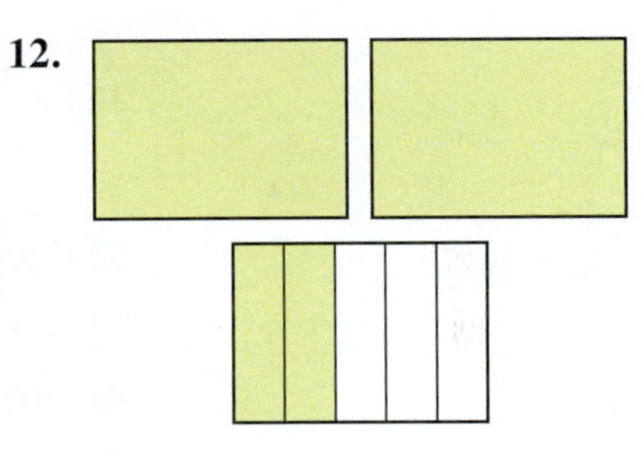

13.

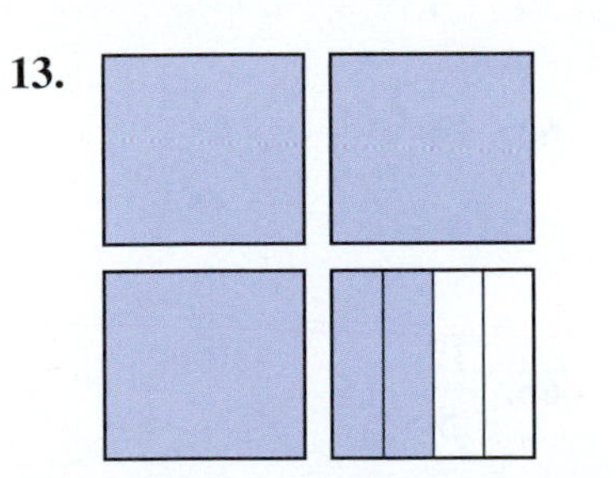

14.

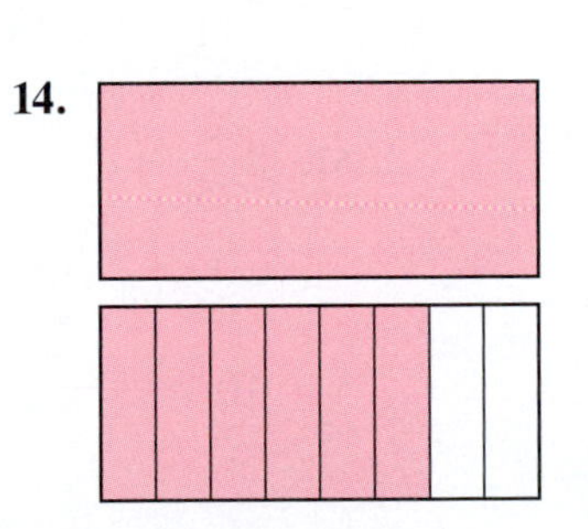

Draw a diagram to represent each fraction or mixed number.

15. $\frac{5}{7}$

16. $\frac{6}{11}$

17. $\frac{2}{9}$

18. $\frac{4}{10}$

19. $\frac{6}{6}$

20. $\frac{11}{11}$

21. $\frac{6}{5}$

22. $\frac{8}{3}$

23. $2\frac{1}{2}$

24. $4\frac{1}{5}$

25. $2\frac{1}{3}$

26. $3\frac{4}{9}$

B *Indicate whether each number is a proper fraction, an improper fraction, or a mixed number.*

27. $\frac{2}{5}$

28. $\frac{7}{12}$

29. $\frac{10}{9}$

30. $\frac{11}{10}$

31. $16\frac{2}{3}$

32. $12\frac{1}{2}$

33. $\frac{5}{5}$

34. $\frac{4}{4}$

35. $\frac{4}{9}$

36. $\frac{5}{6}$

37. $66\frac{2}{3}$

38. $10\frac{3}{4}$

Write each number as an improper fraction.

39. $2\frac{3}{5}$

40. $1\frac{1}{3}$

41. $6\frac{1}{9}$

42. $10\frac{2}{3}$

43. $11\frac{2}{5}$

44. $12\frac{3}{4}$

45. 5

46. 8

47. $7\frac{3}{8}$

48. $6\frac{5}{6}$

49. $9\frac{7}{9}$

50. $10\frac{1}{2}$

51. $13\frac{1}{2}$

52. $20\frac{1}{8}$

53. $19\frac{3}{5}$

54. $11\frac{5}{7}$

55. 14

56. 10

57. $4\frac{10}{11}$

58. $2\frac{7}{13}$

59. $8\frac{3}{14}$

60. $4\frac{1}{6}$

61. $8\frac{2}{25}$

62. $14\frac{1}{10}$

Express each fraction as a mixed or whole number.

63. $\frac{4}{3}$

64. $\frac{6}{5}$

65. $\frac{10}{9}$

66. $\frac{12}{5}$

67. $\frac{9}{3}$

68. $\frac{12}{12}$

69. $\frac{15}{15}$

70. $\frac{100}{100}$

71. $\frac{99}{5}$

72. $\frac{31}{2}$

73. $\frac{82}{9}$

74. $\frac{62}{3}$

75. $\frac{45}{45}$

76. $\frac{40}{3}$

77. $\frac{74}{9}$

78. $\frac{41}{8}$

79. $\frac{27}{2}$ **80.** $\frac{58}{11}$ **81.** $\frac{100}{9}$ **82.** $\frac{38}{3}$

83. $\frac{27}{1}$ **84.** $\frac{72}{9}$ **85.** $\frac{56}{7}$ **86.** $\frac{19}{1}$

C *Find two fractions equivalent to each fraction.*

87. $\frac{1}{8}$ **88.** $\frac{3}{10}$ **89.** $\frac{2}{11}$ **90.** $\frac{1}{10}$

91. $\frac{3}{4}$ **92.** $\frac{5}{6}$ **93.** $\frac{1}{9}$ **94.** $\frac{3}{5}$

Write an equivalent fraction with the given denominator.

95. $\frac{3}{4} = \frac{}{12}$ **96.** $\frac{2}{9} = \frac{}{18}$ **97.** $\frac{5}{8} = \frac{}{24}$ **98.** $\frac{7}{10} = \frac{}{20}$

99. $4 = \frac{}{10}$ **100.** $5 = \frac{}{15}$ **101.** $\frac{3}{5} = \frac{}{60}$ **102.** $\frac{4}{9} = \frac{}{63}$

103. $\frac{5}{8} = \frac{}{64}$ **104.** $\frac{3}{10} = \frac{}{40}$ **105.** $3 = \frac{}{18}$ **106.** $2 = \frac{}{21}$

107. $\frac{4}{9} = \frac{}{81}$ **108.** $\frac{7}{8} = \frac{}{24}$ **109.** $\frac{6}{7} = \frac{}{49}$ **110.** $\frac{5}{6} = \frac{}{48}$

111. $\frac{2}{17} = \frac{}{51}$ **112.** $\frac{1}{3} = \frac{}{90}$ **113.** $\frac{7}{12} = \frac{}{84}$ **114.** $\frac{1}{4} = \frac{}{100}$

115. $\frac{2}{3} = \frac{}{48}$ **116.** $\frac{7}{8} = \frac{}{56}$ **117.** $\frac{3}{10} = \frac{}{100}$ **118.** $\frac{5}{6} = \frac{}{144}$

Simplify, if possible.

119. $\frac{6}{9}$ **120.** $\frac{9}{12}$ **121.** $\frac{10}{10}$ **122.** $\frac{21}{21}$

123. $\frac{5}{15}$ **124.** $\frac{4}{24}$ **125.** $\frac{9}{20}$ **126.** $\frac{25}{49}$

127. $\frac{25}{100}$ **128.** $\frac{75}{100}$ **129.** $\frac{125}{1,000}$ **130.** $\frac{875}{1,000}$

131. $\frac{20}{16}$ **132.** $\frac{15}{9}$ **133.** $\frac{66}{32}$ **134.** $\frac{30}{18}$

135. $\frac{18}{32}$ **136.** $\frac{36}{45}$ **137.** $\frac{7}{24}$ **138.** $\frac{19}{51}$

139. $\frac{27}{9}$ **140.** $\frac{36}{144}$ **141.** $\frac{12}{84}$ **142.** $\frac{21}{36}$

143. $3\frac{38}{57}$ **144.** $11\frac{51}{102}$ **145.** $2\frac{100}{100}$ **146.** $1\frac{144}{144}$

177. In a recent year, the weights of the first six draft picks for the Pittsburgh Steelers football team are given in the table below.

Player	Pouncey	Worilds	Sanders	Gibson	Scott	Butler
Weight (in pounds)	312	240	180	240	346	185

(*Source:* sports.yahoo.com)

What was their average weight?

178. The following chart gives the age of the first six American presidents at the time of their inauguration.

President	Washington	J. Adams	Jefferson	Madison	Monroe	J. Q. Adams
Age (in years)	57	61	57	57	58	57

What was their average age at inauguration? (*Source: Significant American Presidents of the United States*)

• Check your answers on page A-3.

MINDStretchers

Mathematical Reasoning

1. Identify the fraction that the shaded portion of the figure to the right represents.

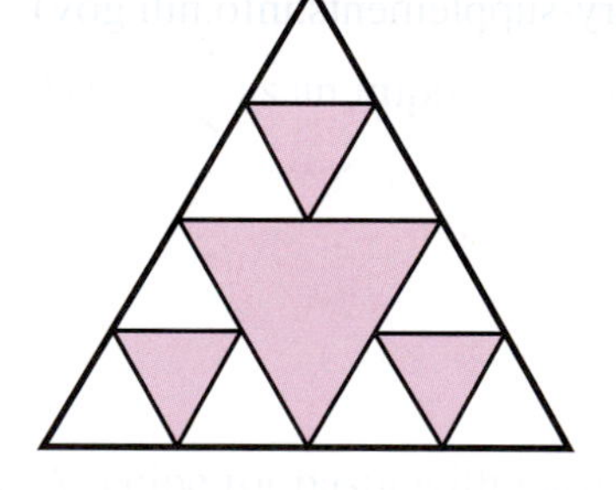

Groupwork

2. Working with a partner, determine how many fractions there are between the numbers 1 and 2.

Critical Thinking

3. Consider the three equivalent fractions shown. Note that the numerators and denominators are made up of the digits 1, 2, 3, 4, 5, 6, 7, 8, and 9—each appearing once.

$$\frac{3}{6} = \frac{7}{14} = \frac{29}{58}$$

a. Verify that these fractions are equivalent by making sure that their cross products are equal.

b. Write another trio of equivalent fractions that use the same nine digits only once.

$$\frac{2}{4} = \frac{\ }{\ } = \frac{\quad}{\quad}$$

2.3 Adding and Subtracting Fractions

OBJECTIVES

- **A** To add or subtract fractions or mixed numbers
- **B** To solve applied problems involving the addition or subtraction of fractions or mixed numbers

In Section 2.2 we examined what fractions mean, how they are written, and how they are compared. In the rest of this chapter, we discuss computations involving fractions, beginning with sums and differences.

Adding and Subtracting Like Fractions

Let's first discuss how to add and subtract like fractions. Suppose that an employee spends $\frac{1}{7}$ of his weekly salary for food and $\frac{2}{7}$ for rent. What part of his salary does he spend for food and rent combined? A diagram can help us understand what is involved. First, we shade one-seventh of the diagram, then another two-sevenths. We see in the diagram that the total shaded area is three-sevenths, $\frac{1}{7} + \frac{2}{7} = \frac{3}{7}$. Note that we added the original numerators to get the numerator of the answer but that *the denominator stayed the same*.

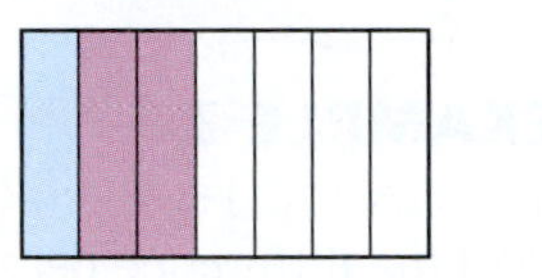

The diagram at the right illustrates the subtraction of like fractions, namely, $\frac{3}{7} - \frac{1}{7}$. If we shade three-sevenths of the diagram and then remove the shading in one-seventh, two-sevenths remain shaded. Therefore, $\frac{3}{7} - \frac{1}{7} = \frac{2}{7}$. Note that we could have gotten this answer simply by subtracting numerators without changing the denominator.

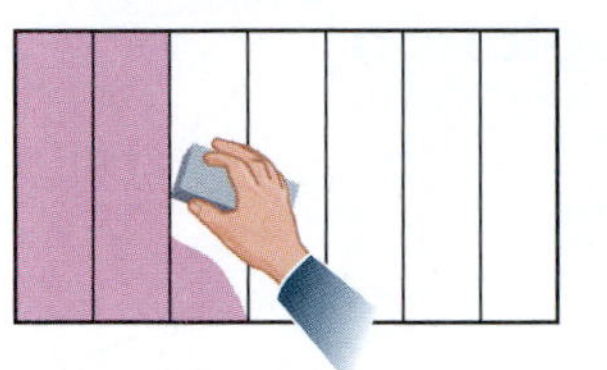

The following rule summarizes how to add or subtract fractions, *provided that they have the same denominator.*

To Add (or Subtract) Like Fractions

- Add (or subtract) the numerators.
- Use the given denominator.
- Write the answer in simplest form.

EXAMPLE 1

Add: $\frac{7}{12} + \frac{2}{12}$

Solution Applying the rule, we get $\frac{7}{12} + \frac{2}{12} = \frac{7+2}{12} = \frac{9}{12}$

Add the numerators. Keep the same denominator.

$= \frac{3 \cdot 3}{4 \cdot 3} = \frac{3 \cdot \cancel{3}^{1}}{4 \cdot \cancel{3}_{1}} = \frac{3}{4}$.

Simplest form

PRACTICE 1

Find the sum of $\frac{7}{15}$ and $\frac{3}{15}$.

TIP Be careful *not* to add the denominators when adding fractions.

EXAMPLE 2

Find the sum of $\frac{12}{16}$, $\frac{3}{16}$, and $\frac{9}{16}$.

Solution

Answer as a mixed number

$$\frac{12}{16} + \frac{3}{16} + \frac{9}{16} = \frac{24}{16} = \frac{3}{2}, \text{ or } 1\frac{1}{2}$$

So the sum of $\frac{12}{16}$, $\frac{3}{16}$, and $\frac{9}{16}$ is $1\frac{1}{2}$.

PRACTICE 2

Add: $\frac{13}{40}$, $\frac{11}{40}$ and $\frac{23}{40}$

EXAMPLE 3

Find the difference between $\frac{11}{7}$ and $\frac{3}{7}$.

Solution

Subtract the numerators.

$$\frac{11}{7} - \frac{3}{7} = \frac{11 - 3}{7} = \frac{8}{7}, \text{ or } 1\frac{1}{7}$$

Keep the same denominator.

PRACTICE 3

Subtract: $\frac{19}{20} - \frac{11}{20}$

EXAMPLE 4

In the following diagram,

a. how far is it from the Administration Building to the Library via the Science Center?

b. which route from the Administration Building to the Library is shorter—via the Science Center or via the Student Center? By how much?

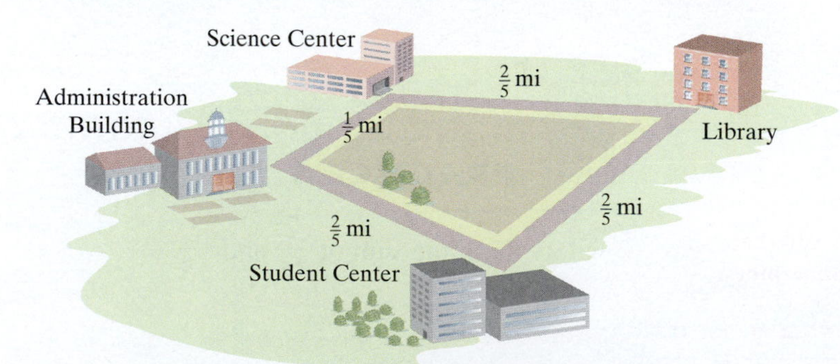

Solution

a. Examining the diagram, we see that

- the distance from the Administration Building to the Science Center is $\frac{1}{5}$ mile, and
- the distance from the Science Center to the Library is $\frac{2}{5}$ mile.

PRACTICE 4

A pediatrician prescribed $\frac{9}{20}$ gram of pain medication for a patient to take every 4 hours.

a. If the dosage were increased by $\frac{3}{20}$ gram, what would the new dosage be?

b. If the original dosage were decreased by $\frac{1}{20}$ gram, find the new dosage.

To find the distance from the Administration Building to the Library via the Science Center, we add.

$$\frac{1}{5} + \frac{2}{5} = \frac{3}{5}$$

So this distance is $\frac{3}{5}$ mile.

b. To find the distance from the Administration Building to the Library via the Student Center, we again add.

$$\frac{2}{5} + \frac{2}{5} = \frac{4}{5}$$

So this distance is $\frac{4}{5}$ mile. Since $\frac{3}{5} < \frac{4}{5}$, the route from the Administration Building to the Library via the Science Center is shorter than the route via the Student Center. Now we find the difference.

$$\frac{4}{5} - \frac{3}{5} = \frac{1}{5}$$

Therefore, the route via the Science Center is $\frac{1}{5}$ mile shorter than the route via the Student Center.

Adding and Subtracting Unlike Fractions

Adding (or subtracting) **unlike fractions** is more complicated than adding (or subtracting) like fractions. An extra step is required: changing the unlike fractions to equivalent like fractions. For instance, suppose that we want to add $\frac{1}{10}$ and $\frac{2}{15}$. Even though we can use any common denominator for these fractions, let's use their *least* common denominator to find equivalent fractions.

$$\begin{aligned} 10 &= 2 \cdot 5 \\ 15 &= 3 \cdot 5 \\ \text{LCD} &= 2 \cdot 3 \cdot 5 = 30 \end{aligned}$$

Let's rewrite the fractions vertically as equivalent fractions with the denominator 30.

$$\begin{aligned} \frac{1}{10} &= \frac{1 \cdot 3}{10 \cdot 3} = \frac{3}{30} \\ +\frac{2}{15} &= \frac{2 \cdot 2}{15 \cdot 2} = +\frac{4}{30} \end{aligned}$$

Now, we add the equivalent like fractions.

$$\begin{array}{r} \frac{3}{30} \\ +\frac{4}{30} \\ \hline \frac{7}{30} \end{array}$$

So $\frac{1}{10} + \frac{2}{15} = \frac{7}{30}$.

We can also add and subtract unlike fractions horizontally.

$$\frac{1}{10} + \frac{2}{15} = \frac{3}{30} + \frac{4}{30} = \frac{3 + 4}{30} = \frac{7}{30}$$

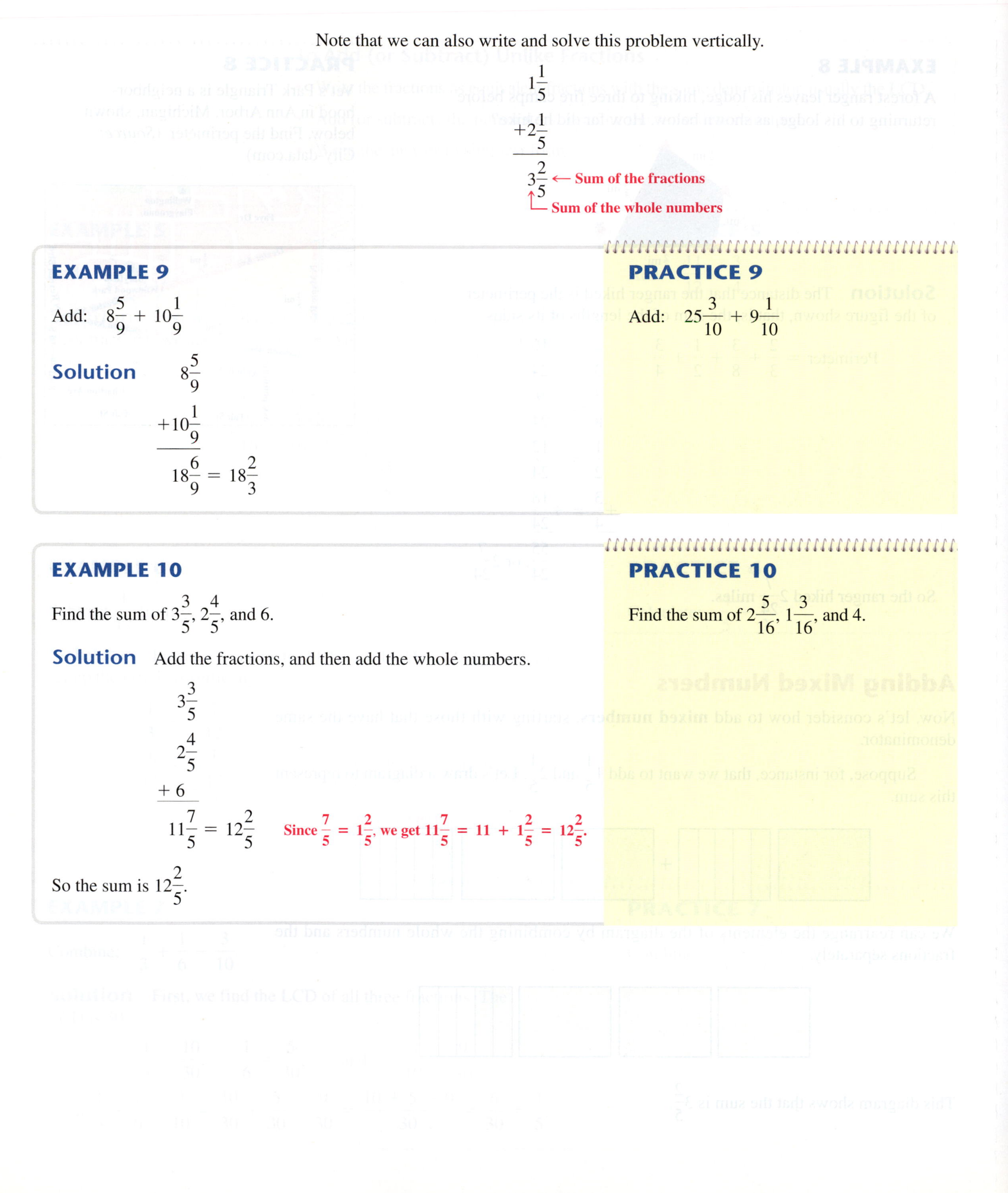

Note that we can also write and solve this problem vertically.

$$\begin{array}{r} 1\frac{1}{5} \\ +2\frac{1}{5} \\ \hline 3\frac{2}{5} \end{array}$$

$\leftarrow$ **Sum of the fractions**

$\uparrow$ **Sum of the whole numbers**

EXAMPLE 9

Add: $8\frac{5}{9} + 10\frac{1}{9}$

Solution

$$\begin{array}{r} 8\frac{5}{9} \\ +10\frac{1}{9} \\ \hline 18\frac{6}{9} \end{array} = 18\frac{2}{3}$$

PRACTICE 9

Add: $25\frac{3}{10} + 9\frac{1}{10}$

EXAMPLE 10

Find the sum of $3\frac{3}{5}$, $2\frac{4}{5}$, and 6.

Solution Add the fractions, and then add the whole numbers.

$$\begin{array}{r} 3\frac{3}{5} \\ 2\frac{4}{5} \\ +\ 6 \\ \hline 11\frac{7}{5} \end{array} = 12\frac{2}{5}$$

Since $\frac{7}{5} = 1\frac{2}{5}$, we get $11\frac{7}{5} = 11 + 1\frac{2}{5} = 12\frac{2}{5}$.

So the sum is $12\frac{2}{5}$.

PRACTICE 10

Find the sum of $2\frac{5}{16}$, $1\frac{3}{16}$, and 4.

EXAMPLE 11

Two movies are shown back-to-back on TV without commercial interruption. The first runs $1\frac{3}{4}$ hours, and the second $2\frac{1}{4}$ hours. How long will it take to watch both movies?

Solution

We need to add $1\frac{3}{4}$ and $2\frac{1}{4}$.

$$\begin{array}{r} 1\frac{3}{4} \\ +2\frac{1}{4} \\ \hline 3\frac{4}{4} \end{array} = 3 + 1 = 4$$

Therefore, it will take 4 hours to watch the two movies.

PRACTICE 11

In a horse race, the winner beat the second-place horse by $1\frac{1}{2}$ lengths, and the second-place horse finished $2\frac{1}{2}$ lengths ahead of the third-place horse. By how many lengths did the third-place horse lose?

We have previously shown that when we add fractions with different denominators, we must first change the unlike fractions to equivalent like fractions. The same applies to adding mixed numbers that have different denominators.

To Add Mixed Numbers

- Write the fractions as equivalent fractions with the same denominator, usually the LCD.
- Add the fractions.
- Add the whole numbers.
- Write the answer in simplest form.

EXAMPLE 12

Find the sum of $3\frac{1}{5}$ and $7\frac{2}{3}$.

Solution

The LCD is 15. Add the fractions, and then add the whole numbers.

$$\begin{array}{rcr} 3\frac{1}{5} & = & 3\frac{3}{15} \\ +7\frac{2}{3} & = & +7\frac{10}{15} \\ \hline & & 10\frac{13}{15} \end{array}$$

The sum of $3\frac{1}{5}$ and $7\frac{2}{3}$ is $10\frac{13}{15}$.

PRACTICE 12

Add $4\frac{1}{8}$ to $3\frac{1}{2}$.

EXAMPLE 16

Subtract $2\frac{7}{100}$ from $5\frac{9}{10}$.

Solution As usual, we use the LCD (which is 100) to find equivalent fractions. Then we subtract the equivalent mixed numbers with the same denominator. Again, let's set up the problem vertically. Subtract the fractions, and then subtract the whole numbers.

$$\begin{array}{rcr} 5\frac{9}{10} & = & 5\frac{90}{100} \\ -2\frac{7}{100} & = & -2\frac{7}{100} \\ \hline & & 3\frac{83}{100} \end{array}$$

The answer is $3\frac{83}{100}$.

PRACTICE 16

Calculate: $8\frac{2}{3} - 4\frac{1}{12}$

EXAMPLE 17

Find the length of the pool shown below.

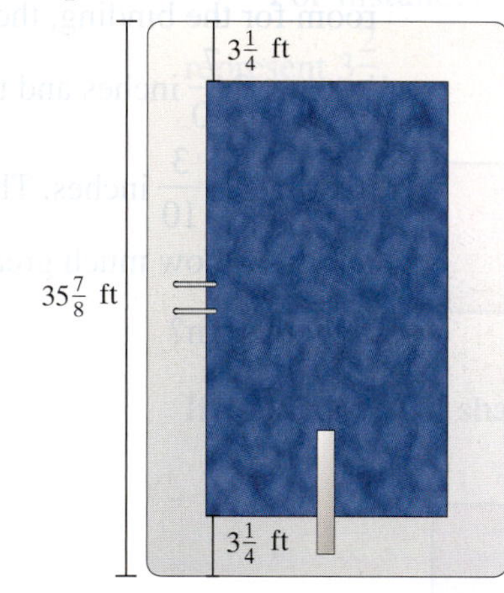

Solution The total length of the pool and the walkway is $35\frac{7}{8}$ feet. To find the length of the pool, we first add $3\frac{1}{4}$ feet and $3\frac{1}{4}$ feet. Then we subtract this sum from $35\frac{7}{8}$ feet.

$$\begin{array}{r} 3\frac{1}{4} \\ +3\frac{1}{4} \\ \hline 6\frac{2}{4} = 6\frac{1}{2} \end{array} \qquad \begin{array}{rcr} 35\frac{7}{8} & = & 35\frac{7}{8} \\ -6\frac{1}{2} & = & -6\frac{4}{8} \\ \hline & & 29\frac{3}{8} \end{array}$$

So the length of the pool is $29\frac{3}{8}$ feet. We can check this answer by adding $3\frac{1}{4}$, $29\frac{3}{8}$, and $3\frac{1}{4}$, getting $35\frac{7}{8}$.

PRACTICE 17

The outline of the state of Nevada is roughly in the shape of a *trapezoid*. The diagram shows the lengths of the borders that Nevada shares with each neighboring state. What is the total length of Nevada's border? (*Source:* econ.umn.edu)

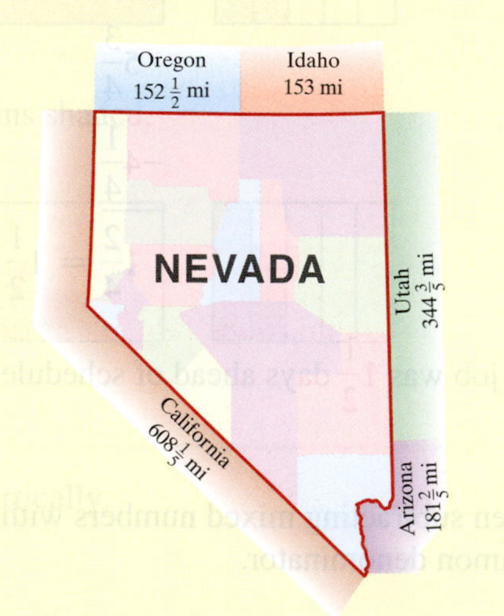

Recall from our discussion of subtracting whole numbers that, in problems in which a digit in the subtrahend is larger than the corresponding digit in the minuend, we need to regroup.

$$\begin{array}{r} \overset{2\,1}{\cancel{3}}\,2\,9 \\ -\;8\,7 \\ \hline 2\,4\,2 \end{array}$$

A similar situation can arise when we are subtracting mixed numbers. If the fraction on the bottom is larger than the fraction on top, we *regroup*, or *borrow from* the whole number on top.

EXAMPLE 18

Subtract: $6 - 1\frac{1}{3}$

Solution Let's rewrite the problem vertically.

$$\begin{array}{r} 6 \\ -1\frac{1}{3} \\ \hline \end{array}$$ There is no fraction on top from which to subtract $\frac{1}{3}$.

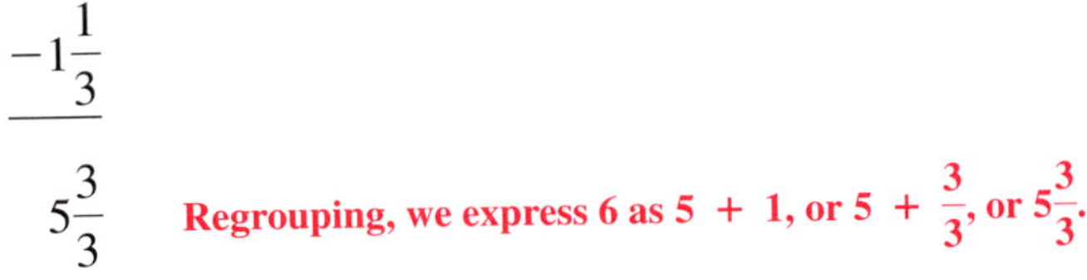

$$\begin{array}{r} 5\frac{3}{3} \\ -1\frac{1}{3} \\ \hline \end{array}$$ Regrouping, we express 6 as $5 + 1$, or $5 + \frac{3}{3}$, or $5\frac{3}{3}$.

$$\begin{array}{r} 5\frac{3}{3} \\ -1\frac{1}{3} \\ \hline 4\frac{2}{3} \end{array}$$ Now subtract.

So $6 - 1\frac{1}{3} = 4\frac{2}{3}$.

As in any subtraction problem, we can check our answer by addition.

$$4\frac{2}{3} + 1\frac{1}{3} = 5\frac{3}{3} = 6$$

PRACTICE 18

Subtract: $9 - 7\frac{5}{7}$

In Example 18, the answer is $4\frac{2}{3}$. Would we get the same answer if we compute $6\frac{1}{3} - 1$?

We have already discussed subtracting mixed numbers without regrouping as well as subtracting a mixed number from a whole number. Now, let's consider the general rule for subtracting mixed numbers.

EXAMPLE 24

Combine and check: $5\frac{1}{3} - \left(2\frac{4}{5} + 1\frac{1}{10}\right)$

Solution Following the order of operations rule, we begin by adding the two mixed numbers in parentheses.

$$\begin{array}{rcr} 2\frac{4}{5} & = & 2\frac{8}{10} \\ +1\frac{1}{10} & = & +1\frac{1}{10} \\ \hline & & 3\frac{9}{10} \end{array}$$

Next, we subtract this sum from $5\frac{1}{3}$.

$$\begin{array}{rcrcr} 5\frac{1}{3} & = & 5\frac{10}{30} & = & 4\frac{40}{30} \\ -3\frac{9}{10} & = & -3\frac{27}{30} & = & -3\frac{27}{30} \\ \hline & & & & 1\frac{13}{30} \end{array}$$

So $5\frac{1}{3} - \left(2\frac{4}{5} + 1\frac{1}{10}\right) = 1\frac{13}{30}$.

Now, let's check this answer by estimating:

$$5\frac{1}{3} - \left(2\frac{4}{5} + 1\frac{1}{10}\right)$$
$$\downarrow \quad \downarrow \quad \downarrow$$
$$5 - (3 + 1) = 5 - 4 = 1$$

The estimate, 1, is close to our answer, $1\frac{13}{30}$.

PRACTICE 24

Calculate and check:

$8\frac{1}{4} - \left(3\frac{2}{5} - 1\frac{9}{10}\right)$

2.3 Exercises

FOR EXTRA HELP MyMathLab Math XL PRACTICE WATCH READ REVIEW

Mathematically Speaking

Fill in each blank with the most appropriate term or phrase from the given list.

denominators	regroup	equivalent
add	numerators	improper

1. To add like fractions, add the __________.

2. To subtract unlike fractions, rewrite them as __________ fractions with the same denominator.

3. When subtracting $2\frac{4}{5}$ from 7, __________ by writing 7 as $6\frac{5}{5}$.

4. Fractions with equal numerators and __________ are equivalent to 1.

A *Add and simplify.*

5. $\frac{5}{8} + \frac{5}{8}$

6. $\frac{7}{10} + \frac{9}{10}$

7. $\frac{11}{12} + \frac{7}{12}$

8. $\frac{71}{100} + \frac{79}{100}$

9. $\frac{1}{5} + \frac{1}{5} + \frac{2}{5}$

10. $\frac{1}{7} + \frac{3}{7} + \frac{2}{7}$

11. $\frac{3}{20} + \frac{1}{20} + \frac{8}{20}$

12. $\frac{1}{10} + \frac{3}{10} + \frac{1}{10}$

13. $\frac{2}{3} + \frac{1}{2}$

14. $\frac{1}{4} + \frac{2}{5}$

15. $\frac{1}{2} + \frac{3}{8}$

16. $\frac{1}{6} + \frac{2}{3}$

17. $\frac{7}{10} + \frac{7}{100}$

18. $\frac{5}{6} + \frac{1}{12}$

19. $\frac{4}{5} + \frac{1}{8}$

20. $\frac{3}{4} + \frac{3}{7}$

21. $\frac{4}{9} + \frac{5}{6}$

22. $\frac{9}{10} + \frac{4}{5}$

23. $\frac{87}{100} + \frac{3}{10}$

24. $\frac{7}{20} + \frac{3}{4}$

25. $\frac{1}{3} + \frac{1}{4} + \frac{1}{6}$

26. $\frac{1}{5} + \frac{1}{6} + \frac{1}{3}$

27. $\frac{3}{8} + \frac{1}{10} + \frac{3}{16}$

28. $\frac{3}{10} + \frac{1}{3} + \frac{1}{9}$

29. $\frac{2}{9} + \frac{5}{8} + \frac{1}{4}$

30. $\frac{1}{2} + \frac{1}{3} + \frac{1}{4}$

31. $\frac{7}{8} + \frac{1}{5} + \frac{1}{4}$

32. $\frac{1}{10} + \frac{2}{5} + \frac{5}{6}$

Add and simplify. Then check by estimating.

33. $1 + 2\frac{1}{3}$

34. $4\frac{1}{5} + 2$

35. $8\frac{1}{10} + 7\frac{3}{10}$

36. $6\frac{1}{12} + 4\frac{1}{12}$

37. $7\frac{3}{10} + 6\frac{9}{10}$

38. $8\frac{2}{3} + 6\frac{2}{3}$

39. $5\frac{1}{6} + 9\frac{5}{6}$

40. $2\frac{3}{10} + 7\frac{9}{10}$

41. $5\frac{1}{4} + 5\frac{1}{6}$

42. $17\frac{3}{8} + 20\frac{1}{5}$

43. $3\frac{1}{3} + \frac{2}{5}$

44. $4\frac{7}{10} + \frac{7}{20}$

45. $8\frac{1}{5} + 5\frac{2}{3}$

46. $4\frac{1}{9} + 20\frac{7}{10}$

47. $\frac{2}{3} + 6\frac{1}{8}$

48. $\frac{1}{6} + 3\frac{2}{5}$

49. $9\frac{2}{3} + 10\frac{7}{12}$

50. $20\frac{3}{5} + 4\frac{1}{2}$

51. $6\frac{1}{10} + 3\frac{93}{100}$

52. $4\frac{8}{9} + 5\frac{1}{3}$

53. $4\frac{1}{2} + 6\frac{7}{8}$ **54.** $10\frac{5}{6} + 8\frac{1}{4}$ **55.** $30\frac{21}{100} + 5\frac{17}{20}$ **56.** $8\frac{3}{10} + 2\frac{321}{1,000}$

57. $80\frac{1}{3} + \frac{3}{4} + 10\frac{1}{2}$ **58.** $\frac{1}{3} + 25\frac{7}{24} + 100\frac{1}{2}$ **59.** $2\frac{1}{3} + 2 + 2\frac{1}{6}$ **60.** $4\frac{1}{8} + 4\frac{3}{16} + \frac{5}{4}$

61. $6\frac{7}{8} + 2\frac{3}{4} + 1\frac{1}{5}$ **62.** $1\frac{2}{3} + 5\frac{5}{6} + 3\frac{1}{4}$ **63.** $2\frac{1}{2} + 5\frac{1}{4} + 3\frac{5}{8}$ **64.** $4\frac{2}{3} + 2\frac{11}{36} + 1\frac{1}{2}$

Subtract and simplify.

65. $\frac{4}{5} - \frac{3}{5}$ **66.** $\frac{7}{9} - \frac{5}{9}$ **67.** $\frac{7}{10} - \frac{3}{10}$ **68.** $\frac{11}{12} - \frac{5}{12}$

69. $\frac{23}{100} - \frac{7}{100}$ **70.** $\frac{3}{2} - \frac{1}{2}$ **71.** $\frac{3}{4} - \frac{1}{4}$ **72.** $\frac{7}{9} - \frac{4}{9}$

73. $\frac{12}{5} - \frac{2}{5}$ **74.** $\frac{1}{8} - \frac{1}{8}$ **75.** $\frac{3}{4} - \frac{2}{3}$ **76.** $\frac{2}{5} - \frac{1}{6}$

77. $\frac{4}{9} - \frac{1}{6}$ **78.** $\frac{9}{10} - \frac{3}{100}$ **79.** $\frac{4}{5} - \frac{3}{4}$ **80.** $\frac{5}{6} - \frac{1}{8}$

81. $\frac{4}{7} - \frac{1}{2}$ **82.** $\frac{2}{5} - \frac{2}{9}$ **83.** $\frac{4}{9} - \frac{3}{8}$ **84.** $\frac{11}{12} - \frac{1}{3}$

85. $\frac{3}{4} - \frac{1}{2}$ **86.** $\frac{5}{6} - \frac{2}{3}$

Subtract and simplify. Then, check either by adding or by estimating.

87. $5\frac{3}{7} - 1\frac{1}{7}$ **88.** $6\frac{2}{3} - 1\frac{1}{3}$ **89.** $3\frac{7}{8} - 2\frac{1}{8}$ **90.** $10\frac{5}{6} - 2\frac{5}{6}$

91. $20\frac{1}{2} - \frac{1}{2}$ **92.** $7\frac{3}{4} - \frac{1}{4}$ **93.** $8\frac{1}{10} - 4$ **94.** $2\frac{1}{3} - 2$

95. $6 - 2\frac{2}{3}$ **96.** $4 - 1\frac{1}{5}$ **97.** $8 - 4\frac{7}{10}$ **98.** $2 - 1\frac{1}{2}$

99. $10 - 3\frac{2}{3}$ **100.** $5 - 4\frac{9}{10}$ **101.** $6 - \frac{1}{2}$ **102.** $9 - \frac{3}{4}$

103. $7\frac{1}{4} - 2\frac{3}{4}$ **104.** $5\frac{1}{10} - 2\frac{3}{10}$ **105.** $6\frac{1}{8} - 2\frac{7}{8}$ **106.** $3\frac{1}{5} - 1\frac{4}{5}$

107. $12\frac{2}{5} - \frac{3}{5}$ **108.** $3\frac{7}{10} - \frac{9}{10}$ **109.** $8\frac{1}{3} - 1\frac{2}{3}$ **110.** $2\frac{1}{5} - \frac{4}{5}$

111. $13\frac{1}{2} - 5\frac{2}{3}$ **112.** $7\frac{1}{10} - 2\frac{1}{7}$ **113.** $9\frac{3}{8} - 5\frac{5}{6}$ **114.** $2\frac{1}{10} - 1\frac{27}{100}$

115. $20\frac{2}{9} - 4\frac{5}{6}$ **116.** $9\frac{13}{100} - 6\frac{7}{10}$ **117.** $3\frac{4}{5} - \frac{5}{6}$ **118.** $1\frac{2}{8} - \frac{2}{6}$

119. $1\frac{3}{4} - 1\frac{1}{2}$ **120.** $2\frac{1}{2} - 1\frac{3}{4}$ **121.** $10\frac{1}{12} - 4\frac{2}{3}$ **122.** $7\frac{1}{4} - 1\frac{5}{16}$

123. $22\frac{7}{8} - 8\frac{9}{10}$ **124.** $9\frac{1}{10} - 3\frac{1}{2}$ **125.** $3\frac{1}{8} - 2\frac{3}{4}$ **126.** $3\frac{1}{4} - 2\frac{5}{16}$

Combine and simplify.

127. $\frac{5}{8} + \frac{9}{10} - \frac{1}{4}$ **128.** $\frac{2}{3} - \frac{1}{5} + \frac{1}{2}$ **129.** $12\frac{1}{6} + 5\frac{9}{10} - 1\frac{3}{10}$ **130.** $7\frac{1}{3} - 2\frac{4}{5} - 1\frac{1}{3}$

131. $15\frac{1}{2} - 3\frac{4}{5} - 6\frac{1}{2}$ **132.** $4\frac{1}{10} + 2\frac{9}{10} - 3\frac{3}{4}$ **133.** $20\frac{1}{10} - \left(\frac{1}{20} + 1\frac{1}{2}\right)$ **134.** $19\frac{1}{6} - \left(8\frac{9}{10} - \frac{1}{5}\right)$

Mixed Practice

Perform the indicated operations and simplify.

135. Subtract $1\frac{7}{8}$ from 6.

136. Add: $6\frac{1}{10} + 3\frac{7}{15}$

137. Calculate: $12\frac{2}{3} - \left(8\frac{5}{6} - 4\frac{1}{2}\right)$

138. Find the sum of $\frac{3}{8}$, $\frac{1}{2}$, and $\frac{1}{3}$.

139. Find the difference between $4\frac{3}{5}$ and $1\frac{2}{3}$.

140. Subtract: $\frac{9}{10} - \frac{1}{4}$

Applications

B *Solve. Write the answer in simplest form.*

141. A $\frac{7}{8}$-inch nail was hammered through a $\frac{3}{4}$-inch door. How far did it extend from the door?

142. A building occupies $\frac{1}{4}$ acre on a $\frac{7}{8}$-acre plot of land. What is the area of the land not occupied by the building?

143. The Kentucky Derby, Belmont Stakes, and the Preakness Stakes are three prestigious horse races that comprise the Triple Crown. (*Source:* http://infoplease.com)

a. Horses run $1\frac{3}{16}$ mile in the Preakness Stakes. If the Preakness Stakes is $\frac{5}{16}$ mile shorter than the Belmont Stakes, how far do horses run in the Belmont Stakes?

b. Horses run $1\frac{1}{4}$ miles in the Kentucky Derby. How much farther do horses run in the Belmont Stakes than in the Kentucky Derby?

144. In the year 2030, the total amount of electricity generated worldwide is projected to be approximately 32 trillion kilowatt-hours. Of this amount, $\frac{1}{32}$ is expected to be generated by liquid fuels, $\frac{1}{8}$ by nuclear power, and $\frac{7}{16}$ by coal. (*Source:* eia.doe.gov)

a. According to these projections, the combined amount of electricity generated by liquid fuels and nuclear power will be what fraction of the total world electricity?

b. As a fraction of the electricity generated worldwide, the amount of electricity generated by coal will be how much greater than the combined amount generated by liquid fuel and nuclear power?

145. The first game of a baseball doubleheader lasted $2\frac{1}{4}$ hours. The second game began after a $\frac{1}{4}$-hour break and lasted $2\frac{1}{2}$ hours. How long did the doubleheader take to play?

146. Three student candidates competed in a student government election. The winner got $\frac{5}{8}$ of the votes, and the second-place candidate got $\frac{1}{4}$ of the votes. If the rest of the votes went to the third candidate, what fraction of the votes did that student get?

147. In 1912, the Titanic, the largest passenger ship in the world, sank on its maiden voyage. The steamship had a length of $882\frac{3}{4}$ feet. Its width at the widest point was $92\frac{1}{2}$ feet. How much greater was its length than its width? (*Source:* titanic-titanic.com)

148. A shopper purchased two boxes of chocolates: one box containing a sugar-free assortment weighs $20\frac{5}{8}$ ounces, and the other box, containing milk toffee sticks, weighs $10\frac{1}{2}$ ounces. Find the total weight of the boxes. (*Source:* Russellstover.com)

149. In testing a new drug, doctors found that $\frac{1}{2}$ of the patients given the drug improved, $\frac{2}{5}$ showed no change in their condition, and the remainder got worse. What fraction got worse?

150. The size of a child's shoe is related to the length of his or her foot. The following table shows the relationship between shoe size and foot length for a variety of sizes.

Size	Foot Length (in inches)	Size	Foot Length (in inches)
4	$5\frac{3}{4}$	8	$6\frac{3}{4}$
5	6	9	7
6	$6\frac{1}{4}$	10	$7\frac{1}{4}$
7	$6\frac{1}{2}$	11	$7\frac{1}{2}$

(*Source:* kidbean.com/sizecharts.html)

Is the difference in foot length greater when comparing sizes 4 and 7 or when comparing sizes 7 and 10?

151. Suppose that four packages are placed on a scale, as shown. If the scale balances, how heavy is the small package on the right?

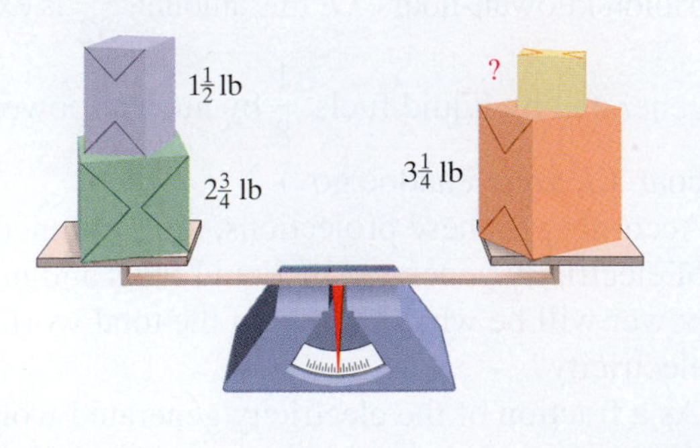

152. If the scale pictured balances, how heavy is the small package on the left?

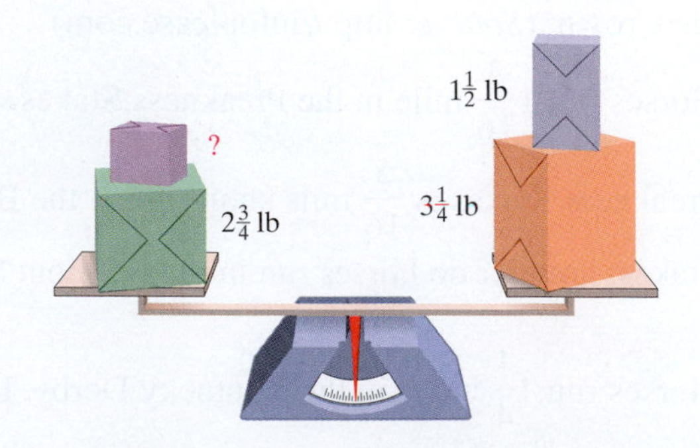

• Check your answers on page A-3.

MINDStretchers

Groupwork

1. Working with a partner, complete the following magic square in which each row, column, and diagonal adds up to the same number.

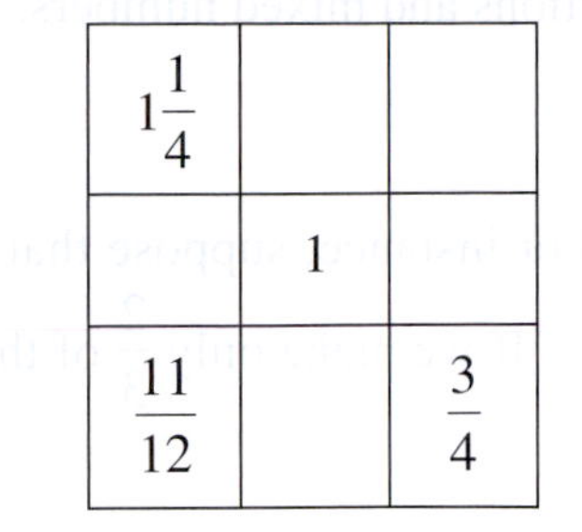

$1\frac{1}{4}$
1
$\frac{11}{12}$
$\frac{3}{4}$

Mathematical Reasoning

2. A fraction with 1 as the numerator is called a **unit fraction**. For example, $\frac{1}{7}$ is a unit fraction. Write $\frac{3}{7}$ as the sum of three unit fractions, using no unit fraction more than once.

$$\frac{3}{7} = \frac{1}{\square} + \frac{1}{\square} + \frac{1}{\square}$$

Writing

3. Consider the following two ways of subtracting $2\frac{4}{5}$ from $4\frac{1}{5}$.

Method 1

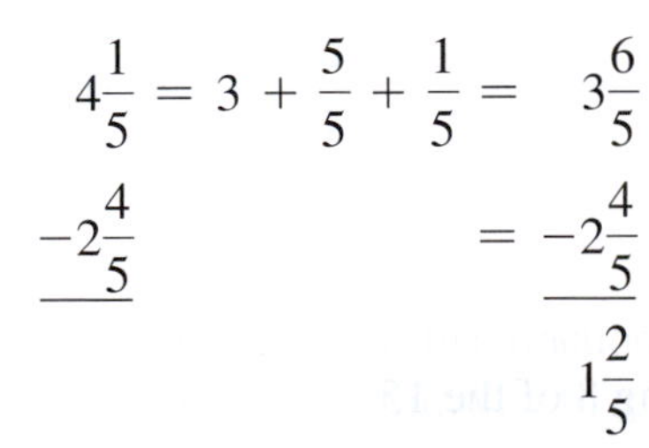

$4\frac{1}{5} = 3 + \frac{5}{5} + \frac{1}{5} = 3\frac{6}{5}$
$-2\frac{4}{5} \qquad = -2\frac{4}{5}$
$1\frac{2}{5}$

Method 2

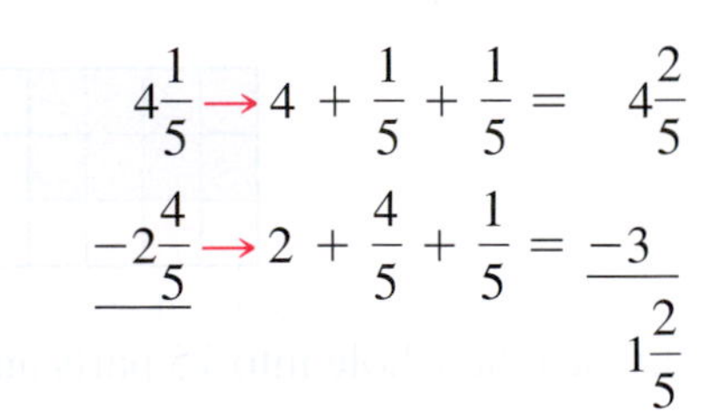

$4\frac{1}{5} \rightarrow 4 + \frac{1}{5} + \frac{1}{5} = 4\frac{2}{5}$
$-2\frac{4}{5} \rightarrow 2 + \frac{4}{5} + \frac{1}{5} = -3$
$1\frac{2}{5}$

a. Explain the difference between the two methods.

b. Explain which method you prefer.

c. Explain why you prefer that method.

EXAMPLE 5

Multiply: $\frac{9}{8} \times \frac{6}{5} \times \frac{7}{9}$

Solution We simplify and then multiply.

$$\frac{9}{8} \times \frac{6}{5} \times \frac{7}{9} = \frac{\overset{1}{\cancel{9}}}{\underset{4}{\cancel{8}}} \times \frac{\overset{3}{\cancel{6}}}{5} \times \frac{7}{\underset{1}{\cancel{9}}}$$

$$= \frac{21}{20}, \text{ or } 1\frac{1}{20}$$

Divide the numerator 9 and the denominator 9 by 9.
Divide the numerator 6 and the denominator 8 by 2.

PRACTICE 5

Multiply: $\frac{7}{27} \cdot \frac{9}{4} \cdot \frac{8}{21}$

EXAMPLE 6

At a college, $\frac{3}{5}$ of the students take a math course. Of these students, $\frac{1}{6}$ take elementary algebra. What fraction of the students in the college take elementary algebra?

Solution We must find $\frac{1}{6}$ of $\frac{3}{5}$.

$$\frac{1}{6} \times \frac{3}{5} = \frac{1}{\underset{2}{\cancel{6}}} \times \frac{\overset{1}{\cancel{3}}}{5} = \frac{1 \times 1}{2 \times 5} = \frac{1}{10}$$

One-tenth of the students in the college take elementary algebra.

PRACTICE 6

A flight from New York to Los Angeles took 7 hours. With the help of the jet stream, the return trip took $\frac{3}{4}$ the time. How long did the trip from Los Angeles to New York take?

EXAMPLE 7

Of the 639 employees at a company, $\frac{4}{9}$ responded to a voluntary survey distributed by the human resources department. How many employees did not respond to the survey?

Solution Apply the strategy of breaking the problem into two parts.

- First, find $\frac{4}{9}$ of 639.
- Then, subtract the result from 639.

In short, we can solve this problem by computing $639 - \left(\frac{4}{9} \times 639\right)$.

$$639 - \left(\frac{4}{9} \times 639\right) = 639 - \left(\frac{4}{\underset{1}{\cancel{9}}} \times \frac{\overset{71}{\cancel{639}}}{1}\right) = 639 - 284 = 355$$

So 355 employees did not respond to the survey.

PRACTICE 7

The state sales tax on a car in Wisconsin is $\frac{1}{20}$ of the price of the car. What is the total amount a consumer would pay for a \$19,780 car? (*Source:* revenue.wi.gov)

Multiplying Mixed Numbers

Some situations require us to multiply mixed numbers. For instance, suppose that your regular hourly wage is $\$7\frac{1}{2}$ and that you make time-and-a-half for working overtime. To find your overtime hourly wage, you need to multiply $1\frac{1}{2}$ by $7\frac{1}{2}$. The key here is to first rewrite each mixed number as an improper fraction.

$$1\frac{1}{2} \times 7\frac{1}{2} = \frac{3}{2} \times \frac{15}{2} = \frac{45}{4}, \text{ or } 11\frac{1}{4}$$

So you make $\$11\frac{1}{4}$ per hour overtime.

To Multiply Mixed Numbers

- Write the mixed numbers as improper fractions.
- Multiply the fractions.
- Write the answer in simplest form.

EXAMPLE 8

Multiply $2\frac{1}{5}$ by $1\frac{1}{4}$.

Solution

$$2\frac{1}{5} \times 1\frac{1}{4} = \frac{11}{5} \times \frac{5}{4}$$ Write each mixed number as an improper fraction.

$$= \frac{11 \times \overset{1}{\cancel{5}}}{\underset{1}{\cancel{5}} \times 4}$$ Simplify and multiply.

$$= \frac{11}{4}, \text{ or } 2\frac{3}{4}$$

PRACTICE 8

Find the product of $3\frac{3}{4}$ and $2\frac{1}{10}$.

EXAMPLE 9

Multiply: $\left(4\frac{3}{8}\right)(4)\left(2\frac{2}{5}\right)$

Solution

$$\left(4\frac{3}{8}\right)(4)\left(2\frac{2}{5}\right) = \left(\frac{35}{8}\right)\left(\frac{4}{1}\right)\left(\frac{12}{5}\right)$$

$$= \left(\frac{\overset{7}{\cancel{35}}}{\underset{\underset{1}{\cancel{2}}}{\cancel{8}}}\right)\left(\frac{\overset{1}{\cancel{4}}}{1}\right)\left(\frac{\overset{6}{\cancel{12}}}{\underset{1}{\cancel{5}}}\right) = 42$$

Note in this problem that, although there are several ways to simplify, the answer always comes out the same.

PRACTICE 9

Multiply: $\left(1\frac{3}{4}\right)\left(5\frac{1}{3}\right)(3)$

EXAMPLE 10

A lawn surrounding a garden is to be installed, as depicted in the following drawing.

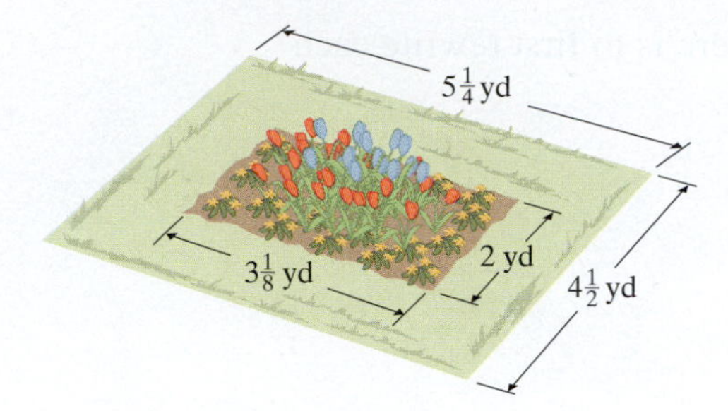

How many square yards of turf will we need to cover the lawn?

Solution Let's break this problem into three steps. First, we find the area of the rectangle with dimensions $5\frac{1}{4}$ yards and $4\frac{1}{2}$ yards. Then, we find the area of the small rectangle whose length and width are $3\frac{1}{8}$ yards and 2 yards, respectively.

$$5\frac{1}{4} \times 4\frac{1}{2} = \frac{21}{4} \times \frac{9}{2}$$

$$= \frac{189}{8}, \text{ or } 23\frac{5}{8}$$ **The area of the large rectangle is $23\frac{5}{8}$ square yards.**

$$3\frac{1}{8} \times 2 = \frac{25}{8} \times \frac{2}{1}$$

$$= \frac{25}{4}, \text{ or } 6\frac{1}{4}$$ **The area of the small rectangle is $6\frac{1}{4}$ square yards.**

Finally, we subtract the area of the small rectangle from the area of the large rectangle.

$$\begin{array}{rcr} 23\frac{5}{8} & = & 23\frac{5}{8} \\ -6\frac{1}{4} & = & -6\frac{2}{8} \\ \hline & & 17\frac{3}{8} \end{array}$$

The area of the lawn is, therefore, $17\frac{3}{8}$ square yards. So we will need $17\frac{3}{8}$ square yards of turf for the lawn.

PRACTICE 10

Two student club posters are shown below. How much greater is the area of the legal-size poster than the letter-size poster?

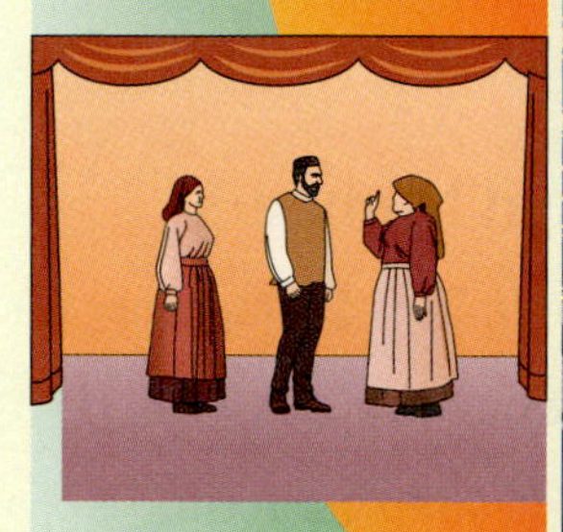

$8\frac{1}{2}$ in. × 11 in.
Letter-size paper

$8\frac{1}{2}$ in. × 14 in.
Legal-size paper

EXAMPLE 11

Simplify: $16\frac{1}{4} - 2 \cdot 4\frac{3}{5}$

Solution We use the order of operations rule, multiplying before subtracting.

$$\begin{aligned}16\frac{1}{4} - 2 \cdot 4\frac{3}{5} &= 16\frac{1}{4} - \frac{2}{1} \cdot \frac{23}{5}\\ &= 16\frac{1}{4} - \frac{46}{5}\\ &= 16\frac{1}{4} - 9\frac{1}{5}\\ &= 16\frac{5}{20} - 9\frac{4}{20}\\ &= 7\frac{1}{20}\end{aligned}$$

PRACTICE 11

Calculate: $6 + \left(3\frac{1}{2}\right)^2$

Dividing Fractions

We now turn to quotients, beginning with dividing a fraction by a whole number. Suppose, for instance, that you want to share $\frac{1}{3}$ of a pizza with a friend, that is, to divide the $\frac{1}{3}$ into two equal parts. What part of the whole pizza will each of you receive?

This diagram shows $\frac{1}{3}$ of a pizza.

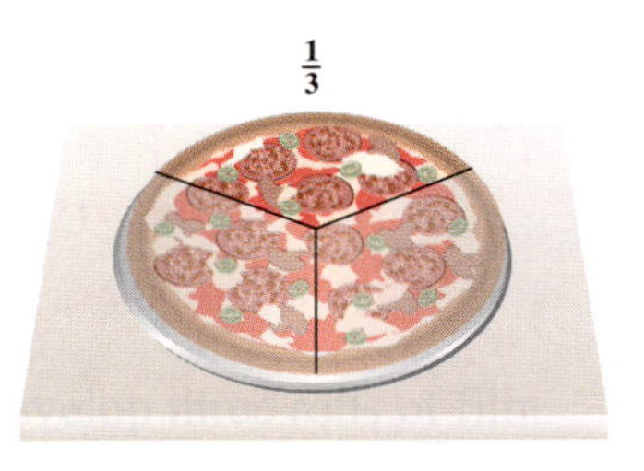

If we split each third into two equal parts, each part is $\frac{1}{6}$ of the pizza.

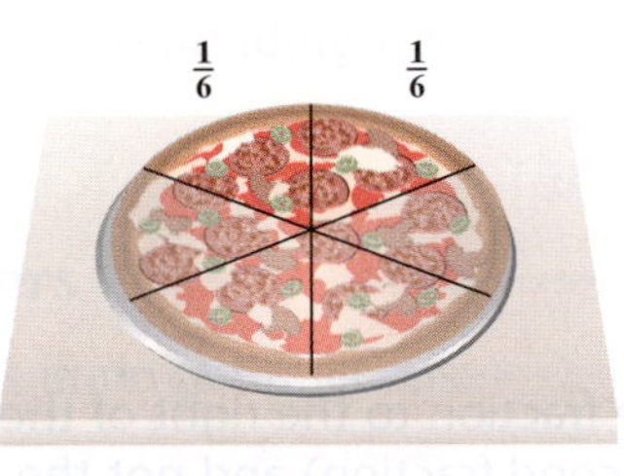

You and your friend will each get $\frac{1}{6}$ of the whole pizza, which you can compute as follows.

$$\frac{1}{3} \div 2 = \frac{1}{6}$$

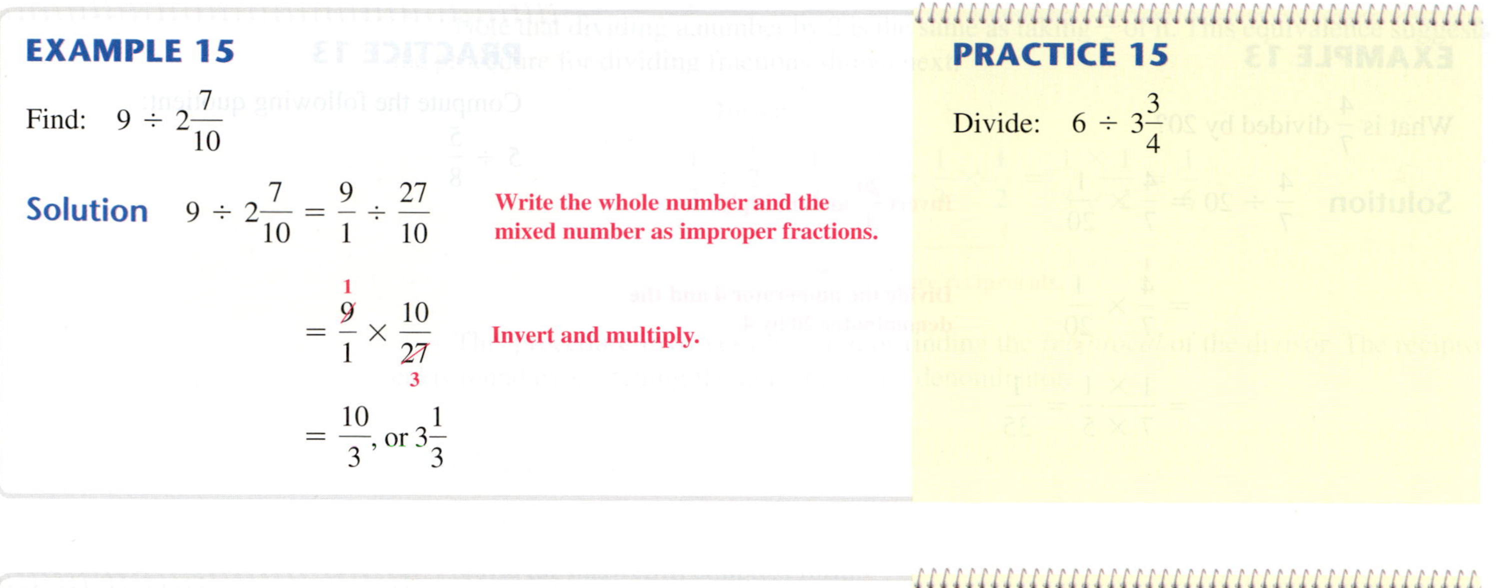

EXAMPLE 15

Find: $9 \div 2\frac{7}{10}$

Solution $9 \div 2\frac{7}{10} = \frac{9}{1} \div \frac{27}{10}$ Write the whole number and the mixed number as improper fractions.

$= \frac{\overset{1}{\cancel{9}}}{1} \times \frac{10}{\underset{3}{\cancel{27}}}$ Invert and multiply.

$= \frac{10}{3}$, or $3\frac{1}{3}$

PRACTICE 15

Divide: $6 \div 3\frac{3}{4}$

EXAMPLE 16

What is $2\frac{1}{2} \div 4\frac{1}{2}$?

Solution $2\frac{1}{2} \div 4\frac{1}{2} = \frac{5}{2} \div \frac{9}{2} = \frac{5}{\underset{1}{\cancel{2}}} \times \frac{\overset{1}{\cancel{2}}}{9} = \frac{5}{9}$

Invert and multiply.

PRACTICE 16

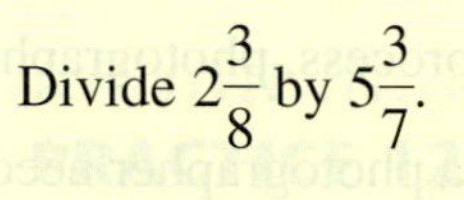

Divide $2\frac{3}{8}$ by $5\frac{3}{7}$.

EXAMPLE 17

There are $6\frac{3}{4}$ yards of silk in a roll. If it takes $\frac{3}{4}$ yard to make one designer tie, how many ties can be made from the roll?

Solution The question is: How many $\frac{3}{4}$'s fit into $6\frac{3}{4}$? It tells us that we must divide.

$6\frac{3}{4} \div \frac{3}{4} = \frac{\overset{9}{\cancel{27}}}{\underset{1}{\cancel{4}}} \times \frac{\overset{1}{\cancel{4}}}{\underset{1}{\cancel{3}}} = 9$

So nine ties can be made from the roll of silk.

PRACTICE 17

According to a newspaper advertisement for a "diet shake," a man lost 33 pounds in $5\frac{1}{2}$ months. How much weight did he lose per month?

Estimating Products and Quotients of Mixed Numbers

As with adding or subtracting mixed numbers, it is important to check our answers when multiplying or dividing. We can check a product or a quotient of mixed numbers by estimating the answer and then confirming that our estimate and answer are reasonably close.

Checking a Product by Estimating

$$2\frac{1}{5} \times 7\frac{2}{3} = \frac{11}{5} \times \frac{23}{3} = \frac{253}{15}, \text{ or } 16\frac{13}{15}$$

$$\downarrow \quad \downarrow$$

$$2 \times 8 = 16$$

Our answer, $16\frac{13}{15}$, is close to 16, the product of the rounded factors.

Because $16\frac{13}{15}$ is near 16, $16\frac{13}{15}$ is a reasonable answer.

Checking a Quotient by Estimating

$$6\frac{1}{4} \div 2\frac{7}{10} = \frac{25}{4} \div \frac{27}{10} = \frac{\overset{}{25}}{\underset{2}{\cancel{4}}} \times \frac{\overset{5}{\cancel{10}}}{27} = \frac{125}{54}, \text{ or } 2\frac{17}{54}$$

$$\downarrow \quad \downarrow$$

$$6 \div 3 = 2$$

Our answer, $2\frac{17}{54}$, is close to 2, the quotient of the rounded dividend and divisor.

Because 2 is near $2\frac{17}{54}$, $2\frac{17}{54}$ is a reasonable answer.

EXAMPLE 18

Simplify and check: $3\frac{3}{4} \times 5\frac{1}{3} \div 2\frac{7}{9}$

Solution Following the order of operations rule, we work from left to right, multiplying the first two mixed numbers.

$$3\frac{3}{4} \times 5\frac{1}{3} = \frac{\overset{5}{\cancel{15}}}{\underset{1}{\cancel{4}}} \times \frac{\overset{4}{\cancel{16}}}{\underset{1}{\cancel{3}}} = 20$$

Then, we divide 20 by $2\frac{7}{9}$ to get the answer.

$$20 \div 2\frac{7}{9} = \frac{20}{1} \div \frac{25}{9} = \frac{\overset{4}{\cancel{20}}}{1} \times \frac{9}{\underset{5}{\cancel{25}}} = \frac{36}{5}, \text{ or } 7\frac{1}{5}$$

Now, let's check by estimating.

$$3\frac{3}{4} \times 5\frac{1}{3} \div 2\frac{7}{9}$$

$$\downarrow \quad \downarrow \quad \downarrow$$

$$4 \times 5 \div 3 = 20 \div 3 \approx 7$$

The answer, $7\frac{1}{5}$, and the estimate, 7, are reasonably close, confirming the answer.

PRACTICE 18

Compute and check:

$$5\frac{3}{5} \div 2\frac{1}{10} \times 2\frac{1}{4}$$

51. $1\frac{1}{7} \times 1\frac{1}{5}$

52. $2\frac{1}{3} \times 1\frac{1}{2}$

53. $\left(2\frac{1}{10}\right)^2$

54. $\left(1\frac{1}{2}\right)^2$

55. $3\frac{9}{10} \cdot 2$

56. $5 \cdot 1\frac{1}{2}$

57. $100 \times 3\frac{3}{4}$

58. $1\frac{5}{6} \times 20$

59. $1\frac{1}{2} \times 5\frac{1}{3}$

60. $5\frac{1}{4} \times 1\frac{1}{9}$

61. $\left(2\frac{1}{2}\right)\left(1\frac{1}{5}\right)$

62. $\left(1\frac{3}{10}\right)\left(2\frac{4}{9}\right)$

63. $12\frac{1}{2} \cdot 3\frac{1}{3}$

64. $5\frac{1}{10} \cdot 1\frac{2}{3}$

65. $66\frac{2}{3} \times 1\frac{7}{10}$

66. $37\frac{1}{2} \times 1\frac{3}{5}$

67. $1\frac{5}{9} \times \frac{3}{8} \times 2$

68. $\frac{1}{8} \times 2\frac{1}{4} \times 6$

69. $\left(\frac{1}{2}\right)^2\left(2\frac{1}{3}\right)$

70. $\left(1\frac{1}{4}\right)^2\left(\frac{1}{5}\right)$

71. $\frac{4}{5} \times \frac{7}{8} \times 1\frac{1}{10}$

72. $8\frac{1}{3} \times \frac{3}{10} \times \frac{5}{6}$

73. $\left(1\frac{1}{2}\right)^3$

74. $\left(2\frac{1}{2}\right)^3$

Divide and simplify.

75. $\frac{3}{5} \div \frac{2}{3}$

76. $\frac{2}{3} \div \frac{3}{5}$

77. $\frac{4}{5} \div \frac{7}{8}$

78. $\frac{7}{8} \div \frac{4}{5}$

79. $\frac{1}{2} \div \frac{1}{7}$

80. $\frac{1}{7} \div \frac{1}{2}$

81. $\frac{5}{9} \div \frac{1}{8}$

82. $\frac{1}{8} \div \frac{5}{9}$

83. $\frac{4}{5} \div \frac{8}{15}$

84. $\frac{3}{10} \div \frac{6}{5}$

85. $\frac{7}{8} \div \frac{3}{8}$

86. $\frac{10}{3} \div \frac{5}{6}$

87. $\frac{9}{10} \div \frac{3}{4}$

88. $\frac{5}{6} \div \frac{1}{3}$

89. $\frac{1}{10} \div \frac{2}{5}$

90. $\frac{3}{4} \div \frac{6}{5}$

91. $\frac{2}{3} \div 7$

92. $\frac{7}{10} \div 10$

93. $\frac{2}{3} \div 6$

94. $\frac{1}{20} \div 2$

95. $8 \div \frac{1}{5}$

96. $8 \div \frac{2}{9}$

97. $7 \div \frac{3}{7}$

98. $10 \div \frac{2}{5}$

99. $4 \div \frac{3}{10}$

100. $10 \div \frac{2}{3}$

101. $1 \div \frac{1}{7}$

102. $3 \div \frac{1}{8}$

103. $2\frac{5}{6} \div \frac{3}{7}$

104. $5\frac{1}{9} \div \frac{2}{3}$

105. $1\frac{1}{3} \div \frac{4}{5}$

106. $7\frac{1}{10} \div \frac{1}{2}$

107. $8\frac{5}{6} \div \frac{9}{10}$

108. $6\frac{1}{2} \div \frac{1}{2}$

109. $20\frac{1}{10} \div \frac{1}{5}$

110. $15\frac{2}{3} \div \frac{5}{6}$

111. $\frac{1}{6} \div 2\frac{1}{7}$

112. $\frac{2}{7} \div 1\frac{1}{3}$

113. $\frac{1}{2} \div 2\frac{3}{5}$

114. $\frac{3}{4} \div 3\frac{1}{9}$

115. $4 \div 1\frac{1}{4}$

116. $7 \div 1\frac{9}{10}$

117. $2\frac{1}{10} \div 20$

118. $5\frac{6}{7} \div 14$

119. $2\frac{1}{2} \div 3\frac{1}{7}$

120. $3\frac{1}{7} \div 2\frac{1}{2}$

121. $8\frac{1}{10} \div 5\frac{3}{4}$

122. $1\frac{7}{10} \div 5\frac{1}{8}$

123. $2\frac{1}{3} \div 4\frac{1}{2}$

124. $8\frac{1}{6} \div 2\frac{1}{2}$

125. $6\frac{3}{8} \div 2\frac{5}{6}$

126. $1\frac{2}{3} \div 1\frac{2}{5}$

Simplify.

127. $\frac{1}{2} + \frac{2}{3} \times 1\frac{1}{3}$

128. $\frac{9}{10} + \frac{4}{5} \cdot 8$

129. $5 - \frac{1}{3} \times \frac{2}{5}$

130. $3 \div \frac{2}{5} - 2\frac{1}{3}$

131. $2\frac{3}{4} \times \frac{1}{8} + \frac{1}{5}$

132. $\frac{3}{8} \cdot \frac{1}{2} - \frac{1}{10}$

133. $4 - \frac{2}{9} \div \frac{3}{4}$

134. $6 \div 5 \times \frac{1}{4}$

135. $3\frac{1}{2} \times 6 \div 5$

136. $4 \cdot \frac{2}{3} - 1\frac{1}{8}$

137. $10 \times \frac{1}{8} \times 2\frac{1}{2}$

138. $\frac{1}{3} \div \frac{1}{6} \times \frac{2}{3}$

139. $8 \div 1\frac{1}{5} + 3 \cdot 1\frac{1}{2}$

140. $3\frac{1}{8} \div 5 + 4 \div 2\frac{1}{2}$

141. $\frac{5}{6} \cdot \frac{9}{10} - \frac{3}{5} \div \frac{6}{7}$

142. $\frac{6}{11} \div \frac{18}{55} - \frac{7}{26} \cdot \frac{13}{14}$

143. $\left(\frac{1}{4}\right)^2 \cdot \left(3 - 1\frac{2}{3}\right)^2$

144. $\left(1 - \frac{2}{5}\right)^2 \div \left(1\frac{1}{2}\right)^2$

145. $\left(1\frac{1}{2} \div \frac{1}{3}\right)^2 + \left(1 - \frac{1}{4}\right)^2$

146. $\left(3\frac{1}{2}\right)^2 + 2\left(1\frac{1}{2} - 1\frac{1}{3}\right)$

Mixed Practice

147. Divide $6\frac{1}{8}$ by $2\frac{3}{4}$.

148. Compute: $14 - 3 \div \left(\frac{4}{5}\right)^2$

149. Find the product of $\frac{3}{5}$ and $\frac{7}{8}$.

150. Find the quotient of $\frac{9}{10}$ and $\frac{2}{5}$.

151. Multiply $\frac{2}{3}$ by 12.

152. Calculate: $\left(4\frac{1}{2}\right)\left(6\frac{2}{3}\right)$

Applications

B *Solve. Write the answer in simplest form.*

153. In a local town, $\frac{5}{6}$ of the voting-age population is registered to vote. If $\frac{7}{10}$ of the registered voters voted in the election for mayor, what fraction of the voting-age population voted?

154. Last year, $\frac{1}{8}$ of the emergency room visits at a hospital were injury related. Of these, $\frac{2}{5}$ were due to motor vehicle accidents. What fraction of the emergency room visits were due to motor vehicle accidents?

155. A couple would like to add on an extension to their house. One of the construction companies that bid on the job would charge \$40,200 with $\frac{1}{30}$ down. How much money does the couple need to put down if they choose to use this company?

156. There is a rule of thumb to not spend more than $\frac{1}{4}$ of one's income on rent. If someone makes \$24,000 a year, what is the most he or she should spend per month on rent according to this rule?

157. A tile store charges $\$46\frac{1}{2}$ per square foot for granite countertops, including installation. Find the cost of buying and installing a granite countertop on the kitchen island shown.

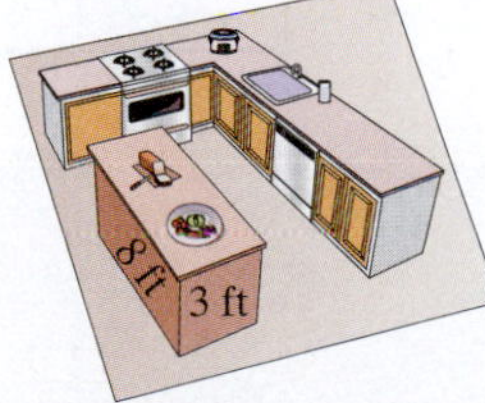

158. Which of these rooms has the larger area?

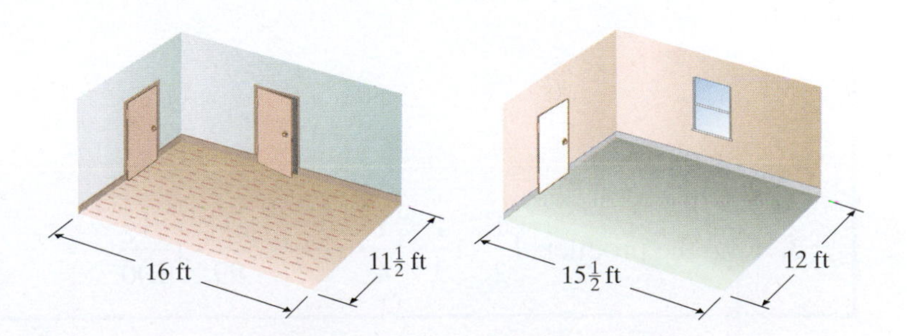

159. Students in an astronomy course learn that a first-magnitude star is $2\frac{1}{2}$ times as bright as a second-magnitude star, which in turn is $2\frac{1}{2}$ times as bright as a third-magnitude star. How many times as bright as a third-magnitude star is a first-magnitude star?

160. Some people believe that gasohol is superior to gasoline as an automotive fuel. Gasohol is a mixture of gasoline $\left(\frac{9}{10}\right)$ and ethyl alcohol $\left(\frac{1}{10}\right)$. How much more gasoline than ethyl alcohol is there in $10\frac{1}{2}$ gallons of gasohol?

161. Because of evaporation, a pond loses $\frac{1}{4}$ of its remaining water each month of summer. If it is full at the beginning of summer, what fraction of the original amount will the pond contain after three summer months?

162. A scientist is investigating the effects of cold on human skin. In one of the scientist's experiments, the temperature starts at 70°F and drops by $\frac{1}{10}$° every 2 minutes. What is the temperature after 6 minutes?

163. A trip to a nearby island takes $3\frac{1}{2}$ hours by boat and $\frac{1}{2}$ hour by airplane. How many times as fast as the boat is the plane?

164. Each dose of aspirin weighs $\frac{3}{4}$ grain. If a hospital pharmacist has 9 grains of aspirin on hand, how many doses can he provide?

165. A store sells two types of candles. The scented candle is 8 inches tall and burns $\frac{1}{2}$ inch per hour, whereas the unscented candle is 10 inches tall and burns $\frac{1}{3}$ inch per hour.

a. In an hour, which candle will burn more?

b. Which candle will last longer?

166. A college-wide fund-raising campaign collected \$3 million in $1\frac{1}{2}$ years for student scholarships.

a. What was the average amount collected per year?

b. By how much would this average increase if an additional \$1 million were collected?

• Check your answers on page A-3.

MINDStretchers

Writing

1. Every number except 0 has a reciprocal. Explain why 0 does not have a reciprocal.

Groupwork

2. In the following magic square, the *product* of every row, column, and diagonal is 1. Working with a partner, complete the square.

$\frac{2}{3}$		$1\frac{1}{2}$
$\frac{1}{2}$		

Patterns

3. Find the product: $1\frac{1}{2} \cdot 1\frac{1}{3} \cdot 1\frac{1}{4} \cdot \cdots \cdot 1\frac{1}{99} \cdot 1\frac{1}{100}$

Key Concepts and Skills

CONCEPT SKILL

Concept/Skill	Description	Example
[2.1] Prime number	A whole number that has exactly two different factors: itself and 1.	2, 3, 5
[2.1] Composite number	A whole number that has more than two factors.	4, 8, 9
[2.1] Prime factorization of a whole number	The number written as the product of its prime factors.	$30 = 2 \cdot 3 \cdot 5$
[2.1] Least common multiple (LCM) of two or more whole numbers	The smallest nonzero whole number that is a multiple of each number.	The LCM of 30 and 45 is 90.
[2.1] To compute the least common multiple (LCM)	• Find the prime factorization of each number. • Identify the prime factors that appear in each factorization. • Multiply these prime factors, using each factor the greatest number of times that it occurs in any of the factorizations.	$20 = 2 \cdot 2 \cdot 5$ $= 2^2 \cdot 5$ $30 = 2 \cdot 3 \cdot 5$ The LCM of 20 and 30 is $2^2 \cdot 3 \cdot 5$, or 60.
[2.2] Fraction	Any number that can be written in the form $\frac{a}{b}$, where a and b are whole numbers and b is nonzero.	$\frac{3}{11}, \frac{9}{5}$
[2.2] Proper fraction	A fraction whose numerator is smaller than its denominator.	$\frac{2}{7}, \frac{1}{2}$
[2.2] Mixed number	A number with a whole-number part and a proper fraction part.	$5\frac{1}{3}, 4\frac{5}{6}$
[2.2] Improper fraction	A fraction whose numerator is greater than or equal to its denominator.	$\frac{9}{4}, \frac{5}{5}$
[2.2] To change a mixed number to an improper fraction	• Multiply the denominator of the fraction by the whole-number part of the mixed number. • Add the numerator of the fraction to this product. • Write this sum over the denominator to form the improper fraction.	$4\frac{2}{3} = \frac{3 \times 4 + 2}{3}$ $= \frac{14}{3}$
[2.2] To change an improper fraction to a mixed number	• Divide the numerator by the denominator. • If there is a remainder, write it over the denominator.	$\frac{14}{3} = 4\frac{2}{3}$
[2.2] To find an equivalent fraction	Multiply the numerator and denominator of $\frac{a}{b}$ by the same whole number; that is, $\frac{a}{b} = \frac{a \cdot n}{b \cdot n}$, where both b and n are nonzero.	$\frac{3}{4} = \frac{3 \cdot 2}{4 \cdot 2} = \frac{6}{8}$

continued

CHAPTER 2 Review Exercises

Say Why *Fill in each blank.*

1. Twenty-seven ________ a composite number
   is/is not
   because ________________________________
   ________________________________.

2. The prime factorization of 180 ________ $3^2 \times 4 \times 5$
   is/is not
   because ________________________________
   ________________________________.

3. The expression $\frac{3}{0}$ ________ a fraction because
   is/is not
   ________________________________.

4. The expression $\frac{12}{11}$ ________ an improper fraction
   is/is not
   because ________________________________
   ________________________________.

5. The expression $\frac{16}{48}$ ________ equivalent to $\frac{1}{3}$ because
   is/is not
   ____
   ________________________________.

6. The expressions $\frac{12}{15}$ and $\frac{12}{16}$ ________ unlike
   are/are not
   fractions because ________________________________
   ________________________________.

7. The least common denominator of $\frac{5}{8}$ and $\frac{7}{12}$ ________
   is/is not
   24 because ________________________________
   ________________________________.

8. The reciprocal of $\frac{6}{8}$ ________ $\frac{3}{4}$ because ________
   is/is not
   ____
   ________________________________.

[2.1] *Find all the factors of each number.*

9. 150 **10.** 180 **11.** 57 **12.** 70

Indicate whether each number is prime or composite.

13. 23 **14.** 33 **15.** 87 **16.** 67

Write the prime factorization of each number, using exponents.

17. 36 **18.** 75 **19.** 99 **20.** 54

Find the LCM.

21. 6 and 14 **22.** 5 and 10 **23.** 18, 24, and 36 **24.** 10, 15, and 20

[2.2] *Identify the fraction or mixed number represented by the shaded portion of each figure.*

25. **26.** **27.** **28.**

Indicate whether each number is a proper fraction, an improper fraction, or a mixed number.

29. $4\frac{1}{8}$ **30.** $\frac{5}{6}$ **31.** $\frac{3}{2}$ **32.** $\frac{7}{1}$

Write each mixed number as an improper fraction.

33. $7\frac{2}{3}$ **34.** $1\frac{4}{5}$ **35.** $9\frac{1}{10}$ **36.** $8\frac{3}{7}$

Write each fraction as a mixed number or a whole number.

37. $\frac{13}{2}$ **38.** $\frac{14}{3}$ **39.** $\frac{11}{4}$ **40.** $\frac{12}{12}$

Write an equivalent fraction with the given denominator.

41. $7 = \frac{}{12}$ **42.** $\frac{2}{7} = \frac{}{14}$ **43.** $\frac{1}{2} = \frac{}{10}$ **44.** $\frac{9}{10} = \frac{}{30}$

Simplify.

45. $\frac{14}{28}$ **46.** $\frac{15}{21}$ **47.** $\frac{30}{45}$ **48.** $\frac{54}{72}$

49. $5\frac{2}{4}$ **50.** $8\frac{10}{15}$ **51.** $6\frac{12}{42}$ **52.** $8\frac{45}{63}$

Insert the appropriate sign: $<$, $=$, *or* $>$.

53. $\frac{5}{8} \quad \frac{3}{8}$ **54.** $\frac{5}{6} \quad \frac{1}{6}$ **55.** $\frac{2}{3} \quad \frac{4}{5}$ **56.** $\frac{9}{10} \quad \frac{7}{8}$

57. $\frac{3}{4} \quad \frac{5}{8}$ **58.** $\frac{7}{10} \quad \frac{5}{9}$ **59.** $3\frac{1}{5} \quad 1\frac{9}{10}$ **60.** $5\frac{1}{8} \quad 5\frac{1}{9}$

Arrange in increasing order.

61. $\frac{2}{7}, \frac{3}{8}, \frac{1}{2}$ **62.** $\frac{1}{5}, \frac{1}{3}, \frac{2}{15}$ **63.** $\frac{4}{5}, \frac{9}{10}, \frac{3}{4}$ **64.** $\frac{7}{8}, \frac{7}{9}, \frac{13}{18}$

[2.3] *Add and simplify.*

65. $\frac{2}{5} + \frac{4}{5}$ **66.** $\frac{7}{20} + \frac{8}{20}$ **67.** $\frac{5}{8} + \frac{7}{8} + \frac{3}{8}$ **68.** $\frac{3}{10} + \frac{1}{10} + \frac{2}{10}$

69. $\frac{1}{3} + \frac{2}{5}$ **70.** $\frac{7}{8} + \frac{5}{6}$ **71.** $\frac{9}{10} + \frac{1}{2} + \frac{2}{5}$ **72.** $\frac{3}{8} + \frac{4}{5} + \frac{3}{4}$

73. $2 + 3\frac{7}{8}$ **74.** $6\frac{1}{4} + 3\frac{1}{4}$ **75.** $8\frac{7}{10} + 1\frac{9}{10}$ **76.** $5\frac{5}{6} + 2\frac{1}{6}$

77. $2\frac{1}{3} + 4\frac{1}{3} + 5\frac{2}{3}$ **78.** $1\frac{3}{10} + \frac{9}{10} + 2\frac{1}{10}$ **79.** $5\frac{2}{5} + \frac{3}{10}$ **80.** $9\frac{1}{6} + 8\frac{3}{8}$

81. $10\frac{2}{3} + 12\frac{3}{4}$ **82.** $20\frac{1}{2} + 25\frac{7}{8}$ **83.** $10\frac{3}{5} + 7\frac{9}{10} + 2\frac{1}{4}$ **84.** $20\frac{7}{8} + 30\frac{5}{6} + 4\frac{1}{3}$

Subtract and simplify.

85. $\frac{3}{8} - \frac{1}{8}$ **86.** $\frac{7}{9} - \frac{1}{9}$ **87.** $\frac{5}{3} - \frac{2}{3}$ **88.** $\frac{4}{6} - \frac{4}{6}$

89. $\frac{3}{10} - \frac{1}{20}$

90. $\frac{1}{2} - \frac{1}{8}$

91. $\frac{3}{5} - \frac{1}{4}$

92. $\frac{1}{3} - \frac{1}{10}$

93. $12\frac{1}{2} - 5$

94. $4\frac{3}{10} - 2$

95. $8\frac{7}{8} - 5\frac{1}{8}$

96. $20\frac{3}{4} - 2\frac{1}{4}$

97. $12 - 5\frac{1}{2}$

98. $4 - 2\frac{3}{10}$

99. $7 - 4\frac{1}{3}$

100. $1 - \frac{4}{5}$

101. $6\frac{1}{10} - 4\frac{3}{10}$

102. $2\frac{5}{8} - 1\frac{7}{8}$

103. $5\frac{1}{4} - 2\frac{3}{4}$

104. $7\frac{1}{6} - 3\frac{5}{6}$

105. $3\frac{1}{10} - 2\frac{4}{5}$

106. $7\frac{1}{2} - 4\frac{5}{8}$

107. $5\frac{1}{12} - 4\frac{1}{2}$

108. $6\frac{2}{9} - 2\frac{1}{3}$

109. $\frac{1}{3} + \frac{5}{6} - \frac{1}{2}$

110. $7\frac{9}{10} - 1\frac{1}{5} + 2\frac{3}{4}$

[2.4] *Multiply and simplify.*

111. $\frac{3}{4} \times \frac{1}{4}$

112. $\frac{1}{2} \times \frac{7}{8}$

113. $\left(\frac{5}{6}\right)\left(\frac{3}{4}\right)$

114. $\left(\frac{2}{3}\right)\left(\frac{1}{4}\right)$

115. $\frac{2}{3} \cdot 8$

116. $\frac{1}{10} \cdot 7$

117. $\left(\frac{1}{5}\right)^3$

118. $\left(\frac{2}{3}\right)^3$

119. $\frac{1}{2} \times \frac{2}{3} \times \frac{3}{4}$

120. $\frac{7}{8} \times \frac{2}{5} \times \frac{1}{6}$

121. $\frac{4}{5} \times 1\frac{1}{5}$

122. $\frac{2}{3} \times 2\frac{1}{3}$

123. $5\frac{1}{3} \cdot \frac{1}{2}$

124. $\frac{1}{10} \cdot 6\frac{2}{3}$

125. $1\frac{1}{3} \cdot 4\frac{1}{2}$

126. $3\frac{1}{4} \cdot 5\frac{2}{3}$

127. $6\frac{3}{4} \times 1\frac{1}{4}$

128. $8\frac{1}{2} \times 2\frac{1}{2}$

129. $\frac{7}{8} \times 1\frac{1}{5} \times \frac{3}{7}$

130. $1\frac{3}{8} \times \frac{10}{11} \times 1\frac{1}{4}$

131. $\left(3\frac{1}{3}\right)^3$

132. $\left(1\frac{1}{2}\right)^3$

133. $\frac{5}{8} + \frac{1}{2} \cdot 5$

134. $1\frac{9}{10} - \left(\frac{2}{3}\right)^2$

135. $4\left(\frac{2}{5}\right) + 3\left(\frac{1}{6}\right)$

136. $6\left(1\frac{1}{2} - \frac{3}{10}\right)$

Find the reciprocal.

137. $\frac{2}{3}$

138. $1\frac{1}{2}$

139. 8

140. $\frac{1}{4}$

Divide and simplify.

141. $\frac{7}{8} \div 5$

142. $\frac{5}{9} \div 9$

143. $\frac{2}{3} \div 5$

144. $\frac{1}{100} \div 2$

145. $\frac{1}{2} \div \frac{2}{3}$

146. $\frac{2}{3} \div \frac{1}{2}$

147. $6 \div \frac{1}{5}$

148. $7 \div \frac{4}{5}$

149. $\frac{7}{8} \div \frac{3}{4}$

150. $\frac{9}{10} \div \frac{1}{2}$

151. $\frac{3}{5} \div \frac{3}{10}$

152. $\frac{2}{3} \div \frac{1}{6}$

153. $3\frac{1}{2} \div 2$

154. $2 \div 3\frac{1}{2}$

155. $6\frac{1}{3} \div 4$

156. $4 \div 6\frac{1}{3}$

157. $8\frac{1}{4} \div 1\frac{1}{2}$

158. $3\frac{2}{5} \div 1\frac{1}{3}$

159. $4\frac{1}{2} \div 2\frac{1}{4}$

160. $7\frac{1}{5} \div 2\frac{2}{5}$

161. $\left(5 - \frac{2}{3}\right) \div \frac{4}{9}$

162. $6\frac{1}{2} \div \left(\frac{1}{2} + 4\frac{1}{2}\right)$

163. $7 \div 2\frac{1}{4} + 5 \div \left(1\frac{1}{2}\right)^2$

164. $\left(1\frac{2}{3}\right)^2 \times 2 + 9 \div 4\frac{1}{2}$

Mixed Applications

Solve. Write the answer in simplest form.

165. The Summer Olympic Games are held during each year divisible by 4. Were the Olympic Games held in 1990?

166. What is the smallest amount of money that you can pay in both all quarters and all dimes?

167. Eight of the 32 human teeth are incisors. What fraction of human teeth are incisors? (*Source:* Ilsa Goldsmith, *Human Anatomy for Children*)

168. The planets in the solar system (including the "dwarf planet" Pluto) consist of Earth, two planets closer to the Sun than Earth, and six planets farther from the Sun than Earth. What fraction of the planets in the solar system are closer than Earth to the Sun? (*Source:* Patrick Moore, *Astronomy for the Beginner*)

169. A Filmworks camera has a shutter speed of $\frac{1}{8{,}000}$ second and a Lensmax camera has a shutter speed of $\frac{1}{6{,}000}$ second. Which shutter is faster? (*Hint:* The faster shutter has the smaller shutter speed.)

170. In a recent year, among Americans who were 65 years of age or older 15 million were male and 21 million female. What fraction of this population was female? (*Source:* U. S. Census Bureau)

171. An insurance company reimbursed a patient \$275 on a dental bill of \$700. Did the patient get more or less than $\frac{1}{3}$ of the money paid back? Explain.

172. A union goes on strike if at least $\frac{2}{3}$ of the workers voting support the strike call. If 23 of the 32 voting workers support a strike, should a strike be declared? Explain.

173. A grand jury has 23 jurors. Sixteen jurors are needed for a quorum, and a vote of 12 jurors is needed to indict.

a. What fraction of the full jury is needed to indict?

b. Suppose that 16 jurors are present. What fraction of those present is needed to indict?

174. In a tennis match, Lisa Gregory went to the net 12 times, winning the point 7 times. By contrast, Monica Yates won the point 4 of the 6 times that she went to the net.

a. Which player went to the net more often?

b. Which player had a better rate of winning points at the net?

175. In a math course, $\frac{3}{5}$ of a student's grade is based on four in-class exams, and $\frac{3}{20}$ of the grade is based on homework. What fraction of a student's grade is based on in-class exams and homework?

176. A metal alloy is made by combining $\frac{1}{4}$ ounce of copper with $\frac{2}{3}$ ounce of tin. Find the alloy's total weight.

177. The weight of a diamond is measured in carats. What is the difference in weight between a $\frac{3}{4}$-carat and a $\frac{1}{2}$-carat diamond?

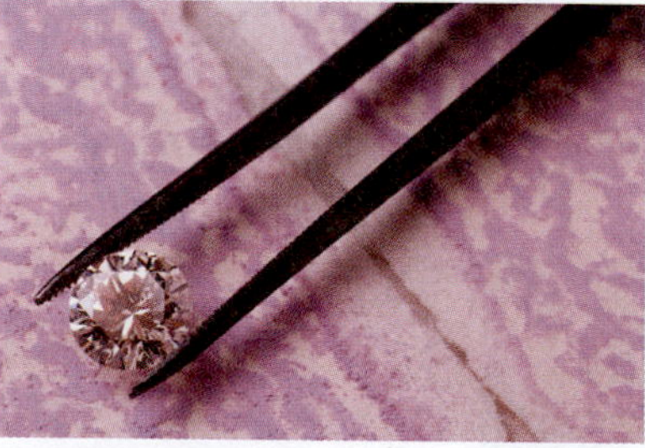

178. During a sale, the price of a sweater was marked $\frac{1}{4}$ off the original price of \$45. Using a coupon, a customer received an additional $\frac{1}{5}$ off the sale price. What fraction of the original price was the final sale price?

179. In a math class, $\frac{3}{8}$ of the students are chemistry majors and $\frac{2}{3}$ of those students are women. If there are 48 students in the math class, how many women are chemistry majors?

180. Three-eighths of the undergraduate students at a two-year college receive financial aid. If the college has 4,296 undergraduate students, how many undergraduate students do not receive financial aid?

181. Of the first 10 artists to receive the Grammy Lifetime Achievement Award, $\frac{4}{5}$ were men. How many of these awardees were women? (*Source: Top 10 of Everything 2010*)

182. A sea otter eats about $\frac{1}{5}$ of its body weight each day. How much will a 35-pound otter eat in a day? (*Source:* Karl W. Kenyon, *The Sea Otter in the Eastern Pacific Ocean*)

183. An investor bought $1,000 worth of a technology stock. At the beginning of last year, it had increased in value by $\frac{2}{5}$. During the year, the value of the stock declined by $\frac{1}{4}$. What was the value of the stock at the end of last year?

184. In Roseville, 40 of every 1,000 people who want to work are unemployed, in contrast to 8 of every 100 people in Georgetown. How many times as great as the unemployment rate in Roseville is the unemployment rate in Georgetown?

185. A brother and sister want to buy as many goldfish as possible for their new fish tank. A rule of thumb is that the total length of fish (in inches) should be less than the capacity of the tank (in gallons). If they have a 10-gallon tank and goldfish average $\frac{1}{2}$ inch in length, how many fish should they buy?

186. A commuter is driving to the city of Denver 15 miles away. If he has already driven $3\frac{1}{4}$ miles, how far is he from Denver?

187. The wingspread of a Boeing 777-300 jet is $199\frac{11}{12}$ feet, whereas the wingspread of a Boeing 747-400 is $211\frac{5}{12}$ feet. How much longer is the wingspread of a Boeing 747-400 jet? (*Source:* boeing.com)

188. A Chicago family plans to take a vacation traveling by express bus either to Kansas City or to Indianapolis. The trip to Kansas City takes $11\frac{3}{4}$ hours, in contrast to a trip to Indianapolis that takes only $4\frac{1}{3}$ hours. How much longer is the first trip? (*Source:* greyhound.com)

189. An airplane is flying $1\frac{1}{2}$ times the speed of sound. If sound travels at about 1,000 feet per second, at what speed is the plane flying?

190. When standing upright, the pressure per square inch on a person's hip joint is about $2\frac{1}{2}$ times his or her body weight. If the person weighs 200 pounds, what is that pressure? (*Source:* pnas.org)

191. A cubic foot of water weighs approximately $62\frac{1}{2}$ pounds. If a basin contains $4\frac{1}{2}$ cubic feet of water, how much does the water weigh?

192. In 2009, the attendance at Fenway Park in Boston was about $1\frac{1}{4}$ times the attendance ten years earlier. If attendance in 1999 was approximately 2,500,000, what was the attendance in 2009? Round to the nearest hundred thousand. (*Source:* ballparks.com)

193. It took the space shuttle Endeavor $1\frac{1}{2}$ hours to orbit Earth. How many orbits did the Endeavor make in 12 hours? (*Source:* NASA)

194. An offshore wind farm measures $10\frac{1}{4}$ miles by $2\frac{1}{2}$ miles. Find the area of the wind farm.

Cumulative Review Exercises

To help you review, solve the following:

1. Write in words: 5,000,315

2. Round 1,876,529 to the nearest hundred thousand.

3. Multiply: $5{,}814 \times 100$

4. Find the quotient: $89\overline{)80{,}812}$

5. Evaluate: $24 \div (2 + 4) - 3$

6. Evaluate: $\left(\dfrac{6-5}{2+3}\right)^2$

7. Write the prime factorization of 84 using exponents.

8. Find the least common multiple of 20 and 24.

9. Write $\dfrac{75}{100}$ in simplest form.

10. Which is larger, $\dfrac{1}{4}$ or $\dfrac{3}{8}$?

11. Add: $\dfrac{5}{8} + \dfrac{5}{6}$

12. Subtract: $8 - 1\dfrac{3}{5}$

13. Find the product: $1\dfrac{1}{2} \cdot 4\dfrac{2}{3}$

14. Divide: $4 \div 2\dfrac{3}{4}$

15. In a recent year, the two U.S. corporations with largest revenues were ExxonMobil and Walmart Stores. How much greater were the ExxonMobil revenues? (*Source: Fortune*)

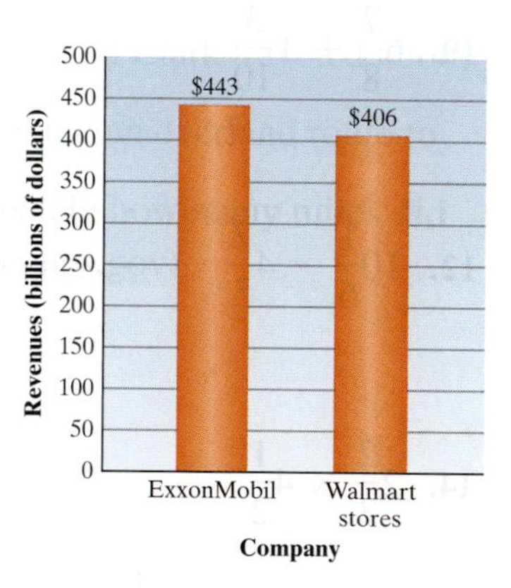

16. A jury decided on punishments in an oil spill case. The jury ordered the captain of the oil barge to pay \$5,000 in punitive damages and the oil company to pay \$5 billion. The amount that the company had to pay is how many times the amount that the captain had to pay?

17. The following table gives the number of named storms (tropical storms, hurricanes, and subtropical storms) that hit the United States in recent years:

Year	Number of Named Storms
2005	28
2006	10
2007	15
2008	18
2009	9

(*Source:* aoml.noaa.gov)

How far above the five-year average was the number of named storms in 2005?

18. A gallon of paint covers approximately 400 square feet. A room has two walls that are 8 feet high and 13 feet wide, and two walls that are 8 feet high and 15 feet wide. The doors and windows along those walls total 78 square feet. Will one gallon of paint cover the walls? Why or why not?

19. A homeowner wants to refinish his basement over three weekends. He completes $\dfrac{1}{4}$ of the job the first weekend and $\dfrac{5}{12}$ of the job the second weekend. What fraction of the job remains to be completed?

20. In a theater program, a student purchases a 7-inch-long piece of trim for costumes. From this purchase, how many $1\dfrac{3}{4}$-inch-long pieces of trim can be made?

• Check your answers on page A-4.

CHAPTER 3

Decimals

Decimals and Blood Tests

Blood tests reveal a great deal about a person's health—whether to reduce the cholesterol level to lower the risk of heart disease, or raise the red blood cell count to prevent anemia. And blood tests identify a variety of diseases, for example, AIDS and mononucleosis.

Blood analyses are typically carried out in clinical laboratories. Technicians in these labs operate giant machines that perform thousands of blood tests per hour.

What these blood tests, known as "chemistries," actually do is to analyze blood for a variety of substances, such as creatinine and calcium.

In any blood test, doctors look for abnormal levels of the substance being measured. For instance, the normal range on the creatinine test is typically from 0.7 to 1.5 milligrams (mg) per unit of blood. A high level may mean kidney disease; a low level, muscular dystrophy.

The normal range on the calcium test may be 9.0 to 10.5 mg per unit of blood. A result outside this range is a clue for any of several diseases.

(*Source:* Dixie Farley, "Top 10 Laboratory Tests: Blood Will Tell," *FDA Consumer*, Vol. 23)

CHAPTER 3 PRETEST

To see if you have already mastered the topics in this chapter, take this test.

1. In the number 27.081, what place does the 8 occupy?

2. Write in words: 4.012

3. Round 3.079 to the nearest tenth.

4. Which is largest: 0.00212, 0.0029, or 0.000888?

Perform the indicated operations.

5. $7.02 + 3.5 + 11$

6. $2.37 + 5.0038$

7. $13.79 - 2.1$

8. $9 - 2.7 + 3.51$

9. $8.3 \times 1{,}000$

10. 8.01×2.3

11. $(0.12)^2$

12. $5 + 3 \times 0.7$

13. $6.05 \div 1{,}000$

14. $\dfrac{9.81}{0.3}$

Express as a decimal.

15. $\dfrac{7}{8}$

16. $2\frac{5}{6}$, rounded to the nearest hundredth

Solve.

17. In a science course, a student learns that an acid is stronger if it has a lower pH value. Which is stronger, an acid with a pH value of 3.7 or an acid with a pH value of 2.95?

18. The following table shows the quarterly revenues for Microsoft Corporation in a recent fiscal year.

	First	Second	Third	Fourth
Revenue (in billions of dollars)	15.06	16.63	13.65	13.1

What was Microsoft's total revenue for the four quarters? (*Source:* microsoft.com)

19. A serving of iceberg lettuce contains 3.6 milligrams (mg) of vitamin C, whereas romaine lettuce contains 11.9 mg of vitamin C. Romaine lettuce is how many times as rich in vitamin C as iceberg lettuce? Round the answer to the nearest whole number. (*Source: The Concise Encyclopedia of Foods and Nutrition*)

20. A tourist visiting Orlando makes a long-distance telephone call that costs \$0.85 for the first 3 minutes and \$0.17 for each additional minute. What is the cost of a 20-minute call?

• Check your answers on page A-4.

3.1 Introduction to Decimals

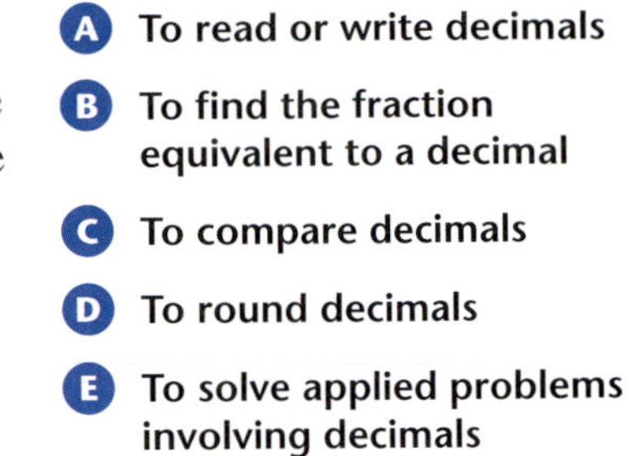

OBJECTIVES

- **A** To read or write decimals
- **B** To find the fraction equivalent to a decimal
- **C** To compare decimals
- **D** To round decimals
- **E** To solve applied problems involving decimals

What Decimals Are and Why They Are Important

Decimal notation is in common use. When we say that the price of a book is $32.75, that the length of a table is 1.8 meters, or that the answer displayed on a calculator is 5.007, we are using decimals.

A number written as a **decimal** has

- a whole-number part, which *precedes* the decimal point, and
- a fractional part, which *follows* the decimal point.

Whole-number part ↓ 4 . 5 1 ↓ Fractional part

↑ Decimal point

A decimal without a decimal point shown is understood to have one at the right end and is the same as a whole number. For instance, 3 and 3. are the same number.

The fractional part of any decimal has as its denominator a power of 10, such as 10, 100, or 1,000. The use of the word *decimal* reminds us of the importance of the number 10 in this notation, just as decade means 10 years or December meant the 10th month of the year (which it was for the early Romans).

In many problems, we can choose to work with either decimals or fractions. Therefore, we need to know how to work with both if we are to use the easier approach to solve a particular problem.

Decimal Places

Each digit in a decimal has a place value. The place value system for decimals is an extension of the place value system for whole numbers.

The places to the right of the decimal point are called **decimal places**. For instance, the number 64.149 is said to have three decimal places.

For a whole number, place values are powers of 10: 1, 10, 100, ... By contrast, each place value for the fractional part of a decimal is the reciprocal of a power of 10: $\frac{1}{10}, \frac{1}{100}, \frac{1}{1{,}000}, \ldots$.

The first decimal place after the decimal point is the tenths place. Working to the right, the next decimal places are the hundredths place, the thousandths place, and so on.

The following table shows the place values in the decimals 0.54 and 0.30716.

Ones	•	Tenths	Hundredths	Thousandths	Ten-thousandths	Hundred-thousandths
1	and	$\frac{1}{10}$	$\frac{1}{100}$	$\frac{1}{1{,}000}$	$\frac{1}{10{,}000}$	$\frac{1}{100{,}000}$
0	.	5	4			
0	.	3	0	7	1	6

EXAMPLE 4

Find the equivalent fraction of each decimal.

a. 3.2 **b.** 3.200

Solution

a. 3.2 represents $3\frac{2}{10}$, or $3\frac{1}{5}$.

b. 3.200 equals $3\frac{200}{1,000}$, or $3\frac{1}{5}$.

PRACTICE 4

Express each decimal in fractional form.

a. 5.6 **b.** 5.6000

TIP Adding zeros in the rightmost decimal places does not change a decimal's value. However, generally decimals can be written without these extra zeros.

EXAMPLE 5

Write each decimal as a mixed number.

a. 1.309 **b.** 1.39

Solution

a. $1.309 = 1\frac{309}{1,000}$

b. $1.39 = 1\frac{39}{100}$

PRACTICE 5

What mixed number is equivalent to each decimal?

a. 7.003 **b.** 4.1

Knowing how to change a decimal to its equivalent fraction also helps us read the decimal.

EXAMPLE 6

Express each decimal in words.

a. 0.319 **b.** 2.71 **c.** 0.08

Solution

a. $0.319 = \frac{319}{1,000}$

We read the decimal as "three hundred nineteen thousandths."

b. $2.71 = 2\frac{71}{100}$

The decimal point is read as "and." We read the decimal as "two and seventy-one hundredths."

c. $0.08 = \frac{8}{100}$

We read the decimal as "eight hundredths." Note that we *do not simplify* the equivalent fraction when reading the decimal.

PRACTICE 6

Express each decimal in words.

a. 0.61

b. 4.923

c. 7.05

EXAMPLE 7

Write each number in decimal notation.

a. Seven tenths **b.** Five and thirty-two thousandths

Solution

a. Since 7 is in the tenths place, the decimal is written as 0.7.

b. **The whole number preceding *and* is in the ones place.** → 5 . 0 3 2 ← **The last digit of 32 is in the thousandths place.**

We replace *and* with the decimal point. **We need a 0 to hold the tenths place.**

The answer is 5.032.

PRACTICE 7

Write each number in decimal notation.

a. Forty-three thousandths

b. Ten and twenty-six hundredths

EXAMPLE 8

For hay fever, an allergy sufferer takes a decongestant pill that has a tablet strength of three hundredths of a gram. Write the equivalent decimal.

Solution "Three hundredths" is written 0.03, with the digit 3 in the hundredths place.

PRACTICE 8

The number pi (usually written π) is approximately three and fourteen hundredths. Write this approximation as a decimal.

Comparing Decimals

Suppose that we want to compare two decimals—say, 0.6 and 0.7. The key is to rethink the problem in terms of fractions.

$$0.6 = \frac{6}{10} \qquad 0.7 = \frac{7}{10}$$

Because $\frac{7}{10} > \frac{6}{10}$, $0.7 > 0.6$.

The following procedure provides another way to compare decimals that is faster than converting the decimals to fractions.

To Compare Decimals

- Rewrite the numbers vertically, lining up the decimal points.
- Working from left to right, compare the digits that have the same place value. At the first place value where the digits differ, the decimal which has the largest digit with this place value is the largest decimal.

119. In terms of land area, North America is 1.36 times as large as South America. Round this decimal to the nearest tenth. (*Source: National Geographic Atlas of the World*)

120. The length of the Panama Canal is 50.7 miles. Round this length to the nearest mile. (*Source: The New Encyclopedia Britannica*)

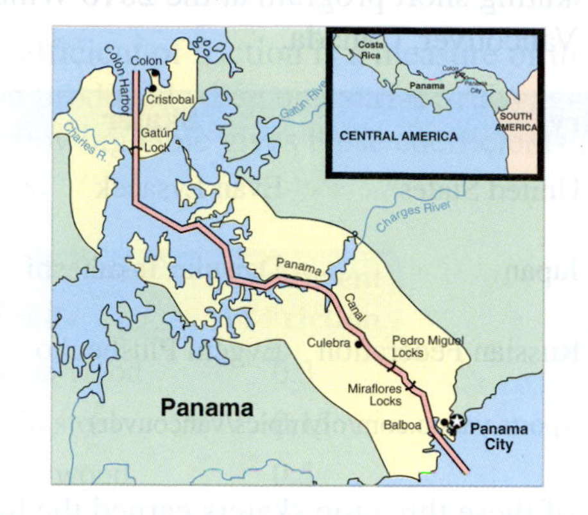

• Check your answers on page A-4.

MINDStretchers

Critical Thinking

1. For each question, either give the answer or explain why there is none.
   - **a.** Find the *smallest* decimal that when rounded to the nearest tenth is 7.5.
   - **b.** Find the *largest* decimal that when rounded to the nearest tenth is 7.5.

Writing

2. The next whole number after 7 is 8. What is the next decimal after 0.7? Explain.

Groupwork

3. Working with a partner, list fifteen numbers between 2.5 and 2.6.

Cultural Note

DISME:
The Art of Tenths,
OR,
Decimall Arithmetike,

In 1585, Simon Stevin, a Dutch engineer, published a book entitled *The Art of Tenths* (*La Disme in French*) in which he presented a thorough account of decimals. Stevin sought to teach everyone "with an ease unheard of, all computations necessary between men by integers without fractions."

Stevin did not invent decimals; their history dates back thousands of years to ancient China, medieval Arabia, and Renaissance Europe. However, Stevin's writings popularized decimals and also supported the notion of decimal coinage—as in American currency, where there are 10 dimes to the dollar.

Source: Morris Kline, *Mathematics, a Cultural Approach* (Reading, Massachusetts: Addison-Wesley Publishing Company, 1962), p. 614.

EXAMPLE 7

Write each number in decimal notation.

a. Seven tenths **b.** Five and thirty-two thousandths

Solution

a. Since 7 is in the tenths place, the decimal is written as 0.7.

b. The whole number preceding *and* is in the ones place. The last digit of 32 is in the thousandths place.

5.032

We replace *and* with the decimal point. We need a 0 to hold the tenths place.

The answer is 5.032.

PRACTICE 7

Write each number in decimal notation.

a. Forty-three thousandths

b. Ten and twenty-six hundredths

EXAMPLE 8

For hay fever, an allergy sufferer takes a decongestant pill that has a tablet strength of three hundredths of a gram. Write the equivalent decimal.

Solution "Three hundredths" is written 0.03, with the digit 3 in the hundredths place.

PRACTICE 8

The number pi (usually written π) is approximately three and fourteen hundredths. Write this approximation as a decimal.

Comparing Decimals

Suppose that we want to compare two decimals—say, 0.6 and 0.7. The key is to rethink the problem in terms of fractions.

$$0.6 = \frac{6}{10} \qquad 0.7 = \frac{7}{10}$$

Because $\frac{7}{10} > \frac{6}{10}$, $0.7 > 0.6$.

The following procedure provides another way to compare decimals that is faster than converting the decimals to fractions.

To Compare Decimals

- Rewrite the numbers vertically, lining up the decimal points.
- Working from left to right, compare the digits that have the same place value. At the first place value where the digits differ, the decimal which has the largest digit with this place value is the largest decimal.

EXAMPLE 9

Which is larger, 0.729 or 0.75?

Solution First, let's line up the decimal point.

↓
0.729
0.75
↑

We see that both decimals have a 0 in the ones place. We next compare the digits in the tenths place and see that, again, they are the same. Looking to the right in the hundredths place, we see that $5 > 2$. Therefore, $0.75 > 0.729$. Note that the decimal with more digits is not necessarily the larger decimal.

PRACTICE 9

Which is smaller, 0.83 or 0.8297?

EXAMPLE 10

Rank from smallest to largest: 2.17, 2.1, and 0.99

Solution First, we position the decimals so that the decimal points are aligned.

↓
2.17
2.1
0.99
↑

Working from left to right, we see that in the ones place, the first two decimals have a 2 and the third decimal has a 0, so the third decimal is the smallest of the three. To decide which of the first two decimals is smaller, we compare the digits in the tenths place. Since both of these decimals have a 1 in the tenths place, we proceed to the hundredths place. A 0 is understood to the right of the 1 in 2.1, so we compare 0 and 7.

↓
2.17
2.1**0**
0.99
↑

Since $0 < 7$, we conclude that $2.10 < 2.17$. Therefore, the three decimals from smallest to largest are 0.99, 2.1, and 2.17.

PRACTICE 10

Rewrite in decreasing order: 3.5, 3.51, and 3.496

EXAMPLE 11

Plastic garbage bags come in three thicknesses (or gauges): 0.003 inch, 0.0025 inch, and 0.002 inch. The three gauges are called lightweight, regular weight, and heavyweight. Which is the lightweight gauge?

Solution To find the smallest of the decimals, we first line up the decimal points.

$$\begin{array}{l} 0.003 \\ 0.0025 \\ 0.002 \end{array}$$

Working from left to right, we see that the three decimals have the same digits until the thousandths place, where $3 > 2$. Therefore, 0.003 must be the heavyweight gauge. To compare 0.0025 and 0.002, we look at the ten-thousandths place. The 5 is greater than the 0 that is understood to be there (0.0020). So 0.0025 inch must be the regular-weight gauge, and 0.002 the lightweight gauge.

PRACTICE 11

The higher the energy efficiency rating (EER) of an air conditioner, the more efficiently it uses electricity. Which of the following air conditioners is least efficient? (*Source:* Consumer Guide)

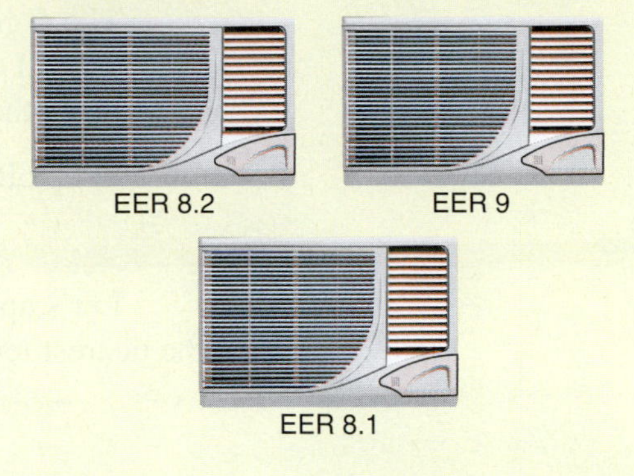

Rounding Decimals

As with whole numbers, we can round decimals to a given place value. For instance, suppose that we want to round the decimal 1.38 to the nearest tenth. The decimal 1.38 lies between 1.3 and 1.4, so one of these two numbers will be our answer—but which? To decide, let's take a look at a number line.

1.31 1.33 1.35 1.37 1.39
1.3 1.32 1.34 1.36 1.38 1.4

Do you see from this diagram that 1.38 is closer to 1.4 than to 1.3?

$$1.38 \approx 1.4$$

Tenths place

Rounding a decimal to the nearest tenth means that the last digit lies in the tenths place.

The following table shows the relationship between the place to which we are rounding and the number of decimal places in our answer.

Rounding to the Nearest	Means That the Rounded Decimal Has
tenth $\left(\frac{1}{10}\right)$	**one** decimal place.
hundredth $\left(\frac{1}{100}\right)$	**two** decimal places.
thousandth $\left(\frac{1}{1,000}\right)$	**three** decimal places.
ten-thousandth $\left(\frac{1}{10,000}\right)$	**four** decimal places.

Note that the number of decimal places is the same as the number of zeros in the corresponding denominator.

The following rule can be used to round decimals.

To Round a Decimal to a Given Decimal Place

- Underline the digit in the place to which the number is being rounded.
- The digit to the right of the underlined digit is called the *critical digit*. Look at the critical digit—if it is 5 or more, add 1 to the underlined digit; if it is less than 5, leave the underlined digit unchanged.
- Drop all decimal places to the right of the underlined digit.

Let's apply this rule to the problem that we just considered—namely, rounding 1.38 to the nearest tenth.

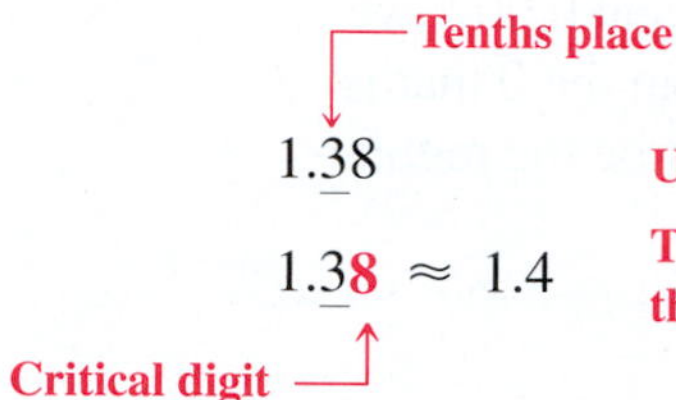

Underline the digit 3, which occupies the tenths place.

The critical digit, 8, is 5 or more, so add 1 to the 3 and then drop all digits to its right.

The following examples illustrate this method of rounding.

EXAMPLE 12

Round 94.735 to

a. the nearest tenth.

b. two decimal places.

c. the nearest thousandth.

d. the nearest ten.

e. the nearest whole number.

Solution

a. First, we underline the digit 7 in the tenths place: 94.$\underline{7}$35. The critical digit, 3, is less than 5, so we do not add 1 to the underlined digit. Dropping all digits to the right of the 7, we get 94.7. Note that our answer has only one decimal place because we are rounding to the nearest tenth.

b. We need to round 94.735 to two decimal places (to the nearest hundredth).

$94.7\underline{3}5 \approx 94.74$

The critical digit is 5 or more. Add 1 to the underlined digit and drop the decimal place to the right.

c. $94.73\underline{5} \approx 94.735$ because the critical digit to the right of the 5 is understood to be 0.

PRACTICE 12

Round 748.0772 to

a. the nearest tenth.

b. the nearest hundredth.

c. three decimal places.

d. the nearest whole number.

e. the nearest hundred.

d. We are rounding 94.735 to the nearest ten (*not tenth*), which is a whole-number place.

$\underline{9}4.735 \approx 90$

Because 4 < 5, keep 9 in the tens place, insert 0 in the ones place and drop all decimal places.

e. Rounding to the nearest whole number means rounding to the nearest 1.

$9\underline{4}.735 \approx 95$

Because 7 > 5, change the 4 to 5 and drop all decimal places.

EXAMPLE 13

Round 3.982 to the nearest tenth.

Solution First, we underline the digit 9 in the tenths place and identify the critical digit: $3.\underline{9}82$. The critical digit, 8, is more than 5, so we add 1 to the 9, get 10, and write down the 0. We add the 1 to 3, getting 4, and drop the 8 and 2.

$$3.\underline{9}\underbrace{82}_{\text{Drop}} \approx 4.0$$

The answer is 4.0. Note that we do not drop the 0 in the tenths place of the answer to indicate that we have rounded to that place.

PRACTICE 13

Round 7.2962 to two decimal places.

EXAMPLE 14

The rate of exchange between currencies varies with time. On a particular day the euro, the currency used in many countries of Western Europe, could have been exchanged for 1.23502 dollars. What was this price to the nearest cent? (*Source:* xe.com)

Solution A cent is one-hundredth of a dollar. Therefore, we need to round 1.23502 to the nearest hundredth.

$$1.2\underline{3}502 \approx 1.24$$

So the price to the nearest cent was $1.24.

PRACTICE 14

Mount Waialeale on the Hawaiian island of Kauai is one of the world's wettest places, with an average annual rainfall of 11.68 meters. What is the amount of rainfall to the nearest tenth of a meter? (*Source:* wikipedia.org)

3.1 Exercises

FOR EXTRA HELP MyMathLab MathXL PRACTICE WATCH READ REVIEW

Mathematically Speaking

Fill in each blank with the most appropriate term or phrase from the given list.

less	greater	increasing	left
decreasing	ten	hundredths	multiple
thousandths	power	right	tenth

1. A decimal place is a place to the ________ of the decimal point.

2. The fractional part of a decimal has as its denominator a ________ of 10.

3. The decimal 0.17 is equivalent to the fraction seventeen ____________.

4. The decimal 209.95 rounded to the nearest ________ is 210.0.

5. The decimal 0.371 is ________ than the decimal 0.3499.

6. The decimals 0.48, 0.4, and 0.371 are written in ________ order.

A *Underline the digit that occupies the given place.*

7. 2.78 Tenths place

8. 6.835 Tenths place

9. 9.01 Hundredths place

10. 0.772 Hundredths place

11. 2.00175 Ten-thousandths place

12. 4.00189 Ten-thousandths place

13. 823.001 Thousandths place

14. 829.006 Thousandths place

Identify the place occupied by the underlined digit.

15. $25.\underline{7}1$

16. $3.00\underline{2}$

17. $8.1\underline{8}3$

18. $\underline{4}9.771$

19. $1{,}077.04\underline{2}$

20. $2.8371\underline{0}7$

21. $\$25\underline{3}.72$

22. $\$7{,}571.3\underline{9}$

Write each decimal in words.

23. 0.53

24. 0.72

25. 0.305

26. 0.849

27. 0.6

28. 0.3

29. 5.72

30. 3.89

31. 24.002

32. 370.081

Write each number in decimal notation.

33. Eight tenths

34. Six tenths

35. One and forty-one thousandths

36. Eighteen and four thousandths

37. Sixty and one hundredth

38. Ninety-two and seven hundredths

39. Four and one hundred seven thousandths

40. Five and sixty-three thousandths

41. Three and two tenths meters

42. Ninety-eight and six tenths degrees

B *For each decimal, find the equivalent fraction or mixed number, written in lowest terms.*

43. 0.6

44. 0.8

45. 0.39

46. 0.27

47. 1.5

48. 9.8

49. 8.000

50. 6.700

51. 5.012

52. 20.304

C *Between each pair of numbers, insert the appropriate sign, <, =, or >, to make a true statement.*

53. 3.21 2.5 **54.** 8.66 4.952 **55.** 0.71 0.8 **56.** 1.2 1.38

57. 9.123 9.11 **58.** 0.72 0.7 **59.** 4 4.000 **60.** 7.60 7.6

61. 8.125 feet 8.2 feet **62.** 2.45 pounds 2.5 pounds

Rearrange each group of numbers from smallest to largest.

63. 7.1, 7, 7.07 **64.** 0.002, 0.2, 0.02 **65.** 5.001, 4.9, 5.2

66. 3.85, 3.911, 2 **67.** 9.6 miles, 9.1 miles, 9.38 miles **68.** 2.7 seconds, 2.15 seconds, 2 seconds

D *Round as indicated.*

69. 17.36 to the nearest tenth

70. 8.009 to two decimal places

71. 3.5905 to the nearest thousandth

72. 3.5902 to the nearest thousandth

73. 37.08 to one decimal place

74. 3.08 to one decimal place

75. 0.396 to the nearest hundredth

76. 0.978 to the nearest hundredth

77. 7.0571 to two decimal places

78. 3.038 to one decimal place

79. 8.7 miles to the nearest mile

80. $3.57 to the nearest dollar

Round to the indicated place.

81.

To the Nearest	8.0714	0.9916
Tenth		
Hundredth		
Ten		

82.

To the Nearest	0.8166	72.3591
Tenth		
Hundredth		
Ten		

Mixed Practice

Solve.

83. In the decimal 0.024, underline the digit in the tenths place.

84. What is the equivalent fraction of 3.8?

85. Round 870.062 to the nearest hundredth.

86. Write four and thirty-one thousandths in decimal notation.

87. Write in increasing order: 2.14 meters, 2.4 meters, and 2.04 meters.

88. Write 0.05 in words.

Applications

The following statements involve decimals. Write all decimals in words.

89. It takes the Earth 23.934 hours to rotate once about its axis. (*Source:* NASA)

90. Male Rufous hummingbirds weigh an average of 0.113 ounce. (*Source:* Lanny Chambers, *Facts about Hummingbirds*)

91. Over two years, the average score on a college admissions exam increased from 18.7 to 18.8.

92. A chemistry text gives 55.85 as the atomic mass of iron and 63.55 as the atomic mass of copper.

93. The following table shows the average amount of time spent on selected daily household activities each year by U.S. civilians.

Activity	Average Amount of Time (in number of hours per year)
Housework	211.7
Lawn and garden care	69.4
Food preparation and cleanup	189.8
Caring for household members	193.5
Household management	47.5

(*Source:* bls.gov)

94. The coefficient of friction is a measure of the amount of friction produced when one surface rubs against another. The following table gives these coefficients for various surfaces.

Materials	Coefficient of Friction
Wood on wood	0.3
Steel on steel	0.15
Steel on wood	0.5
A rubber tire on a dry concrete road	0.7
A rubber tire on a wet concrete road	0.5

(*Source: CRC Handbook of Chemistry and Physics*)

95. Bacteria are single-celled organisms that typically measure from 0.00001 inch to 0.00008 inch across.

96. In one month, the consumer confidence index rose from 71.9 points to 80.2 points.

The following statements involve numbers written in words. Write each number in decimal notation.

97. The area of a plot of land is one and two tenths acres.

98. The lead in many mechanical pencils is seven tenths millimeter thick.

99. At the first Indianapolis 500 auto race in 1911, the winning speed was seventy-four and fifty-nine hundredths miles per hour. (*Source:* Jack Fox, *The Indianapolis 500*)

100. In 1796, there was a U.S. coin in circulation, the half cent, worth five thousandths of a dollar.

101. At sea level, the air pressure on each square inch of surface area is fourteen and seven tenths pounds.

102. A doctor prescribed a dosage of one hundred twenty-five thousandths milligram of Prolixin.

103. According to the owner's manual, the voltage produced by a camcorder battery is nine and six tenths volts (V).

104. In preparing an injection, a nurse measured out one and eight tenths milliliters of sterile water.

105. The electrical usage in a tenant's apartment last month amounted to three hundred fifty-two and one tenth kilowatt hours (kWh).

106. In one day, the Dow Jones Industrial Average fell by three and sixty-three hundredths points.

Solve.

107. The following table shows the three medalists in the men's skating short program at the 2010 Winter Olympics in Vancouver, Canada.

Country	Skater	Score
United States	Evan Lysacek	90.3
Japan	Daisuke Takahashi	90.25
Russian Federation	Evgeni Plushenko	90.85

(*Source:* sports.yahoo.com/olympics/vancouver)

Which of these three top skaters earned the highest score for the short program?

108. The more powerful an earthquake is, the higher its magnitude is on the Richter scale. Great earthquakes, such as the 1906 San Francisco earthquake, have magnitudes of 8.0 or higher. Is an earthquake with magnitude 7.8 considered to be a great earthquake? (*Source: The New Encyclopedia Britannica*)

109. Last winter, a homeowner's average daily heating bill was for 8.75 units of electricity. This winter, it was for 8.5 units. During which winter was the average higher?

110. In order to qualify for the dean's list at a community college, a student's grade point average (GPA) must be 3.5 or above. Did a student with a GPA of 3.475 make the dean's list?

111. The following table shows estimates of the lead emissions in the United States for two given years.

Year	Amount of Lead (in millions of tons)
1985	0.022
2005	0.003

(*Source:* Environmental Protection Agency)

In which year was the amount of lead emissions less?

112. As part of her annual checkup, a patient had a blood test. The normal range for a particular substance is 1.1 to 2.3. If she scored 0.95, was her blood in the normal range?

113. The following table shows the amount of money that a jury awarded a husband and wife who were plaintiffs in a lawsuit.

Plaintiff	Award (in millions of dollars)
Husband	1.875
Wife	1.91

Whose award was less than the $1.9 million that each plaintiff had demanded?

114. The table below shows the amount of toxic emissions released into the air from three factories during the same time in a recent year. Which of the factories released the most toxic emissions?

	Electronics Factory	Food Factory	Chemical Factory
Toxic emissions (in millions of pounds)	1.5	1.4	1.48

Round to the indicated place.

115. A bank pays interest on all its accounts to the nearest cent. If the interest on an account is $57.0285, how much interest does the bank pay?

116. A city's sales tax rate, expressed as a decimal, is 0.0825. What is this rate to the nearest hundredth?

117. According to the organizers of a lottery, the probability of winning the lottery is 0.0008. Round this probability to three decimal places.

118. One day last week, a particular foreign currency was worth $0.7574. How much is this currency worth to the nearest tenth of a dollar?

119. In terms of land area, North America is 1.36 times as large as South America. Round this decimal to the nearest tenth. (*Source: National Geographic Atlas of the World*)

120. The length of the Panama Canal is 50.7 miles. Round this length to the nearest mile. (*Source: The New Encyclopedia Britannica*)

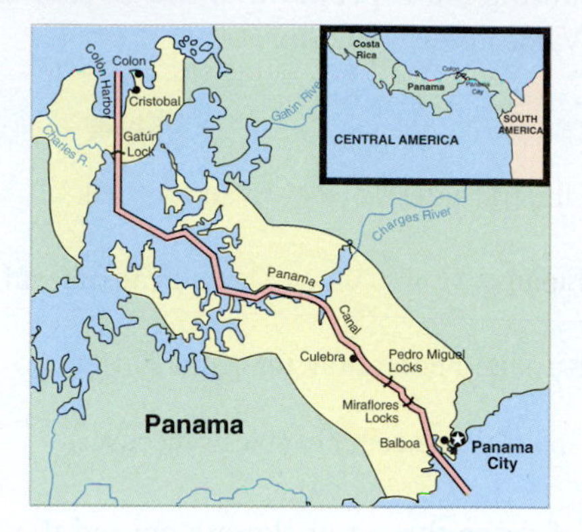

• Check your answers on page A-4.

MIND*Stretchers*

Critical Thinking

1. For each question, either give the answer or explain why there is none.

   a. Find the *smallest* decimal that when rounded to the nearest tenth is 7.5.

   b. Find the *largest* decimal that when rounded to the nearest tenth is 7.5.

Writing

2. The next whole number after 7 is 8. What is the next decimal after 0.7? Explain.

Groupwork

3. Working with a partner, list fifteen numbers between 2.5 and 2.6.

Cultural Note

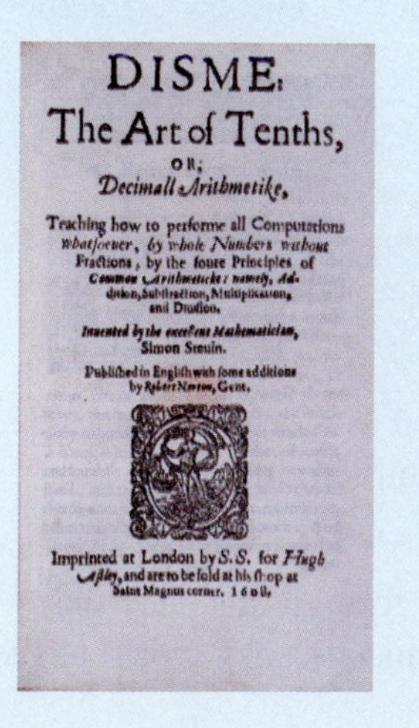
DISME:
The Art of Tenths,
OR;
Decimall Arithmetike,
Teaching how to performe all Computations whatsoever, by whole Numbers without Fractions, by the foure Principles of Common Arithmeticke: namely, Addition, Subtraction, Multiplication, and Division.
Invented by the excellent Mathematician, Simon Stevin.
Published in English with some additions by Robert Norton, Gent.
Imprinted at London by S. S. for Hugh Astley, and are to be sold at his shop at Saint Magnus corner. 1608.

In 1585, Simon Stevin, a Dutch engineer, published a book entitled *The Art of Tenths* (*La Disme in French*) in which he presented a thorough account of decimals. Stevin sought to teach everyone "with an ease unheard of, all computations necessary between men by integers without fractions."

Stevin did not invent decimals; their history dates back thousands of years to ancient China, medieval Arabia, and Renaissance Europe. However, Stevin's writings popularized decimals and also supported the notion of decimal coinage—as in American currency, where there are 10 dimes to the dollar.

Source: Morris Kline, *Mathematics, a Cultural Approach* (Reading, Massachusetts: Addison-Wesley Publishing Company, 1962), p. 614.

3.2 Adding and Subtracting Decimals

A To add or subtract decimals

B To solve applied problems involving the addition or subtraction of decimals

In Section 3.1 we discussed the meaning of decimals and how to compare and round them. Now, we turn our attention to computing with decimals, starting with addition and subtraction.

Adding Decimals

Adding decimals is similar to adding whole numbers: We add the digits in each place value position, regrouping when necessary. Suppose that we want to find the sum of two decimals: 1.2 + 3.5. First, we rewrite the problem vertically, lining up the decimal points in the addends. Then, we add as usual, inserting the decimal point below the other decimal points.

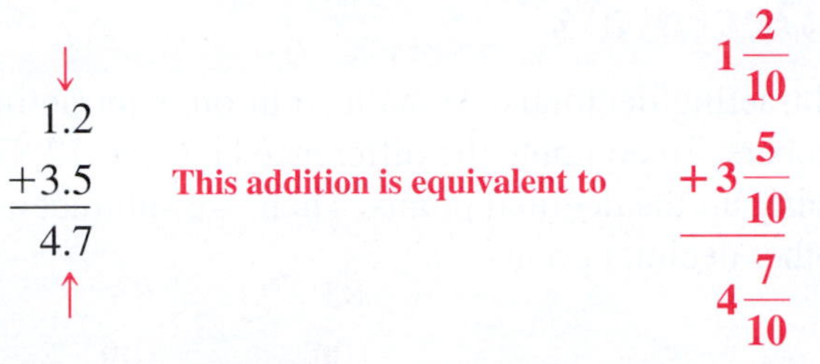

Note that when we added the mixed numbers corresponding to the decimals, we got $4\frac{7}{10}$, which is equivalent to 4.7. This example suggests the following rule:

To Add Decimals

- Rewrite the numbers vertically, lining up the decimal points.
- Add.
- Insert a decimal point in the sum below the other decimal points.

EXAMPLE 1

Find the sum: 2.7 + 80.13 + 5.036

Solution

$$\begin{array}{r} 2.7 \\ 80.13 \\ +5.036 \\ \hline 87.866 \end{array}$$

Rewrite the addends with decimal points lined up vertically.

Add.

Insert the decimal point in the sum.

PRACTICE 1

Add: 5.12 + 4.967 + 0.3

EXAMPLE 2

Compute: 2.367 + 5 + 0.143

Solution Recall that 5 and 5. are equivalent.

$$\begin{array}{r} 2.367 \\ 5. \\ +0.143 \\ \hline 7.510 \end{array} = 7.51$$

Line up the decimal points and add.

Insert the decimal point in the sum.

We can drop the extra 0 at the right end.

PRACTICE 2

What is the sum of 7.31, 8, and 23.99?

EXAMPLE 3

A runner's time was 0.06 second longer than the world record of 21.71 seconds. What was the runner's time?

Solution We need to compute the sum of the two numbers. The runner's time was 21.77 seconds.

$$\begin{array}{r} 0.06 \\ +21.71 \\ \hline 21.77 \end{array}$$

PRACTICE 3

A child has the flu. This morning, his body temperature was 99.4°F. What was his temperature after it went up by 2.7°?

Subtracting Decimals

Now, let's discuss subtracting decimals. As with addition, subtracting decimals is similar to subtracting whole numbers. To compute the difference between 12.83 and 4.2, we rewrite the problem vertically, lining up the decimal points. Then, we subtract as usual, inserting a decimal point below the other decimal points.

$$\begin{array}{r} 12.83 \\ -4.2 \\ \hline 8.63 \end{array} \quad \text{is equivalent to} \quad \begin{array}{r} 12\frac{83}{100} = 12\frac{83}{100} \\ -4\frac{2}{10} = -4\frac{20}{100} \\ \hline 8\frac{63}{100}, \text{ or } 8.63 \end{array}$$

Again, note that when we subtracted the equivalent mixed numbers, we got the same difference.

As in any subtraction problem, we can check this answer by adding the subtrahend (4.2) to the difference (8.63), confirming that we get the original minuend (12.83). This example suggests the following rule:

$$\begin{array}{r} 8.63 \\ +4.2 \\ \hline 12.83 \end{array}$$

To Subtract Decimals

- Rewrite the numbers vertically, lining up the decimal points.
- Subtract, inserting extra zeros in the minuend if necessary for regrouping.
- Insert a decimal point in the difference, below the other decimal points.

EXAMPLE 4

Subtract and check: $5.038 - 2.11$

Solution

$$\begin{array}{r} 5.038 \\ -2.11 \\ \hline 2.928 \end{array}$$

Rewrite the problem with decimal points lined up vertically.

Subtract. Regroup when necessary.

Insert the decimal point in the answer.

Check To verify that our difference is correct, we check by addition.

$$\begin{array}{r} 2.928 \\ +2.11 \\ \hline 5.038 \end{array}$$

PRACTICE 4

Find the difference and check: $71.3825 - 25.17$

EXAMPLE 5

65 is how much larger than 2.04?

Solution Recall that 65 and 65. are equivalent.

Insert the zeros needed for regrouping.

$$\begin{array}{r} 65.00 \\ -2.04 \\ \hline 62.96 \end{array}$$

Line up the decimal points.
Subtract.
Insert the decimal point in the answer.

Check

$$\begin{array}{r} 62.96 \\ +\ 2.04 \\ \hline 65.00 \end{array} = 65$$

PRACTICE 5

How much greater is \$735 than \$249.57?

EXAMPLE 6

A Burger King Whopper Jr.® contains 1.02 grams of sodium, whereas a McDonald's Quarter Pounder ® contains 0.73 grams. How much more sodium does the Whopper Jr.® contain? (*Sources:* nutrition.mcdonalds.com and bk.com)

Solution We need to find the difference between 1.02 and 0.73.

$$\begin{array}{r} 1.02 \\ -0.73 \\ \hline 0.29 \end{array}$$

So a Burger King Whopper Jr.® contains 0.29 grams more sodium.

PRACTICE 6

The New York Marathon has timing mats located every 3.1 miles throughout its 26.2-mile course. A runner passes one of the mats 9.3 miles into the race. How much further does the runner have to go to finish the race? (*Source*: nycmarathon.org)

EXAMPLE 7

Suppose that a part-time employee's salary is \$350 a week, less deductions. The following table shows these deductions.

Deduction	Amount
Federal, state, and city taxes	\$100.80
Social Security	26.78
Union dues	8.88

What is the employee's take-home pay?

Solution Let's use the strategy of breaking the question into two simpler questions.

- *How much money is deducted per week?* The weekly deductions (\$100.80, \$26.78, and \$8.88) add up to \$136.46.
- *How much of the salary is left after subtracting the total deductions?* The difference between \$350 and \$136.46 is \$213.54, which is the employee's take-home pay.

PRACTICE 7

A sales rep, working in Ohio, wants to drive from Circleville to Columbus. How much shorter is it to drive directly to Columbus instead of going by way of Lancaster? (*Source:* mapquest.com)

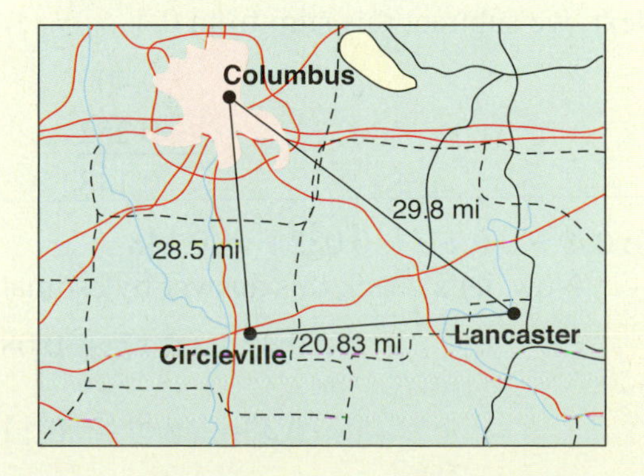

Estimating Sums and Differences

Being able to estimate in your head the sum or difference between two decimals is a useful skill, for either checking or approximating an exact answer. To estimate, simply round the numbers to be added or subtracted and then carry out the operation on the rounded numbers.

EXAMPLE 8

Compute the sum 0.17 + 0.4 + 0.083. Use estimation to check.

Solution First, we add. Then, to check, we round the addends—say, to the nearest tenth—and add the rounded numbers.

$$\begin{array}{rcr} 0.17 & \approx & 0.2 \\ 0.4 & \approx & 0.4 \\ +0.083 & \approx & +0.1 \\ \hline \text{Exact sum} \longrightarrow 0.653 & & 0.7 \longleftarrow \text{Estimated sum} \end{array}$$

Our exact sum is close to our estimated sum, and in fact, rounds to it.

PRACTICE 8

Add 0.093, 0.008, and 0.762. Then, check by estimating.

EXAMPLE 9

Subtract 0.713 − 0.082. Then check by estimating.

Solution First, we find the exact answer and then round the given numbers to get an estimate.

$$\begin{array}{rcr} 0.713 & \approx & 0.7 \\ -0.082 & \approx & -0.1 \\ \hline \text{Exact difference} \longrightarrow 0.631 & & 0.6 \longleftarrow \text{Estimated difference} \end{array}$$

Our exact answer, 0.631, is close to 0.6.

PRACTICE 9

Compute: 0.17 − 0.091. Use estimation to check.

EXAMPLE 10

Combine and check: 0.4 − (0.17 + 0.082)

Solution Following the order of operations rule, we begin by adding the two decimals in parentheses.

$$\begin{array}{r} 0.17 \\ +0.082 \\ \hline 0.252 \end{array}$$

Next, we subtract this sum from 0.4.

$$\begin{array}{r} 0.400 \\ -0.252 \\ \hline 0.148 \end{array}$$

So 0.4 − (0.17 + 0.082) = 0.148.

Now, let's check this answer by estimating:

$$\begin{array}{l} 0.4 - (0.17 + 0.082) \\ \downarrow \quad\quad \downarrow \quad\quad \downarrow \\ 0.4 - (0.2 + 0.1) = 0.4 - 0.3 = 0.1 \end{array}$$

The estimate, 0.1, is close to 0.148.

PRACTICE 10

Calculate and check:
0.813 − (0.29 − 0.0514)

In the following examples, we estimate a sum or difference to approximate the correct answer, not to check it.

EXAMPLE 11

A movie budgeted at $7.25 million ended up costing $1.655 million more. Estimate the final cost of the movie.

Solution Let's round each number to the nearest million dollars.

$$\begin{array}{rcr} 1.655 & \approx & 2 \text{ million} \\ 7.25 & \approx & +7 \text{ million} \\ \hline & & 9 \text{ million} \end{array}$$

Adding the rounded numbers, we see that the movie cost approximately $9 million.

PRACTICE 11

From the deposit ticket shown below, estimate the total amount deposited.

DEPOSIT SLIP

Lonnie Chavis
9 West Drive
Petersburg, VA 23805

DATE June 3, 2011
Deposits may not be available for immediate withdrawal

Sign here for cash received (if required)

OMNI BANK 701 Halifax Street Petersburg, VA 23805

	DOLLARS	CENTS
	76	35
	312	95
	42	30
	49	
TOTAL $		

:002582910: 56024036" 02

Estimate: ______________

EXAMPLE 12

When the underwater tunnel connecting the United Kingdom and France was built, French and British construction workers dug from their respective countries. They met at the point shown on the map.

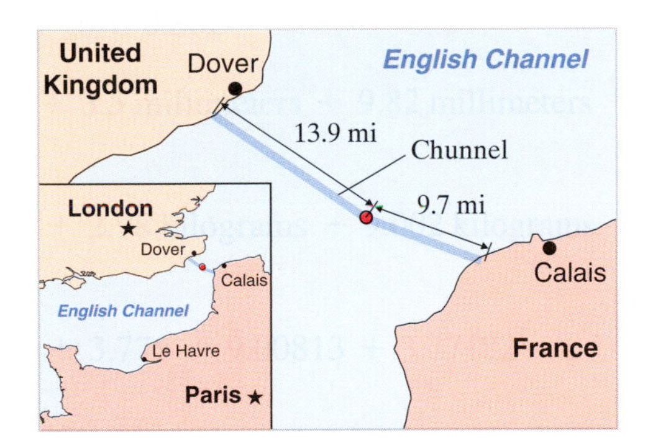

Estimate how much farther the British workers had dug than the French workers. (*Source: The New York Times*)

Solution We can round 13.9 to 14 and 9.7 to 10. The difference between 14 and 10 is 4, so the British workers dug about 4 miles farther than the French workers.

PRACTICE 12

An art collector bought a painting for $2.3 million. A year later, she sold the painting for $4.1 million. Estimate her profit on the sale.

Compute.

55. $35.2 - 2.86 + 9.07 - 1.658$

56. $10 - 2.38 + 4.92 - 6.02$

57. 30 milligrams − 0.5 milligram − 1.6 milligrams

58. $20.93 + $1.07 − $19.58

59. $5.21 - (1.03 + 0.975)$

60. $6.953 - (4.09 + 0.008)$

61. $41.075 - 2.87104 - 17.005$

62. $0.00661 + 1.997 - 0.05321$

In each group of three computations, one answer is wrong. Use estimation to identify which answer is incorrect.

63. a. $\begin{array}{r} 0.059 \\ 0.00234 \\ +0.036 \\ \hline 0.09734 \end{array}$ b. $\begin{array}{r} 0.1903 \\ 0.074 \\ +0.2051 \\ \hline 0.4694 \end{array}$ c. $\begin{array}{r} 0.00441 \\ 0.06882 \\ +0.0103 \\ \hline 0.8353 \end{array}$

64. a. $\begin{array}{r} \$32.71 \\ 43.09 \\ +\ 8.27 \\ \hline \$74.07 \end{array}$ b. $\begin{array}{r} \$19.37 \\ 2. \\ +\ 7.22 \\ \hline \$28.59 \end{array}$ c. $\begin{array}{r} \$139.26 \\ 82.87 \\ +\ 3.01 \\ \hline \$225.14 \end{array}$

65. a. $\begin{array}{r} 0.35 \\ -0.1007 \\ \hline 0.2493 \end{array}$ b. $\begin{array}{r} 0.072 \\ -0.0056 \\ \hline 0.664 \end{array}$ c. $\begin{array}{r} 0.03 \\ -0.008 \\ \hline 0.022 \end{array}$

66. a. $\begin{array}{r} 8.551 \\ -2.9995 \\ \hline 5.5515 \end{array}$ b. $\begin{array}{r} 78.328 \\ -\ 5.5 \\ \hline 7.2828 \end{array}$ c. $\begin{array}{r} 65 \\ -\ 2.778 \\ \hline 62.222 \end{array}$

Mixed Practice

Solve.

67. Calculate: $4.78 + 13 - 10.009$

68. Find the difference between 90.1 and 12.58.

69. Add: 0.5 pound + 3 pounds + 4.25 pounds

70. Subtract: $20 − $6.95

71. Compute: $8 - 2.4 + 6.0013$

72. What is the sum of 1.265, 7, and 0.14?

Applications

B *Solve and check.*

73. A paperback book that normally sells for $13 is now on sale for $11.97. What is the discount in dollars and cents?

74. During a drought, the mayor of a city attempted to reduce daily water consumption to 3.1 million gallons. If daily water consumption fell to 1.948 million gallons above that goal, estimate the city's consumption.

75. A skeleton was found at an archaeological dig. Radiocarbon dating—a technique used for estimating age—indicated that the skeleton was 56 centuries old, plus or minus 0.8 centuries. According to this estimate, what is the greatest possible age of the skeleton?

76. A college launched a campaign to collect $3 million to build a new technology complex. If $1.316 million has been collected so far, how much more money, to the nearest million dollars, is needed?

77. As an investment, a couple bought an apartment house for $2.3 million. Two years later, they sold the apartment house for $4 million. What was their profit?

78. A woman sued her business partner and was awarded $1.5 million. On appeal, however, the award was reduced to $0.75 million. By how much was the award reduced?

79. In setting up a page in word processing, the margins of a page are usually expressed in decimal parts of an inch. How long is each typed line on the page?

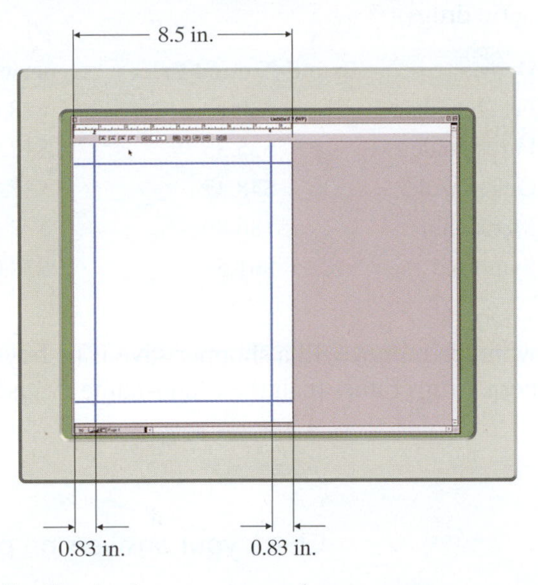

80. In the picture below, how much clearance will there be between the top of the roof cargo carrier and the top of the garage door?

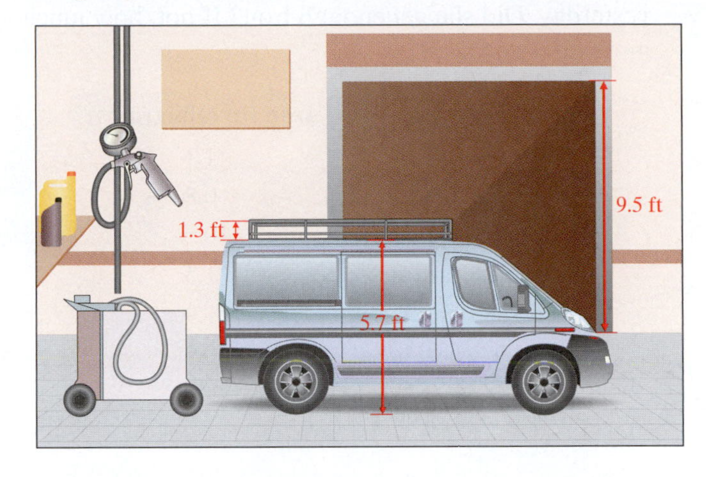

81. A radio disc jockey wants to choose among compact disc tracks that last 3.5, 2.8, 2.9, 2.6, and 1.6 minutes. Can he select tracks so as to get between 9.8 and 10 minutes of music? Explain.

82. A shopper plans to buy three items that cost \$4.99, \$7.99, and \$2.99 each. If she has \$15 with her, will she have enough money to pay for all three items? Explain.

83. When gymnasts compete, they receive scores in four separate events: vault (VT), uneven bar (UB), balance beam (BB), and floor exercises (FX). The total of these four event scores is called the all-around score (AA). The following chart shows the qualifying scores earned by three gymnasts at the Beijing, China 2008 Summer Olympics.

Gymnast	VT	UB	BB	FX	AA
Nastia Liukin (U.S.)	15.1	15.95	15.975	15.35	
Yang Yilin (China)	15.2	16.65	15.5	15	
Shawn Johnson (U.S.)	16	15.325	15.975	15.425	

(*Source*: wikipedia.org)

a. Calculate the all-around scores for these three competitors.

b. Which competitor had the highest all-around score?

84. The graph shown gives the number of households that watched particular TV programs in a given week, according to the Nielsen Top 20 ratings.

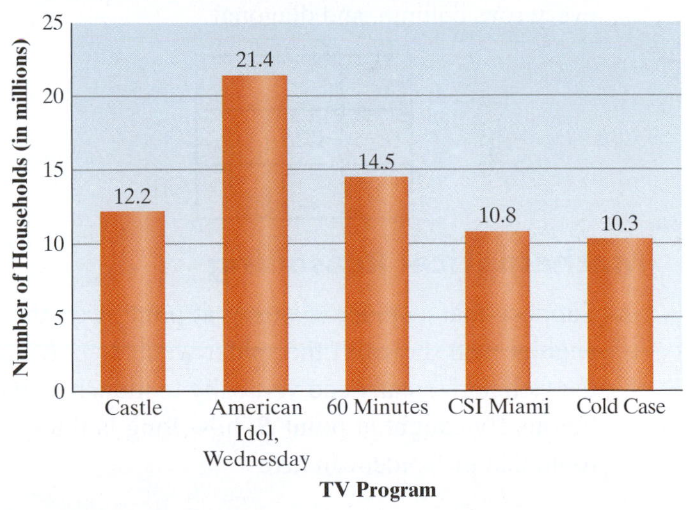

(*Source:* usatoday.com)

a. How many more households watched *American Idol-Wednesday* than *60 Minutes*?

b. Write the names of the five programs in increasing order of viewership.

85. To prevent anemia, a doctor advises his patient to take at least 18 milligrams of iron each day. The following table shows the amount of iron in the food that the patient ate yesterday. Did she get enough iron? If not, how much more does she need?

Food	Iron (in milligrams)
Chicken, breast	1.1
Canned tuna	0.8
Fortified oatmeal	10
Seedless raisins	1.5
Frozen spinach	1.9
Whole wheat bread	0.9

(*Source:* ods.od.nih.gov)

86. When filling a prescription, buying a generic drug rather than a brand-name drug can often save money. The following table shows the prices of various brand-name and generic drugs.

Drug	Brand-Named Price	Generic Price
Fexofenadine	$30	$8.80
Furosemide	$35.35	$17.94
Omeprazole	$48.47	$37.44
Metformin	$186.89	$78.39
Synthroid	$61.56	$38.02

How much money will a shopper save if he buys all five generic drugs rather than the brand-name drugs?

• Check your answers on page A-4.

MINDStretchers

Groupwork

1. Working with a partner, find the missing entries in the following magic square, in which 3.75 is the sum of every row, column, and diagonal.

0.75	1.25	
2.		

Mathematical Reasoning

2. Suppose that a spider is sitting at point *A* on the rectangular web shown. If the spider wants to crawl along the web horizontally and vertically to munch on the delicious fly caught at point *B*, how long is the shortest route that the spider can take?

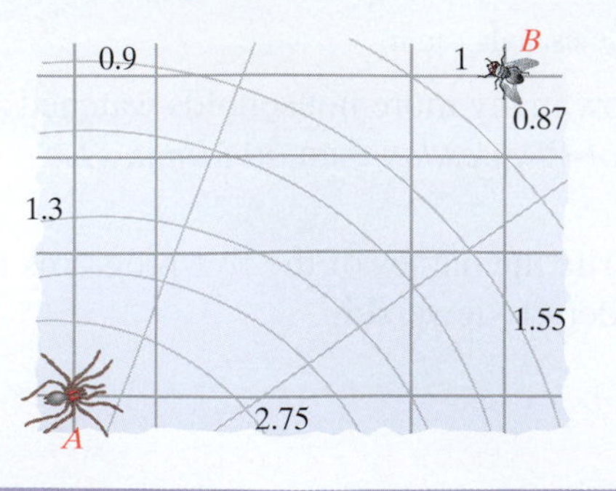

Writing

3. a. How many pairs of whole numbers are there whose sum is 7?

b. How many pairs of decimals are there whose sum is 0.7?

c. Explain why (a) and (b) have different answers.

3.3 Multiplying Decimals

OBJECTIVES

- **A** To multiply decimals
- **B** To solve applied problems involving the multiplication of decimals

In this section, we discuss how to multiply two or more decimals, finding both the exact product and an estimated product.

Multiplying Decimals

To find the product of two decimals—say, 1.02 and 0.3—we multiply the same way we multiply whole numbers. But with decimals we need to know where the decimal point goes in the product. To find out, let's change each decimal to its fractional equivalent.

$$\begin{array}{r} 1.02 \\ \times\ 0.3 \\ \hline 306 \end{array} \text{ is equivalent to } \rightarrow \quad 1\frac{2}{100} \times \frac{3}{10}$$

Where should we place the decimal point?

$$= \frac{102}{100} \times \frac{3}{10} = \frac{306}{1{,}000}, \text{ or } 0.306$$

The product is in thousandths, so it has three decimal places.

Looking at the multiplication problem with decimals, note that *the product has as many decimal places as the total number of decimal places in the factors*. This example illustrates the following rule for multiplying decimals.

$$\begin{array}{r} 1.0\,2 \\ 0.3 \\ \hline 0.3\,0\,6 \end{array}$$

To Multiply Decimals

- Multiply the factors as if they were whole numbers.
- Find the total number of decimal places in the factors.
- Count that many places from the right end of the product, and insert a decimal point.

EXAMPLE 1

Multiply: 6.1 × 3.7

Solution First, multiply 61 by 37, ignoring the decimal points.

$$\begin{array}{r} 6.1 \\ \times\ 3.7 \\ \hline 4\,2\,7 \\ 1\,8\,3\ \ \\ \hline 2\,2\,5\,7 \end{array}$$

Then, count the total number of decimal places in the factors.

$$\begin{array}{rl} 6.1 & \leftarrow \text{One decimal place} \\ \times\ 3.7 & \leftarrow \text{One decimal place} \\ \hline 4\,2\,7 & \\ 1\,8\,3\ \ & \\ \hline 2\,2.5\,7 & \leftarrow \text{Two decimal places in the product} \end{array}$$

Insert the decimal point two places from the right end.
So, 22.57 is the product.

PRACTICE 1

Find the product: 2.81 × 3.5

Estimating Products

Being able to estimate mentally the product of two decimals is useful for either checking or approximating an exact answer. To estimate, round each factor so that it has only one nonzero digit. Then, multiply the rounded factors.

EXAMPLE 9

Multiply 0.703 by 0.087 and check the answer by estimating.

Solution First, we multiply the factors to find the exact product. Then, we round each factor and multiply them.

$$\begin{array}{rrl} 0.703 & \approx \quad 0.7 & \leftarrow \text{Rounded to have one nonzero digit} \\ \times\ 0.087 & \approx \times\ 0.09 & \leftarrow \text{Rounded to have one nonzero digit} \\ \hline \text{Exact product} \rightarrow 0.061161 & 0.063 & \leftarrow \text{Estimated product} \end{array}$$

We see that the exact product and the estimated product are fairly close.

PRACTICE 9

Find the product of 0.0037×0.092, estimating to check.

EXAMPLE 10

Calculate and check: $(4.061)(0.72) + (0.91)(0.258)$

Solution Following the order of operations rule, we begin by finding the two products.

$$(4.061)(0.72) = 2.92392 \qquad (0.91)(0.258) = 0.23478$$

Then, we add these two products.

$$2.92392 + 0.23478 = 3.1587$$

So $(4.061)(0.72) + (0.91)(0.258) = 3.1587$.

Now, let's check this answer by estimating.

$$\begin{array}{c} (4.061)(0.72) + (0.91)(0.258) \\ \downarrow \qquad \downarrow \qquad\quad \downarrow \qquad \downarrow \\ (4) \quad (0.7) + (0.9) \quad (0.3) = 2.8 + 0.27 = 3.07 \approx 3 \end{array}$$

The estimate 3 is close to 3.1587.

PRACTICE 10

Compute and check:
$(0.488)(9.1) - (3.5)(0.227)$

EXAMPLE 11

The sound waves of an elephant call can travel through both the ground and the air. Through the air, the waves may travel 6.63 miles. If they travel 1.5 times as far through the ground, what is the estimated ground distance? (*Source:* wikipedia.org)

Solution We know that the waves may travel 6.63 miles through the air and 1.5 times as far through the ground. To find the estimated ground distance, we compute this product.

$$\begin{array}{rcr} 6.63 & \approx & 7 \\ \times 1.5 & \approx & \times 2 \\ \hline & & 14 \end{array}$$

So the estimated ground distance of the sound waves of an elephant call is about 14 miles.

PRACTICE 11

Earth travels through space at a speed of 18.6 miles per second. Estimate how far Earth travels in 60 seconds. (*Source:* The Diagram Group, *Comparisons*)

Multiplying Decimals on a Calculator

Multiply decimals on a calculator by entering each decimal as you would enter a whole number, but insert a decimal point as needed. If there are too many decimal places in your answer to fit in the display, investigate how your calculator displays the answer.

EXAMPLE 12

Compute 8,278.55 × 0.875, rounding your answer to the nearest hundredth. Then, check the answer by estimating.

Solution

Press	Display
8278.55 [×] 0.875 [ENTER]	8278.55 * 0.875 7243.73125

Now, 7,243.73125 rounded to the nearest hundredth is 7,243.73. Checking by estimating, we get 8,000 × 0.9, or 7,200, which is close to our exact answer.

PRACTICE 12

Find the product of 2,471.66 and 0.33, rounding to the nearest tenth. Check the answer.

EXAMPLE 13

Find $(1.9)^2$

Solution

Press	Display
1.9 [^] 2 [ENTER]	1.9 ^ 2 3.61

Now, let's check by estimating. Since 1.9 rounded to the nearest whole number is 2, $(1.9)^2$ should be close to 2^2, or 4, which is close to our exact answer, 3.61.

PRACTICE 13

Calculate: $(2.1)^3$

3.3 Exercises

FOR EXTRA HELP MyMathLab Math XL PRACTICE WATCH READ REVIEW

Mathematically Speaking

Fill in each blank with the most appropriate term or phrase from the given list.

add	three	factors	five
first factor	four	multiplication	
square	two	division	

1. The operation understood in the expression (3.4) (8.9) is____________.

2. When multiplying decimals, the number of decimal places in the product is equal to the total number of decimal places in the ______.

3. To multiply a decimal by 100, move the decimal point ______ places to the right.

4. The product of 0.27 and 8.18 has ______ decimal places.

5. To compute the expression $(8.5)^2 + 2.1$, first ______.

6. To multiply a decimal by 1,000, move the decimal point ______ places to the right.

A *Insert a decimal point in each product. Check by estimating.*

7. $2.356 \times 1.27 = 299212$
8. $97.26 \times 5.3 = 515478$
9. $3{,}144 \times 0.065 = 204360$
10. $837 \times 0.15 = 12555$
11. $71.2 \times 35 = 24920$
12. $0.002 \times 37 = 0074$
13. $0.0019 \times 0.051 = 969$
14. $0.0089 \times 0.0021 = 1869$
15. $2.87 \times 1{,}000 = 287000$
16. $492.31 \times 10 = 492310$
17. $\$4.25 \times 0.173 = \$73525$
18. 11.2 feet $\times$ 0.75 = 8400 feet

Find the product. Check by estimating.

19. 0.6×0.9
20. 0.8×0.7
21. 0.5×0.8
22. 0.6×0.8
23. 0.1×0.2
24. 0.9×0.5
25. 0.04×0.07
26. 0.03×0.01
27. 2.55×0.3
28. 8.07×0.6
29. 0.96×2.1
30. 0.87×3.1
31. 38.01×0.2
32. 12.02×0.05
33. 125×0.004
34. 135×0.006
35. 3.8×1.54
36. 9.51×2.7
37. 13.74×11
38. 12.45×11
39. 12.459×0.3
40. 72.558×0.2
41. $(0.675)(2.66)$
42. $(4.003)(0.59)$

43. 83.127×100

44. 49.247×100

45. $0.0023 \times 10{,}000$

46. $0.0135 \times 10{,}000$

47. $(1.5)(0.6)(0.1)$

48. $(12)(3.5)(0.2)$

49. $(0.03)(1.4)(25)$

50. $(2.6)(0.5)(0.9)$

51. $(0.001)^3$

52. $(0.1)^4$

53. 17 feet × 2.5

54. 15 hours × 7.5

55. 3.5 miles × 0.4

56. 9.1 meters × 1,000

57. $\begin{array}{r} 43.87 \\ \times\ 0.975 \\ \hline \end{array}$

58. $\begin{array}{r} 18{,}275.33 \\ \times\ \ \ 0.39 \\ \hline \end{array}$

59. $\begin{array}{r} 99{,}125 \\ \times\ \ 2.75 \\ \hline \end{array}$

60. $\begin{array}{r} 3.512 \\ \times\ 1.47 \\ \hline \end{array}$

Simplify.

61. 0.7×10^2

62. 0.6×10^4

63. $30 - 2.5 \times 1.7$

64. $18 - 3.4 \times 1.6$

65. $1 + (0.3)^2$

66. $6 + (1.2)^2$

67. $0.8(1.3 + 2.9) - 0.5$

68. $4 - 2.1(3.5 - 1.8)$

69. $(5.2 - 3.9)(0.9 + 2.14)$

70. $(8 + 4.5)(8 - 4.5)$

71. $0.4(3 - 2.9)(2 + 1.5)$

72. $0.5(1 + 0.2)(1 - 0.2)$

Complete each table.

73.

Input	Output
1	3.8 × **1** − 0.2 =
2	3.8 × **2** − 0.2 =
3	3.8 × **3** − 0.2 =
4	3.8 × **4** − 0.2 =

74.

Input	Output
1	7.5 × **1** + 0.4 =
2	7.5 × **2** + 0.4 =
3	7.5 × **3** + 0.4 =
4	7.5 × **4** + 0.4 =

Each product is rounded to the nearest hundredth. In each group of three products, one is wrong. Use estimation to explain which product is incorrect.

75. a. $51.6 \times 0.813 \approx 419.51$ **b.** $2.93 \times 7.283 \approx 21.34$ **c.** $(5.004)^2 \approx 25.04$

76. a. $0.004 \times 3.18 \approx 0.01$ **b.** $2.99 \times 0.287 \approx 0.86$ **c.** $(1.985)^3 \approx 10.82$

77. a. $4.913 \times 2.18 \approx 10.71$ **b.** $0.023 \times 0.71 \approx 0.16$ **c.** $(8.92)(1.0027) \approx 8.94$

78. a. \$138.28 × 0.075 ≈ \$10.37 **b.** 0.19 × \$487.21 ≈ \$92.57 **c.** 0.77 × \$6,005.79 ≈ \$462.45

Mixed Practice

Solve.

79. Simplify: $9 - (0.5)^2$

80. Compute: $2.1 + 5 \times 0.6$

81. Multiply 0.75 and 0.09, rounding the answer to the nearest thousandth.

82. Multiply: $(2.3)(4.5)(0.6)$

83. Find the product of 0.56 and 8.

84. Find the product: $3.01 \times 1{,}000$

Applications

B *Solve. Check by estimating.*

85. Sound travels at approximately 1,000 feet per second (fps). If a jet is flying at Mach 2.9 (that is, 2.9 times the speed of sound), what is its speed?

86. If insurance premiums of $323.50 are paid yearly for 10 years for a life insurance policy, how much did the policy holder pay altogether in premiums?

87. The planet in the solar system closest to the Sun is Mercury. The average distance between these two bodies is 57.9 million kilometers. Express this distance in standard form. (*Source:* Jeffrey Bennett et al., *The Cosmic Perspective*)

88. According to the first American census in 1790, the population of the United States was approximately 3.9 million. Write this number in standard form. (*Source: The Statistical History of the United States*)

89. A construction company builds custom-designed swimming pools, including the circular pool shown below. The area of the bottom of this pool is approximately 3.14×9^2 square feet. Find this area to the nearest tenth. (*Source:* Pritchett Construction Co.)

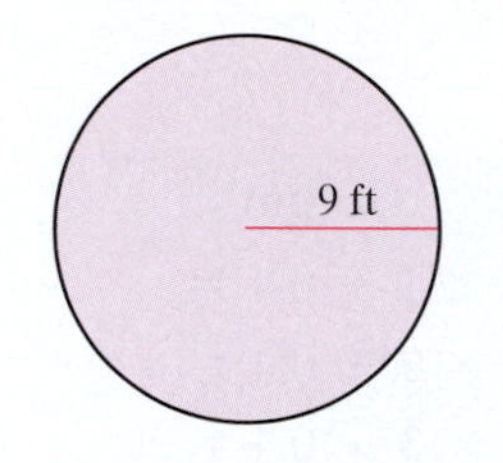

90. Find the area (in square meters) of the floor pictured.

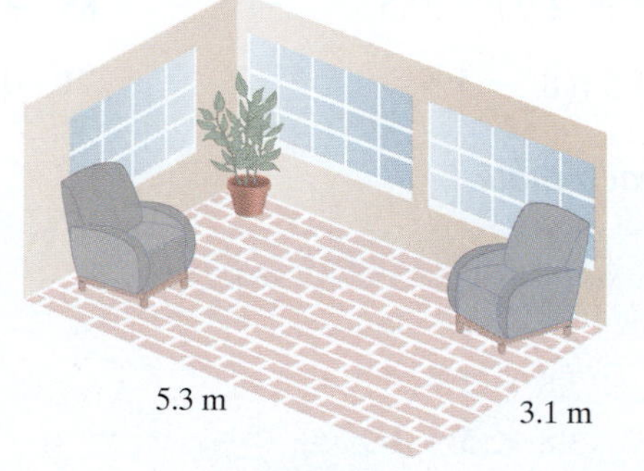

91. Over a 5-day period, a nurse administered 10 tablets to a patient. If each tablet contained 0.125 milligram of the drug Digoxin, how much Digoxin did the nurse administer?

92. Water weighs approximately 62.5 pounds per cubic foot (lb/ft^3). If a bathtub contains about 30 ft^3 of water, how much does the water in the bathtub weigh?

93. A tennis player weighing 180 pounds burns 10.9 calories per minute while playing singles tennis. How many calories would he burn in 2 hours? (*Source:* caloriesperhour.com)

94. A plumber is paid $37.50 per hour for the first 40 hours worked. She gets time and a half, $56.25, for any time over her 40-hour week. If she works 49 hours in a week, how much is her pay?

95. The sales receipt for a shopper's purchases is as follows:

Purchase	Quantity	Unit Price	Price
Belt	1	$11.99	$___.__
Shirt	3	$16.95	$___.__
Total Price			$___.__

a. Complete the table.

b. If the shopper pays for these purchases with four $20 bills, how much change should he get?

96. On an electric bill, *usage* is the difference between the meter's *current reading* and the *previous reading* in kilowatt hours (kWh). The *amount due* is the product of the usage and the *rate per kWh*. Find the two missing quantities in the table, rounding to the nearest hundredth.

Previous Reading	750.07 kWh
Current Reading	1,115.14 kWh
Usage	kWh
Rate per kWh	$0.10
Amount Due	$

97. Scientists have discovered a relationship between the length of a person's bones and the person's overall height. For instance, an adult male's height (in inches) can be predicted from the length of his femur bone by using the formula $(1.9 \times \text{femur}) + 32.0$. With this formula, find the height of the German giant Constantine, whose femur measured 29.9 inches. (*Source: Guinness World Records*)

98. In order to buy a $125,000 house, a couple puts down $25,000 and takes out a mortgage on the balance. To pay off the mortgage, they pay $877.57 per month for the following 360 months. How much more will they end up paying for the house than the original price of $125,000?

• Check your answers on page A-4.

MINDStretchers

Patterns

1. When $(0.001)^{100}$ is multiplied out, how many decimal places will it have?

Mathematical Reasoning

2. Give an example of two decimals

a. whose sum is greater than their product, and

b. whose product is greater than their sum.

Groupwork

3. In the product to the right, each letter stands for a different digit. Working with a partner, identify all the digits.

$$\begin{array}{r} A.B \\ \times B.A \\ \hline C\,D \end{array}$$

3.4 Dividing Decimals

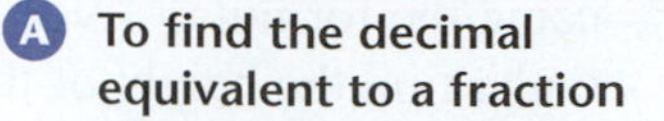

OBJECTIVES

- **A** To find the decimal equivalent to a fraction
- **B** To divide decimals
- **C** To solve applied problems involving the division of decimals

In this section, we first consider changing a fraction to its decimal equivalent, which involves both division and decimals. We then move on to our main concern—the division of decimals.

Changing a Fraction to the Equivalent Decimal

Earlier in this chapter, we discussed how to change a decimal to its equivalent fraction. Now let's consider the opposite problem—how to change a fraction to its equivalent decimal.

When the denominator of a fraction is already a power of 10, the problem is simple. For example, the decimal equivalent of $\frac{43}{100}$ is 0.43.

But what about the more difficult problem in which the denominator is *not* a power of 10? A good strategy is to find an equivalent fraction that does have a power of 10 as its denominator. Consider, for instance, the fraction $\frac{3}{4}$. Since 4 is a factor of 100, which is a power of 10, we can easily find an equivalent fraction having a denominator of 100.

$$\frac{3}{4} = \frac{3 \cdot 25}{4 \cdot 25} = \frac{75}{100} = 0.75$$

So 0.75 is the decimal equivalent of $\frac{3}{4}$.

There is a faster way to show that $\frac{3}{4}$ is the same as 0.75, without having to find an equivalent fraction. Because $\frac{3}{4}$ can mean $3 \div 4$, we divide the numerator (3) by the denominator (4). Note that if we continue to divide to the hundredths place, there is no remainder.

```
     0.75      ← Insert the decimal point directly above
   ______        the decimal point in the dividend.
  4)3.00       ← The decimal point is after the 3.
    0            Insert enough 0's to continue
    -            dividing as far as necessary.
    3 0
    2 8
    ---
      20
      20
      --
```

So this division also tells us that $\frac{3}{4}$ equals 0.75.

To Change a Fraction to the Equivalent Decimal

- Divide the denominator of the fraction into the numerator, inserting to its right both a decimal point and enough zeros to get an answer either without a remainder or rounded to a given decimal place.
- Place a decimal point in the quotient directly above the decimal point in the dividend.

EXAMPLE 1

Express $\frac{1}{2}$ as a decimal.

Solution To find the decimal equivalent, we divide the fraction's numerator by its denominator.

$$\begin{array}{r} 0.5 \\ 2\overline{)1.0} \\ 0 \\ \hline 1\,0 \\ 1\,0 \\ \hline \end{array}$$

Add a decimal point and a 0 to the right of the 1.

So 0.5 is the decimal equivalent of $\frac{1}{2}$.

Check We verify that the fractional equivalent of 0.5 is $\frac{1}{2}$.

$$0.5 = \frac{5}{10} = \frac{1}{2}$$

The answer checks.

PRACTICE 1

Write the fraction $\frac{3}{8}$ as a decimal.

EXAMPLE 2

Convert $2\frac{3}{5}$ to a decimal.

Solution Let's first change this mixed number to an improper fraction: $2\frac{3}{5} = \frac{13}{5}$. We can then change this improper fraction to a decimal by dividing its numerator by its denominator.

$$\begin{array}{r} 2.6 \\ 5\overline{)13.0} \\ 10 \\ \hline 3\,0 \\ 3\,0 \\ \hline \end{array}$$

So 2.6 is the decimal form of $2\frac{3}{5}$.

Check We convert this answer back from a decimal to its mixed number form.

$$2.6 = 2\frac{6}{10} = 2\frac{3}{5}$$ The answer checks.

PRACTICE 2

Write $7\frac{5}{8}$ as a decimal.

When converting some fractions to decimal notation, we keep getting a remainder as we divide. In this case, we round the answer to a given decimal place.

EXAMPLE 3

Convert $4\frac{8}{9}$ to a decimal, rounded to the nearest hundredth.

Solution First, we change this mixed number to an improper fraction. Then, we convert it to a decimal.

$$4\frac{8}{9} = \frac{44}{9} = 9\overline{)44.000}$$

```
     4.888
 9)44.000   ← In order to round to the nearest hundredth, we must
   36         continue to divide to the thousandths place. So we
    8 0       insert three 0's.
    7 2
      80
      72
       80
       72
```

Finally, we round to the nearest hundredth: $4.888 \approx 4.89$

PRACTICE 3

Express $83\frac{1}{3}$ as a decimal, rounded to the nearest tenth.

In Example 3, note that if instead of rounding we had continued to divide we would have gotten as our answer the **repeating decimal** 4.88888 . . . (also written $4.\overline{8}$). Can you think of any other fraction that is equivalent to a repeating decimal?

Let's now consider some word problems in which we need to convert fractions to decimals.

EXAMPLE 4

Reformulated gasoline (RFG) is a "cleaner" gasoline required to be used in nine major metropolitan areas of the United States with the worst ozone air pollution problem. In a particular week, the average retail regular gas price for RFG areas was $\$2\frac{7}{8}$. Express this amount in dollars and cents, to the nearest cent. (*Source:* epa.gov and eia.doe.gov)

Solution To solve, we must convert the mixed number $2\frac{7}{8}$ to a decimal.

$$2\frac{7}{8} = \frac{23}{8} = 8\overline{)23.000}$$

```
    2.875  ≈ 2.88
 8)23.000
   16
    70
    64
     60
     56
      40
      40
```

So the average price for RFG was $2.88, to the nearest cent.

PRACTICE 4

About $\frac{13}{125}$ of the world's land surface is covered with ice. Express this fraction as a decimal, rounded to the nearest tenth. (*Source:* enotes.com)

Dividing Decimals

Before we turn our attention to dividing one decimal by another, let's consider simpler problems in which we are dividing a decimal by a whole number. An example of such a problem is $0.6 \div 2$. We can write this expression as the fraction $\frac{0.6}{2}$, which can be rewritten as the quotient of two whole numbers by multiplying the numerator and denominator by 10 as follows:

$$\frac{0.6}{2} = \frac{0.6 \times 10}{2 \times 10} = \frac{6}{20}$$

We then convert the fraction $\frac{6}{20}$ to the equivalent decimal, as we have previously discussed.

$$\begin{array}{r} 0.3 \\ 20\overline{)6.0} \\ \underline{0} \\ 6\,0 \\ \underline{6\,0} \end{array}$$

So $\frac{0.6}{2} = 0.3$

Note that we get the same quotient if we divide the number in the original problem as follows:

$$\begin{array}{r} 0.3 \leftarrow \textbf{Quotient} \\ \textbf{Divisor} \rightarrow 2\overline{)0.6} \leftarrow \textbf{Dividend} \\ \underline{0} \\ 6 \\ \underline{6} \end{array}$$

It is important to write the decimal point in the quotient directly above the decimal point in the dividend.

Next, let's consider a division problem in which we are dividing one decimal by another: $0.006 \div 0.02$. Writing this expression as a fraction, we get $\frac{0.006}{0.02}$. Since we have already discussed how to divide a decimal by a whole number, the goal here is to find a fraction equivalent to $\frac{0.006}{0.02}$ where the denominator is a whole number. Multiplying the numerator and denominator by 100 will do just that.

$$0.006 \div 0.02 = \frac{0.006}{0.02} = \frac{0.006 \times 100}{0.02 \times 100} = \frac{0.6}{2}$$

We know from the previous problem that $\frac{0.6}{2} = 0.3$. Since $\frac{0.006}{0.02} = \frac{0.6}{2}$, we conclude $\frac{0.006}{0.02} = 0.3$.

A shortcut to multiplying by 100 in both the divisor and the dividend is to move the decimal point two places to the right.

So $0.02\overline{)0.006}$ (quotient 0.3) is equivalent to $2\overline{)0.6}$ (quotient 0.3)

As in any division problem, we can check our answer by confirming that the product of the quotient and the *original divisor* equals the *original dividend.*

Division Problem	**Check**
$0.6 \div 2 = 0.3$	$\begin{array}{r} 0.3 \\ \underline{\times 2} \\ 0.6 \end{array}$
$0.006 \div 0.02 = 0.3$	$\begin{array}{r} 0.3 \\ \underline{\times 0.02} \\ 0.006 \end{array}$

These examples suggest the following rule:

To Divide Decimals

- Move the decimal point in the divisor to the right end of the number.
- Move the decimal point in the dividend the same number of places to the right as in the divisor.
- Insert a decimal point in the quotient directly above the decimal point in the dividend.
- Divide the new dividend by the new divisor, inserting zeros at the right end of the dividend as necessary.

EXAMPLE 5

What is 0.035 divided by 0.25?

Solution **Move the decimal point to the right end, making the divisor a whole number.**

$$0.25\overline{)0.035}$$

Move the decimal point in the dividend the same number of places.

Now, we divide 3.5 by 25, which gives us 0.14.

$$\begin{array}{r} 0.14 \\ 25\overline{)3.50} \\ \underline{2\,5} \\ 1\,00 \\ \underline{1\,00} \end{array}$$

Check We see that the product of the quotient and the original divisor is equal to the original dividend.

$$\begin{array}{r} 0.14 \\ \underline{\times 0.25} \\ 070 \\ \underline{028} \\ 0.0350 \end{array} = 0.035$$

PRACTICE 5

Divide and check: $2.706 \div 0.15$

EXAMPLE 6

Find the quotient: $6 \div 0.0012$. Check the answer.

Solution

The decimal point is moved four places to the right.

$$0.0012\overline{)6.0000}$$

To move the decimal point four places to the right, we must add four 0's as placeholders.

$$\begin{array}{r} 5{,}000 \\ 12\overline{)60{,}000} \end{array}$$

Check

$$\begin{array}{r} 5{,}000 \\ \underline{\times\ 0.0012} \\ 6.0000 \end{array} = 6$$

The answer checks.

PRACTICE 6

Divide $\dfrac{8.2}{0.004}$ and then check.

EXAMPLE 7

Divide and round to the nearest hundredth: $0.7\overline{)40.2}$
Then, check.

Solution $0.7\overline{)40.2}$

$$\begin{array}{r} 57.428 \approx 57.43 \text{ to the nearest hundredth} \\ 7\overline{)402.000} \\ \underline{35} \\ 52 \\ \underline{49} \\ 3\,0 \\ \underline{2\,8} \\ 20 \\ \underline{14} \\ 60 \\ \underline{56} \\ 4 \end{array}$$

Check

$$\begin{array}{r} 57.43 \\ \underline{\times\ 0.7} \\ 40.201 \approx 40.2 \end{array}$$

Because we rounded our answer, the check gives us a product only approximately equal to the original dividend.

PRACTICE 7

Find the quotient of 8.07 and 0.11, rounded to the nearest tenth.

EXAMPLE 8

Compute and check: $8.319 \div 1{,}000$

Solution

$$\begin{array}{r} 0.008319 \\ 1{,}000\overline{)8.319000} \\ \underline{8\,000} \\ 3190 \\ \underline{3000} \\ 1900 \\ \underline{1000} \\ 9000 \\ \underline{9000} \end{array}$$

Check

$0.008319 \times 1{,}000 = 0008.319 = 8.319$

Note that the divisor (1,000) is a power of 10 ending in three zeros, and that the quotient is identical to the dividend except that the decimal point is moved to the left three places.

$$\frac{8.319}{1{,}000} = 0.008319$$

PRACTICE 8

Divide: $100\overline{)3.41}$

Example 8 suggests the following shortcut.

TIP To divide a decimal by a power of 10, move the decimal point to the left the same number of places as the power of 10 has zeros.

Can you explain the difference between the shortcuts for multiplying and for dividing by a power of 10?

EXAMPLE 9

Compute: $\frac{7.2}{100}$

Solution Since we are dividing by 100, a power of 10 with two zeros, we can find this quotient simply by moving the decimal point in 7.2 to the left two places.

$$\frac{7.2}{100} = .072, \quad \text{or} \quad 0.072$$

So the quotient is 0.072.

PRACTICE 9

Calculate: $0.86 \div 1{,}000$

Now, let's try using these skills in some applications.

EXAMPLE 10

The following table gives the area of each of the world's three largest oceans.

Ocean	Area (in millions of square kilometers)
Pacific Ocean	155.6
Atlantic Ocean	76.8
Indian Ocean	68.6

(*Source:* cia.gov)

The area of the Pacific Ocean is how many times as great as the area of the Atlantic Ocean, rounded to the nearest tenth?

Solution The area of the Pacific Ocean is 155.6 and that of the Atlantic Ocean is 76.8 (both in millions of square kilometers). To find how many times as great 155.6 is when compared to 76.8, we find the quotient of these numbers.

$$76.8\overline{)155.6} = 768\overline{)1556}$$

$$\begin{array}{r} 2.02 \\ 768\overline{)1556.00} \\ \underline{1536} \\ 20\;0 \\ \underline{0} \\ 2000 \\ \underline{1536} \\ 464 \end{array}$$

Rounding to the nearest tenth, we find that the area of the Pacific Ocean is 2.0 times that of the Atlantic Ocean.

PRACTICE 10

The table gives the amount of selected foods consumed per capita in the United States in a recent year.

Food	Annual Per Capita Consumption (in pounds)
Red meat	112.0
Poultry	72.7
Fish and shellfish	16.5

The amount of red meat consumed was how many times as great as the amount of poultry, rounded to the nearest tenth? (*Source:* U.S. Department of Agriculture)

Estimating Quotients

As we have shown, one way to check the quotient of two decimals is by multiplying. Another way is by estimating.

To check a decimal quotient by estimating, we can round each decimal to one nonzero digit and then mentally divide the rounded numbers. But we must be careful to position the decimal point correctly in our estimate.

Mental estimation is also a useful skill for approximating a quotient.

EXAMPLE 11

Divide and check by estimating: $3.36 \div 0.021$

Solution $0.021\overline{)3.360}$

We compute the exact answer.

$$\begin{array}{r} 160 \\ 21\overline{)3{,}360} \\ \underline{2\,1} \\ 1\,26 \\ \underline{1\,26} \\ 00 \\ \underline{00} \end{array}$$

So 160 is our quotient.

Now, let's check by estimating. Because $3.36 \approx 3$ and $0.021 \approx 0.02$, $3.36 \div 0.021 \approx 3 \div 0.02$. We mentally divide to get the estimate.

$$\begin{array}{r} 150 \\ 0.02\overline{)3.00} \end{array}$$

Our estimate 150 is reasonably close to our exact answer, 160.

PRACTICE 11

Compute and check by estimating: $8.229 \div 0.39$

EXAMPLE 12

Calculate and check: $(9.13) \div (0.2) + (4.6)^2$

Solution Following the order of operations rule, we begin by finding the square and then the quotient.

$$(4.6)^2 = 21.16$$
$$(9.13) \div (0.2) = 45.65$$

Then we add these two results.

$$21.16 + 45.65 = 66.81$$

So $(9.13) \div (0.2) + (4.6)^2 = 66.81$.

Now, let's check this answer by estimating.

$$(9.13) \div (0.2) + (4.6)^2$$
$$\underbrace{9 \div 0.2}_{45} + 25 \approx 45 + 25 \approx 70$$

The estimate 70 is close to our answer, 66.81.

PRACTICE 12

Compute and check: $13.07 + (8.4 \div 0.5)^2$

EXAMPLE 13

The water in a filled aquarium weighs 638.25 pounds. If 1 cubic foot of water weighs 62.5 pounds, estimate how many cubic feet of water there are in the aquarium.

Solution We know that the water in the aquarium weighs 638.25 pounds. Since 1 cubic foot of water weighs 62.5 pounds, we can estimate the number of cubic feet of water in the aquarium by computing the quotient $638.25 \div 62.5$, which is approximately $600 \div 60$, or 10. So a reasonable estimate for the amount of water in the aquarium is 10 cubic feet.

PRACTICE 13

The following graph shows the number of farms, in a recent year, in five states.

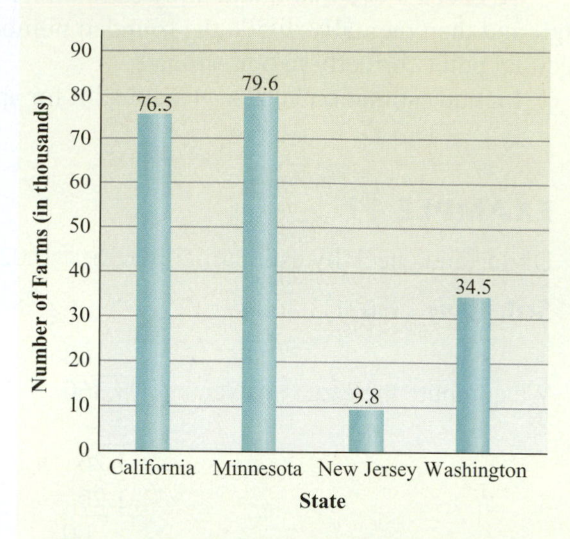

The number of farms in Minnesota is about how many times as great as the number in New Jersey? (*Source:* U.S. Department of Agriculture)

Dividing on a Calculator

When dividing decimals on a calculator, be careful to enter the dividend first and then the divisor. Note that when the dividend is larger than the divisor, the quotient is greater than 1 and when the dividend is smaller than the divisor, the quotient is less than 1.

EXAMPLE 14

Calculate $8.6 \div 1.6$ and round to the nearest tenth.

Solution

Press	Display
8.6 ÷ 1.6 ENTER	8.6/1.6 5.375

The answer, when rounded to the nearest tenth, is 5.4. As expected, the answer is greater than 1, because $8.6 > 1.6$.

PRACTICE 14

Compute the quotient $8.6\overline{)1.6}$ and round to the nearest tenth.

EXAMPLE 15

Divide $0.3\overline{)0.07}$, rounding to the nearest hundredth.

Solution

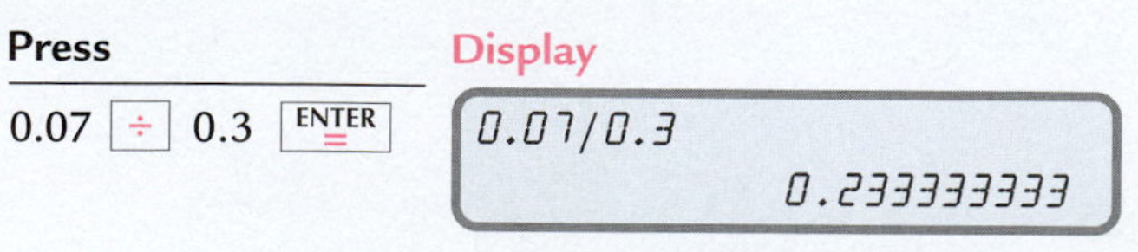

The answer, when rounded to the nearest hundredth, is 0.23. As expected, the answer is less than 1, because $0.07 < 0.3$.

PRACTICE 15

Find the quotient, rounding to the nearest hundredth: $0.3 \div 0.07$

3.4 Exercises

FOR EXTRA HELP MyMathLab Math XL PRACTICE WATCH READ REVIEW

Mathematically Speaking

Fill in each blank with the most appropriate term or phrase from the given list.

quotient	three	divisor	decimal
dividend	right	terminating	four
fraction	product	left	repeating

1. To change a fraction to the equivalent ________, divide the numerator of the fraction by its denominator.
2. An example of a(n) ________ decimal is 0.3333 . . . .
3. When dividing decimals, move the decimal point in the divisor to the ________ end.
4. To divide a decimal by 1,000, move the decimal point ________ places to the left.
5. To estimate the ________ of 0.813 and 0.187, divide 0.8 by 0.2.
6. When dividing a decimal by a whole number, the decimal point in the quotient is placed above the decimal point in the ________.

A *Change to the equivalent decimal. Then, check.*

7. $\frac{1}{2}$ **8.** $\frac{3}{5}$ **9.** $\frac{3}{8}$ **10.** $\frac{1}{8}$

11. $\frac{37}{10}$ **12.** $\frac{57}{10}$ **13.** $1\frac{5}{8}$ **14.** $2\frac{7}{8}$

15. $6\frac{1}{5}$ **16.** $8\frac{2}{5}$ **17.** $21\frac{3}{100}$ **18.** $60\frac{17}{100}$

Change to the equivalent decimal. Round to the nearest hundredth.

19. $\frac{2}{3}$ **20.** $\frac{5}{6}$ **21.** $\frac{7}{9}$ **22.** $\frac{1}{3}$

23. $3\frac{1}{9}$ **24.** $2\frac{4}{7}$ **25.** $5\frac{1}{16}$ **26.** $10\frac{11}{32}$

B *Divide and check.*

27. $4\overline{)17}$ **28.** $2\overline{)35}$ **29.** $5\overline{)21}$ **30.** $6\overline{)33}$

31. $8\overline{)11}$ **32.** $6\overline{)9}$ **33.** $18\overline{)153}$ **34.** $14\overline{)217}$

Divide. Express any remainder as a decimal rounded to the nearest thousandth.

35. $7\overline{)23}$ **36.** $9\overline{)41}$ **37.** $11\overline{)3}$ **38.** $13\overline{)2}$

39. $7\overline{)46}$ **40.** $6\overline{)82}$ **41.** $13\overline{)911}$ **42.** $12\overline{)208}$

Insert the decimal point in the appropriate place.

43. $\begin{array}{r} 5\,8\,8\,2 \\ 0.7\overline{)4\,1.1\,7\,4} \end{array}$ **44.** $\begin{array}{r} 5\,7 \\ 3\overline{)0.0\,1\,7\,1} \end{array}$ **45.** $\begin{array}{r} 6\,6\,3 \\ 0.5\,8\overline{)0.0\,3\,8\,4\,5\,4} \end{array}$ **46.** $\begin{array}{r} 6\,9 \\ 3.9\overline{)2\,6.9\,1} \end{array}$

Divide. Check, either by multiplying or by estimating.

47. $8\overline{)23.1}$ **48.** $8\overline{)24.6}$ **49.** $7\overline{)2.002}$ **50.** $6\overline{)4.002}$

51. $\frac{17.2}{4}$ 52. $\frac{18.5}{5}$ 53. $\frac{0.003}{2}$ 54. $\frac{0.009}{5}$

55. $8.65 \div 5$ 56. $7.74 \div 6$ 57. $11.5 \div 4$ 58. $16.5 \div 4$

59. $0.2\overline{)0.8}$ 60. $0.3\overline{)0.6}$ 61. $0.05\overline{)3.52}$ 62. $0.04\overline{)1.92}$

63. $\frac{47}{0.5}$ 64. $\frac{86}{0.2}$ 65. $\frac{5}{0.4}$ 66. $\frac{9}{0.6}$

67. $0.03 \div 0.1$ 68. $0.04 \div 0.2$ 69. $0.38 \div 1.9$ 70. $0.75 \div 2.5$

71. $95.2 \div 100$ 72. $81.6 \div 100$ 73. $0.082 \div 100$ 74. $0.093 \div 100$

Divide, rounding to the nearest hundredth. Check, either by multiplying or by estimating.

75. $0.8\overline{)307.1}$ 76. $0.6\overline{)401.8}$ 77. $0.9\overline{)0.0057}$ 78. $0.2\overline{)0.0063}$

79. $\frac{3.69}{0.4}$ 80. $\frac{3.98}{0.8}$ 81. $\frac{87}{0.009}$ 82. $\frac{23}{0.006}$

83. $41 \div 0.021$ 84. $91 \div 0.071$ 85. $35.77 \div 0.11$ 86. $29.11 \div 0.17$

87. $49.071 \div 0.728$ 88. $18.032 \div 0.796$ 89. $3 \div 0.0721$ 90. $4 \div 0.0826$

Perform the indicated operations.

91. $\frac{3.06}{4} + 2$ 92. $\frac{2.04}{3} + 1$ 93. $\frac{18.27 - 8.4}{0.3}$ 94. $\frac{26.77 - 10.1}{0.4}$

95. $\frac{13.05}{7.27 - 7.02}$ 96. $\frac{14.07}{8.41 - 8.01}$ 97. $\frac{8.1 \times 0.2}{0.4}$ 98. $\frac{4.7 \times 5.6}{0.8}$

99. $(82.9 - 3.6) \div (0.21 - 0.01)$ 100. $(3.21 - 0.207) \div (2.08 - 2.072)$

101. $8.73 \div 0.2 + (2.5)^2$ 102. $4.86 \div 0.2 + (3.1)^2$

Complete each table.

103.

Input	Output
1	**1** ÷ 5 − 0.2 =
2	**2** ÷ 5 − 0.2 =
3	**3** ÷ 5 − 0.2 =
4	**4** ÷ 5 − 0.2 =

104.

Input	Output
1	**1** ÷ 4 + 0.4 =
2	**2** ÷ 4 + 0.4 =
3	**3** ÷ 4 + 0.4 =
4	**4** ÷ 4 + 0.4 =

Each of the following quotients is rounded to the nearest hundredth. In each group of three quotients, one is wrong. Use estimation to identify which quotient is incorrect.

105. **a.** $5.7 \div 89 \approx 0.06$ **b.** $0.77 \div 0.0019 \approx 405.26$ **c.** $31.5 \div 0.61 \approx 516.39$

106. **a.** $\frac{9.83}{4.88} \approx 0.20$ **b.** $\frac{2.771}{0.452} \approx 6.13$ **c.** $\frac{389.224}{1.79} \approx 217.44$

107. **a.** $61.27 \div 0.057 \approx 1{,}074.91$ **b.** $0.614 \div 2.883 \approx 2.13$ **c.** $0.0035 \div 0.00481 \approx 0.73$

108. **a.** $\$365 \div \$4.89 \approx 7.46$ **b.** $\$17{,}358.27 \div \$365 \approx 47.56$ **c.** $\$3{,}000 \div \$2.54 \approx 1{,}181.10$

Mixed Practice

Solve.

109. Express $\frac{4}{5}$ as a decimal.

110. Divide $1.6\overline{)8.5}$ and round to the nearest tenth.

111. Change $1\frac{1}{6}$ to a decimal, rounded to the nearest hundredth.

112. Simplify: $81.5 - \frac{32}{0.4}$

113. What is 0.063 divided by 0.14?

114. Find the quotient: $9 \div 0.0072$

Applications

C *Solve and check.*

115. A stalactite is an icicle-shaped mineral deposit that hangs from the roof of a cave. If it took a thousand years for a stalactite to grow to a length of 3.7 inches, how much did it grow per year?

116. In a strong storm, the damage to 100 houses was estimated at $12.7 million. What was the average damage per house?

117. The women's softball team won 21 games and lost 14. The men's softball team won 22 games and lost 18.

a. The women's team won what fraction of the games that it played, expressed as a decimal?

b. The men's team won what fraction of its games, expressed as a decimal?

c. Which team has a better record? Explain.

118. Yesterday, 0.08 inches of rain fell. Today, $\frac{1}{4}$ inch of rain fell.

a. How much rain fell today, expressed as a decimal?

b. Which day did more rain fall? Explain.

119. The table shown gives the highest gasoline mileage for three small SUVs.

a. For each SUV, compute how many miles it gets per gallon, rounded to the nearest whole number.

b. Which SUV gives the highest mileage?

SUV	Distance Driven (in miles)	Gasoline Used (in gallons)	Gasoline Mileage (in miles per gallon)
Honda CR-V	40.5	1.9	
Ford Escape Hybrid	62.4	2.4	
GMC Terrain	42.6	2.4	

(Source: Consumer Reports)

120. The table shown gives the total number of games played and assists made by three basketball players over the same period of time.

a. Compute the average number of assists per game for each player, expressed as a decimal rounded to the nearest tenth.

b. Decide which player had the highest average.

Player	Number of Games	Number of Assists	Average
Steve Nash	236	2,507	
Chris Paul	203	2,266	
Deron Williams	226	2,385	

(*Source:* nba.com/statistics)

121. On the stock market, shares of Citigroup, Inc. common stock were traded at $3.63 per share. How many shares could have been bought for $7,260? (*Source:* finance.yahoo.com)

122. A light microscope can distinguish two points separated by 0.0005 millimeter, whereas an electron microscope can distinguish two points separated by 0.0000005 millimeter. The electron microscope is how many times as powerful as the light microscope?

123. Typically, the heaviest organ in the body is the skin, weighing about 9 pounds. By contrast, the heart weighs approximately 0.7 pound. About how many times the weight of the heart is that of the skin? (*Source: World of Scientific Discovery*)

124. At a community college, each student enrolled pays a $19.50 student fee per semester. In a given semester, if the college collected $39,000 in student fees, how many students were enrolled?

125. A shopper buys four organic chickens. The chickens weigh 3.2 pounds, 3.5 pounds, 2.9 pounds, and 3.6 pounds. How much less than the average weight of the four chickens was the weight of the lightest one?

126. A dieter joins a weight-loss club. Over a 5-month period, she loses 8 pounds, 7.8 pounds, 4 pounds, 1.5 pounds, and 0.8 pound. What was her average monthly weight loss, to the nearest tenth of a pound?

127. Babe Ruth got 2,873 hits in 8,399 times at bat, resulting in a batting average of $\frac{2,873}{8,399}$, or approximately .342. Another great player, Ty Cobb, got 4,189 hits out of 11,434 times at bat. What was his batting average, expressed as a decimal rounded to the nearest thousandth? (Note that batting averages don't have a zero to the left of the decimal point because they can never be greater than 1.) (*Source:* baseball-reference.com)

128. On January 23, 1960, the Trieste became the only manned deep-diving research vessel ever to reach the bottom of the Mariana Trench. This trench is the deepest part of any ocean on Earth, with a depth of 10,911 meters. At a descent rate of 3,290.4 meters per hour, approximately how long did it take for the Trieste to reach the bottom? Round the answer to the nearest tenth of an hour. (*Source:* absoluteastronomy.com)

• Check your answers on page A-5.

MINDStretchers

Patterns

1. In the *repeating decimal* 0.142847142847142847 . . . , identify the 994th digit to the right of the decimal point.

Groupwork

2. In the following magic square, 3.375 is the *product* of the numbers in every row, column, and diagonal. Working with a partner, fill in the missing numbers.

	0.25	3
		0.5

Writing

3. a. $0.5 \div 0.8 = ?$

b. $0.8 \div 0.5 = ?$

c. Find the product of your answers in parts (a) and (b). Explain how you could have predicted this product.

Key Concepts and Skills

CONCEPT SKILL

Concept/Skill	Description	Example
[3.1] **Decimal**	A number written with two parts: a whole number, which precedes the decimal point, and a fractional part, which follows the decimal point.	Whole-number part → 3.721 ← Fractional part; Decimal point (points to the decimal point in 3.721)
[3.1] **Decimal place**	A place to the right of the decimal point.	Decimal places: 8.**035** (0 = Tenths, 3 = Hundredths, 5 = Thousandths)
[3.1] **To change a decimal to the equivalent fraction or mixed number**	• Copy the nonzero whole-number part of the decimal and drop the decimal point. • Place the fractional part of the decimal in the numerator of the equivalent fraction. • Make the denominator of the equivalent fraction 1 followed by as many zeros as the decimal has decimal places. • Simplify the resulting fraction, if possible.	The decimal 3.25 is equivalent to the mixed number $3\frac{25}{100}$ or $3\frac{1}{4}$.
[3.1] **To compare decimals**	• Rewrite the numbers vertically, lining up the decimal points. • Working from left to right, compare the digits that have the same place value. At the first place value where the digits differ, the decimal which has the largest digit with this place value is the largest decimal.	1.073 1.06999 In the ones place and the tenths place, the digits are the same. But in the hundredths place, $7 > 6$, so $1.073 > 1.06999$.
[3.1] **To round a decimal to a given decimal place**	• Underline the digit in the place to which the number is being rounded. • The digit to the right of the underlined digit is called the *critical digit*. Look at the critical digit—if it is 5 or more, add 1 to the underlined digit; if it is less than 5, leave the underlined digit unchanged. • Drop all decimal places to the right of the underlined digit.	23.9<u>3</u>81 ≈ 23.94 (Critical digit: 8)
[3.2] **To add decimals**	• Rewrite the numbers vertically, lining up the decimal points. • Add. • Insert a decimal point in the sum below the other decimal points.	0.035 0.08 C+ 0.00813 0.12313
[3.2] **To subtract decimals**	• Rewrite the numbers vertically, lining up the decimal points. • Subtract, inserting extra zeros in the minuend if necessary for regrouping. • Insert a decimal point in the difference below the other decimal points.	0.90370 C− 0.17052 0.73318

continued

CONCEPT SKILL

Concept/Skill	Description	Example
[3.3] **To multiply decimals**	• Multiply the factors as if they were whole numbers. • Find the total number of decimal places in the factors. • Count that many places from the right end of the product, and insert a decimal point.	21.07 ← Two decimal places × 0.18 ← Two decimal places 3.7926 ← Four decimal places
[3.4] **To change a fraction to the equivalent decimal**	• Divide the denominator of the fraction into the numerator, inserting to its right both a decimal point and enough zeros to get an answer either without a remainder or rounded to a given decimal place. • Place a decimal point in the quotient directly above the decimal point in the dividend.	$\frac{7}{8} = 8\overline{)7.000}$ = 0.875
[3.4] **To divide decimals**	• Move the decimal point in the divisor to the right end of the number. • Move the decimal point in the dividend the same number of places to the right as in the divisor. • Insert a decimal point in the quotient directly above the decimal point in the dividend. • Divide the new dividend by the new divisor, inserting zeros at the right end of the dividend as necessary.	$3.5\overline{)71.05}$ = 20.3 $35\overline{)710.5}$ 70 10 0 10 5 10 5

CHAPTER 3 Review Exercises

Say Why *Fill in each blank.*

1. The expression $7\frac{3}{10}$ ________ (is/is not) considered to be a decimal because ______________________________________.

2. In the decimal 2.781, the 2 ________ (is/is not) in a decimal place because ______________________________________.

3. The number 48.726 rounded to the nearest hundredth ________ (is/is not) 48.72 because ______________________________________.

4. We ________ (can/cannot) add, subtract, or compare decimals by rewriting them vertically and then lining up the decimal points because ______________________________________.

5. The number 9.1313 ________ (is/is not) a repeating decimal because ______________________________________.

6. Without calculating, we know that $0.04 \div 0.23$ ________ (is/is not) less than 1 because ______________________________________.

[3.1] *Name the place that each underlined digit occupies.*

7. $8.3\underline{5}9$ **8.** $13.\underline{0}05$ **9.** $8{,}024.\underline{5}$ **10.** $0.000\underline{3}$

Express each number as a fraction, mixed number, or whole number.

11. 0.35 **12.** 8.2 **13.** 4.007 **14.** 10.000

Write each decimal in words.

15. 0.72 **16.** 5.6

17. 3.0009 **18.** 510.036

Write each decimal in decimal notation.

19. Seven thousandths **20.** Two and one tenth

21. Nine hundredths **22.** Seven and forty-one thousandths

Between each pair of numbers, insert the appropriate sign, <, =, or >, to make a true statement.

23. 0.037 0.04 **24.** 2.031 2.0301 **25.** 5.12 4.71932 **26.** 2 1.8

Rearrange each group of numbers from largest to smallest.

27. 0.72, 0.8, 1.002 **28.** 0.003, 0.00057, 0.004

Round as indicated.

29. 7.31 to the nearest tenth **30.** 0.0387 to the nearest thousandth

31. 4.3868 to two decimal places **32.** $899.09 to the nearest dollar

[3.2] *Perform the indicated operations. Check by estimating.*

33. $8.2 + 3.91$ **34.** $50 + 2.7 + 0.05$ **35.** $8 + $3.25 + $12.88 **36.** 8.4 m + 3.6 m

37. $30.7 - 1.92$

38. $93 - 5.248$

39. $2.5 - (0.72 - 0.054)$

40. $54.17 - (8 - 2.731)$

41. $5.398 + 8.72 + 92.035 + 0.7723 - 3.714 - 5.008$

42. $\$87{,}259.39 + \$2{,}098.35 + \$1{,}387.92 + \$203.14$

[3.3] *Find the product. Check by estimating.*

43. 7.28×0.4

44. $(288)(3.5)$

45. 0.005×0.002

46. $(3.7)^2$

47. $2.71 \cdot 1{,}000$

48. 0.0034×10

49. $8 - (1.5)^2$

50. $3(2.4) + 7(0.9)$

51. $18{,}772.35 \times 0.0836$

52. $(74.862)(5.901)$

[3.4] *Change to the equivalent decimal.*

53. $\frac{5}{8}$

54. $90\frac{1}{5}$

55. $4\frac{1}{16}$

56. $\frac{45}{1{,}000}$

Express each fraction as a decimal. Round to the nearest hundredth.

57. $\frac{1}{6}$

58. $\frac{2}{7}$

59. $8\frac{1}{3}$

60. $11\frac{2}{9}$

Divide and check.

61. $2\overline{)1.3}$

62. $\frac{4.8}{3}$

63. $0.7 \div 4$

64. $\frac{2.77}{10}$

Divide. Round to the nearest tenth.

65. $4.67 \div 0.9$

66. $\frac{2.35}{0.73}$

67. $\frac{7.11}{0.3}$

68. $0.06\overline{)981.5}$

69. $18.74 \div 9.7$

70. $220 \div 0.61$

71. $81.37\overline{)247.062}$

72. $247.062\overline{)81.37}$

Simplify.

73. $\frac{(1.3)^2 - 1.1}{0.5}$

74. $\frac{2.5 - (0.4)^2}{0.02}$

75. $\frac{13.75}{9.6 - 9.2}$

76. $(2.5)(3.5) \div 6.25$

Mixed Applications

Solve.

77. Recently, a champion swimmer swam 50 meters in 25.2 seconds and then swam 100 meters farther in 29.29 seconds. How long did she take to swim the 150 meters?

78. On a certain day, the closing price of one share of Home Depot was \$14.57. If the closing price for the day was \$0.43 higher than the opening price, what was the opening price? (*Source:* finance.yahoo.com)

79. The venom of a certain South American frog is so poisonous that 0.0000004 ounce of the venom can kill a person. How is this decimal read?

80. A carpenter is constructing steps leading to a terrace. The cross-section of the steps is shown below. Find the missing dimension.

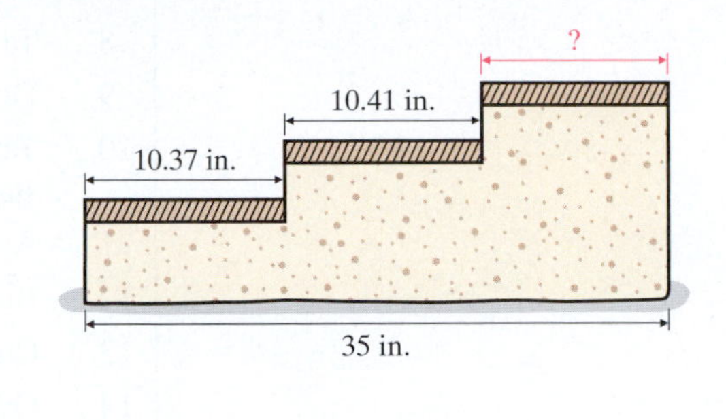

81. Astronomers use the term *astronomical unit* (or AU) for the average distance between Earth and the Sun. The distance 1 AU is about 93,000,000 miles. The average distance from the Sun to Mars is about 140,000,000 miles. Express in astronomical units, rounded to the nearest tenth, the average distance between Mars and the Sun. (*Source*: Jeffrey Bennett et al., *The Cosmic Perspective*)

82. In the United States Congress, there are 100 senators and 435 representatives. How many times as many representatives as senators are there?

83. A supermarket sells a 4-pound package of ground meat for $5.20 and a 5-pound package of ground meat for $6.20. What is the difference between the two prices per pound?

84. A homeowner is making plans to build a patio in her backyard and would like to save money by doing the work herself. The dimensions of the proposed patio are 9.67 yards by 5.33 yards. The labor costs would be about $5 per square yard. Estimate how much money the homeowner would save in labor by building the patio herself.

85. In a chemistry lab, a student weighs a compound three times, getting 7.15 grams, 7.18 grams, and 7.23 grams. What is the average of these weights, to the nearest hundredth of a gram?

86. A team of geologists scaled a mountain. At the base of the mountain, the temperature had been 11°C. The temperature fell 0.75 degrees for every 100 meters the team climbed. After they climbed 1,000 meters, what was the temperature?

87. The following form was adapted from the *U.S. Individual Income Tax Return*. Find the total income in line 22.

7	Wages, salaries, tips, etc.	7	28,774.71
8	Taxable interest income	8	
9	Dividend income	9	232.55
10	Taxable refunds, credits, or offsets of state and local income taxes	10	349.77
11	Alimony received	11	
12	Business income or (loss)	12	
13	Capital gain or (loss)	13	511.74
14	Other gains or (losses)	14	5,052.71
15	IRA distributions: taxable amount	15	
16	Pensions and annuities: taxable amount	16	
17	Rents, royalties, partnerships, estates, trusts, etc.	17	1,240.97
18	Farm income or (loss)	18	
19	Unemployment compensation	19	
20	Social Security benefits	20	
21	Other income	21	
22	Add the amounts shown in the far right column	22	

88. The following table shows the quarterly revenues, in billions of dollars, for Google and Yahoo! for a recent year.

Ending	Google	Yahoo!
June 30	5.523	1.573
September 30	5.945	1.575
December 31	6.674	1.732
March 31	6.775	1.597

(*Source:* finance.yahoo.com)

How much more were Google's earnings than Yahoo!'s for the year?

• Check your answers on page A-5.

CHAPTER 3 Posttest

FOR EXTRA HELP

CHAPTER Test Prep VIDEOS

The Chapter Test Prep Videos with test solutions are available on DVD, in MyMathLab, and on YouTube (search "AkstBasicMath" and click on "Channels").

To see if you have mastered the topics in this chapter, take this test.

1. In the number 0.79623, which digit occupies the thousandths place?

2. Write in words: 5.102

3. Round 320.1548 to the nearest hundredth.

4. Which is smallest, 0.04, 0.0009, or 0.00028?

5. Express 3.04 as a mixed number.

6. Write as a decimal: four thousandths

Perform the indicated operations.

7. $2.3 + 0.704 + 1.35$

8. $\$5.27 + \$9 - \$8.61$

9. 2.09×10

10. 5.2×1.1

11. $(0.1)^3$

12. $\frac{3.52}{2} + \frac{4.8}{3}$

13. $2.9 \div 1{,}000$

14. $\frac{9.81}{0.3}$

Express as a decimal.

15. $\frac{3}{8}$

16. $4\frac{1}{6}$, rounded to the nearest hundredth

Solve.

17. The element hydrogen is so light that 1 cubic foot of hydrogen weighs only 0.005611 pound. Round this weight to the nearest hundredth of a pound.

18. Historically, a mile was the distance that a Roman soldier covered when he took 2,000 steps. If a mile is 5,280 feet, how many feet, to the nearest tenth of a foot, was a Roman's step?

19. The Triple Crown consists of three horse races—the Kentucky Derby (1.25 miles), the Preakness Stakes (1.1875 miles), and the Belmont Stakes (1.5 miles). Which race is longest? (*Source: World Almanac 2010*)

20. A part of the real estate tax in Berkeley, California provides funds for the maintenance and servicing of traffic signals and other public lighting. This amount of tax (in dollars) on a house is 0.0108 times the area of the house (in square feet). What is this tax amount on a house that is 3,000 square feet in area? (*Source:* ci.berkeley.ca.us)

• Check your answers on page A-5.

CHAPTER 4 PRETEST

To see if you have already mastered the topics in this chapter, take this test.

Write each algebraic expression in words.

1. $t - 4$

2. $\frac{y}{3}$

Translate each phrase to an algebraic expression.

3. 8 more than m

4. Twice n

Evaluate each algebraic expression.

5. $\frac{x}{4}$, for $x = 16$

6. $5 - y$, for $y = 3\frac{1}{2}$

Translate each sentence to an equation.

7. The sum of x and 3 equals 5.

8. The product of 4 and y is 12.

Solve and check.

9. $x + 4 = 10$

10. $t - 1 = 9$

11. $2n = 26$

12. $\frac{a}{4} = 3$

13. $8 = m + 1.9$

14. $15 = 0.5n$

15. $m - 3\frac{1}{2} = 10$

16. $\frac{n}{10} = 1.5$

Write an equation. Solve and check.

17. The planet Jupiter has 36 more moons than the planet Uranus. If Jupiter has 63 moons, how many does Uranus have? (*Source:* NASA)

18. Tickets for all movies shown before 5:00 P.M. at a local movie theater qualify for the bargain matinee price, which is $2.75 less than the regular ticket price. If the bargain price is $6.75, what is the regular price?

19. In Michigan, about two-fifths of the area is covered with water. This portion of the state represents about 39,900 square miles. What is the area of Michigan? (*Source: The New York Times Almanac, 2010*)

20. An 8-ounce cup of regular tea has about 40 milligrams of caffeine, which is 10 times the amount of caffeine in a cup of decaffeinated tea. How much caffeine is in a cup of decaffeinated tea?

• Check your answers on page A-6.

4.1 Introduction to Basic Algebra

OBJECTIVES

- **A** To translate phrases to algebraic expressions and vice versa
- **B** To evaluate an algebraic expression for a given value of the variable
- **C** To solve applied problems involving algebraic expressions

What Algebra Is and Why It Is Important

In this chapter, we discuss some of the basic ideas in algebra. These ideas will be important throughout the rest of this book.

In algebra, we use letters to represent unknown numbers. The expression $2 + 3$ is arithmetic, whereas the expression $x + y$ is algebraic, since x and y represent numbers whose values are not known. With *algebraic expressions*, such as $x + y$, we can make general statements about numbers and also find the value of unknown numbers.

We can think of algebra as a *language*: The idea of translating ordinary words to algebraic notation and vice versa is the key. Often, just writing a problem algebraically makes the problem much easier to solve. We present ample proof of this point repeatedly in the chapters that follow.

We begin our discussion of algebra by focusing on what algebraic expressions mean and how to translate and evaluate them.

Translating Phrases to Algebraic Expressions and Vice Versa

To apply mathematics to a real-world situation, we often need to be able to express that situation algebraically. Consider the following example of this kind of translation.

Suppose that you are enrolled in a college course that meets 50 minutes a day for 3 days a week. The course therefore meets for $50 \cdot 3$, or 150 minutes, a week.

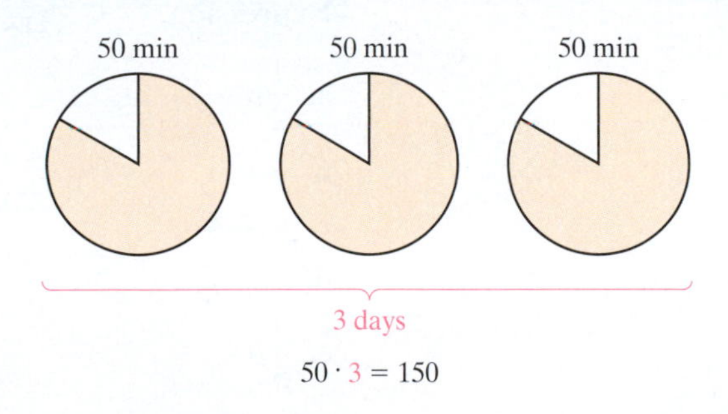

Now, suppose that in a semester the 50-minute class meets d days but that we do not know what number the letter d represents. How many minutes per semester does the class meet? Do you see that we can express the answer as $50d$, that is, 50 times d days?

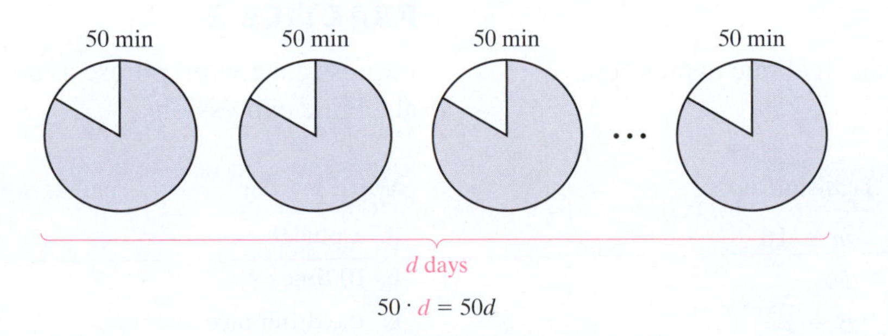

In algebra, a *variable* is a letter, or other symbol, used to represent an unknown number. In the algebraic expression $50d$, for instance, d is a variable and 50 is a *constant.* Note that in writing an algebraic expression, we usually omit any multiplication symbol: $50d$ means $50 \cdot d$.

DEFINITIONS

A **variable** is a letter that represents an unknown number.

A **constant** is a known number.

An **algebraic expression** is an expression that combines variables, constants, and arithmetic operations.

There are many translations of an algebraic expression to words, as the following table indicates.

$x + 4$ translates to	$n - 3$ translates to	$\frac{3}{4} \cdot y$ or $\frac{3}{4}y$ translates to	$z \div 5$ or $\frac{z}{5}$ translates to
• x plus 4 • x increased by 4 • the sum of x and 4 • 4 more than x • 4 added to x	• n minus 3 • n decreased by 3 • the difference between n and 3 • 3 less than n • 3 subtracted from n	• $\frac{3}{4}$ times y • the product of $\frac{3}{4}$ and y • $\frac{3}{4}$ of y	• z divided by 5 • the quotient of z and 5 • z over 5

EXAMPLE 1

Translate each algebraic expression in the table to words.

Solution

Algebraic Expression	Translation
a. $\frac{p}{3}$	p divided by 3
b. $x - 4$	4 less than x
c. $5f$	5 times f
d. $2 + y$	the sum of 2 and y
e. $\frac{2}{3}a$	$\frac{2}{3}$ of a

PRACTICE 1

Translate each algebraic expression to words.

Algebraic Expression	Translation
a. $\frac{1}{2}p$	
b. $5 - x$	
c. $y \div 4$	
d. $n + 3$	
e. $\frac{3}{5}b$	

EXAMPLE 2

Translate each word phrase in the table to an algebraic expression.

Solution

Word Phrase	Translation
a. 16 more than m	$m + 16$
b. the product of 5 and b	$5b$
c. the quotient of 6 and z	$6 \div z$
d. a decreased by 4	$a - 4$
e. $\frac{3}{8}$ of t	$\frac{3}{8}t$

PRACTICE 2

Express each word phrase as an algebraic expression.

Word Phrase	Translation
a. x plus 9	
b. 10 times y	
c. the difference between n and 7	
d. p divided by 5	
e. $\frac{2}{5}$ of v	

4.1 Exercises

FOR EXTRA HELP MyMathLab MathXL PRACTICE WATCH READ REVIEW

Mathematically Speaking

Fill in each blank with the most appropriate term or phrase from the given list.

arithmetic	constant	evaluate
translate	variable	algebraic

1. A(n) __________ is a letter that represents an unknown number.
2. A(n) __________ is a known number.
3. A(n) __________ expression combines variables, constants, and arithmetic operations.
4. To __________ an algebraic expression, replace each variable with the given number, and carry out the computation.

A *Translate each algebraic expression to two different word phrases.*

5. $t + 9$
6. $8 + r$
7. $c - 12$
8. $x - 5$
9. $c \div 3$
10. $\frac{z}{7}$
11. $10s$
12. $11t$
13. $y - 10$
14. $w - 1$
15. $7a$
16. $4x$
17. $x \div 6$
18. $\frac{y}{5}$
19. $x - \frac{1}{2}$
20. $x - \frac{1}{3}$
21. $\frac{1}{4}w$
22. $\frac{4}{5}y$
23. $2 - x$
24. $8 - y$
25. $1 + x$
26. $n + 7$
27. $3p$
28. $2x$
29. $n - 1.1$
30. $x - 6.5$
31. $y \div 0.9$
32. $\frac{n}{2.4}$

Translate each word phrase to an algebraic expression.

33. x plus 10
34. d plus 12
35. 1 less than n
36. 9 less than b
37. the sum of y and 5
38. the sum of x and 11

As we have seen, any letter or symbol can be used to represent a variable. For example, *five less than a number* can be translated to $n - 5$, where n represents the number.

Let's consider the following example.

EXAMPLE 3

Express each phrase as an algebraic expression.

Solution

Word Phrase	Translation
a. 2 less than a number	$n - 2$, where n represents the number
b. an amount divided by 10	$\frac{a}{10}$, where a represents the amount
c. $\frac{3}{8}$ of a price	$\frac{3}{8}p$, where p represents the price

PRACTICE 3

Translate each word phrase to an algebraic expression.

Word Phrase	Translation
a. a quantity increased by 12	
b. the quotient of 9 and an account balance	
c. a cost multiplied by $\frac{2}{7}$	

Now, let's look at word problems that involve translations.

EXAMPLE 4

Suppose that p partners share equally in the profits of a business. What is each partner's share if the profit was $2,000?

Solution Each partner should get the quotient of 2,000 and p, which can be written algebraically as $\frac{2{,}000}{p}$ dollars.

PRACTICE 4

Next weekend, a student wants to study for his four classes. If he has h hours to study in all and he wants to devote the same amount of time to each class, how much time will he study per class?

EXAMPLE 5

At registration, n out of 100 classes are closed. How many classes are not closed?

Solution Since n classes are closed, the remainder of the 100 classes are not closed. So we can represent the number of classes that are not closed by the algebraic expression $100 - n$.

PRACTICE 5

Of s shrubs in front of a building, 3 survived the winter. How many shrubs died over the winter?

Evaluating Algebraic Expressions

In this section, we look at how to evaluate algebraic expressions. Let's begin with a simple example.

Suppose that the balance in a savings account is $200. If d dollars is then deposited, the balance will be $(200 + d)$ dollars.

To evaluate the expression $200 + d$ for a particular value of d, we replace d with that number. If $50 is deposited, we replace d by 50:

$$200 + d = 200 + 50 = 250$$

So the new balance will be $250.

The following rule is helpful for evaluating expressions:

To Evaluate an Algebraic Expression

- Substitute the given value for each variable.
- Carry out the computation.

Now, let's consider some more examples.

EXAMPLE 6

Evaluate each algebraic expression.

Solution

Algebraic Expression	Value
a. $n + 8$, if $n = 15$	$15 + 8 = 23$
b. $9 - z$, if $z = 7.89$	$9 - 7.89 = 1.11$
c. $\frac{2}{3}r$, if $r = 18$	$\frac{2}{3} \cdot 18 = 12$
d. $y \div 4$, if $y = 3.6$	$3.6 \div 4 = 0.9$

PRACTICE 6

Find the value of each algebraic expression.

Algebraic Expression	Value
a. $\frac{s}{4}$, if $s = 100$	
b. $0.2y$, if $y = 1.9$	
c. $x - 4.2$, if $x = 9$	
d. $25 + z$, if $z = 1.6$	

The following examples illustrate how to write and evaluate expressions to solve word problems.

EXAMPLE 7

Power consumption for a period of time is measured in watt-hours, where a watt-hour means 1 watt of power for 1 hour. How many watt-hours of energy will a 60-watt bulb consume in h hours? In 3 hours?

Solution The expression that represents the number of watt-hours used in h hours is $60h$. So for $h = 3$, the number of watt-hours is $60 \cdot 3$, or 180. Therefore, 180 watt-hours of energy will be consumed in 3 hours.

PRACTICE 7

When deciding how much money to spend on a new car, a good rule of thumb to follow is to budget about one-fifth of your monthly net income for a car payment. How much should you set aside for the monthly car payment if your net income is n dollars per month? If your net income is $3,750? (*Source:* automotive.com)

EXAMPLE 8

Suppose that there are 180 days in the local school year. How many days was a student present at school if she was absent d days? 9 days?

Solution If d represents the number of days that the student was absent, the expression $180 - d$ represents the number of days that she was present. If she was absent 9 days, we substitute 9 for d in the expression:

$$180 - d = 180 - 9 = 171$$

So the student was present 171 days.

PRACTICE 8

At a coffee shop, a lunch bill was $18.45 plus the tip. What was the total amount of the lunch, including a tip of t dollars? A tip of $3?

39. t divided by 6

40. r divided by 2

41. the product of 10 and y

42. the product of 5 and p

43. the difference between w and 5

44. the difference between n and 5

45. n increased by $\frac{4}{5}$

46. x increased by $\frac{2}{3}$

47. the quotient of z and 3

48. The quotient of n and 10

49. $\frac{2}{7}$ of x

50. $\frac{2}{3}$ of y

51. 6 subtracted from k

52. 8 subtracted from z

53. 12 more than a number

54. 18 more than a number

55. the difference between a number and 5.1

56. the difference between a number and 8.2

B ***Evaluate each algebraic expression.***

57. $y + 7$, if $y = 19$

58. $3 + n$, if $n = 2.9$

59. $7 - x$, if $x = 4.5$

60. $19 - y$, if $y = 6.7$

61. $\frac{3}{4}p$, if $p = 20$

62. $\frac{4}{5}n$, if $n = 30$

63. $x \div 2$, if $x = 2\frac{1}{3}$

64. $\frac{n}{3}$, if $n = 7.5$

65. $p - 7.9$, if $p = 9$

66. $y - 20.1$, if $y = 30$

67. $x \div \frac{5}{6}$, if $x = \frac{1}{6}$

68. $\frac{1}{3}y$, if $y = \frac{1}{2}$

Complete each table.

69.

x	$x + 8$
1	
2	
3	
4	

70.

x	$x + 10$
1	
2	
3	
4	

71.

n	$n - 0.2$
1	
2	
3	
4	

72.

b	$b - 0.4$
1	
2	
3	
4	

73.

x	$\frac{3}{4}x$
4	
8	
12	
16	

74.

n	$\frac{2}{3}n$
3	
6	
9	
12	

75.

z	$\frac{z}{2}$
2	
4	
6	
8	

76.

y	$\frac{y}{5}$
5	
10	
15	
20	

Mixed Practice

Solve.

77. Translate the phrase "7 less than x" to an algebraic expression.

78. Evaluate the algebraic expression $0.5t$, if $t = 8$.

79. Translate the algebraic expression $\frac{n}{2}$ to two different phrases.

80. Evaluate the algebraic expression $\frac{1}{4}y$, if $y = \frac{2}{3}$.

81. Translate the phrase "the product of 3.5 and t" to an algebraic expression.

82. Evaluate the algebraic expression $x + 1$, if $x = 4$.

83. Translate the algebraic expression $x + 6$ to two different phrases.

84. Evaluate the algebraic expression $n - 20$, if $n = 30$.

Applications

C *Solve.*

85. A patient receives m milligrams of medication per dose. Her doctor orders her medication to be decreased by 25 milligrams. How much medication will she then receive per dose?

86. When a borrower takes out a mortgage, each monthly payment has two parts. One part goes toward the principal and the other toward the interest. If the principal payment is \$344.86 and the interest payment is i, write an algebraic expression for the total payment.

87. The top of the Flatiron Building in New York City, so called because it is shaped like a clothing iron, is in a form similar to the triangle pictured below. Write an expression for the sum of the measures of the three angles. (*Source:* flatironbid.org)

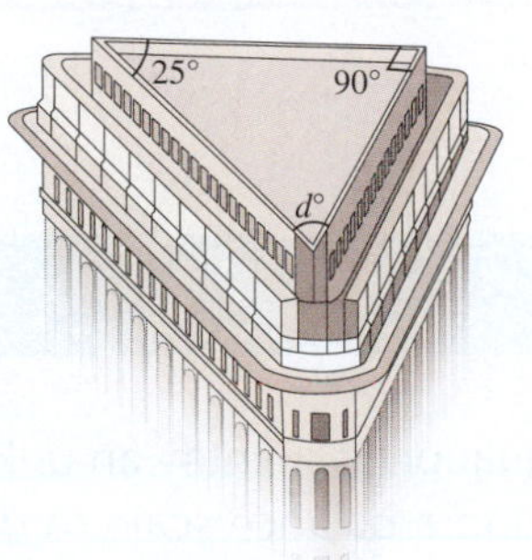

88. Professional land surveyors establish official land, air space, and water boundaries. Below is a sample of a typical lot survey. Write an expression for the sum of the lengths of the sides of the lot. (*Source:* lsrp.com)

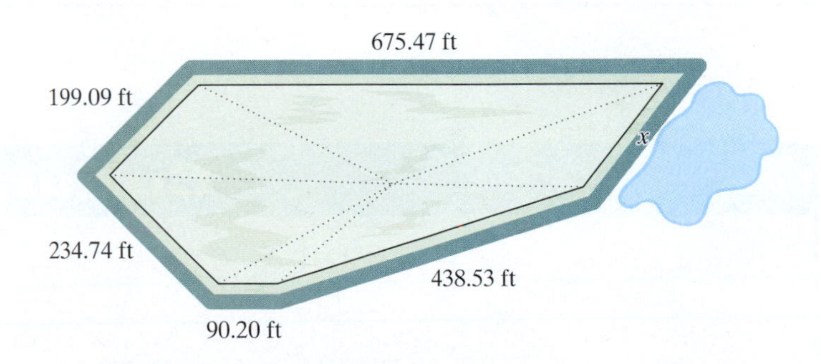

89. If a long-distance trucker drives at a speed of r miles per hour for t hours, she will travel a distance of $r \cdot t$ miles. How far will she travel at a speed of 55 miles per hour in 4 hours?

90. If a basketball player makes b baskets in a attempts, his field goal average is defined to be $\frac{b}{a}$. Find the field goal average of a player who made 12 baskets in 25 attempts.

91. A bank charges customers a fee of \$2.50 for each withdrawal made at its ATMs.

a. Write an expression for the total fee charged to a customer for w of these withdrawals

b. Find the total fee if the customer makes 9 withdrawals.

92. A computer network technician charges \$80 per hour for labor.

a. Write an expression for the cost of h hours of work.

b. Find the cost of a networking job that takes $2\frac{1}{2}$ hours.

• Check your answers on page A-6.

MIND*Stretchers*

Mathematical Reasoning

1. Consider the expression $x + x$.

   a. Why does this expression mean the same as the expression $2x$?

   b. What does the expression $\underbrace{x + x + x + \cdots + x}_{n \text{ times}}$ mean in terms of multiplication?

Groupwork

2. Working with a partner, consider the areas of the following rectangles. For some values of x, the rectangle on the left has a larger area; for other values of x, the rectangle on the right is larger.

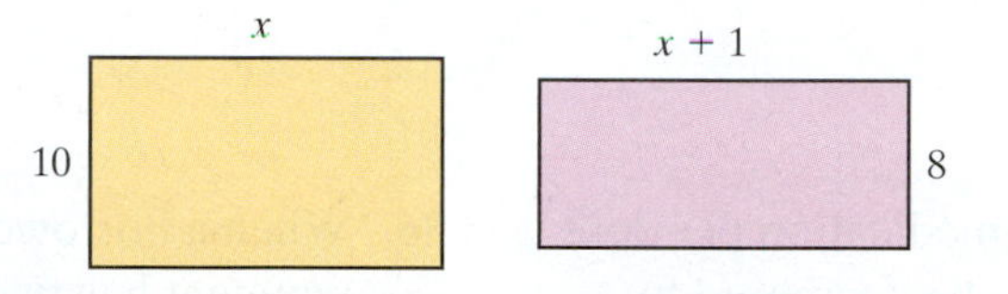

   a. Find a value of x for which the rectangle on the left has the larger area.

   b. Find a value of x for which the area of the rectangle on the right is larger.

Writing

3. Algebra is universal; that is, it is used in all countries of the world regardless of the language spoken. If you know how to speak a language other than English, translate each of the following algebraic expressions to that language.

   a. $7x$ **b.** $x - 2$ **c.** $3 + x$ **d.** $\frac{x}{3}$

Cultural Note

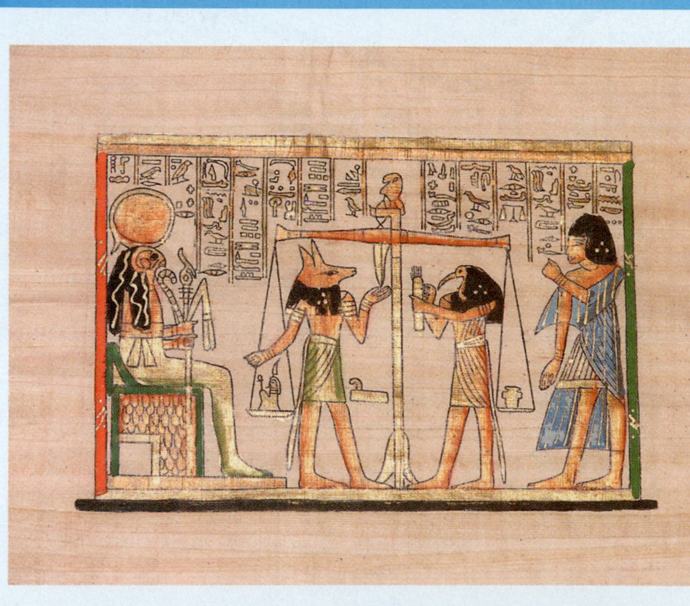

Solving an equation to identify an unknown number is similar to using a balance scale to determine an unknown weight. Egyptians 3,400 years ago used balance scales to weigh objects such as gold rings.

The balance scale is an ancient measuring device. These scales were used by Sumerians for weighing precious metals and gems at least 9,000 years ago.

Source: O. Dilke, *Mathematics and Measurement* (Berkeley: University of California Press/British Museum, 1987).

4.2 Solving Addition and Subtraction Equations

OBJECTIVES

- **A** To translate sentences involving addition or subtraction to equations
- **B** To solve addition or subtraction equations
- **C** To solve applied problems involving addition or subtraction equations

What an Equation Is

An equation contains two expressions separated by an equal sign.

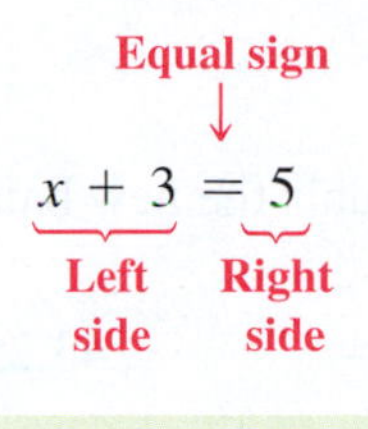

DEFINITION
An **equation** is a mathematical statement that two expressions are equal.

For example,

$$1 + 2 = 3$$
$$x - 5 = 6$$
$$2 + 7 + 3 = 12$$
$$3x = 9$$

are all equations.

Equations are used to solve a wide range of problems. A key step in solving a problem is to translate the sentences that describe the problem to an equation that models the problem. In this section, we focus on equations that involve either addition or subtraction. In the next section, we consider equations involving multiplication or division.

Translating Sentences to Equations

In translating sentences to equations, certain words and phrases mean the same as the equal sign:

- equals
- is
- is equal to
- is the same as
- yields
- results in

Let's look at some examples of translating sentences to equations that involve addition or subtraction and vice versa.

EXAMPLE 1

Translate each sentence in the table to an equation.

Solution

Sentence	Equation
a. The sum of y and 3 is equal to $7\frac{1}{2}$.	$y + 3 = 7\frac{1}{2}$
b. The difference between x and 9 is the same as 14.	$x - 9 = 14$
c. Increasing a number by 1.5 yields 3.	$n + 1.5 = 3$
d. 6 less than a number is 10.	$n - 6 = 10$

PRACTICE 1

Write an equation for each word phrase or sentence.

Sentence	Equation
a. n decreased by 5.1 is 9.	
b. y plus 2 is equal to 12.	
c. The difference between a number and 4 is the same as 11.	
d. 5 more than a number is $7\frac{3}{4}$.	

EXAMPLE 2

In a savings account, the previous balance P plus a deposit of \$7.50 equals the new balance of \$43.25. Write an equation that represents this situation.

Solution

The previous balance plus the deposit equals the new balance.

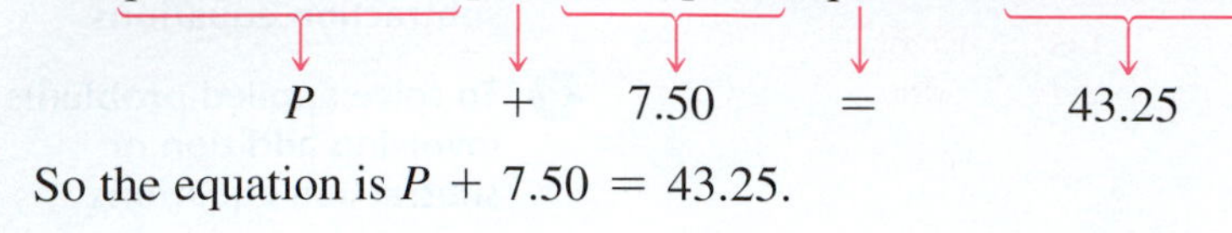

So the equation is $P + 7.50 = 43.25$.

PRACTICE 2

The sale price of a jacket is \$49.95. This amount is \$6 less than the regular price p. Write an equation that represents this situation.

Equations Involving Addition and Subtraction

Suppose that you are told that five *more than some number* is equal to seven. You can find that number by solving the addition equation $x + 5 = 7$. To solve this equation means to find a number that, when substituted for the variable x, makes the equation a true statement. Such a number is called a *solution* of the equation.

To solve the equation $x + 5 = 7$, we can think of a balance scale like the one shown.

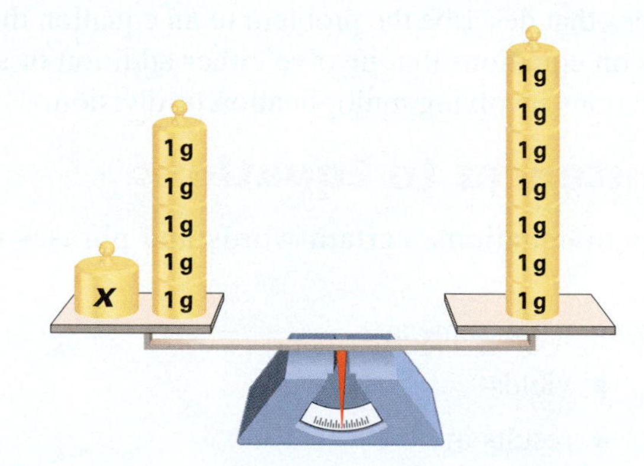

For the balance to remain level, whatever we do to one side, we must also do to the other side. In this case, if we subtract 5 grams from each side of the balance, we can conclude that the unknown weight, x, must be 2 grams, as shown below. So 2 is the solution of the equation $x + 5 = 7$.

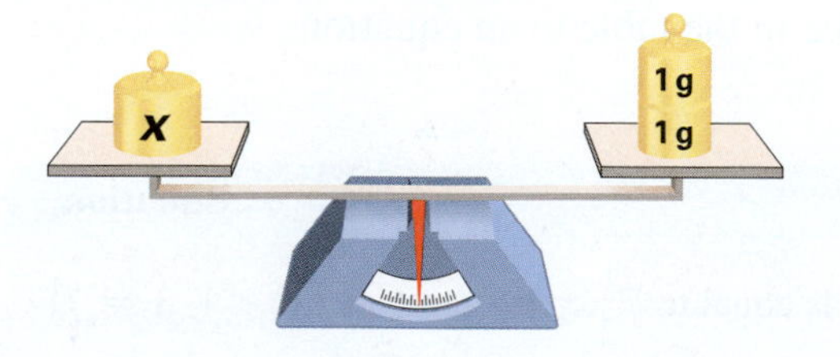

Similarly in the *subtraction equation* $x - 5 = 7$, if we add 5 to each side of the equation, we find that x equals 12.

In solving these and other equations, the key is to **isolate the variable**, that is, to get the variable alone on one side of the equation.

These examples suggest the following rule:

To Solve Addition or Subtraction Equations

- For an addition equation, *subtract* the same number from each side of the equation in order to isolate the variable on one side.
- For a subtraction equation, *add* the same number to each side of the equation in order to isolate the variable on one side.
- In either case, check the solution by substituting the value of the unknown in the original equation to verify that the resulting equation is true.

Because addition and subtraction are **opposite operations**, one operation "undoes" the other. The following examples illustrate how to perform an opposite operation to each side of an equation when you are solving for the unknown.

EXAMPLE 3

Solve and check: $y + 9 = 17$

Solution

$$y + 9 = 17$$

$$y + 9 - 9 = 17 - 9 \quad \text{Subtract 9 from each side of the equation.}$$

$$y + 0 = 8$$

$$y = 8 \quad \text{Any number added to 0 is the number.}$$

The solution is 8.

Check

$$y + 9 = 17$$

$$8 + 9 \stackrel{?}{=} 17 \quad \text{Substitute 8 for } y \text{ in the original equation.}$$

$$17 \stackrel{\checkmark}{=} 17 \quad \text{The equation is true, so 8 is the solution to the equation.}$$

Note that, because 9 was added to y in the original equation, we solved by subtracting 9 from both sides of the equation in order to isolate the variable.

PRACTICE 3

Solve and check: $x + 5 = 14$

EXAMPLE 4

Solve and check: $n - 2.5 = 0.7$

Solution

$$n - 2.5 = 0.7$$

$$n - 2.5 + 2.5 = 0.7 + 2.5 \quad \text{Add 2.5 to each side of the equation.}$$

$$n - 0 = 3.2$$

$$n = 3.2$$

The solution is 3.2.

Check

$$n - 2.5 = 0.7$$

$$3.2 - 2.5 \stackrel{?}{=} 0.7 \quad \text{Substitute 3.2 for } n \text{ in the original equation.}$$

$$0.7 \stackrel{\checkmark}{=} 0.7$$

Can you explain why checking an answer is important?

PRACTICE 4

Solve and check: $t - 0.9 = 1.8$

4.2 Exercises

FOR EXTRA HELP MyMathLab Math XL PRACTICE WATCH READ REVIEW

Mathematically Speaking

Fill in each blank with the most appropriate term or phrase from the given list.

constant	subtract	equation
translates	simplifies	variable
add	sentence	

1. A(n) __________ is a mathematical statement that two expressions are equal.

2. A solution of an equation is a number that, when substituted for the __________, makes the equation a true statement.

3. In the equation $x + 2 = 5$, __________ from each side of the equation in order to isolate the variable.

4. The equation $x - 1 = 6$ __________ to the sentence "The difference between x and 1 is 6."

A *Translate each sentence to an equation.*

5. z minus 9 is 25.

6. x minus 7 is 29.

7. The sum of 7 and x is 25.

8. The sum of m and 19 is 34.

9. t decreased by 3.1 equals 4.

10. r decreased by 5.1 equals 6.4.

11. $\frac{3}{2}$ increased by a number yields $\frac{9}{2}$.

12. $\frac{8}{3}$ increased by a number yields $\frac{13}{3}$.

13. $3\frac{1}{2}$ less than a number is equal to 7.

14. $1\frac{1}{2}$ less than a number is equal to $7\frac{1}{4}$.

B *By answering yes or no, indicate whether the value of x shown is a solution of the given equation.*

15.

Equation	Value of x	Solution?
a. $x + 1 = 9$	8	
b. $x - 3 = 4$	5	
c. $x + 0.2 = 5$	4.8	
d. $x - \frac{1}{2} = 1$	$\frac{1}{2}$	

16.

Equation	Value of x	Solution?
a. $x - 39 = 5$	44	
b. $x - 2 = 6$	4	
c. $x + 2.8 = 4$	1.2	
d. $x - \frac{2}{3} = 1$	$1\frac{2}{3}$	

Identify the operation to perform on each side of the equation to isolate the variable.

17. $x + 4 = 6$

18. $x + 10 = 17$

19. $x - 6 = 9$

20. $x - 11 = 4$

21. $x - 7 = 24$

22. $10 = x - 3$

23. $x + 21 = 25$

24. $3 = x + 2$

Solve and check.

25. $a - 7 = 24$

26. $x - 9 = 13$

27. $y + 19 = 21$

28. $z + 23 = 31$

29. $x - 2 = 10$

30. $t - 4 = 19$

31. $n + 9 = 13$

32. $d + 12 = 12$

33. $5 + m = 7$

34. $17 + d = 20$

35. $39 = y - 51$

36. $44 = c - 3$

37. $z + 2.4 = 5.3$

38. $t + 2.3 = 6.7$

39. $n - 8 = 0.9$

40. $c - 0.7 = 6$

41. $y + 8.1 = 9$

42. $a + 0.7 = 2$

43. $x + \frac{1}{3} = 9$

44. $z + \frac{2}{5} = 11$

45. $m - 1\frac{1}{3} = 4$

46. $s - 4\frac{1}{2} = 8$

47. $x + 3\frac{1}{4} = 7$

48. $t + 1\frac{1}{2} = 5$

49. $c - 14\frac{1}{5} = 33$

50. $a - 9\frac{7}{10} = 27\frac{2}{3}$

51. $x - 3.4 = 9.6$

52. $m - 12.5 = 13.7$

53. $5 = y - 1\frac{1}{4}$

54. $3 = t - 1\frac{2}{3}$

55. $5\frac{3}{4} = a + 2\frac{1}{3}$

56. $4\frac{1}{3} = n + 3\frac{1}{2}$

57. $2.3 = x - 5.9$

58. $4.1 = d - 6.9$

59. $y - 7.01 = 12.9$

60. $x - 3.2 = 5.23$

61. $x + 3.443 = 8$

62. $x + 0.035 = 2.004$

63. $2.986 = y - 7.265$

64. $3.184 = y - 1.273$

Translate each sentence to an equation. Solve and check.

65. 3 more than n is 11.

66. 15 more than x is 33.

67. 6 less than y equals 7.

68. 4 less than t equals 1.

69. If 10 is added to n, the sum is 19.

70. 25 added to a number m gives a result of 53.

71. x increased by 3.6 is equal to 9.

72. n increased by 3.5 is equal to 7.

73. A number minus $4\frac{1}{3}$ is the same as $2\frac{2}{3}$.

74. A number minus $5\frac{1}{2}$ is the same as $2\frac{1}{2}$.

Choose the equation that best describes the situation.

75. After 6 months of dieting and exercising, an athlete lost $8\frac{1}{2}$ pounds. If she now weighs 135 pounds, what was her original weight?

a. $w + 8\frac{1}{2} = 135$ **b.** $w - 126\frac{1}{2} = 8\frac{1}{2}$

c. $w - 8\frac{1}{2} = 135$ **d.** $w + 135 = 143\frac{1}{2}$

76. A teenager has d dollars. After buying an Xbox 360 Elite for \$299.99, he has \$6.01 left. How many dollars did he have at first?

a. $d + 299.99 = 306$ **b.** $d - 299.99 = 6.01$

c. $d - 299.99 = 306$ **d.** $d + 6.01 = 299.99$

77. According to a 30-day sample, the two most downloaded English-language authors are the British novelist Charles Dickens and the American humorist Mark Twain. In the sample, there were 37,541 downloads of Dickens and 5,268 fewer of Twain. How many downloads of Twain were there? (*Source:* Project Gutenberg)

a. $x + 5{,}268 = 37{,}541$ **b.** $x - 5{,}268 = 37{,}541$

c. $x + 5{,}268 = 32{,}273$ **d.** $x - 37{,}541 = 5{,}268$

78. The CN Tower in Canada and Canton Tower in China are two of the tallest telecommunications towers in the world. Of these structures, the CN Tower is 555 meters tall, which is 55 meters shorter than the tower in China. How tall is the Canton Tower? (*Source:* gztvtower.info.com)

a. $x + 55 = 555$ **b.** $x + 555 = 610$

c. $x - 55 = 555$ **d.** $x - 555 = 55$

Mixed Practice *Solve and check.*

79. $10 = a - 4.5$

80. $x + \frac{1}{2} = 6$

Solve.

81. The life expectancy in the United States of a female born in the year 2000 was 79.3 years. For a female born two decades later, it is projected to be 2.6 years greater. Choose the equation to find the life expectancy of a female born in the year 2020. (*Source:* U.S. Census Bureau)

a. $x + 79.3 = 2.6$ **b.** $x - 2.6 = 79.3$
c. $x + 2.6 = 79.3$ **d.** $x - 20 = 2.6$

82. The hygienist at a dentist's office cleaned a patient's teeth. The total bill came to \$125, which was partially covered by dental insurance. If the patient paid \$60 out of pocket toward the bill, choose the equation to find how much of the bill was covered by insurance.

a. $x + 60 = 125$ **b.** $x - 60 = 125$
c. $x + 125 = 60$ **d.** $x - 125 = 60$

83. Is 3 a solution to the equation $10 - x = 7$?

84. Is 6 a solution to the equation $x + 4.5 = 7.5$?

85. Translate the sentence "The sum of 4.2 and n is 8" to an equation.

86. Translate the sentence "x decreased by 4 is 10" to an equation.

87. Identify the operation to perform on each side of the equation $y - 1.9 = 5$ to isolate the variable.

88. Identify the operation to perform on each side of the equation $n + 2 = 10$ to isolate the variable.

Applications

C *Write an equation. Solve and check.*

89. A local community college increased the cost of a credit hour by \$12 for this year. If the cost of a credit hour for this year is \$106, what was the cost last year?

90. The first algebra textbook was written by the Arab mathematician Muhammad ibn Musa al-Khwarazmi. The title of that book, which gave rise to the word *algebra*, was *Aljabr wa'lmuqabalah*, meaning "the art of bringing together unknowns to match a known quantity." If the book appeared in the year 825, how many years ago was this? (*Source:* R.V. Young, *Notable Mathematicians*)

91. In the triangle shown, angles A and B are complementary, that is, the sum of their measures is 90°. Find x, the number of degrees in angle B.

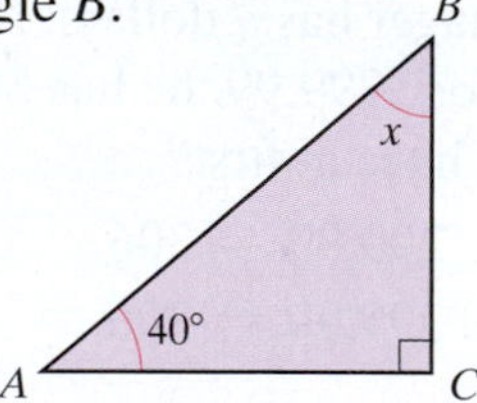

92. In the following diagram, angles ABD and CBD are supplementary, that is, the sum of their measures is 180°. Find y.

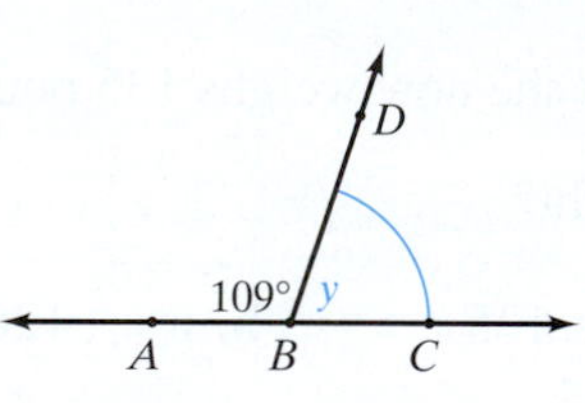

93. An article on Broadway shows reported that this week the box office receipts for a particular show were \$621,000. If that amount was \$13,000 less than last week's, how much money did the show take in last week?

94. Mount Kilimanjaro, the highest elevation on the continent of Africa, is 299 meters lower than Mount McKinley, the highest elevation on the continent of North America. If Mount Kilimanjaro is 5,895 meters high, how high is Mount McKinley? (*Source: The World Factbook, 2010*)

95. On a state freeway, the minimum speed limit is 45 miles per hour. This is 20 miles per hour lower than the maximum speed limit. What is the maximum speed limit?

96. The melting point of silver is 1,763 degrees Fahrenheit. This is 185 degrees less than the melting point of gold. What is the melting point of gold? (*Source: The New York Times Almanac, 2010*)

97. In a recent year, the U.S. charities that received the greatest private support were two organizations headquartered in Alexandria, Virginia—United Way Worldwide ($4,023,362,895) and the Salvation Army ($1,876,674,000). How much more money did United Way Worldwide receive? (*Source:* philanthropy.com)

98. During a recession, an automobile company laid off 18,578 employees, reducing its workforce to 46,894. How many employees did the company have before the recession?

• Check your answers on page A-6.

MINDStretchers

Groupwork

1. Working with a partner, compare the equations $x - 4 = 6$ and $x - a = b$.

a. Use what you know about the first equation to solve the second equation for x.

b. What are the similarities and the differences between the two equations?

Writing

2. Equations often serve as models for solving word problems.
Write two different word problems corresponding to each of the following equations.

a. $x + 4 = 9$

•

•

b. $x - 1 = 5$

•

•

Critical Thinking

3. In the magic square at the right, the sum of each row, column, and diagonal is the same. Find that sum and write and solve equations to get the values of f, g, h, r, and t.

f	6	11
g	10	h
r	14	t

EXAMPLE 4 (continued)

Check $5 = \frac{y}{2}$

$5 \stackrel{?}{=} \frac{10}{2}$ Substitute 10 for y in the original equation.

$5 \stackrel{\checkmark}{=} 5$

EXAMPLE 5

Solve and check: $0.2n = 4$

Solution $0.2n = 4$

$\frac{0.2n}{0.2} = \frac{4}{0.2}$ Divide each side by 0.2: $0.2\overline{)4.0}$ = 20. or 20.

$n = 20$

The solution is 20.

Check $0.2n = 4$

$0.2(20) \stackrel{?}{=} 4$ Substitute 20 for n in the original equation.

$4.0 \stackrel{?}{=} 4$

$4 \stackrel{\checkmark}{=} 4$

PRACTICE 5

Solve and check: $1.5x = 6$

EXAMPLE 6

Solve and check: $\frac{m}{0.5} = 1.3$

Solution $\frac{m}{0.5} = 1.3$

$(0.5)\frac{m}{0.5} = (0.5)(1.3)$ Multiply each side by 0.5.

$m = 0.65$

The solution is 0.65.

Check $\frac{m}{0.5} = 1.3$

$\frac{0.65}{0.5} \stackrel{?}{=} 1.3$ Substitute 0.65 for m in the original equation.

$1.3 \stackrel{\checkmark}{=} 1.3$

PRACTICE 6

Solve and check: $\frac{a}{2.4} = 1.2$

EXAMPLE 7

Solve and check: $\frac{2}{3}n = 6$

Solution

$$\frac{2}{3}n = 6$$

$$\frac{2}{3}n \div \frac{2}{3} = 6 \div \frac{2}{3} \quad \text{Divide each side by } \frac{2}{3}.$$

$$\left(\frac{2}{3}n\right)\left(\frac{3}{2}\right) = 6\left(\frac{3}{2}\right)$$

$$\left(\frac{2}{3}\right)\left(\frac{3}{2}\right)n = 6\left(\frac{3}{2}\right)$$

$$n = 9$$

The solution is 9.

Check

$$\frac{2}{3}n = 6$$

$$\frac{2}{3}(9) \stackrel{?}{=} 6 \quad \text{Substitute 9 for } n \text{ in the original equation.}$$

$$\frac{2}{3}\left(\frac{9}{1}\right) \stackrel{?}{=} 6$$

$$6 \stackrel{\checkmark}{=} 6$$

PRACTICE 7

Solve and check: $\frac{3}{4}x = 12$

As in the case of addition and subtraction equations, multiplication and division equations can be useful mathematical models of real-world situations. To derive these models, we translate word sentences to algebraic equations and solve.

EXAMPLE 8

Write each sentence as an algebraic equation. Then, solve and check.

Solution

Sentence	Equation	Check
a. Thirty-five is equal to the product of 5 and x.	$35 = 5x$ $\frac{35}{5} = \frac{5x}{5}$ $7 = x$, or $x = 7$	$35 = 5x$ $35 \stackrel{?}{=} 5(7)$ $35 \stackrel{\checkmark}{=} 35$
b. One equals p divided by 3.	$1 = \frac{p}{3}$ $3 \cdot 1 = 3 \cdot \frac{p}{3}$ $3 = p$, or $p = 3$	$1 = \frac{p}{3}$ $1 \stackrel{?}{=} \frac{3}{3}$ $1 \stackrel{\checkmark}{=} 1$

PRACTICE 8

Translate each sentence to an equation. Then, solve and check.

Sentence	Equation	Check
a. Twelve is equal to the quotient of z and 6.		
b. Sixteen equals twice x.		

[4.2] *Solve and check.*

39. $x + 11 = 20$
40. $y + 15 = 24$
41. $n - 19 = 7$
42. $b - 12 = 8$
43. $a + 2.5 = 6$
44. $c + 1.6 = 9.1$
45. $x - 1.8 = 9.2$
46. $y - 1.4 = 0.6$
47. $w + 1\frac{1}{2} = 3$
48. $s + \frac{2}{3} = 1$
49. $c - 1\frac{1}{4} = 5\frac{1}{2}$
50. $p - 6 = 5\frac{2}{3}$
51. $7 = m + 2$
52. $10 = n + 10$
53. $39 = c - 39$
54. $72 = y - 18$
55. $38 + n = 49$
56. $37 + x = 62$
57. $4.0875 + x = 35.136$
58. $24.625 = m - 1.9975$

[4.2–4.3] *Translate each sentence to an equation.*

59. n decreased by 19 is 35.
60. 37 less than a equals 234.
61. 9 increased by a number is equal to $15\frac{1}{2}$.
62. 26 more than a number is $30\frac{1}{3}$.
63. Twice y is 16.
64. The product of t and 25 is 175.
65. 34 is equal to n divided by 19.
66. 17 is the quotient of z and 13.
67. $\frac{1}{3}$ of a number equals 27.
68. $\frac{2}{5}$ of a number equals 4.

By answering yes or no, indicate whether the value of x shown is a solution to the given equation.

69.

Equation	Value of x	Solution?
a. $0.3x = 6$	2	
b. $x - \frac{1}{2} = 1\frac{2}{3}$	$2\frac{1}{6}$	
c. $\frac{x}{0.5} = 7$	3.5	
d. $x + 0.1 = 3$	3.1	

70.

Equation	Value of x	Solution?
a. $0.2x = 6$	30	
b. $x + \frac{1}{2} = 1\frac{2}{3}$	$\frac{5}{6}$	
c. $\frac{x}{0.2} = 4.1$	8.2	
d. $x + 0.5 = 7.4$	6.9	

[4.3] *Solve and check.*

71. $2x = 10$
72. $8t = 16$
73. $\frac{a}{7} = 15$
74. $\frac{n}{6} = 9$
75. $9y = 81$
76. $10r = 100$
77. $\frac{w}{10} = 9$
78. $\frac{x}{100} = 1$
79. $1.5y = 30$
80. $1.2a = 144$
81. $\frac{1}{8}n = 4$
82. $\frac{1}{2}b = 16$
83. $\frac{m}{1.5} = 2.1$
84. $\frac{z}{0.3} = 1.9$
85. $100x = 40$
86. $10t = 5$
87. $0.3 = \frac{m}{4}$
88. $1.4 = \frac{b}{7}$
89. $0.866x = 10.825$
90. $\frac{x}{0.707} = 2.1$

Mixed Applications

Write an algebraic expression for each problem. Then, evaluate the expression for the given amount.

91. The temperature increases 2 degrees an hour. By how many degrees will the temperature increase in h hours? In 3 hours?

92. During the fall term, a math tutor works 20 hours per week. What is the tutor's hourly wage if she earns d dollars per week? $191 per week?

93. The local supermarket sells a certain fruit for 89¢ per pound. How much will p pounds cost? 3 pounds?

94. After having borrowed $3,000 from a bank, a customer must pay the amount borrowed plus a finance charge. How much will he pay the bank if the finance charge is d dollars? $225?

Write an equation. Then, solve and check.

95. After depositing $238 in a checking account, the balance will be $517. What was the balance before the deposit?

96. A bowler's final score is the sum of her handicap and scratch score (actual score). If a bowler has a final score of 225 and a handicap of 50, what was her scratch score?

97. Drinking bottled water is more popular in some countries than in others. In a recent year, the per capita consumption of bottled water in the United States was about 100 liters, or approximately 2.9 times as much as it was in the United Kingdom. Find the per capita consumption in the United Kingdom, to the nearest liter. (*Source:* britishbottledwater.org)

98. Hurricane Gilbert was one of the strongest storms to hit the Western Hemisphere in the twentieth century. A newspaper reported that the hurricane left 500,000 people, or about one-fourth of the population of Jamaica, homeless. Approximately what was the population of Jamaica? (*Source:* J. B. Elsner and A. B. Kara, *Hurricanes of the North Atlantic*)

99. On the Moon, a person weighs about one-sixth of his or her weight on Earth. What is the weight on Earth of an astronaut who weighs 30 pounds on the Moon?

100. During an economic recession, a U.S. senator proposed a recovery plan that would cost $3 trillion. Opponents criticized the plan for being too expensive and said that it was 2.5 times the cost of an alternative plan. What was the cost of the alternative plan?

101. The normal body temperature is 98.6 degrees Fahrenheit. An ill patient had a temperature of 101°F. This temperature is how many degrees above normal?

102. This year, a community college received 8,957 applications for admission, which amounts to 256 fewer than were received last year. How many applications did the community college receive last year?

• Check your answers on page A-6.

CHAPTER 4 Posttest

FOR EXTRA HELP

The Chapter Test Prep Videos with test solutions are available on DVD, in MyMathLab, and on YouTube (search "AkstBasicMath" and click on "Channels").

To see if you have mastered the topics in this chapter, take this test.

Write each algebraic expression in words.

1. $x + \frac{1}{2}$

2. $\frac{a}{3}$

Translate each word phrase to an algebraic expression.

3. 10 less than a number

4. The quotient of 8 and p

Evaluate each algebraic expression.

5. $a - 1.5$, for $a = 1.5$

6. $\frac{b}{9}$, for $b = 2\frac{1}{4}$

Translate each sentence to an equation.

7. The difference between x and 6 is $4\frac{1}{2}$.

8. The quotient of y and 8 is 3.2.

Solve and check.

9. $x + 10 = 10$

10. $y - 6 = 6$

11. $81 = 3n$

12. $82 = \frac{a}{9}$

13. $m - 1.8 = 6$

14. $1.5n = 75$

15. $10x = 5\frac{1}{2}$

16. $\frac{n}{100} = 7.6$

Write an equation. Then, solve and check.

17. A recipe for seafood stew requires $2\frac{1}{4}$ pounds of fish. After buying $1\frac{3}{4}$ pounds of bluefish, a chef decides to fill out the recipe with codfish. How many pounds of codfish should he buy?

18. In a recent year, the federal government removed the Rocky gray wolf from the list of endangered species, resulting in the first Rocky gray wolf hunting season in decades. During this season, about one-fourth of the population, or an estimated 500 wolves, were killed. Approximately how many Rocky gray wolves had lived in the wild before the hunt began? (*Source:* treehugger.com)

19. According to official estimates, the world population in 2045 is projected to be about one-and-a-half times what it was in 1999. If the projected population is 9 billion, what was the population in 1999? (*Source:* census.gov)

20. In chemistry, an *endothermic reaction* is one that absorbs heat. As a result of an endothermic reaction, the temperature of a solution dropped by 19.8 degrees Celsius to 7.6°C. What was the temperature of the solution before the reaction took place? (*Source:* Timberlake, *Chemistry: An Introduction to General, Organic, and Biological Chemistry*)

• Check your answers on page A-7.

Cumulative Review Exercises

To help you review, solve the following:

1. Round 314,159 to the nearest hundreds.

2. One of the three differences below is wrong. Use estimation to identify which difference is incorrect.

   a. $\begin{array}{r} 675{,}029 \\ -126{,}384 \\ \hline 548{,}645 \end{array}$ b. $\begin{array}{r} 539{,}324 \\ -126{,}384 \\ \hline 412{,}940 \end{array}$ c. $\begin{array}{r} 954{,}736 \\ -365{,}976 \\ \hline 488{,}760 \end{array}$

3. Multiply: 804×29

4. Find the quotient and check: $35{,}020 \div 34$

5. Write the equivalent fraction with the given denominator: $\frac{7}{8} = \frac{\quad}{96}$

6. Write as a mixed number in simplified form: $\frac{56}{40}$

7. Subtract: $8\frac{1}{4} - 2\frac{7}{8}$

8. Write the decimal 5.239 in words.

9. Insert the appropriate sign, $<$, $=$, or $>$, to make a true statement. 6.356 6.36

10. Compute: $12 - (3.2 + 4.91)$

11. Find the quotient: $7.5 \div 1{,}000$

12. Decide whether 2 is a solution of the equation $w + 3 = 5$

13. Solve and check: $n - 3.8 = 4$

14. Solve and check: $\frac{x}{2} = 16$

15. In animating a cartoon, artists had to draw 24 images to appear during 1 second of screen time. How many images did they have to draw to produce a 5-minute cartoon?

16. Dental insurance reimbursed a patient \$200 on a bill of \$700. Did the patient get less or more than $\frac{1}{3}$ of his money back? Explain.

17. In a recent review of world energy, the annual oil production of the United States was approximately 300 million tons, or three-fifths of Saudi Arabia's production. Write an equation to find Saudi Arabia's annual oil production. Then solve and check. (*Source: Top 10 of Everything 2010*)

18. According to the Environmental Protection Agency, U.S. residents produce about 4.4 pounds of garbage per person per day. At this rate, how much garbage does a family of 3 produce in a week? (*Source:* epa.gov)

19. Farmers depend on bees to pollinate many crop plants, such as apples and cherries. In the American Midwest, the acreage of crops is large as compared with the number of bees, so farmers are especially concerned if the number of beehives declines. When the number of beehives in the state of Illinois dropped from 101,000 to 46,000, how big a drop was this? (*Source:* ag.uiuc.edu)

20. Due to World War I, the number of U.S. military personnel on active duty in 1918—roughly 2.9 million—was about 4.5 times what it had been the year before. (*Source: The New York Times Almanac 2010*)

a. Write an equation to find the number of personnel in 1917.

b. Solve, rounding the solution to the nearest hundred thousand.

• Check your answers on page A-7.

CHAPTER 5

Ratio and Proportion

Ratio, Proportion, and the Connecticut Compromise

One of the most contested issues at the 1787 Constitutional Convention in Philadelphia was how the U.S. Congress was to be structured. The convention delegates from large states, such as Massachusetts, supported the Virginia Plan in which the number of representatives from a state would be *proportional* to its population so that a state with double the population of another would have twice as many representatives. By contrast, the delegates from smaller states, such as Delaware, favored the New Jersey Plan, in which every state has the same number of representatives in Congress.

Finally, after months of heated debate, the convention delegates agreed to the Connecticut Compromise under which Congress would have two chambers whose approval would be required for legislation to pass. In one chamber, the Senate, big and small states alike would have the same number of representatives. In the other chamber, the House of Representatives, representation would be proportional so that for all states the *ratio* of the population to the number of House representatives would be roughly the same. To this day, the Constitution provides that seats in the House are apportioned among the states by population, as determined by the census conducted every ten years.

(*Source:* Douglas Brinkley, *American Heritage History of the United States*, Viking, 1998)

CHAPTER 5 PRETEST

To see if you have already mastered the topics in this chapter, take this test.

Write each ratio or rate in simplest form.

1. 6 to 8

2. 40 to 100

3. \$30 to \$18

4. 19 feet to 51 feet

5. 48 gallons of water in 15 minutes

6. 10 milligrams every 6 hours

Find the unit rate.

7. 12 dental assistants for every 6 dentists

8. 35 calculators for 35 students

Determine the unit price.

9. \$690 for 3 boxes of ceramic tiles

10. 12 bottles of lemon iced tea for \$6.00

Determine whether each proportion is true or false.

11. $\frac{2}{3} = \frac{16}{24}$

12. $\frac{32}{20} = \frac{8}{3}$

Solve and check.

13. $\frac{6}{8} = \frac{x}{12}$

14. $\frac{21}{x} = \frac{2}{3}$

15. $\frac{\frac{1}{2}}{4} = \frac{2}{x}$

16. $\frac{x}{6} = \frac{8}{0.3}$

Solve.

17. A contractor combines 80 pounds of sand with 100 pounds of gravel. In this mixture, what is the ratio of sand to gravel?

18. A machine at a potato chip factory can peel 12,000 pounds of potatoes in 60 minutes. At this rate, how many pounds of potatoes can it peel per minute?

19. The *aspect ratio* of an image is the ratio of the image's width to its height. In digital camera photos, a common aspect ratio is 4 to 3. With this ratio, how high is a photo that is 6 inches wide? (*Source:* wikipedia.org)

20. Suppose the scale on a Louisiana map is 3 inches to 31 miles. If two cities, Baton Rouge and New Orleans, are 7.4 inches apart on the map, what is the actual distance, to the nearest mile, between the two cities? (*Source:* ersys.com)

• Check your answers on page A-7.

5.1 Introduction to Ratios

OBJECTIVES

- **A** To write ratios of like quantities in simplest form
- **B** To write rates in simplest form
- **C** To solve applied problems involving ratios

What Ratios Are and Why They Are Important

We frequently need to compare quantities. Sports, medicine, and business are just a few areas where we use **ratios** to make comparisons. Consider the ratios in the following examples.

- The volleyball team won 4 games for every 3 they lost.
- A physician assistant prepared a 1-to-25 boric acid solution.
- The stock's price-to-earnings ratio was 13 to 1.

Can you think of other examples of ratios in your daily life?

The preceding examples illustrate the following definition of a ratio.

DEFINITION

A **ratio** is a comparison of two quantities expressed as a quotient.

There are several ways to write a ratio. For instance, we can write the ratio 3 to 10 as

$$3 \text{ to } 10 \qquad 3{:}10 \qquad \frac{3}{10}$$

No matter which notation we use for this ratio, it is read "3 to 10."

Simplifying Ratios

Because a ratio can be written as a fraction, we can say that, as with any fraction, a ratio is in simplest form (or written in lowest terms) when 1 is the only common factor of the numerator and denominator.

Let's consider some examples of writing ratios in simplest form.

EXAMPLE 1

Write the ratio 10 to 5 in simplest form.

Solution The ratio 10 to 5 expressed as a fraction is $\frac{10}{5}$.

$$\frac{10}{5} = \frac{10 \div 5}{5 \div 5} = \frac{2}{1}$$

So the ratio 10 to 5 is the same as the ratio 2 to 1. Note that the ratio 2 to 1 means that the first number is twice as large as the second number.

PRACTICE 1

Write the ratio 8:12 in simplest form.

Frequently, we deal with quantities that have units, such as months or feet. When both quantities in a ratio have the same unit, they are called **like quantities**. In a ratio of like quantities, the units drop out.

EXAMPLE 2

Express the ratio 5 months to 3 months in simplest form.

Solution The ratio 5 months to 3 months expressed as a fraction is $\frac{5 \text{ months}}{3 \text{ months}}$. Simplifying, we get $\frac{5}{3}$, which is already in lowest terms.

PRACTICE 2

Express in simplest form the ratio 9 feet to 5 feet.

TIP Note that with ratios we do not rewrite improper fractions as mixed numbers because our answer must be a comparison of *two* numbers. So in Example 2, we write the ratio as $\frac{5}{3}$ rather than as $1\frac{2}{3}$.

EXAMPLE 3

24-karat gold is pure gold. By contrast, 14-karat gold, commonly used to make jewelry, consists of 14 parts out of 24 parts pure gold; the rest is another metal such as copper or silver added for hardness. In 14-karat gold, what is the ratio of gold to the other metal? (*Source:* essortment.com)

Solution In 14-karat gold, 14 parts of 24 parts are pure gold. First, we calculate the number of parts that are not pure gold.

$$24 - 14 = 10$$

So 10 of 24 parts are the other metal. Next, let's write the ratio of pure gold to the other metal.

$$\frac{\text{Number of parts of pure gold}}{\text{Number of parts of the other metal}} = \frac{14}{10} = \frac{7}{5}$$

We conclude that the ratio of gold to the other metal is 7 to 5.

PRACTICE 3

The Waist to Hip Ratio (WHR) is a ratio, commonly expressed as a decimal, that has been shown to be a good predictor of possible cardiovascular problems in both men and women. If a male has a WHR greater than 1, then he is considered to be at high risk for these problems. Calculate the WHR of a male with a waist measurement of 40 inches and a hip measurement 2 inches less. Is he at high risk? (*Source: The Medical Journal of Australia*)

Now, let's compare **unlike quantities**, that is, quantities that have different units or are different kinds of measurement. Such a comparison is called a **rate**.

DEFINITION
A **rate** is a ratio of unlike quantities.

For instance, suppose that your rate of pay is \$52 for each 8 hours of work. Simplifying this rate, we get:

$$\frac{\$52}{8 \text{ hours}} = \frac{\$13}{2 \text{ hours}}$$

So you are paid \$13 for every 2 hours that you worked. Note that the units are expressed as part of the answer.

EXAMPLE 4

Simplify each rate.

a. 350 miles to 18 gallons of gas

b. 18 trees to produce 2,000 pounds of paper

Solution

a. 350 miles to 18 gallons $= \frac{350 \text{ miles}}{18 \text{ gallons}} = \frac{175 \text{ miles}}{9 \text{ gallons}}$

b. 18 trees to 2,000 pounds $= \frac{18 \text{ trees}}{2{,}000 \text{ pounds}} = \frac{9 \text{ trees}}{1{,}000 \text{ pounds}}$

PRACTICE 4

Express each rate in simplest form.

a. 150 milliliters of medication infused every 60 minutes

b. 18 pounds lost in 12 weeks

Examples 1, 2, 3, and 4 illustrate the following rule for simplifying a ratio or rate:

To Simplify a Ratio or Rate

- Write the ratio or rate as a fraction.
- Express the fraction in simplest form.
- If the quantities are like, drop the units. If the quantities are unlike, keep the units.

Frequently, we want to find a particular kind of rate called a *unit rate*. In the rate $\frac{\$13}{2 \text{ hours}}$, for instance, it would be useful to know what is earned for each hour (that is, the hourly wage). We need to rewrite $\frac{\$13}{2 \text{ hours}}$ so that the denominator is 1 hour.

$$\frac{\$13}{2 \text{ hours}} = \frac{\$13 \div 2}{2 \text{ hours} \div 2} = \frac{\$6.50}{1 \text{ hour}} = \$6.50 \text{ per hour, or } \$6.50/\text{hr}$$

Note that "per" means "divided by."

Here, we divided the numbers in both the numerator and denominator by the number in the denominator.

DEFINITION

A **unit rate** is a rate in which the number in the denominator is 1.

EXAMPLE 5

Write as a unit rate.

a. 275 miles in 5 hours

b. $3,453 for 6 weeks

Solution First, we write each rate as a fraction. Then, we divide numbers in the numerator and denominator by the number in the denominator, getting 1 in the denominator.

PRACTICE 5

Express as a unit rate.

a. a fall of 192 feet in 4 seconds

b. 15 hits in 40 times at bat

EXAMPLE 5 (continued)

a. 275 miles in 5 hours $= \dfrac{275 \text{ miles}}{5 \text{ hours}} = \dfrac{275 \text{ miles} \div 5}{5 \text{ hours} \div 5} = \dfrac{55 \text{ miles}}{1 \text{ hour}}$, or 55 mph

b. \$3,453 for 6 weeks $= \dfrac{\$3{,}453}{6 \text{ weeks}} = \dfrac{\$3{,}453 \div 6}{6 \text{ weeks} \div 6} = \dfrac{\$575.50}{1 \text{ week}}$, or \$575.50 per week

EXAMPLE 6

To reduce expenses, a commuter buys a fuel-efficient car. If the car goes 60 miles on 2.5 gallons of gas, what is its fuel economy, that is, how many miles per gallon does it get?

Solution To find the car's fuel economy, we calculate the ratio of the distance that it travels (60 miles) to the amount of gas that it uses (2.5 gallons).

$$\text{Fuel economy} = \frac{60 \text{ miles}}{2.5 \text{ gallons}}$$

Next, we simplify by dividing both the numerator and denominator by the number in the denominator, 2.5:

$$\frac{60 \text{ miles}}{2.5 \text{ gallons}} = \frac{24 \text{ miles}}{1 \text{ gallon}}$$

So the car gets 24 miles per gallon.

PRACTICE 6

Because of heavy traffic, a bus took 30 minutes to cover a distance of 20 city blocks. How many minutes per city block did the bus move?

In order to get the better buy, we sometimes compare prices by computing the price of a single item. This **unit price** is a type of unit rate.

DEFINITION

A **unit price** is the price of one item, or one unit.

To find a unit price, we write the ratio of the total price of the units to the number of units and then, simplify.

$$\text{Unit price} = \frac{\text{Total price}}{\text{Number of units}}$$

Let's consider some examples of unit pricing.

EXAMPLE 7

Find the unit price.

a. \$300 for 12 months of membership

b. 6 credits for \$234

c. 10-ounce box of wheat flakes for \$2.70

PRACTICE 7

Determine the unit price.

a. 4 supersaver flights for \$696

b. \$22 for 8 hours of parking

c. \$19.80 for 20 song downloads

Solution

a. $\dfrac{\$300}{12 \text{ months}} = \$25/\text{month}$

b. $\dfrac{\$234}{6 \text{ credits}} = \$39/\text{credit}$

c. $\dfrac{\$2.70}{10 \text{ ounces}} = \$0.27/\text{ounce}$

In the next example, we apply the concept of unit price to determine which is the better deal.

EXAMPLE 8

For the following two boxes of bandages, which is the better buy?

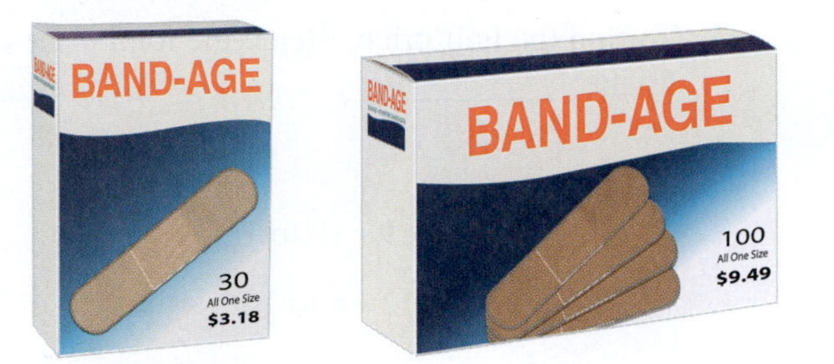

Solution First, we find the unit price for each box of bandages. Then, we round to the nearest cent and compare the prices.

$$\text{Unit Price} = \frac{\text{Total price}}{\text{Number of units}} = \frac{\$3.18}{30} = \$0.106 \approx \$0.11 \text{ per bandage}$$

$$\text{Unit Price} = \frac{\text{Total price}}{\text{Number of units}} = \frac{\$9.49}{100} = \$0.0949 \approx \$0.09 \text{ per bandage}$$

Since $\$0.09 < \$0.11$, the better buy is the box with 100 bandages.

PRACTICE 8

Which bottle of vitamin C has the lower unit price?

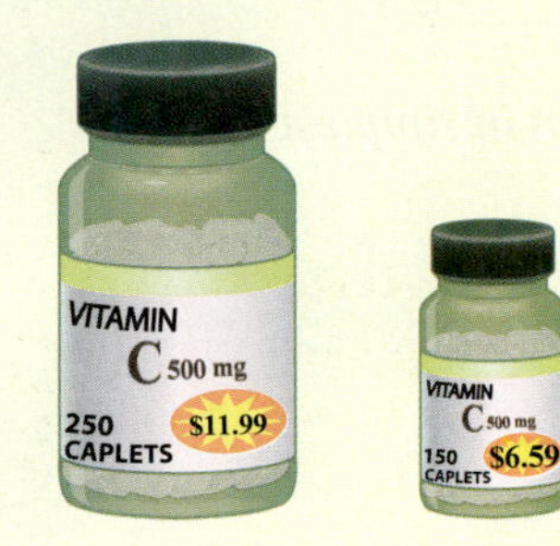

Cultural Note

The shape of a grand piano is dictated by the length of its strings. When a stretched string vibrates, it produces a particular pitch, say C. A second string of comparable tension will produce another pitch, which depends on the ratio of the string lengths. For instance, if the ratio of the second string to the first string is 18 to 16, then plucking the second string will produce the pitch B.

Around 500 B.C., the followers of the mathematician Pythagoras learned to adjust string lengths in various ratios so as to produce an entire scale. Thus the concept of ratio is central to the construction of pianos, violins, and many other musical instruments.

Sources: John R. Pierce, *The Science of Musical Sound* (New York: Scientific American Library, 1983)

David Bergamini, *Mathematics* (New York: Time-Life Books, 1971)

5.1 Exercises

FOR EXTRA HELP MyMathLab MathXL PRACTICE WATCH READ REVIEW

Mathematically Speaking

Fill in each blank with the most appropriate term or phrase from the given list.

weight of a unit	difference	number of units
like	fractional form	denominator
simplest form	unlike	
numerator	quotient	

1. A ratio is a comparison of two quantities expressed as a(n) ____________.

2. A rate is a ratio of ____________ quantities.

3. A ratio is said to be in ____________ when 1 is the only common factor of the numerator and denominator.

4. Quantities that have the same units are called ____________ quantities.

5. A unit rate is a rate in which the number in the ____________ is 1.

6. To find the unit price, divide the total price of the units by the ____________.

A *Write each ratio in simplest form.*

7. 6 to 9
8. 9 to 12
9. 10 to 15
10. 21 to 27
11. 55 to 35
12. 8 to 10
13. 2 to $1\frac{1}{3}$
14. 25 to $1\frac{1}{4}$
15. 2.5 to 10
16. 1.25 to 100
17. 60 minutes to 45 minutes
18. $40 to $25
19. 10 feet to 10 feet
20. 75 tons to 75 tons
21. 30¢ to 18¢
22. 66 years to 32 years
23. 7 miles per hour to 24 miles per hour
24. 21 gallons to 20 gallons
25. 1,000 acres to 50 acres
26. 2,000 miles to 25 miles
27. 8 grams to 7 grams
28. 19 ounces to 51 ounces
29. 24 seconds to 30 seconds
30. 28 milliliters to 42 milliliters

B *Write each rate in simplest form.*

31. 25 telephone calls in 10 days
32. 42 gallons in 4 minutes
33. 288 calories burned in 40 minutes
34. 190 e-mails in 25 days
35. 2 million hits on a website in 6 months
36. 50 million troy ounces of gold produced in 12 months
37. 68 baskets in 120 attempts
38. 18 boxes of cookies for $45
39. 296 points in 16 games
40. 12 knockouts in 16 fights
41. 500 square feet of carpeting for $1,645
42. 300 full-time students to 200 part-time students

43. 48 males for every 9 females

44. 3 case workers for every 80 clients

45. 40 Democrats for every 35 Republicans

46. $12,500 in 6 months

47. 2 pounds of zucchini for 16 servings

48. 57 hours of work in 9 days

49. 1,535 flights in 15 days

50. 25 pounds of plaster for 2,500 square feet of wall

51. 3 pounds of grass seeds for 600 square feet of lawn

52. 684 parts manufactured in 24 hours

Determine the unit rate.

53. 3,375 revolutions in 15 minutes

54. 3,000 houses to 1,500 acres of land

55. 120 gallons of heating oil for 15 days

56. 48 yards in 8 carries

57. 3 tanks of gas to cut 10 acres of lawn

58. 192 meters in 6 seconds

59. 8 yards of material for 5 dresses

60. 648 heartbeats in 9 minutes

61. 20 hours of homework in 10 days

62. $200 for 8 hours of work

63. A run of 5 kilometers in 20 minutes

64. 56 calories in 4 ounces of orange juice

65. 140 fat calories in 2 tablespoons of peanut butter

66. 60 children for every 5 adults

Find the unit price, rounding to the nearest cent if necessary.

67. 12 bars of soap for $5.40

68. 4 credit hours for $200

69. 6 rolls of film that cost $17.70

70. 2 notebooks that cost $6.90

71. 3 plants for $200

72. $240,000 for a 30-second prime time television commercial spot

73. 5 nights in a hotel for $495

74. 60 minutes of Internet access for $3

Complete each table, rounding if necessary. Determine which is the better buy.

75. Honey lemon cough drops

Number of Units	Total Price	Unit Price
30	$1.69	
100	$5.49	

76. Stretchable disposable diapers

Number of Units	Total Price	Unit Price
36	$8.69	
60	$14.99	

77. Staples® Bright White Inkjet paper

Number of Units (Sheets)	Total Price	Unit Price
500	$9.69	
2,500	$42.99	

78. Colgate® Total toothpaste

Number of Units (Ounces)	Total Price	Unit Price
6	$3.59	
7	$4.19	

Fill in the table. Which is the best buy?

79. Glad® trash bags, large, with drawstring

Number of Units	Total Price	Unit Price
14	$8.49	
25	$11.49	
28	$7.49	

80. CVS® AA alkaline batteries

Number of Units	Total Price	Unit Price
4	$2.57	
10	$3.89	
24	$6.89	

Mixed Practice

Solve.

81. To the nearest cent, find the unit price of an 18-ounce jar of creamy peanut butter that costs $2.89.

82. Complete the table. Then, find the best buy. Starbucks® Cappuccino

Number of units (Fluid ounces)	Total Price	Unit Price
12	$3.15	
16	$3.95	
24	$4.25	

83. Simplify the rate: 4 tutors for every 30 students.

84. Write as a unit rate: 50 lots to 0.2 square mile.

85. Write the ratio 20 to 4 in simplest form.

86. Express $\frac{30 \text{ centimeters}}{45 \text{ centimeters}}$ in simplest form.

Applications

C *Solve. Simplify if possible.*

87. The number line shown is marked off in equal units. Find the ratio of the length of the distance x to the distance y.

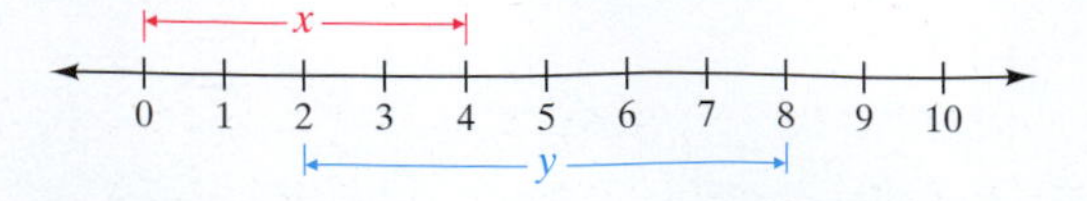

88. In the following rectangle, what is the ratio of the width to the length?

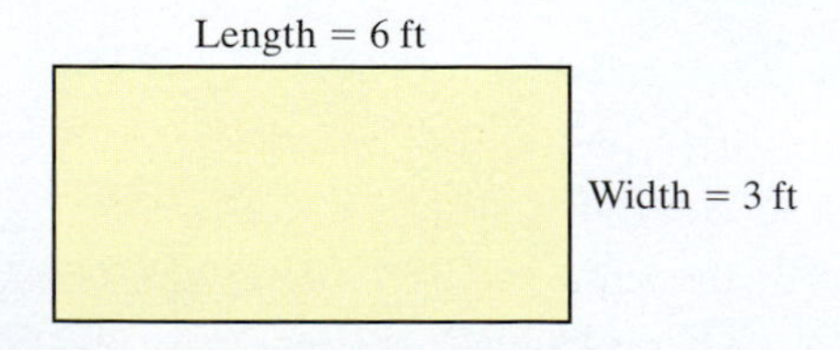

89. In 10 ounces of cashew nuts, there are 1,700 calories. How many calories are there per ounce?

90. For a building valued at $200,000 the property tax is $4,000. Find the ratio of the tax to the building's value.

91. The *cartridge yield* of a computer printer is the number of pages that it will print before the toner runs out. A cartridge that sells for $69.99 has a yield of 2,000 pages. Rounded to the nearest cent, what is the cost per page of printing? (*Source:* smartcomputing.com)

92. A bathtub contains 20 gallons of water. If the tub empties in 4 minutes, what is the rate of flow of the water per minute?

93. In a student government election, 1,000 students cast a vote for the incumbent, 900 voted for the opponent, and 100 cast a protest vote. What was the ratio of the incumbent's vote to the total number of votes?

94. At a college, 4,500 of the 7,500 students are female. What is the ratio of females to males at the college?

95. In finance, the *return on investment* (ROI) is the ratio of profit or loss on an investment relative to the amount of money invested. ROI is commonly calculated to compare the performance of one investment relative to another. Find this ratio for an investment of $9,000 with a profit of $1,500. (*Sources:* investopedia.com and ehow.com)

96. About 15,600 people can ride El Toro, a roller coaster at Six Flags Great Adventure in New Jersey, in 12 hours. Approximately how many people per hour can ride El Toro? (*Source:* wikipedia.org)

97. At the beginning of the 112th U.S. Congress, there were 193 Democrats in the House of Representatives and 51 Democrats in the Senate. For Republicans, there were 242 in the House and 47 in the Senate. Was the ratio of Democrats to Republicans higher in the House or in the Senate? (*Source:* wikipedia.org)

98. The table below shows the breakdown of the number of patients in two hospital units at a local city hospital. Is the ratio of nurses to patients in the intensive care unit higher or lower than the ratio of nurses to patients in the medical unit?

	Intensive Care Unit	Medical Unit
Patients	25	65
Nurses	8	11

99. The following table deals with five of the longest-reigning monarchs in history.

Monarch	Country	Reign	Length of Reign (in years)
King Louis XIV	France	1643–1715	72
King John II	Liechtenstein	1858–1929	71
Emperor Franz-Josef	Austria-Hungary	1848–1916	68
Queen Victoria	United Kingdom	1837–1901	64
Emperor Hirohito	Japan	1926–1989	63

(*Source: The Top 10 of Everything 2006*)

a. What is the ratio of Emperor Hirohito's length of reign to that of Emperor Franz-Josef?

b. What is the ratio of Queen Victoria's length of reign to that of King Louis XIV?

100. The following bar graph deals with popular singers and the number of their albums that "went platinum," that is, sold more than 1 million copies.

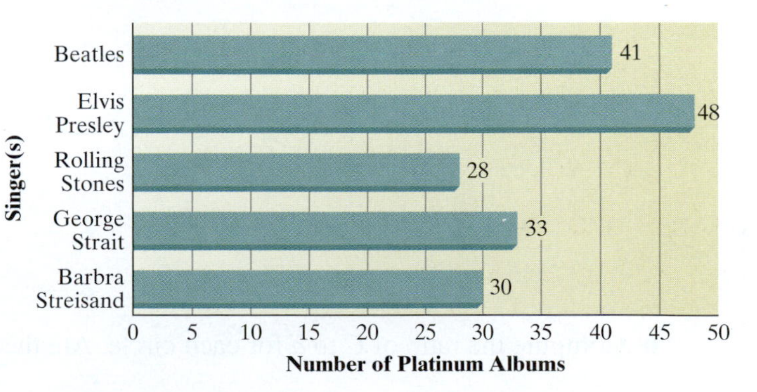

(*Source:* riaa.com)

a. Find the ratio of the number of platinum albums for Elvis Presley as compared to George Strait.

b. What is the ratio of the number of platinum albums for the Rolling Stones as compared to Barbra Streisand?

101. In the insurance industry, a loss ratio is the ratio of total losses paid out by an insurance company to total premiums collected for a given time period.

$$\text{Loss ratio} = \frac{\text{Losses paid}}{\text{Premiums collected}}$$

In 2 months, a certain insurance company paid losses of \$6,400,000 and collected premiums of \$12,472,000. What is the loss ratio, rounded to the nearest hundredth?

102. Analysts for a brokerage firm prepare research reports on companies with stocks traded in various stock markets. One statistic that an analyst uses is the price-to-earnings (P.E.) ratio.

$$\text{P.E. ratio} = \frac{\text{Market price per share}}{\text{Earnings per share}}$$

Find the P.E. ratio, rounded to the nearest hundredth, for a stock that had a per-share market price of \$70.75 and earnings of \$5.37 per share.

• Check your answers on page A-7.

MINDStretchers

History

1. For a **golden rectangle**, the ratio of its length to its width is approximately 1.618 to 1 (the **golden ratio**).

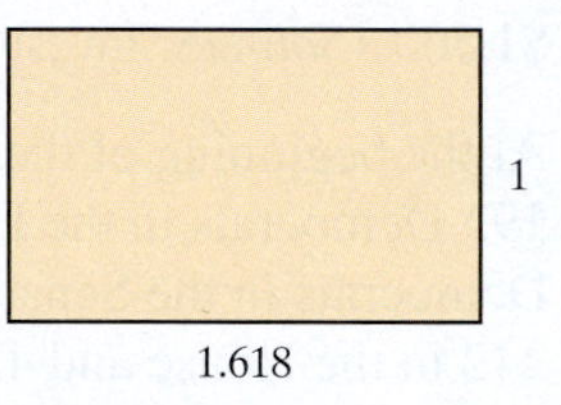

To the ancient Egyptians and Greeks, a golden rectangle was considered to be the ratio most pleasing to the eye. Show that index cards in either of the two standard sizes (3×5 and 5×8) are close approximations to the golden rectangle.

Investigation

2. The distance around a circle is called its **circumference** (C). The distance across the circle through its center is called its **diameter** (d).

a. Use a string and ruler to measure C and d for both circles shown.

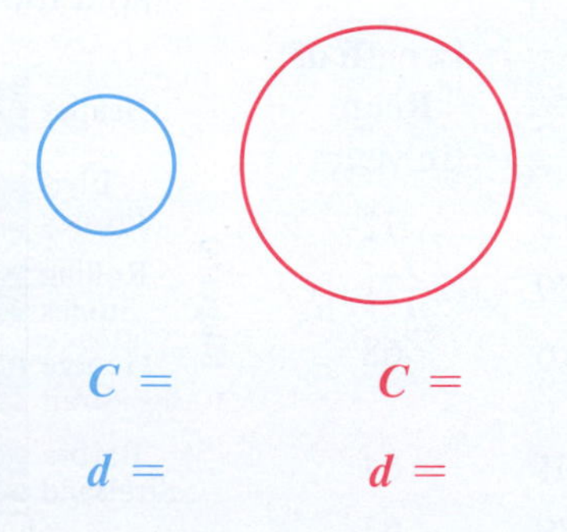

$C =$ $\qquad$ $C =$

$d =$ $\qquad$ $d =$

b. Compute the ratio of C to d for each circle. Are the ratios approximately equal?

$\frac{C}{d} =$ $\qquad$ $\frac{C}{d} =$

Writing

3. Sometimes we use *differences* rather than *quotients* to compare two quantities. Give an example of each kind of comparison and any advantages and disadvantages of each approach.

5.2 Solving Proportions

OBJECTIVES

A To write or solve proportions

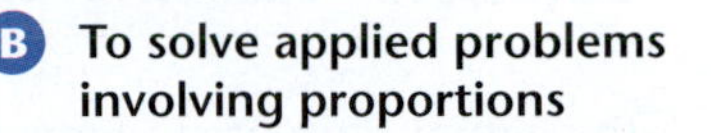

B To solve applied problems involving proportions

Writing Proportions

When two ratios—for instance, 1 to 2 and 4 to 8—are equal, they are said to be *in proportion*. We can write "1 is to 2 as 4 is to 8" as $\frac{1}{2} = \frac{4}{8}$. Such an equation is called a **proportion**.

Proportions are common in daily life and are used in many situations, such as finding the distance between two cities from a map with a given scale or the amount of a worker's pay for four weeks given the weekly pay.

DEFINITION

A **proportion** is a statement that two ratios are equal.

One way to see if a proportion is true is to determine whether the *cross products* of the ratios are equal. For example, we see that the proportion

$$\frac{1}{2} = \frac{4}{8}$$

is true, because $2 \cdot 4 = 1 \cdot 8$, or $8 = 8$. However, the proportion $\frac{3}{5} = \frac{9}{10}$ is not true, since $5 \cdot 9 \neq 3 \cdot 10$.

EXAMPLE 1

Determine whether the proportion 4 is to 3 as 16 is to 12 is true.

Solution First, we write the ratios in fractional form: $\frac{4}{3} = \frac{16}{12}$

$$\frac{4}{3} = \frac{16}{12}$$

$3 \cdot 16 \stackrel{?}{=} 4 \cdot 12$ Set the cross products equal.

$48 \stackrel{\checkmark}{=} 48$

So the proportion 4 is to 3 as 16 is to 12 is true.

PRACTICE 1

Are the ratios 10 to 4 and 15 to 6 in proportion?

EXAMPLE 2

Is $\frac{15}{9} = \frac{8}{5}$ a true proportion?

Solution $\frac{15}{9} \stackrel{?}{=} \frac{8}{5}$

$9 \cdot 8 \stackrel{?}{=} 15 \cdot 5$ Set the cross products equal.

$72 \neq 75$

The cross products are not equal. So the proportion is not true.

PRACTICE 2

Determine whether $\frac{15}{6} = \frac{8}{3}$ is a true proportion.

EXAMPLE 3

A college claims that the student-to-faculty ratio is 13 to 1. If there are 96 faculty for 1,248 students, is the college's claim true?

Solution The college claims a student-to-faculty ratio of $\frac{13}{1}$, and the actual ratio of students to faculty is $\frac{1{,}248}{96}$. We want to know if these two ratios are equal.

$$\begin{array}{l} \text{Students} \rightarrow \\ \text{Faculty} \rightarrow \end{array} \frac{13}{1} \stackrel{?}{=} \frac{1{,}248}{96} \begin{array}{l} \leftarrow \text{Students} \\ \leftarrow \text{Faculty} \end{array}$$

$$1 \cdot 1{,}248 \stackrel{?}{=} 13 \cdot 96 \quad \text{Set the cross products equal.}$$

$$1{,}248 \stackrel{\checkmark}{=} 1{,}248$$

Since the cross products are equal, the college's claim is true.

PRACTICE 3

A company has a policy making the compensation of its CEO proportional to the dividends that are paid to shareholders. If the dividends increase from $72 to $80 and the CEO's compensation is increased from $360,000 to $420,000, was the company's policy followed?

Solving Proportions

Suppose that you make $840 for working 4 weeks in a book shop. At this rate of pay, how much money will you make in 10 weeks? To solve this problem, we can write a proportion in which the rates compare the amount of pay to the time worked. We want to find the amount of pay corresponding to 10 weeks, which we call x.

$$\begin{array}{l} \text{Pay} \rightarrow \\ \text{Time} \rightarrow \end{array} \frac{840}{4} = \frac{x}{10} \begin{array}{l} \leftarrow \text{Pay} \\ \leftarrow \text{Time} \end{array}$$

After setting the cross products equal, we find the missing value.

$$\frac{840}{4} = \frac{x}{10}$$

$$4 \cdot x = 840 \cdot 10$$

$$4x = 8{,}400$$

$$\frac{4x}{4} = \frac{8{,}400}{4} \quad \text{Divide each side of the equation by 4.}$$

$$x = 2{,}100$$

So you will make $2,100 in 10 weeks.

We can check our solution by substituting 2,100 for x in the original proportion.

$$\frac{840}{4} = \frac{x}{10}$$

$$\frac{840}{4} \stackrel{?}{=} \frac{2{,}100}{10}$$

$$4 \cdot 2{,}100 \stackrel{?}{=} 840 \cdot 10 \quad \text{Set the cross products equal.}$$

$$8{,}400 \stackrel{\checkmark}{=} 8{,}400$$

Our solution checks.

To Solve a Proportion

- Find the cross products, and set them equal.
- Solve the resulting equation.
- Check the solution by substituting the value of the unknown in the original equation to be sure that the resulting proportion is true.

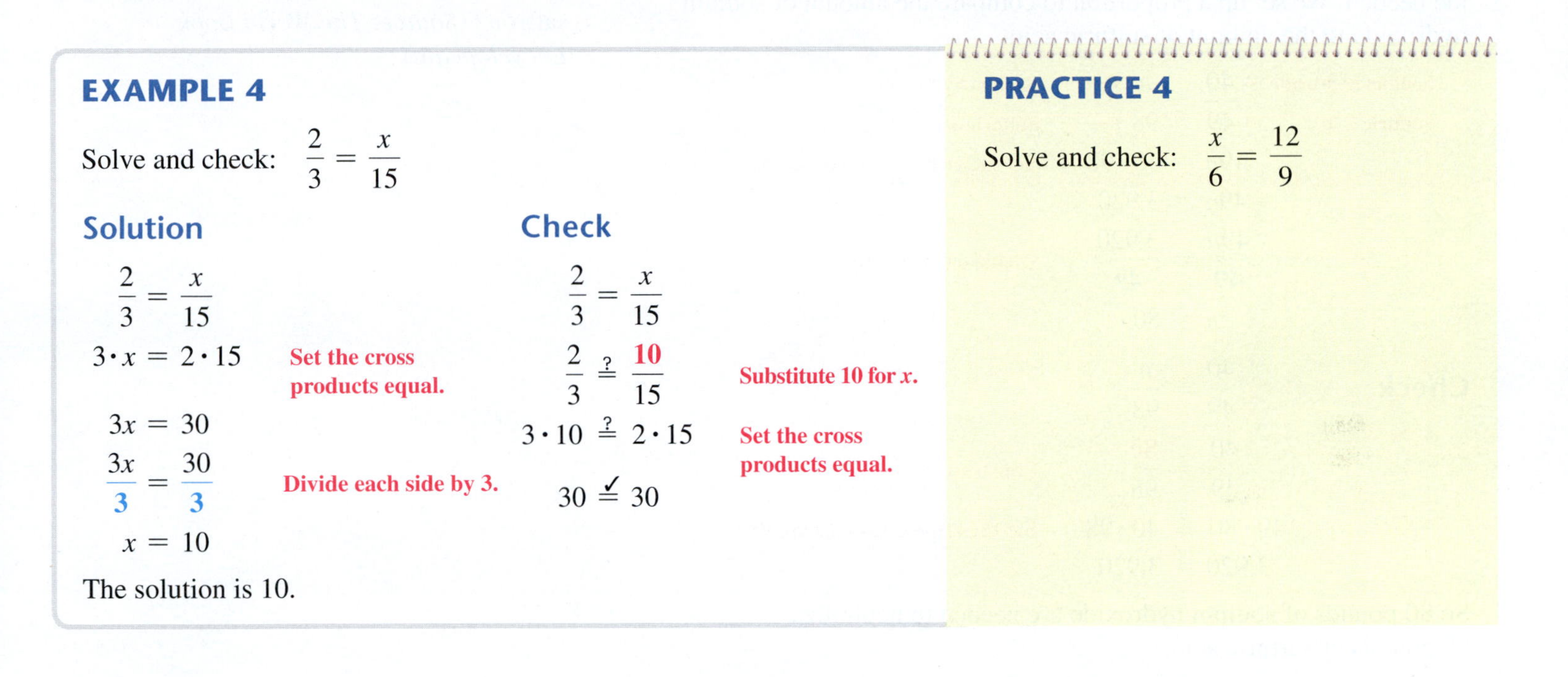

EXAMPLE 4

Solve and check: $\frac{2}{3} = \frac{x}{15}$

Solution

$$\frac{2}{3} = \frac{x}{15}$$

$$3 \cdot x = 2 \cdot 15 \qquad \text{Set the cross products equal.}$$

$$3x = 30$$

$$\frac{3x}{3} = \frac{30}{3} \qquad \text{Divide each side by 3.}$$

$$x = 10$$

The solution is 10.

Check

$$\frac{2}{3} = \frac{x}{15}$$

$$\frac{2}{3} \stackrel{?}{=} \frac{10}{15} \qquad \text{Substitute 10 for } x.$$

$$3 \cdot 10 \stackrel{?}{=} 2 \cdot 15 \qquad \text{Set the cross products equal.}$$

$$30 \stackrel{\checkmark}{=} 30$$

PRACTICE 4

Solve and check: $\frac{x}{6} = \frac{12}{9}$

EXAMPLE 5

Solve and check: $\frac{\frac{1}{4}}{12} = \frac{x}{96}$

Solution

$$\frac{\frac{1}{4}}{12} = \frac{x}{96}$$

$$12 \cdot x = \frac{1}{4} \cdot 96 \qquad \text{Set the cross products equal.}$$

$$12x = 24$$

$$\frac{12x}{12} = \frac{24}{12} \qquad \text{Divide each side by 12.}$$

$$x = 2$$

The solution is 2.

Check

$$\frac{\frac{1}{4}}{12} = \frac{x}{96}$$

$$\frac{\frac{1}{4}}{12} \stackrel{?}{=} \frac{2}{96} \qquad \text{Substitute 2 for } x.$$

$$12(2) \stackrel{?}{=} \frac{1}{4} \cdot (96) \qquad \text{Set the cross products equal.}$$

$$24 \stackrel{\checkmark}{=} 24$$

PRACTICE 5

Solve and check: $\frac{\frac{1}{2}}{2} = \frac{3}{x}$

EXAMPLE 6

Forty pounds of sodium hydroxide are needed to neutralize 49 pounds of sulfuric acid. At this rate, how many pounds of sodium hydroxide are needed to neutralize 98 pounds of sulfuric acid? (*Source:* Peter Atkins and Loretta Jones, *Chemistry*)

Solution Let n represent the number of pounds of sodium hydroxide needed. We set up a proportion to compare the amount of sodium hydroxide to the amount of sulfuric acid.

Sodium hydroxide → 40, n ← Sodium hydroxide
Sulfuric acid → 49, 98 ← Sulfuric acid

$$\frac{40}{49} = \frac{n}{98}$$

$$49n = 40 \cdot 98 \quad \text{Set the cross products equal.}$$

$$49n = 3{,}920$$

$$\frac{49n}{49} = \frac{3{,}920}{49} \quad \text{Divide each side by 49.}$$

$$n = 80$$

Check

$$\frac{40}{49} = \frac{n}{98}$$

$$\frac{40}{49} \stackrel{?}{=} \frac{80}{98} \quad \text{Substitute 80 for } n.$$

$$49 \cdot 80 \stackrel{?}{=} 40 \cdot 98 \quad \text{Set the cross products equal.}$$

$$3{,}920 \stackrel{\checkmark}{=} 3{,}920$$

So 80 pounds of sodium hydroxide are needed to neutralize 98 pounds of sulfuric acid.

PRACTICE 6

Saffron is a powder made from crocus flowers and is used in the manufacture of perfume. Some 8,000 crocus flowers are required to make 2 ounces of saffron. How many flowers are needed to make 16 ounces of saffron? (*Source: The World Book Encyclopedia*)

TIP A good way to set up a proportion is to write quantities of the same kind in the numerators and their corresponding quantities of the other kind in the denominators.

EXAMPLE 7

If St. Louis and Cincinnati on the map below are 1.6 inches apart, what is the actual distance between them?

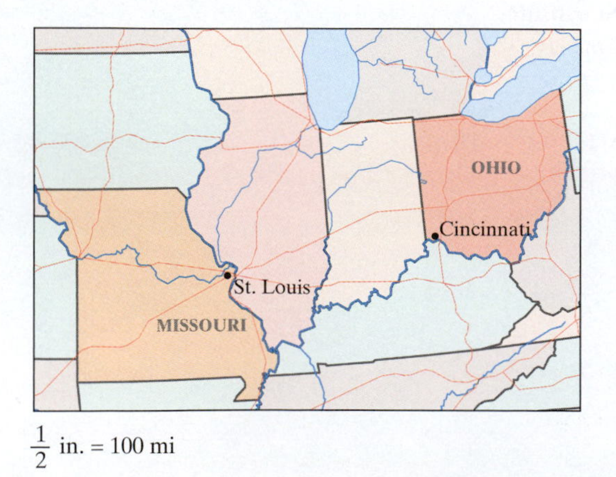

$\frac{1}{2}$ in. = 100 mi

PRACTICE 7

The first aircraft to fly faster than the speed of sound was a research plane piloted by Major Charles E. Yeager of the U.S. Air Force on Oct. 14, 1947. Aircraft speed, especially supersonic, is commonly expressed as a Mach number—the ratio of the speed of the aircraft to the speed of sound (1,066 kilometers per hour). If Yeager's plane reached a top speed of 1,126 kilometers per hour exceeding the speed of sound, what was the Mach speed of the plane, rounded to the nearest hundredth? (*Source:* concorde-jet.com)

Solution We know that $\frac{1}{2}$ inch corresponds to 100 miles. Let's set up a proportion that compares inches to miles, letting m represent the unknown number of miles.

$$\frac{\frac{1}{2}\text{ inch}}{100\text{ miles}} = \frac{1.6\text{ inches}}{m\text{ miles}}$$

$$\frac{\frac{1}{2}}{100} = \frac{1.6}{m}$$

$$\frac{1}{2}m = (100)(1.6) \quad \text{Set the cross products equal.}$$

$$\frac{1}{2}m = 160$$

$$\frac{1}{2}m \div \frac{1}{2} = 160 \div \frac{1}{2} \quad \text{Divide each side by } \frac{1}{2}.$$

$$\frac{1}{2}m \times \frac{2}{1} = 160 \times \frac{2}{1}$$

$$m = 320$$

Check

$$\frac{\frac{1}{2}}{100} = \frac{1.6}{m}$$

$$\frac{\frac{1}{2}}{100} \stackrel{?}{=} \frac{1.6}{320}$$

$$100(1.6) \stackrel{?}{=} \frac{1}{2}\cdot(320)$$

$$160 \stackrel{\checkmark}{=} 160$$

So the cities are 320 miles apart.

EXAMPLE 8

In the following diagram, the heights and shadow lengths of the two objects shown are in proportion. Find the height of the tree h.

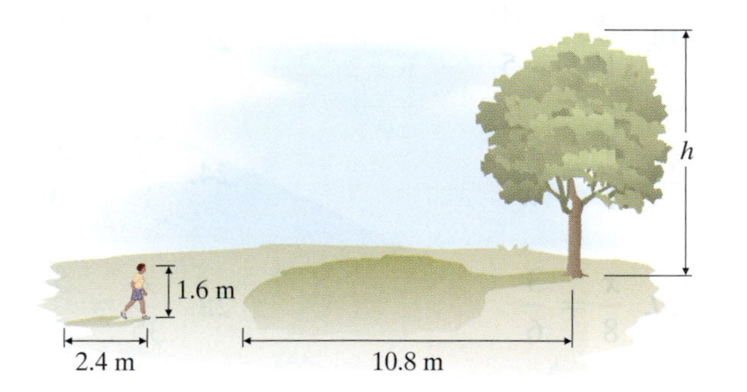

Solution The heights and shadow lengths are in proportion, so we write the following:

$$\begin{matrix}\text{Height} \rightarrow \\ \text{Shadow} \rightarrow\end{matrix} \frac{h\text{ meters}}{10.8\text{ meters}} = \frac{1.6\text{ meters}}{2.4\text{ meters}} \begin{matrix}\leftarrow \text{Height} \\ \leftarrow \text{Shadow}\end{matrix}$$

$$\frac{h}{10.8} = \frac{1.6}{2.4}$$

$$2.4h = (10.8)(1.6)$$

$$\frac{2.4h}{2.4} = \frac{17.28}{2.4}$$

$$h = 7.2$$

Check

$$\frac{h}{10.8} = \frac{1.6}{2.4}$$

$$\frac{7.2}{10.8} \stackrel{?}{=} \frac{1.6}{2.4}$$

$$(10.8)(1.6) \stackrel{?}{=} (7.2)(2.4)$$

$$17.28 \stackrel{\checkmark}{=} 17.28$$

So the height of the tree is 7.2 meters.

PRACTICE 8

A community college in Hawaii has a student-to-faculty ratio of 14 to 1. How many faculty members are at the college if it has 4,200 students? (*Source:* hawaii.edu)

Key Concepts and Skills

CONCEPT SKILL

Concept/Skill	Description	Example
[5.1] Ratio	A comparison of two quantities expressed as a quotient.	3 to 4, $\frac{3}{4}$, or 3:4
[5.1] Rate	A ratio of unlike quantities.	$\frac{10 \text{ students}}{3 \text{ tutors}}$
[5.1] To simplify a ratio	• Write the ratio as a fraction. • Express the fraction in simplest form. • If the quantities are alike, drop the units. If the quantities are unlike, keep the units.	9:27 is the same as 1:3, because $\frac{9}{27} = \frac{1}{3}$ 21 hours to 56 hours $= \frac{21 \text{ hours}}{56 \text{ hours}} = \frac{21}{56} = \frac{3}{8}$ 175 miles per 7 gallons $= \frac{175 \text{ miles}}{7 \text{ gallons}} = \frac{25 \text{ miles}}{1 \text{ gallon}}$, or 25 mpg
[5.1] Unit rate	A rate in which the number in the denominator is 1.	$\frac{180 \text{ calories}}{1 \text{ ounce}}$, or 180 calories per ounce, or 180 cal/oz
[5.1] Unit price	The price of one item, or one unit.	\$0.69 per can, or \$0.69/can
[5.2] Proportion	A statement that two ratios are equal.	$\frac{5}{8} = \frac{15}{24}$
[5.2] To solve a proportion	• Find the cross products, and set them equal. • Solve the resulting equation. • Check the solution by substituting the value of the unknown in the original equation to verify that the resulting proportion is true.	$\frac{6}{9} = \frac{2}{x}$ $6x = 18$ $x = 3$ **Check** $\frac{6}{9} = \frac{2}{x}$ $\frac{6}{9} \stackrel{?}{=} \frac{2}{3}$ $6 \cdot 3 \stackrel{?}{=} 9 \cdot 2$ $18 = 18$ ✓

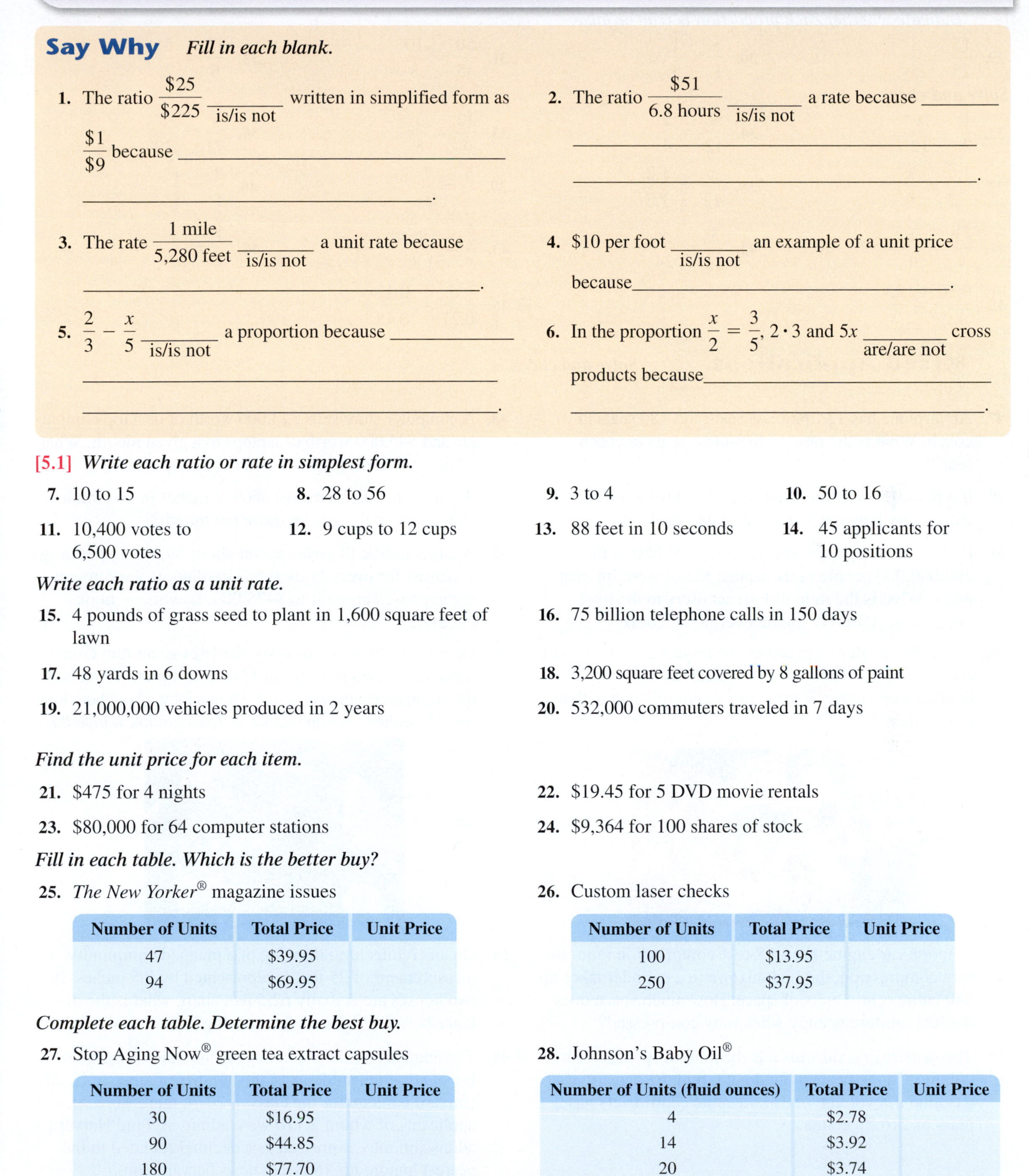

CHAPTER 5 Review Exercises

Say Why *Fill in each blank.*

1. The ratio $\frac{\$25}{\$225}$ ________ (is/is not) written in simplified form as $\frac{\$1}{\$9}$ because ____.

2. The ratio $\frac{\$51}{6.8 \text{ hours}}$ ________ (is/is not) a rate because ____.

3. The rate $\frac{1 \text{ mile}}{5{,}280 \text{ feet}}$ ________ (is/is not) a unit rate because ____.

4. $10 per foot ________ (is/is not) an example of a unit price because____.

5. $\frac{2}{3} - \frac{x}{5}$ ________ (is/is not) a proportion because ____.

6. In the proportion $\frac{x}{2} = \frac{3}{5}$, $2 \cdot 3$ and $5x$ ________ (are/are not) cross products because____.

[5.1] *Write each ratio or rate in simplest form.*

7. 10 to 15
8. 28 to 56
9. 3 to 4
10. 50 to 16
11. 10,400 votes to 6,500 votes
12. 9 cups to 12 cups
13. 88 feet in 10 seconds
14. 45 applicants for 10 positions

Write each ratio as a unit rate.

15. 4 pounds of grass seed to plant in 1,600 square feet of lawn
16. 75 billion telephone calls in 150 days
17. 48 yards in 6 downs
18. 3,200 square feet covered by 8 gallons of paint
19. 21,000,000 vehicles produced in 2 years
20. 532,000 commuters traveled in 7 days

Find the unit price for each item.

21. $475 for 4 nights
22. $19.45 for 5 DVD movie rentals
23. $80,000 for 64 computer stations
24. $9,364 for 100 shares of stock

Fill in each table. Which is the better buy?

25. *The New Yorker®* magazine issues

Number of Units	Total Price	Unit Price
47	$39.95	
94	$69.95	

26. Custom laser checks

Number of Units	Total Price	Unit Price
100	$13.95	
250	$37.95	

Complete each table. Determine the best buy.

27. Stop Aging Now® green tea extract capsules

Number of Units	Total Price	Unit Price
30	$16.95	
90	$44.85	
180	$77.70	

28. Johnson's Baby Oil®

Number of Units (fluid ounces)	Total Price	Unit Price
4	$2.78	
14	$3.92	
20	$3.74	

[5.2] *Indicate whether each proportion is true or false.*

29. $\frac{15}{25} = \frac{3}{5}$ **30.** $\frac{3}{1} = \frac{1}{3}$ **31.** $\frac{50}{45} = \frac{10}{8}$ **32.** $\frac{15}{6} = \frac{5}{2}$

Solve and check.

33. $\frac{1}{2} = \frac{x}{12}$ **34.** $\frac{9}{12} = \frac{x}{4}$ **35.** $\frac{12}{x} = \frac{3}{8}$ **36.** $\frac{x}{72} = \frac{5}{12}$

37. $\frac{1.6}{7.2} = \frac{x}{9}$ **38.** $\frac{x}{12} = \frac{1.2}{1.8}$ **39.** $\frac{5}{\frac{1}{2}} = \frac{7}{x}$ **40.** $\frac{3}{5} = \frac{x}{\frac{2}{3}}$

41. $\frac{2\frac{1}{4}}{x} = \frac{1}{30}$ **42.** $\frac{3}{1\frac{3}{5}} = \frac{x}{24}$ **43.** $\frac{\frac{5}{6}}{x} = \frac{2}{1.8}$ **44.** $\frac{\frac{2}{3}}{4} = \frac{x}{0.9}$

45. $\frac{0.36}{4.2} = \frac{2.4}{x}$ **46.** $\frac{x}{0.21} = \frac{0.12}{0.18}$

Mixed Applications

Solve and check.

47. An airplane has 12 first-class seats and 180 seats in coach. What is the ratio of first-class seats to coach seats?

48. A computer store sells $23,000 worth of desktop computers and $45,000 worth of laptops in a given month. What is the ratio of desktop computer to laptop sales?

49. If a personal care attendant earns $540 for a 6-day workweek, how much does she earn per day?

50. A glacier in Alaska moves about 2 inches in 16 months. How far does the glacier move per month?

51. In a recent year, approximately 200,000,000 of the 300,000,000 people in the United States were Internet users. What is the ratio of Internet users to the total population? (*Source:* internetworldstats.com)

52. A city's public libraries spend about $9.50 in operating expenses for every book they circulate. If their operating expenses amount to $475,000, how many books circulate?

53. In a college's day-care center, the required staff-to-child ratio is 2 to 5. If there are 60 children and 12 staff in the day-care center, is the center in compliance with the requirement?

54. Despite the director's protests, the 1924 silent film *Greed* was edited down from about 42 reels of film to 10 reels. If the original version was about 9 hours long, about how long was the edited version? (*Source: The Film Encyclopedia*)

55. A sports car engine has an 8-to-1 compression ratio. Before compression, the fuel mixture in a cylinder takes up 440 cubic centimeters of space. How much space does the fuel mixture occupy when fully compressed?

56. On an architectural drawing of a planned community, a measurement of 25 feet is represented by 0.5 inches. If two houses are actually 62.5 feet apart, what is the distance between them on the drawing?

57. The *density of a substance* is the ratio of its mass to its volume. To the nearest hundredth, find the density of gasoline if a volume of 317.45 cubic centimeters has a mass of 216.21 grams.

58. The admission rate at a college is the ratio of the number of admitted students to the number of applicants. At Harvard College for the class of 2014, there were 30,489 applicants of whom 2,110 were admitted. Find Harvard's admission rate, expressed as a decimal rounded to the nearest hundredth. (*Source:* news.harvard.edu)

• Check your answers on page A-7.

CHAPTER 5 Posttest

FOR EXTRA HELP

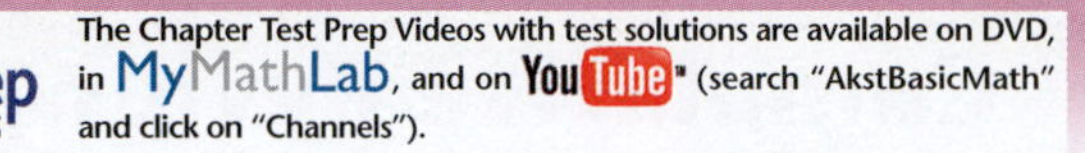

CHAPTER
Test Prep
VIDEOS
The Chapter Test Prep Videos with test solutions are available on DVD, in MyMathLab, and on YouTube (search "AkstBasicMath" and click on "Channels").

To see if you have mastered the topics in this chapter, take this test.

Write each ratio or rate in simplest form.

1. 8 to 12

2. 15 to 42

3. 55 ounces to 31 ounces

4. 180 miles to 15 miles

5. 65 revolutions in 60 seconds

6. 3 centimeters for every 75 kilometers

Find the unit rate.

7. 340 miles in 5 hours

8. 200-meter dash in 25 seconds

Determine the unit price.

9. $4,080 for 30 days

10. 25 greeting cards for $20

Determine whether each proportion is true or false.

11. $\frac{8}{21} \stackrel{?}{=} \frac{16}{40}$

12. $\frac{7}{3} \stackrel{?}{=} \frac{63}{27}$

Solve and check.

13. $\frac{15}{x} = \frac{6}{10}$

14. $\frac{102}{17} = \frac{36}{x}$

15. $\frac{0.9}{36} = \frac{0.7}{x}$

16. $\frac{\frac{1}{3}}{4} = \frac{x}{12}$

Solve.

17. To advertise his business, an owner can purchase 3 million e-mail addresses for $120 or 5 million e-mail addresses for $175. Which is the better buy?

18. The Association of American Medical Colleges has called for increasing the number of students attending medical schools so as to reduce projected physician shortages. In a recent year, the entering class of medical schools was about 18,000 students as compared to 16,000 students five years earlier. What is the ratio of the later enrollment to the earlier enrollment? (*Source:* aamc.org)

19. A man $6\frac{1}{4}$ feet tall casts a 5-foot shadow. A nearby tree casts a 20-foot shadow. If the heights and shadow lengths of the man and tree are proportional, how tall is the tree?

20. A nurse takes his patient's pulse. What is the patient's pulse per minute if it beats 12 times in 15 seconds?

• Check your answers on page A-8.

Cumulative Review Exercises

To help you review, solve the following:

1. Add: $\$93{,}281 + \$8{,}429 + \$6{,}701$

2. Divide: $\dfrac{5{,}103}{27}$

3. Calculate: $7 \cdot 2^3 - \dfrac{21 - 13}{2}$

4. Write the prime factorization of 168.

5. Find the difference: $3\dfrac{1}{10} - 2\dfrac{7}{10}$

6. Divide: $\dfrac{1}{3} \div 3\dfrac{1}{4}$

7. Simplify: $\dfrac{2}{3} \times 1\dfrac{1}{2} - \dfrac{1}{4}$

8. Multiply: $8.2 \times 1{,}000$

9. Estimate: $12\dfrac{1}{7} \div 3\dfrac{9}{10}$

10. Solve and check: $x + 6.5 = 9$

11. Solve and check: $\dfrac{3}{10}n = 21$

12. Simplify the ratio: 2.5 to 10

13. Find the unit price: 3 yards for $12

14. Solve and check: $\dfrac{\frac{1}{2}}{4} = \dfrac{x}{6}$

15. What is the area of the singles tennis court shaded in the diagram?

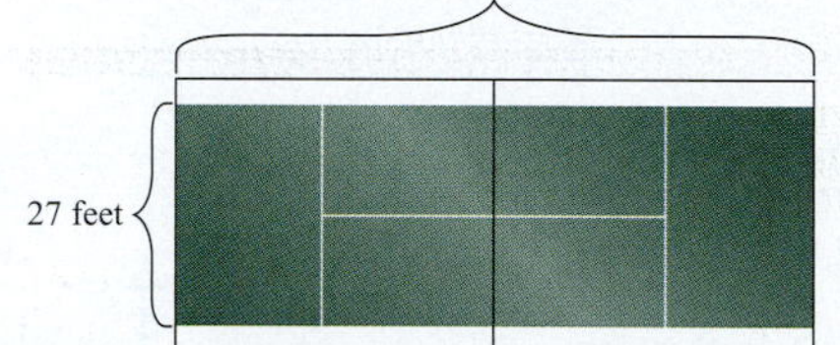

16. A college graduate looking for a teaching position takes a temporary substitute-teaching job. He works two days and makes $178.35. At this rate of pay, how much would he make, to the nearest cent, for teaching five days? (*Source:* okaloosaschools.com)

17. The barometric pressure fell from 30.02 inches to 29.83 inches. By how many inches did it fall?

18. Write the algebraic expression for the number of miles a driver travels if she drives at a speed of r miles per hour for t hours. How far will she travel in 4 hours at a speed of 65 miles per hour?

19. A rule of thumb for growing lily bulbs is to plant them 3 times as deep as they are wide. How deep should a gardener plant a lily bulb that is 2.5 inches wide?

20. In 2010, air traffic between the United States and Western Europe was disrupted because of ash spewing from a volcano in Iceland. On one day alone, 1,000 of 29,000 scheduled flights were cancelled. What fraction of the flights were cancelled? (*Source:* cnn.com)

• Check your answers on page A-8.

CHAPTER 5 Posttest

To see if you have mastered the topics in this chapter, take this test.

Write each ratio or rate in simplest form.

1. 8 to 12

2. 15 to 42

3. 55 ounces to 31 ounces

4. 180 miles to 15 miles

5. 65 revolutions in 60 seconds

6. 3 centimeters for every 75 kilometers

Find the unit rate.

7. 340 miles in 5 hours

8. 200-meter dash in 25 seconds

Determine the unit price.

9. $4,080 for 30 days

10. 25 greeting cards for $20

Determine whether each proportion is true or false.

11. $\frac{8}{21} \stackrel{?}{=} \frac{16}{40}$

12. $\frac{7}{3} \stackrel{?}{=} \frac{63}{27}$

Solve and check.

13. $\frac{15}{x} = \frac{6}{10}$

14. $\frac{102}{17} = \frac{36}{x}$

15. $\frac{0.9}{36} = \frac{0.7}{x}$

16. $\frac{\frac{1}{3}}{4} = \frac{x}{12}$

Solve.

17. To advertise his business, an owner can purchase 3 million e-mail addresses for $120 or 5 million e-mail addresses for $175. Which is the better buy?

18. The Association of American Medical Colleges has called for increasing the number of students attending medical schools so as to reduce projected physician shortages. In a recent year, the entering class of medical schools was about 18,000 students as compared to 16,000 students five years earlier. What is the ratio of the later enrollment to the earlier enrollment? (*Source:* aamc.org)

19. A man $6\frac{1}{4}$ feet tall casts a 5-foot shadow. A nearby tree casts a 20-foot shadow. If the heights and shadow lengths of the man and tree are proportional, how tall is the tree?

20. A nurse takes his patient's pulse. What is the patient's pulse per minute if it beats 12 times in 15 seconds?

• Check your answers on page A-8.

Cumulative Review Exercises

To help you review, solve the following:

1. Add: $\$93,281 + \$8,429 + \$6,701$

2. Divide: $\frac{5,103}{27}$

3. Calculate: $7 \cdot 2^3 - \frac{21 - 13}{2}$

4. Write the prime factorization of 168.

5. Find the difference: $3\frac{1}{10} - 2\frac{7}{10}$

6. Divide: $\frac{1}{3} \div 3\frac{1}{4}$

7. Simplify: $\frac{2}{3} \times 1\frac{1}{2} - \frac{1}{4}$

8. Multiply: $8.2 \times 1,000$

9. Estimate: $12\frac{1}{7} \div 3\frac{9}{10}$

10. Solve and check: $x + 6.5 = 9$

11. Solve and check: $\frac{3}{10}n = 21$

12. Simplify the ratio: 2.5 to 10

13. Find the unit price: 3 yards for $12

14. Solve and check: $\frac{\frac{1}{2}}{4} = \frac{x}{6}$

15. What is the area of the singles tennis court shaded in the diagram?

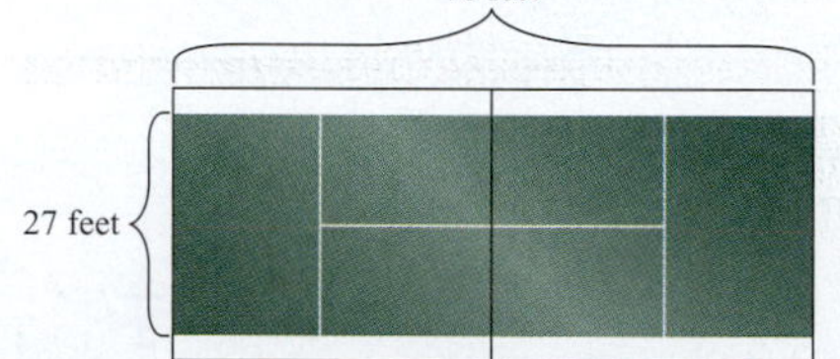

16. A college graduate looking for a teaching position takes a temporary substitute-teaching job. He works two days and makes $178.35. At this rate of pay, how much would he make, to the nearest cent, for teaching five days? (*Source:* okaloosaschools.com)

17. The barometric pressure fell from 30.02 inches to 29.83 inches. By how many inches did it fall?

18. Write the algebraic expression for the number of miles a driver travels if she drives at a speed of r miles per hour for t hours. How far will she travel in 4 hours at a speed of 65 miles per hour?

19. A rule of thumb for growing lily bulbs is to plant them 3 times as deep as they are wide. How deep should a gardener plant a lily bulb that is 2.5 inches wide?

20. In 2010, air traffic between the United States and Western Europe was disrupted because of ash spewing from a volcano in Iceland. On one day alone, 1,000 of 29,000 scheduled flights were cancelled. What fraction of the flights were cancelled? (*Source:* cnn.com)

• Check your answers on page A-8.

CHAPTER 6

Percents

Percents and Political Polls

In 2007, Barack Obama and Hillary Clinton competed in Iowa for the Democratic nomination for President of the United States. When asked which candidate they would vote for if the Democratic caucus were held that day, 30% of interviewees said Mr. Obama and 26% Mrs. Clinton. But every poll and survey has a *margin of error*. The Iowa poll's margin of error was plus or minus 5%, meaning that support for Mr. Obama was likely between 25% and 35%, and support for Mrs. Clinton between 21% and 31%. The results in Iowa were therefore too close to call—there was essentially a tie. Barack Obama went on to be nominated and to win the election.

(*Source*: *New York Times*, November 28, 2007)

CHAPTER 6 PRETEST

To see if you have already mastered the topics in this chapter, take this test.

Rewrite.

1. 5% as a fraction

2. $37\frac{1}{2}\%$ as a fraction

3. 250% as a decimal

4. 3% as a decimal

5. 0.007 as a percent

6. 8 as a percent

7. $\frac{2}{3}$ as a percent, rounded to the nearest whole percent

8. $1\frac{1}{10}$ as a percent

Solve.

9. What is 75% of 50 feet?

10. Find 110% of 50.

11. 80% of what number is 25.6?

12. 2% of what number is 5?

13. What percent of 10 is 4?

14. What percent of 4 is 10?

15. In a municipal savings account, a city employee earned 3% interest on $350. How much money did the employee earn in interest for 1 year?

16. The number of students enrolled at a community college rose from 2,475 last year to 2,673 this year. What was the percent increase in the college's enrollment?

17. In the depths of the Great Depression, 24% of the U.S. civilian labor force was unemployed. Write this percent as a simplified fraction. (*Source:* census.gov)

18. In a chemistry lab, a student dissolved 10 milliliters of acid in 30 milliliters of water. What percent of the solution was acid?

19. For parties of 8 or more, a restaurant automatically adds an 18% tip to the restaurant check. What tip would be added to a dinner check for a party of 10 if the total bill was $339.50?

20. A patient's health insurance covered 80% of the cost of her operation. She paid the remainder, which came to $2,000. Find the total cost of the operation.

• Check your answers on page A-8.

6.1 Introduction to Percents

OBJECTIVES

- **A** To find the fraction or the decimal equivalent of a given percent
- **B** To find the percent equivalent of a given fraction or decimal
- **C** To solve applied problems involving percent conversions

What Percents Are and Why They Are Important

Percent means divided by 100. So 50% (read "fifty percent") means 50 divided by 100 (or 50 out of 100).

A percent can also be thought of as a ratio or a fraction with denominator 100. For example, we can look at 50% either as the ratio of 50 parts to 100 parts or as the fraction $\frac{50}{100}$, or $\frac{1}{2}$. Since a fraction can be written as a decimal, we can also think of 50% as 0.50, or 0.5.

In the diagram at the right, 50 of the 100 squares are shaded. This shaded portion represents 50%.

We can use diagrams to represent other percents.

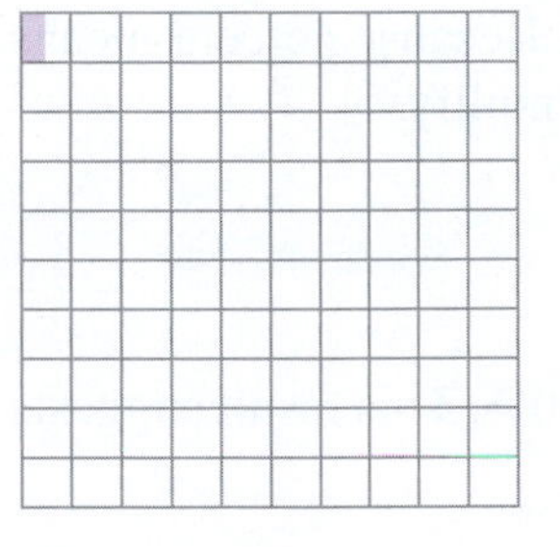

In the diagram to the left, $\frac{1}{2}$% is equivalent to the shaded portion,

$$\frac{\frac{1}{2}}{100}, \text{ or } \frac{1}{200}.$$

The entire diagram at the right is shaded, so 100% means $\frac{100}{100}$, or 1.

We can express 105% as $\frac{105}{100}$, or $1\frac{1}{20}$, as shown by the shaded portions of the diagrams.

Percents are commonly used, as the following statements taken from a single page of a newspaper illustrate.

- About 10% of the city's budget goes to sanitation.
- Blanket Sale—30% to 40% off!
- The number of victims of the epidemic increased by 125% in just 6 months.

A key reason for using percents so frequently is that they are easy to compare. For instance, we can tell right away that a discount of 30% is larger than a discount of 22%, simply by comparing the whole numbers 30 and 22.

To see how percents relate to fractions and decimals, let's consider finding equivalent fractions, decimals, and percents. In Chapter 3, we discussed two of the six types of conversions:

- changing a decimal to a fraction, and
- changing a fraction to a decimal.

Here, we consider the remaining four types of conversions:

- changing a percent to a fraction,
- changing a percent to a decimal,
- changing a decimal to a percent, and
- changing a fraction to a percent.

Note that each type of conversion changes the way the number is written—but not the number itself.

Changing a Percent to a Fraction

Suppose that we want to rewrite a percent—say, 30%—as a fraction. Because percent means divided by 100, we simply drop the % sign, place 30 over 100, and simplify.

$$30\% = \frac{30}{100} = \frac{3}{10}$$

Therefore, the fraction $\frac{3}{10}$ is just another way of writing the percent 30%. This result suggests the following rule:

To Change a Percent to the Equivalent Fraction

- Drop the % sign from the given percent and place the number over 100.
- Simplify the resulting fraction, if possible.

EXAMPLE 1

Write 7% as a fraction.

Solution To change this percent to a fraction, we drop the percent sign and write the 7 over 100. The fraction is already in lowest terms.

$$7\% = \frac{7}{100}$$

PRACTICE 1

Find the fractional equivalent of 21%.

EXAMPLE 2

Express 150% as a fraction.

Solution $150\% = \frac{150}{100} = \frac{3}{2}$, or $1\frac{1}{2}$

Note that the answer is larger than 1 because the original percent was more than 100%.

PRACTICE 2

What is the fractional equivalent of 225%?

EXAMPLE 3

Express $\frac{1}{10}\%$ as a fraction.

Solution To find the equivalent fraction, we drop the % sign and then put the number over 100.

$$\frac{\frac{1}{10}}{100} = \frac{1}{10} \div 100 = \frac{1}{10} \div \frac{100}{1} = \frac{1}{10} \times \frac{1}{100} = \frac{1}{1{,}000}$$

So $\frac{1}{10}\%$ expressed as a fraction is $\frac{1}{1{,}000}$.

PRACTICE 3

Change $\frac{2}{3}\%$ to a fraction.

EXAMPLE 4

Express $33\frac{1}{3}\%$ as a fraction.

Solution To find the equivalent fraction, we first drop the % sign and then put the number over 100.

$$\frac{33\frac{1}{3}}{100} = 33\tfrac{1}{3} \div 100 = 33\tfrac{1}{3} \div \frac{100}{1} = \frac{100}{3} \div \frac{100}{1} = \frac{\overset{1}{\cancel{100}}}{3} \times \frac{1}{\underset{1}{\cancel{100}}} = \frac{1}{3}$$

So $33\frac{1}{3}\%$ expressed as a fraction is $\frac{1}{3}$.

PRACTICE 4

Change $12\frac{1}{2}\%$ to a fraction.

EXAMPLE 5

The Ring of Fire contains 75% of the volcanoes on Earth. Express this percent as a fraction. (*Source:* nationalgeographic.com)

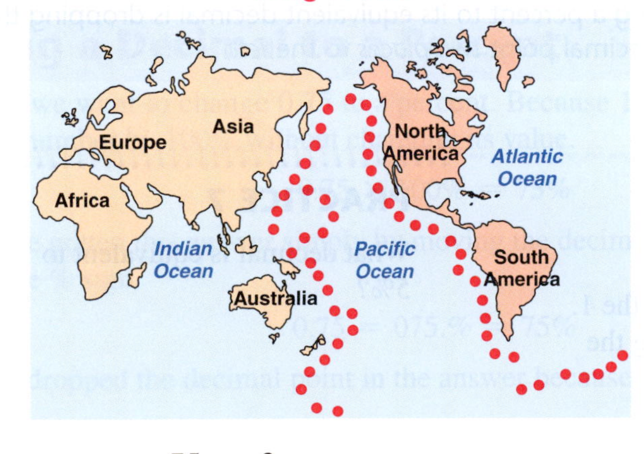

Solution $75\% = \frac{75}{100} = \frac{3}{4}$

So $\frac{3}{4}$ of the volcanoes on Earth are located in the Ring of Fire.

PRACTICE 5

Rwanda is the first country where more than half of the members of the legislature (56%) are women. Express this percent as a fraction. (*Source: Top 10 of Everything, 2010*)

EXAMPLE 11

Write 0.425 as a percent.

Solution We multiply 0.425 by 100 and add a % sign.

$$0.425 = 0.425 \times 100\% = 42.5\%, \text{ or } 42\frac{1}{2}\%$$

PRACTICE 11

What percent is equivalent to the decimal 0.025?

TIP A shortcut for changing a decimal to its equivalent percent is inserting a % sign and moving the decimal point *two places* to the *right*.

EXAMPLE 12

Convert 0.03 to a percent.

Solution $0.03 = 003.\% = 3\%$

PRACTICE 12

Change 0.09 to a percent.

EXAMPLE 13

Express 0.1 as a percent.

Solution In the given number, only a single digit is to the right of the decimal point. So to move the decimal point two places to the right, we need to insert a 0 as a placeholder.

$$0.1 = 0.10 = 10.\% = 10\%$$

PRACTICE 13

What percent is equivalent to 0.7?

EXAMPLE 14

What percent is equivalent to 2?

Solution Recall that a whole number such as 2 has a decimal point understood to its right. We move the decimal point two places to the right.

$$2 = 2. = 2.00 = 200.\% = 200\%$$

So the answer is 200%, which makes sense: 200% is double 100%, just as 2 is double 1.

PRACTICE 14

Rewrite 3 as a percent.

EXAMPLE 15

Express 0.2483 as a percent, rounded to the nearest whole percent.

Solution First, we obtain the exact percent equivalent.

$$0.2483 = 24.83\%$$

To round this number to the nearest whole percent, we underline the digit 4. Then, we check the critical digit immediately to its right. This digit is 8, so we round up.

$$2\underline{4}.83\% \approx 25.\% = 25\%$$

PRACTICE 15

Convert 0.714 to a percent, rounded to the nearest whole percent.

EXAMPLE 16

Red blood cells make up about 0.4 of the total blood volume in the human body, whereas 55% of the total blood volume is plasma. Which makes up more of the blood volume—red blood cells or plasma? Explain. (*Source:* Mayo Clinic)

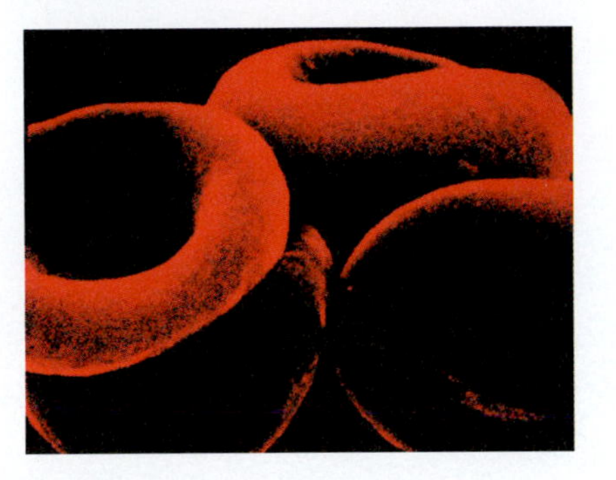

Solution We want to compare the decimal 0.4 and the percent 55%. One way is to change the decimal to a percent.

$$0.4 = 0.40 = 40.\% = 40\%$$

Since 40% is less than 55%, we conclude that plasma makes up more of the blood volume.

PRACTICE 16

Air is a mixture of many gases. For example, 0.78 of air is nitrogen, and 0.93% is argon. Is there more nitrogen or argon in air? Explain.

Changing a Fraction to a Percent

Now, let's change a fraction to a percent. Consider, for instance, the fraction $\frac{1}{5}$. To convert this fraction to a percent, multiply $\frac{1}{5}$ by 100%, which is equal to 1.

$$\frac{1}{5} = \frac{1}{5} \times 100\% = \frac{1}{\cancelto{1}{5}} \times \frac{\cancelto{20}{100}}{1}\% = 20\%$$

To Change a Fraction to the Equivalent Percent

- Multiply the fraction by 100%.

EXAMPLE 17

Rewrite $\frac{7}{20}$ as a percent.

Solution To change the given fraction to a percent, we multiply by 100%.

$$\frac{7}{20} = \frac{7}{20} \times 100\% = \frac{7}{\cancel{20}_{1}} \times \frac{\cancel{100}^{5}}{1}\% = 35\%$$

PRACTICE 17

Convert $\frac{4}{25}$ to a percent.

EXAMPLE 18

Which is larger: 130% or $1\frac{3}{8}$?

Solution To compare, let's express $1\frac{3}{8}$ as a percent.

$$1\frac{3}{8} = 1\frac{3}{8} \times 100\% = \frac{11}{8} \times \frac{100}{1}\%$$

$$= \frac{11}{\cancel{8}_{2}} \times \frac{\cancel{100}^{25}}{1}\% = \frac{275}{2}\% = 137\frac{1}{2}\%$$

Because $137\frac{1}{2}\%$ is larger than 130%, so is $1\frac{3}{8}$.

PRACTICE 18

True or false: $\frac{2}{3} > 60\%$. Justify your answer.

EXAMPLE 19

A student got 28 of 30 questions correct on a test. If all the questions were equal in value, what was the student's grade, rounded to the nearest whole percent?

Solution The student answered $\frac{28}{30}$ of the questions right. To find the student's grade, we change this fraction to a percent.

$$\frac{28}{30} = \frac{28}{30} \times 100\% = \frac{28}{\cancel{30}_{3}} \times \frac{\cancel{100}^{10}}{1}\%$$

$$= \frac{280}{3}\% = 93\frac{1}{3}\% = 93.3\ldots\% \approx 93\%$$

Note that the critical digit is 3, so we round down. The rounded grade was therefore 93%.

PRACTICE 19

An administrative assistant spends $490 out of her monthly salary of $1,834 on rent. What percent of her monthly salary is spent on rent, rounded to the nearest whole percent?

6.1 Exercises

FOR EXTRA HELP MyMathLab MathXL PRACTICE WATCH READ REVIEW

Mathematically Speaking

Fill in each blank with the most appropriate term or phrase from the given list.

right	fraction	percent	divide
decimal	left	whole number	multiply

1. A(n) __________ is a ratio or fraction with denominator 100.

2. To change a percent to the equivalent __________, drop the % sign from the given percent, and place the number over 100.

3. To change a percent to the equivalent decimal, move the decimal point two places to the __________ and drop the % sign.

4. To change a fraction to the equivalent percent, __________ the fraction by 100 and insert a % sign.

A *Change each percent to a fraction or mixed number. Simplify.*

5. 8% **6.** 3% **7.** 250% **8.** 110%

9. 33% **10.** 41% **11.** 18% **12.** 44%

13. 14% **14.** 45% **15.** 65% **16.** 92%

17. $\frac{3}{4}\%$ **18.** $\frac{1}{10}\%$ **19.** $\frac{3}{10}\%$ **20.** $\frac{1}{5}\%$

21. $7\frac{1}{2}\%$ **22.** $2\frac{1}{2}\%$ **23.** $14\frac{2}{7}\%$ **24.** $28\frac{4}{7}\%$

Convert each percent to a decimal.

25. 6% **26.** 9% **27.** 72% **28.** 25%

29. 0.1% **30.** 0.2% **31.** 102% **32.** 113%

33. 42.5% **34.** 10.5% **35.** 500% **36.** 400%

37. $106\frac{9}{10}\%$ **38.** $201\frac{1}{10}\%$ **39.** $3\frac{1}{2}\%$ **40.** $2\frac{4}{5}\%$

41. $\frac{9}{10}\%$ **42.** $\frac{7}{10}\%$ **43.** $\frac{3}{4}\%$ **44.** $\frac{1}{4}\%$

B *Express each decimal as a percent.*

45. 0.31 **46.** 0.37 **47.** 0.17 **48.** 0.18

49. 0.3 **50.** 0.4 **51.** 0.04 **52.** 0.05

53. 0.125 **54.** 0.875 **55.** 1.29 **56.** 1.07

57. 2.9 **58.** 3.5 **59.** 2.87 **60.** 3.62

61. 1.016 **62.** 1.003 **63.** 9 **64.** 7

Change each fraction to a percent.

65. $\frac{3}{10}$ **66.** $\frac{1}{2}$ **67.** $\frac{1}{10}$ **68.** $\frac{3}{20}$

69. $\frac{4}{25}$ **70.** $\frac{6}{25}$ **71.** $\frac{9}{10}$ **72.** $\frac{7}{10}$

73. $\frac{3}{50}$ **74.** $\frac{1}{50}$ **75.** $\frac{5}{9}$ **76.** $\frac{2}{9}$

77. $\frac{1}{9}$ **78.** $\frac{4}{7}$ **79.** 6 **80.** 8

81. $1\frac{1}{2}$ **82.** $2\frac{3}{5}$ **83.** $2\frac{1}{6}$ **84.** $1\frac{1}{3}$

Replace ■ *with* $<$ *or* $>$.

85. $2\frac{1}{4}$ ■ 240% **86.** $3\frac{5}{6}$ ■ 380% **87.** $\frac{1}{2}\%$ ■ 50% **88.** $\frac{1}{40}$ ■ $\frac{1}{4}\%$

Express as a percent, rounded to the nearest whole percent.

89. $\frac{4}{9}$ **90.** $\frac{3}{7}$ **91.** 2.2469 **92.** 1.1633

Complete each table.

93.

Fraction	Decimal	Percent
		$33\frac{1}{3}$
	0.666 …	
	0.25	
		75%
		20%
$\frac{2}{5}$		
	0.6	

94.

Fraction	Decimal	Percent
	0.8	
$\frac{1}{6}$		
$\frac{5}{6}$		
		12.5%
	0.375	
		$62\frac{1}{2}\%$
	0.875	

Mixed Practice

Solve.

95. Change 104% to a mixed number.

96. What percent is equivalent to $\frac{2}{5}$?

97. Express $3\frac{1}{6}$ as a percent.

98. Express $62\frac{1}{2}\%$ as a fraction.

99. Convert 27.5% to a decimal.

100. Find the decimal equivalent to $\frac{3}{8}\%$.

101. What percent is equivalent to 3.1?

102. Change 0.003 to a percent.

103. Which is smaller, $2\frac{5}{9}$ or 254%?

104. Express 1.2753 to the nearest whole percent.

Applications

C *Solve.*

105. It is estimated that 96% of all e-mail messages received are *spam* (unsolicited junk e-mail). Express this percent as a decimal. (*Source:* govtech.com)

106. According to a recent study, 65% of children have had an imaginary companion by age 7. Express this percent as a fraction. (*Source:* uwnews.org)

107. The following graph shows the percent of people in the United States who get their local news regularly from various sources.

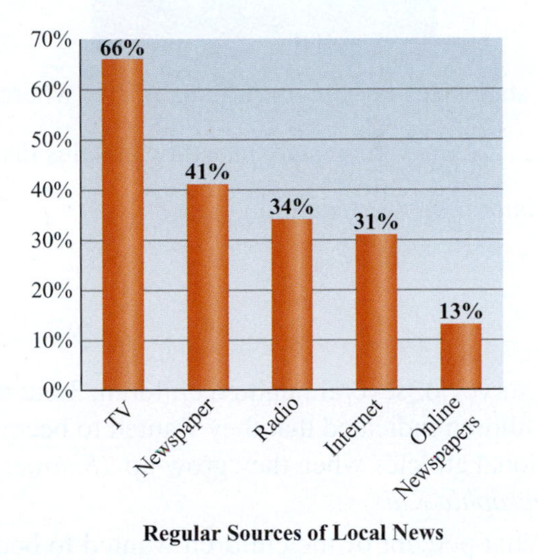

Regular Sources of Local News
(*Source:* people-press.org)

What fraction of people in the U.S. do *not* get their news regularly from TV?

108. The following graph shows the distribution of investments for a retiree. Express as a decimal the percent of investments that are in equities.

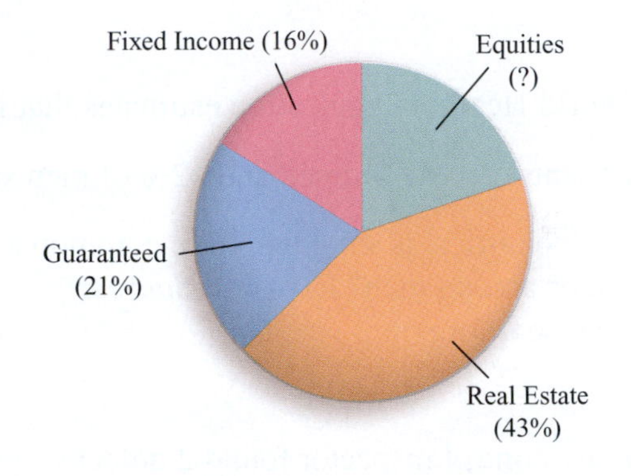

109. According to the nutrition label, one large egg contains 6 grams of protein. This is 10% of the daily value (DV) for protein. Express this percent as a fraction.

110. A bank offers a Visa credit card with a fixed annual percentage rate (APR) of 16.99%. Express the APR as a decimal.

111. Los Angeles has an area 10.05 times that of San Francisco. Express this decimal as a percent. (*Source:* U.S. Census Bureau)

112. In France, the main source of electricity is nuclear power. In a recent year, more than $\frac{3}{4}$ of the country's total electrical production was nuclear. Write this fraction as a percent. (*Source:* International Atomic Energy Agency)

113. When the recession ended, the factory's output grew by 135%. Write this percent as a simplified mixed number.

114. According to a survey, 78% of the arguments that couples have are about money. Express this percent as a decimal.

115. According to a recent U.S. Bureau of Labor Statistics report, 51.5% of all union members are government employees. Convert this percent to a decimal. (*Source:* rsc.tomprice.house.gov)

116. After an oil spill, 15% of the wildlife survived. Express this percent as a fraction.

117. The World Health Organization estimates that in developed countries, $\frac{1}{4}$ of women and 42% of men smoke. In these countries, is smoking more common among women or among men? Explain. (*Source:* americanheart.org)

118. The state sales tax rate in Indiana is 7%, whereas in Iowa it is $\frac{3}{50}$. Which state has a lower sales tax rate? Explain. (*Source:* taxadmin.org)

119. A quality control inspector found 2 defective machine parts out of 500 manufactured.

a. What percent of the machine parts manufactured were defective?

b. What percent of the machine parts manufactured were not defective?

120. In a survey of several hundred children, 3 out of every 25 children indicated that they wanted to become professional athletes when they grow up. (*Source: National Geographic Kids*)

a. What percent of the children wanted to become professional athletes?

b. What percent of the children did not want to become professional athletes?

121. In the 2008 U.S. presidential election, 131,257,328 people turned out to vote. At the time, there were 189,844,867 registered voters. What percent of the registered voters, to the nearest whole percent, voted? (*Sources:* fec.gov and eac.gov)

122. The first Social Security retirement benefits were paid in 1940 to Ida May Fuller of Vermont. She had paid in a total of $24.85 and got back $20,897 before her death in 1975. Express the ratio of what she got back to what she put in as a percent, rounded to the nearest whole percent. (*Source:* James Trager, *The People's Chronology*)

• Check your answers on page A-8.

MINDStretchers

Mathematical Reasoning

1. By mistake, you move the decimal point to the right instead of to the left when changing a percent to a decimal. Your answer is how many times as large as the correct answer?

Writing

2. A study of the salt content of seawater showed that the average salt content varies from 33‰ to 37‰, where the symbol ‰ (read "per mil") means "for every thousand." Explain why you think the scientist who wrote this study did not use the % symbol.

Critical Thinking

3. What percent of the region shown is shaded in?

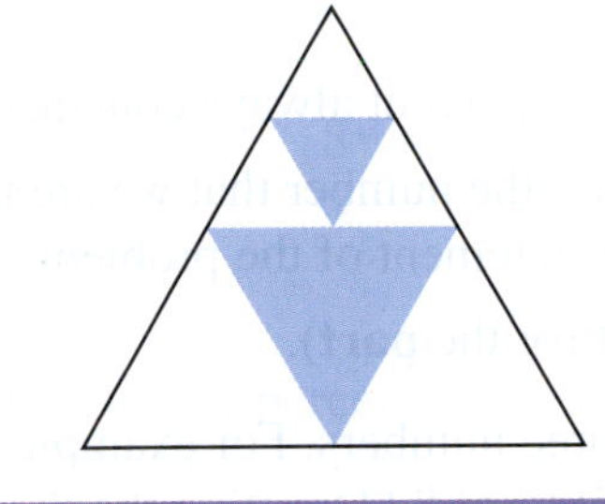

Cultural Note

Throughout history, the concepts of percent and taxation have been interrelated. At the peak of the Roman Empire, the Emperor Augustus instituted an inheritance tax of 5% to provide retirement funds for the military. Another emperor, Julius Caesar, imposed a 1% sales tax on the population. And in Roman Asia, tax collectors exacted a tithe of 10% on crops. If landowners could not pay, the collectors offered to lend them funds at interest rates that ranged from 12% up to 48%.

Roman taxation served as a model for modern countries when these countries developed their own systems of taxation many centuries later.

Sources: Frank J. Swetz, *Capitalism and Arithmetic: The New Math of the 15th Century,* Open Court, 1987; Carolyn Webber and Aaron Wildavsky, *A History of Taxation and Expenditure in the Western World,* Simon and Schuster, 1986.

TIP When the percent is less than 100%, the amount is *less* than the base. When the percent is more than 100%, the amount is *more* than the base.

EXAMPLE 4

What is $66\frac{2}{3}\%$ of 15?

Solution First, let's change the percent to a fraction:

$$66\frac{2}{3}\% = \frac{66\frac{2}{3}}{100} = 66\frac{2}{3} \div \frac{100}{1} = \frac{\overset{2}{\cancel{200}}}{3} \times \frac{1}{\underset{1}{\cancel{100}}} = \frac{2}{3}$$

Translating the question to an equation, we get:

$$\begin{array}{ccccc} \text{What} & \text{is} & 66\frac{2}{3}\% & \text{of} & 15? \\ \downarrow & \downarrow & \downarrow & \downarrow & \downarrow \\ x & = & \frac{2}{3} & \cdot & 15 \end{array}$$

$$= \frac{2}{\underset{1}{\cancel{3}}} \cdot \overset{5}{\cancel{15}} = \frac{10}{1} = 10$$

So $66\frac{2}{3}\%$ of 15 is 10.

PRACTICE 4

Find $33\frac{1}{3}\%$ of 600.

EXAMPLE 5

A marketing account manager has 3.5% of her monthly salary put into a 401(k) plan. How much did she put into the 401(k) plan if her monthly salary is $3,200?

Solution We are looking for the monthly amount placed into the 401(k) plan, which is 3.5% of $3,200.

$$\begin{array}{ccccc} \text{What} & \text{is} & 3.5\% & \text{of} & \$3{,}200? \\ \downarrow & \downarrow & \downarrow & \downarrow & \downarrow \\ x & = & (0.035) & \cdot & (3{,}200) \\ & = & 112 & & \end{array}$$

So she has $112 per month put into the 401(k) plan. Note that this amount has the same unit (dollars) as the base.

PRACTICE 5

Of the 600 workers at a factory, 8.5% belong to a union. How many workers are in the union?

Finding a Base

Now, let's consider some examples of using the translation method to find the base when we know the percent and the amount.

EXAMPLE 6

4% of what number is 8?

Solution We begin by writing the appropriate equation.

4%	of	what number	is	8?
↓	↓	↓	↓	↓
0.04	·	x	=	8

Next, we solve this equation.

$$0.04x = 8$$

$$\frac{0.04}{0.04}x = \frac{8}{0.04} \quad \text{Divide each side by 0.04.}$$

$$x = \frac{8}{0.04} = 200 \qquad 0.04\overline{)8.00} = 4\overline{)800.}\ \ 200.$$

So 4% of 200 is 8.

PRACTICE 6

6 is 12% of what number?

EXAMPLE 7

108 is 120% of what number?

Solution We consider the following question:

120%	of	what number	is	108?
↓	↓	↓	↓	↓
1.2	·	x	=	108

Solving, we get:

$$1.2x = 108$$

$$\frac{1.2}{1.2}x = \frac{108}{1.2}$$

$$x = 90$$

So 108 is 120% of 90.

PRACTICE 7

250% of what number is 18?

EXAMPLE 8

A college awarded financial aid to 3,843 students, which was 45% of the total number of students enrolled at the college. What was the student enrollment at the college?

Solution We must answer the following question:

45%	of	what number	is	3,843?
↓	↓	↓	↓	↓
0.45	·	x	=	3,843

Next, we solve the equation.

$$0.45x = 3{,}843$$

$$\frac{0.45x}{0.45} = \frac{3{,}843}{0.45}$$

$$x = 8{,}540$$

So 8,540 students were enrolled at the college.

PRACTICE 8

There was a glut of office space in a city, with 400,000 square feet, or 16% of the total office space, vacant. How much office space did the city have?

Finding a Percent

Finally, let's look at the third type of percent problem, in which we are given the base and the amount and are looking for the percent.

EXAMPLE 9

What percent of 80 is 60?

Solution We begin by writing the appropriate equation.

$$\begin{array}{ccccc} \text{What percent} & \text{of} & 80 & \text{is} & 60? \\ \downarrow & \downarrow & \downarrow & \downarrow & \downarrow \\ x & \cdot & 80 & = & 60 \end{array}$$

$$80x = 60 \quad \text{Write the equation in standard form.}$$

$$\frac{80}{80}x = \frac{60}{80} \quad \text{Divide each side by 80.}$$

$$x = \frac{\overset{3}{\cancel{60}}}{\underset{4}{\cancel{80}}} = \frac{3}{4} \quad \text{Simplify.}$$

Since we are looking for a percent, we change $\frac{3}{4}$ to a percent. So 75% of 80 is 60.

$$x = \frac{3}{4} = \frac{3}{\underset{1}{\cancel{4}}} \cdot \frac{\overset{25}{\cancel{100}}}{1}\% = 75\%$$

PRACTICE 9

What percent of 6 is 5?

EXAMPLE 10

What percent of 60 is 80?

Solution We begin by writing the appropriate equation, as shown to the right.

$$\begin{array}{ccccc} \text{What percent} & \text{of} & 60 & \text{is} & 80? \\ \downarrow & \downarrow & \downarrow & \downarrow & \downarrow \\ x & \cdot & 60 & = & 80 \end{array}$$

$$60x = 80$$

$$\frac{60}{60}x = \frac{80}{60}$$

$$x = \frac{\overset{4}{\cancel{80}}}{\underset{3}{\cancel{60}}} = \frac{4}{3}$$

Finally, we want to change $\frac{4}{3}$ to a percent.

$$x = \frac{4}{3} = \frac{4}{3} \cdot \frac{100}{1}\% = \frac{400}{3}\% = 133\frac{1}{3}\%$$

So $133\frac{1}{3}\%$ of 60 is 80.

PRACTICE 10

What percent of 8 is 9?

EXAMPLE 11

A young couple buys a house for $125,000, making a down payment of $25,000 and paying the difference over time with a mortgage. What percent of the cost of the house was the down payment?

Solution We write the question as shown to the right.

What percent of $125,000 is $25,000?

$$x \cdot 125{,}000 = 25{,}000$$

$$125{,}000x = 25{,}000$$

$$\frac{125{,}000}{125{,}000}x = \frac{25{,}000}{125{,}000}$$

$$x = \frac{25}{125} = \frac{1}{5}$$

Next, we change $\frac{1}{5}$ to a percent. $x = \frac{1}{5} = \frac{1}{5} \cdot \frac{100}{1}\% = 20\%$

So the down payment was 20% of the total cost of the house.

PRACTICE 11

Of the 400 acres on a farm, 120 were used to grow corn. What percent of the total acreage was used to grow corn?

The Proportion Method

So far, we have used the translation method to solve percent problems. Now, let's consider an alternative approach, the proportion method.

Using the proportion method, we view a percent relationship in the following way.

$$\frac{\text{Amount}}{\text{Base}} = \frac{\text{Percent}}{100}$$

If we are given two of the three quantities, we set up this proportion and then solve it to find the third quantity.

EXAMPLE 12

What is 60% of 35?

Solution The base (the number after the word *of*) is 35. The percent (the number followed by the % sign) is 60. The amount is unknown. We set up the proportion, substitute into it, and solve.

$$\frac{\text{Amount}}{\text{Base}} = \frac{\text{Percent}}{100}$$

$$\frac{x}{35} = \frac{60}{100}$$

$$100x = 60 \cdot 35 \quad \text{Set cross products equal.}$$

$$\frac{100}{100}x = \frac{2{,}100}{100} \quad \text{Divide each side by 100.}$$

$$x = 21$$

So 60% of 35 is 21.

PRACTICE 12

Find 108% of 250.

EXAMPLE 13

15% of what number is 21?

Solution Here, the number after the word *of* is missing, so we are looking for the base. The amount is 21, and the percent is 15. We set up the proportion, substitute into it, and solve.

$$\frac{\text{Amount}}{\text{Base}} = \frac{\text{Percent}}{100}$$

$$\frac{21}{x} = \frac{15}{100}$$

$$15x = 2{,}100 \quad \text{Set cross products equal.}$$

$$\frac{\cancel{15}}{\cancel{15}}x = \frac{2{,}100}{15} \quad \text{Divide each side by 15.}$$

$$x = 140$$

So 15% of 140 is 21.

PRACTICE 13

2% of what number is 21.6?

EXAMPLE 14

What percent of \$45 is \$30?

Solution We know that the base is 45, the amount is 30, and we are looking for the percent.

$$\frac{30}{45} = \frac{x}{100}$$

$$45x = 3{,}000$$

$$\frac{\cancel{45}}{\cancel{45}}x = \frac{3{,}000}{45}$$

$$x = 66\frac{2}{3}$$

So we conclude that $66\frac{2}{3}\%$ of \$45 is \$30.

PRACTICE 14

What percent of 63 is 21?

EXAMPLE 15

A car depreciated, that is, dropped in value, by 20% during its first year. By how much did the value of the car decline if it cost \$30,500 new?

Solution The question here is: What is 20% of \$30,500? So the percent is 20, the base is \$30,500 and we are looking for the amount. We set up the proportion and solve.

$$\frac{x}{30{,}500} = \frac{20}{100}$$

$$100x = 610{,}000$$

$$\frac{\cancel{100}}{\cancel{100}}x = \frac{610{,}000}{100}$$

$$x = 6{,}100$$

So the value of the car depreciated by \$6,100.

PRACTICE 15

A credit card company requires a minimum payment of 4% of the balance. What is the minimum payment if the credit card balance is \$2,450?

EXAMPLE 16

Each day, an adult takes tablets containing 24 milligrams of zinc. If this amount is 160% of the recommended daily allowance, how many milligrams are recommended? (*Source: Podiatry Today*)

Solution Here, we are looking for the base. The question is: 160% of what amount is 24 milligrams? We set up the proportion and solve.

$$\frac{24}{x} = \frac{160}{100}$$

$$160x = 2{,}400$$

$$\frac{\cancel{160}}{\cancel{160}}x = \frac{2{,}400}{160}$$

$$x = 15$$

Therefore, the recommended daily allowance of zinc is 15 milligrams. Note that this base is less than the amount (24 milligrams). Why must that be true?

PRACTICE 16

According to a newspaper article, a Nobel Prize winner had to pay the Internal Revenue Service $129,200—or 38% of his prize—in taxes. How much was his Nobel Prize worth?

EXAMPLE 17

A college accepted 1,620 of the 4,500 applicants for admission. What was the acceptance rate, expressed as a percent?

Solution The question is: What percent of 4,500 is 1,620?

$$\frac{1{,}620}{4{,}500} = \frac{x}{100}$$

$$4{,}500x = 162{,}000$$

$$\frac{\cancel{4{,}500}}{\cancel{4{,}500}}x = \frac{162{,}00\cancel{0}}{4{,}50\cancel{0}}$$

$$x = 36$$

So the college's acceptance rate was 36%.

PRACTICE 17

A bookkeeper's annual salary was raised from $38,000 to $39,900. What percent of her original annual salary is her new annual salary?

Percents on a Calculator

Many calculators have a percent key (%), sometimes used with the 2nd function (2nd). However, the percent key functions differently on different models. Check to see if the following approach works on your machine. If it does not, experiment to find an approach that does.

EXAMPLE 18

Use a calculator to find 50% of 8.

Solution

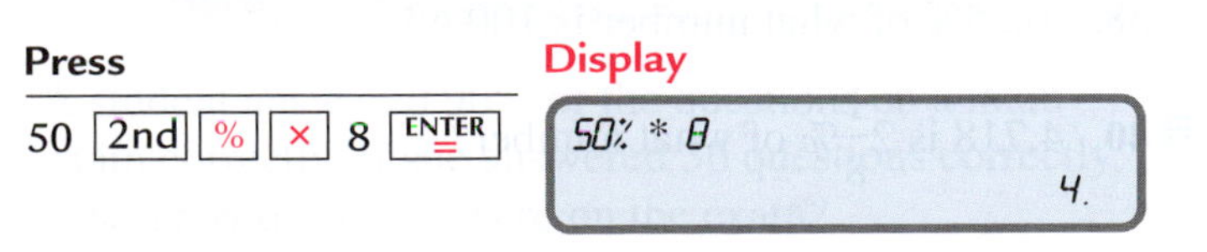

PRACTICE 18

What is 8.25% of $72.37, to the nearest cent?

75. Payroll deductions comprise 40% of the gross income of a student working part-time. If his deductions total $240, what is his gross income?

76. In 1862, the U.S. Congress enacted the nation's first income tax, at the rate of 3%. How much in income tax would you have paid if you made $2,500? (*Source:* U.S. Bureau of the Census)

77. According to the report on a country's economic conditions, 1.5 million people, or 8% of the workforce, were unemployed. How large was the workforce?

78. A recipe for cattle feed calls for 1,200 pounds of corn, 400 pounds of oats, 200 pounds of protein, 100 pounds of beet pulp, 75 pounds of cottonseed hulls, and 25 pounds of molasses. What percent of this mixture is corn? (*Source:* cattlepages.com)

79. In basketball, a foul shot is called a *free throw*. The recipient of the most valuable player award on a college basketball team made 75% of 96 free throw attempts. How many free throws did he make? (*Source:* wikipedia.org)

80. According to a recent telephone survey, 424 thousand out of 673 thousand adults interviewed in the U.S. were either overweight or obese. What percent of interviewees, to the nearest whole percent, were either overweight or obese? (*Source:* webmd.com)

81. Flexible-fuel vehicles run on E85, an alternative fuel that is a blend of ethanol and gasoline containing 85% ethanol. How much ethanol is in 12 gallons of E85?

82. Of the 80 classrooms on a community college campus, 75% are equipped with whiteboards. How many classrooms have whiteboards?

83. A company's profits amounted to 10% of its sales. If the profits were $3 million, compute the company's sales.

84. In a recent survey of U.S. colleges, the average tuition and fees at private four-year colleges was approximately $20 thousand, in contrast to about $2 thousand for public two-year colleges. The second figure is what percent of the first figure? (*Source: The Chronicle of Higher Education Almanac*)

85. A math lab coordinator is willing to spend up to 25% of her income on housing. What is the most she can spend if her annual income is $36,000?

86. An office supply warehouse shipped 648 cases of copy paper. If this represents 72% of the total inventory, how many cases of paper did the warehouse have in its inventory?

87. A lab technician mixed 36 milliliters of alcohol with 84 milliliters of water to make a solution. What percent of the solution was alcohol?

88. A shopper lives in a town where the sales tax is 5%. Across the river, the tax is 4%. If it costs her $6 to make the round trip across the river, should she cross the river to buy a $250 television set?

89. The following graph shows the breakdown of the projected U.S population by gender in the year 2020. If the population is expected to be 340 million people, how many more women than men will there be in 2020? (*Source:* census.gov)

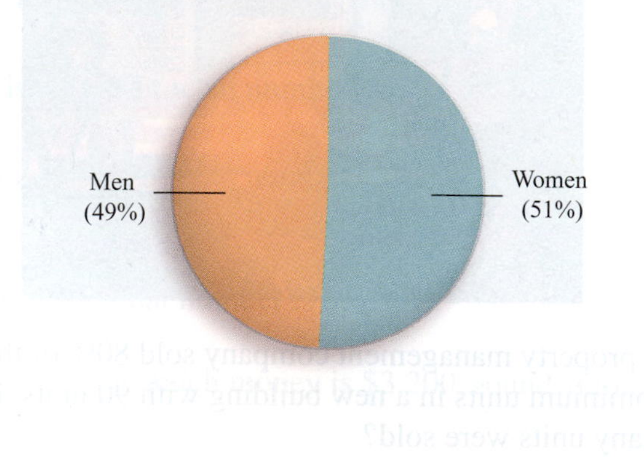

90. The graph gives the use of the Internet by U.S. adults, according to a recent national survey. All percents are rounded to the nearest whole percent.

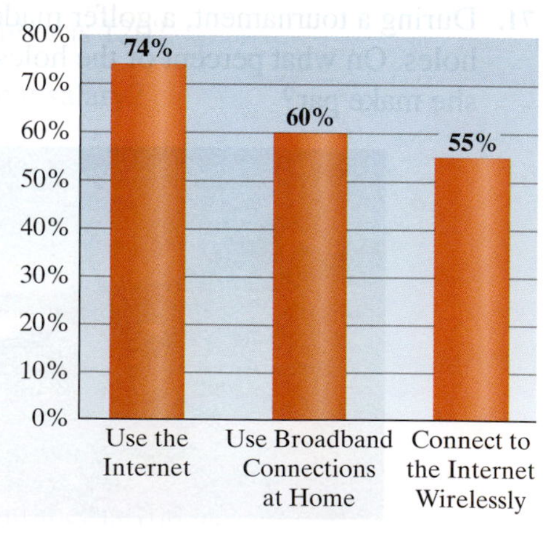

The survey involved tracking a sample of 2,258 adults. How many more adults in this sample use broadband connections at home than those who connect to the internet wirelessly? (*Source:* pewinternet.org)

91. In a company, 85% of the employees are female. If 765 males work for the company, what is the total number of employees?

92. A quarterback completed 15 passes or 20% of his attempted passes. How many of his attempted passes did he *not* complete?

93. The state of Michigan has an urban population of about 8 million people, with the remaining population of 2 million people living in rural areas. (*Source:* ers.usda.gov)

a. Approximately what is the total population of Michigan?

b. Approximately what percent of Michigan's population is urban?

94. A homeowner builds a family room addition on his 1,650-square-foot house, increasing the area of the house by 495 square feet.

a. Calculate the total area of the house with the addition.

b. What percent of the original area is the total area?

• Check your answers on page A-8.

MINDStretchers

Writing

1. Do you prefer solving percent problems using the translation method or the proportion method? In a few sentences, explain why.

Critical Thinking

2. At a college, 20% of the women commute, in contrast to 30% of the men. Yet more women than men commute. Explain how this result is possible.

Technology

3. On the web, go to the U.S. Bureau of the Census home page (census.gov). Write a percent problem of interest to you involving data from the site, and solve the problem.

6.3 More on Percents

OBJECTIVES

- **A** To solve percent increase or decrease problems
- **B** To solve percent problems involving taxes, commissions, or discounts
- **C** To solve simple or compound interest problems
- **D** To solve applied problems involving percents

Finding a Percent Increase or Decrease

Next, let's consider a type of "what percent" problem that deals with a *changing quantity*. If the quantity is increasing, we speak of a *percent increase*; if it is decreasing, of a *percent decrease*.

Here is an example: Last year, a family paid $2,000 in health insurance, and this year, their health insurance bill was $2,500. By what percent did this expense increase?

Note that this problem states the value of a quantity at two points in time. We are asked to find the percent increase between these two values.

To solve, we first compute the difference between the values, that is, between the *new value* and the *original value*.

$$\underset{\text{New value}}{2{,}500} - \underset{\text{Original value}}{2{,}000} = \underset{\text{Change in value}}{500}$$

The question posed is expressed as follows:

$$\begin{array}{ccccc} \text{What percent} & \text{of} & 2{,}000 & \text{is} & 500? \\ \downarrow & \downarrow & \downarrow & \downarrow & \downarrow \\ x & \cdot & 2{,}000 & = & 500 \end{array}$$

It is important to note that the *base* here—as in all percent change problems—is the original value of the quantity.

Next, we solve the equation.

$$2{,}000x = 500$$
$$\frac{2{,}000}{2{,}000}x = \frac{500}{2{,}000}$$
$$x = \frac{1}{4} = 0.25, \text{ or } 25\%$$

So we conclude that the family's health insurance expense *increased* by 25%.

To Find a Percent Increase or Decrease

- Compute the difference between the new and the original values.
- Compute what percent this difference is of the original value.

EXAMPLE 1

The cost of a marriage license had been $10. Later it rose to $15. What percent increase was this?

PRACTICE 1

To accommodate a flood of tourists, businesses in town boosted the number of hotel beds from 25 to 100. What percent increase is this?

Solution The original cost of the license was \$10, and the new cost was \$15. The change in cost is, therefore, \$15 − \$10, or \$5. So the question is as follows:

What percent	of	\$10	is	\$5?
↓	↓	↓	↓	↓
x	$\cdot$	10	=	5

$$10x = 5$$

$$\frac{\cancel{10}}{\cancel{10}}x = \frac{\overset{1}{\cancel{5}}}{\underset{2}{\cancel{10}}}$$

$$x = \frac{1}{2} = 0.5, \text{ or } 50\%$$

So the cost of the license increased by 50%.

EXAMPLE 2

Suppose that an animal species is considered to be endangered if its population drops by more than 60%. If a species' population fell from 40 to 18, should we consider the animal endangered?

Solution The population dropped from 40 to 18, that is, by 22. The question is how the percent decrease compares with 60%. We compute.

What percent	of	40	is	22?
↓	↓	↓	↓	↓
x	$\cdot$	40	=	22

$$40x = 22$$

$$x = \frac{22}{40} = \frac{11}{20} = 0.55, \text{ or } 55\%$$

Since the population decreased by less than 60%, the species is not considered to be endangered.

PRACTICE 2

Major financial crashes took place on both Tuesday, October 29, 1929, and Monday, October 19, 1987. On the earlier date, the stock index dropped from 300 to 230. On the latter date, it dropped from 2,250 to 1,750. As a percent, did the stock index drop more in 1929 or in 1987? (*Source: The Wall Street Journal*)

Business Applications of Percent

The idea of percent is fundamental to business and finance. Percent applications are part of our lives whenever we buy or sell merchandise, pay taxes, and borrow or invest money.

Taxes

Governments levy taxes to pay for a variety of services, from supporting schools to paving roads. There are many kinds of taxes, including sales, income, property, and import taxes.

In general, the amount of a tax that we pay is a percent of a related value. For instance, sales tax is usually computed as a percent of the price of merchandise sold. Thus, in a town where the sales tax rate is 7%, we could compute the tax on any item sold by computing 7% of the price of that item.

Similarly, property tax is commonly computed by taking a given percent (the tax rate) of the property's assessed value. And an import tax is calculated by taking a specified percent of the market value of the imported item.

EXAMPLE 3

The sales tax on a $950 digital camcorder is $71.25. What is the sales tax rate, expressed as a percent?

Solution We must consider the following question:

71.25 is what percent of 950?

$$71.25 = x \cdot 950$$

$$950x = 71.25$$

$$\frac{950x}{950} = \frac{71.25}{950}$$

$$x = 0.075, \text{ or } 7.5\%$$

So the rate of the sales tax is 7.5%, or $7\frac{1}{2}\%$.

PRACTICE 3

When registering a new car, the owner paid a 2.5% import tax on the purchase price of $18,500. How much import tax did he pay? (*Source:* justlanded.com)

Commission

To encourage salespeople to make more sales, many of them, instead of receiving a fixed salary, are paid on *commission*. Working on commission means that the amount of money that they earn is a specified percent—say, 10%—of the total sales for which they are responsible. Often salespeople make a flat fee in addition to a commission based on sales.

EXAMPLE 4

The owner of a condo in San Diego sold it for $222,000. On this amount, she paid a real estate agent a commission of 6%.

a. Find the amount of the commission.

b. How much money did the owner make from the sale after paying the agent's fee?

Solution

a. The commission is 6% of $222,000.

What is 6% of $222,000?

$$x = 0.06 \cdot 222{,}000 = 13{,}320$$

So the commission amounted to $13,320.

b. The seller made $222,000 − $13,320, or $208,680.

PRACTICE 4

A sales associate at a furniture store is paid a base monthly salary of $1,500. In addition, she earns a 9% commission on her monthly sales. If her total sales this month is $12,500, calculate

a. her commission, and

b. her total monthly income.

Discount

In buying or selling merchandise, the term *discount* refers to a reduction on the merchandise's original price. The rate of discount is usually expressed as a percent of the original price.

EXAMPLE 5

A drugstore gives senior citizens a 10% discount. If some pills normally sell for \$16 a bottle, how much will a senior citizen pay?

Solution Note that, because senior citizens get a discount of 10%, they pay 100% − 10%, or 90%, of the normal price.

The question then becomes:

What is 90% of \$16?

$$x = 0.9 \cdot 16 = 14.4$$

So a senior citizen will pay \$14.40 for a bottle of the pills.

Note that another way to solve this problem is first to compute the amount of the discount (10% of \$16) and then to subtract this discount from the original price. With this approach, do we get the same answer?

PRACTICE 5

Find the sale price.

FAMOUS DESIGNER
JEANS
REGULARLY $87
20%
OFF
TODAY ONLY

Simple Interest

Anyone who has been late in paying a credit card bill or who has deposited money in a savings account knows about *interest*. When we lend or deposit money, we make interest. When we borrow money, we pay interest.

Interest depends on the amount of money borrowed (the *principal*), the annual rate of interest (usually expressed as a percent), and the length of time the money is borrowed (usually expressed in years). We can compute the amount of interest by multiplying the principal by the rate of interest and the number of years. This type of interest is called *simple interest* to distinguish it from *compound interest* (which we discuss later).

EXAMPLE 6

How much simple interest is earned in 1 year on a principal of \$900 at an annual interest rate of 6.5%?

Solution To compute the interest, we multiply the principal by the rate of interest and the number of years.

Principal ↓ Rate of Interest ↓ Number of Years ↓

$$\text{Interest} = 900 \times 0.065 \times 1$$
$$= 58.5$$

So \$58.50 in interest is earned.

PRACTICE 6

What is the simple interest on an investment of \$20,000 for 1 year at an annual interest rate of 7.25%?

EXAMPLE 7

A customer deposited \$825 in a savings account that each year pays 5% in simple interest, which is credited to his account. What is the account balance after 2 years?

Solution To solve this problem, let's break it into two questions:

- How much interest did the customer make after 2 years?
- What is the sum of the original deposit and that interest?

First, let's find the interest. To do this, we multiply the principal by the rate of interest and the number of years.

Principal → (825), Rate of Interest → (0.05), Number of Years → (2)

$$\text{Interest} = (825)\,(0.05)\,(2)$$
$$= 82.50$$

The customer made \$82.50 in interest.

Now, let's find the account balance by adding the amount of the original deposit to the interest made.

Original Deposit → 825, Interest → 82.50

$$\text{Account Balance} = 825 + 82.50$$
$$= 907.50$$

So the account balance after 2 years is \$907.50

PRACTICE 7

A bank account pays 6% simple interest on \$1,600 for 2 years. Compute the account balance after 2 years.

Compound Interest

As we have seen, simple interest is based on the principal. Most banks, however, pay their customers *compound interest*, which is based on both the principal and the previous interest generated.

For instance, suppose that a bank customer has \$1,000 deposited in a savings account that pays 5% interest compounded annually. There were no withdrawals or other deposits. Let's compute the balance in the account at the end of the third year.

The following table shows the account balance after the customer has left the money in the account for 3 years. After 1 year, the account will contain \$1,050 (that is, 100% of the original \$1,000 added to 5% of \$1,000, giving us 105% of \$1,000).

Year	Balance at the End of the Year
0	\$1,000
1	\$1,000 + 0.05 × \$1,000 = \$1,050.00
2	\$1,050 + 0.05 × \$1,050 = \$1,102.50
3	\$1,102.50 + 0.05 × \$1,102.50 ≈ \$1,157.63

The balance in the account after the third year is \$1,157.63, rounded to the nearest cent.

Alternatively, for each year we can multiply the account balance by 1.05 to compute the balance at the end at the next year.

Year	Balance at the End of the Year
0	\$1,000
1	$1.05 \times \$1{,}000 = \$1{,}050.00$
2	$(1.05)^2 \times \$1{,}000 = \$1{,}102.50$
3	$(1.05)^3 \times \$1{,}000 \approx \$1{,}157.63$

Note that the balance at the end of the third year is $(1.05)^3 \times \$1{,}000$, or \$1,157.63, in agreement with our previous computation. What would the balance be at the end of the fourth year? What is the relationship between the number of years the money has been invested and the power of 1.05?

In computing the preceding answer, we needed to raise the number 1.05 to a power. Before scientific calculators became available, compound interest problems were commonly solved by use of a compound interest table that contained information such as the following:

Number of Years	4%	5%	6%	7%
1	1.04000	1.05000	1.06000	1.07000
2	1.08160	1.10250	1.12360	1.14490
3	1.12486	1.15763	1.19102	1.22504

When using such a table to calculate a balance, we simply multiply the principal by the number in the table corresponding to the rate of interest and the number of years for which the principal is invested. For instance, after 3 years a principal of \$1,000 compounded at 5% per year results in a balance of $1.15763 \times 1{,}000$, or \$1,157.63, as we previously noted.

Today, problems of this type are generally solved on a calculator.

EXAMPLE 8

A couple deposited \$7,000 in a bank account and did not make any withdrawals or deposits in the account for 3 years. The interest is compounded annually at a rate of 3.5%. What will be the amount in their account at the end of this period?

Solution Each year, the amount in the account is 100% + 3.5%, or 1.035 times the previous year's balance. So at the end of 3 years, the number of dollars in the account is calculated as follows:

Principal ↓ 7,000 × First Year ↓ 1.035 × Second Year ↓ 1.035 × Third Year ↓ 1.035

$$7{,}000 \times 1.035 \times 1.035 \times 1.035$$

It makes sense to use a calculator to carry out this computation. One way to key in this computation on a calculator is as follows.

Press	Display
7000 [×] 1.035 [^] 3 [ENTER]	7000 * 1.035 ^ 3 7761.025125

So at the end of 3 years, they have \$7,761.03 in the account, rounded to the nearest cent.

PRACTICE 8

Find the balance after 4 years on a principal amount of \$2,000 invested at a rate of 6% compounded annually.

47. At a home goods store, a customer bought a down comforter that originally cost $180.

a. What was the sale price of the comforter?
b. Calculate the total amount the customer paid after 6% sales tax was added to the purchase.

48. During a sale, a shoe store marked down the price of a pair of sneakers that originally cost $80 by 40%.

a. What was the sale price of the sneakers?
b. After two weeks, the store marked down the sale price by another 60%. What percent off the original price was the sale price after the second discount was applied?

49. An investor put $3,000 in an account that pays 4% interest, compounded annually. Find the amount in the account after 2 years.

50. A bank pays 4.5% interest, compounded annually, on a 2-year certificate of deposit (CD) that initially costs $500. What is the value of the CD at the end of the 2 years, rounded to the nearest cent?

51. A city had a population of 4,000. If the city's population increased by 10% per year, what was the population 4 years later?

52. An art dealer bought a painting for $10,000. If the value of the painting increased by 50% per year, what was its value 4 years later?

• Check your answers on page A-8.

MINDStretchers

Writing

1. Explain the difference between simple interest and compound interest.

Technology

2. Using a spreadsheet, construct a three-column table showing the original price, the 10% discount, and the selling price for items with an original price of any whole number of dollars between $1 and $100.

Mathematical Reasoning

3. If a quantity increases by a given percent and then decreases by the same percent, will the final value be the same as the original value? Explain.

Key Concepts and Skills

CONCEPT SKILL

Concept/Skill	Description	Example
[6.1] **Percent**	A ratio or fraction with denominator 100. It is written with the % sign, which means divided by 100.	$7\% = \frac{7}{100}$ ↑ Percent
[6.1] **To change a percent to the equivalent fraction**	• Drop the % sign from the given percent and place the number over 100. • Simplify the resulting fraction, if possible.	$25\% = \frac{25}{100} = \frac{1}{4}$
[6.1] **To change a percent to the equivalent decimal**	• Drop the % sign from the given percent and divide the number by 100.	$23.5\% = .235$, or 0.235
[6.1] **To change a decimal to the equivalent percent**	• Multiply the number by 100 and insert a % sign.	$0.125 = 12.5\%$
[6.1] **To change a fraction to the equivalent percent**	• Multiply the fraction by 100%.	$\frac{1}{5} = \frac{1}{5} \times 100\% = \frac{1}{\cancel{5}_1} \times \frac{\cancel{100}^{20}}{1}\%$ $= 20\%$
[6.2] **Base**	The number that we are taking the percent of. It always follows the word *of* in the statement of a percent problem.	50% of 8 is 4. ↑ Base (8)
[6.2] **Amount**	The result of taking the percent of the base.	50% of 8 is 4. ↑ Amount (4)
[6.2] **To solve a percent problem using the translation method**	• Translate as follows: What number, what percent → x is → $=$ of → $\times$ or $\cdot$ % → decimal or fraction • Set up the equation. **The percent of the base is the amount.** • Solve.	What is 50% of 8? ↓ ↓ ↓ ↓ ↓ $x = 0.5 \cdot 8$ $x = 4$ 30% of what number is 6? ↓ ↓ ↓ ↓ ↓ $0.3 \cdot x = 6$ $\frac{\cancel{0.3}x}{\cancel{0.3}} = \frac{6}{0.3}$ $x = \frac{6}{0.3} = 20$ What percent of 8 is 2? ↓ ↓ ↓ ↓ ↓ $x \cdot 8 = 2$ $8x = 2$ $x = \frac{2}{8} = \frac{1}{4} = 25\%$

continued

[6.2] *Solve.*

11. What is 40% of 30?

12. What percent of 5 is 6?

13. 2 feet is what percent of 4 feet?

14. 30% of what number is 6?

15. What percent of 8 is 3.5?

16. Find 55% of 10.

17. $12 is 200% of what amount of money?

18. 2 is what percent of 10?

19. What is 1.2% of 25?

20. Find 115% of 400.

21. 35% of $200 is what?

22. $\frac{1}{2}$% of what number is 5?

23. 15 is what percent of 0.75?

24. 4.5 is what percent of 18?

25. Calculate $33\frac{1}{3}$% of $600.

26. What percent of $9 is $4?

27. Find 60% of $20.

28. 2.5% of how much money is $40?

29. What percent of $7.99 is $1.35, to the nearest whole percent?

30. 3.5 is $8\frac{1}{4}$% of what number, to the nearest hundredth?

[6.3] *Complete the following tables.*

31.

Original Value	New Value	Percent Decrease
24	16	

32.

Original Value	New Value	Percent Decrease
360 mi	300 mi	

33.

Selling Price	Rate of Sales Tax	Sales Tax
$50	6%	

34.

Sales	Rate of Commission	Commission
$600	4%	

35.

Original Price	Rate of Discount	Discount	Sale Price
$200	15%		

36.

Principal	Interest Rate	Time (in years)	Simple Interest	Final Balance
$200	4%	2		

Mixed Applications

Solve.

37. A compact fluorescent light bulb (CFL) will last up to 8,000 hours. If another CFL lasts 25% longer, what is the life of this bulb? (*Sources:* smarthome.com and bulbs.com)

38. Jonas Salk developed the polio vaccine in 1954. The number of reported polio cases in the United States dropped from 29,000 to 15,000 between 1955 and 1956. What was the percent drop, to the nearest whole percent? (*Source:* census.gov)

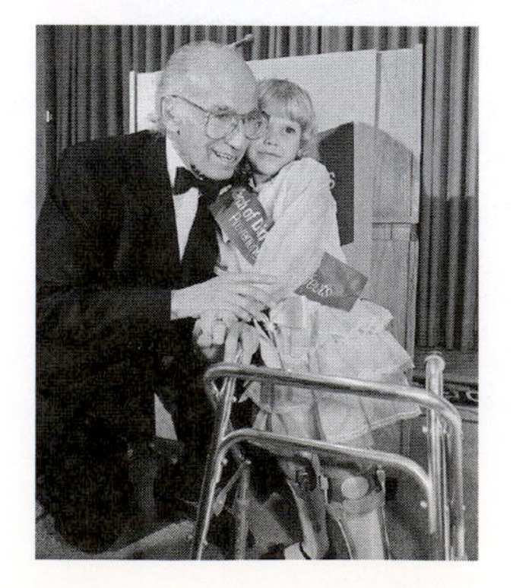

39. For their fees, one real estate agent charges 11% of a year's rent and another charges the first month's rent. Which agent charges more?

40. A particular community bank makes available loans with simple interest. How much interest is due on a five-year car loan of $24,000 based on a simple interest rate of 6%?

41. According to a city survey, 49% of respondents approve of how the mayor is handling his job and 31% disapprove. What percent neither approved nor disapproved?

42. Plastics make up about 11% and paper makes up $\frac{9}{25}$ of the solid municipal waste in the United States. Which makes up more of the solid municipal waste? (*Source:* Energy Information Administration)

43. According to a study, 25% of employees do not take all of their vacation time due to the demands of their jobs. Express this percent as a fraction. (*Source:* Families and Work Institute)

44. In a recent year, 16 out of every 25 high school graduates in Ohio took the ACT college entrance exam. Express this fraction as a percent. (*Source: The World Almanac, 2010*)

45. It takes a worker 50 minutes to commute to work. If he has been traveling for 20 minutes, what percent of his trip has been completed?

46. Among left-handed people are a number of U.S. presidents, including Ronald Reagan, George H.W. Bush, Bill Clinton, and Barack Obama. About 3 out of every 20 people are lefties. Express as a percent. (*Sources:* indiana.edu and scientificamerican.com)

47. A couple financed a 30-year mortgage at a fixed interest rate of 6.29%. Express this rate as a decimal.

48. The length of a person's thigh bone is usually about 27% of his or her height. Estimate someone's height whose thigh bone is 20 inches long. (*Source: American Journal of Physical Anthropology*)

49. The following table deals with estimates of the recoverable coal reserves in two states that are leading coal producers.

State	Coal Reserves (in billions of short tons)
West Virginia	17
Illinois	38

The size of West Virginia's reserves is what percent of the size of Illinois' reserves rounded to the nearest percent? (*Source*: nma.org)

50. A clothing store places the following ad in a local newspaper:

At the store, what is the sale price of a suit that regularly sells for $230?

51. The salary of an executive assistant had been $30,000 before she got a raise of $1,000. If the rate of inflation is 5%, has her salary kept pace with inflation?

52. In a scientific study that relates weight to health, people are considered overweight if their actual weight is at least 20% above their ideal weight. If you weigh 160 pounds and have an ideal weight of 130 pounds, are you considered overweight?

53. An airline oversold a flight to Los Angeles by nine seats, or 5% of the total number of seats available on the airplane. How many seats does the airplane have?

54. When an assistant became editor, the magazine's weekly circulation increased from 50,000 to 60,000. By what percent did the circulation increase?

55. The winner of a men's U.S. Open tennis match got $87\frac{1}{2}\%$ of his first serves in. If he had 72 first serves, how many went in?

56. According to a recent survey, there are approximately 78 million owned dogs and 94 million owned cats in the United States. The number of dogs is what percent of the number of cats, to the nearest percent? (*Source:* humanesociety.org)

57. At an auction, a bidder bought a table for \$150. The auction house also charged a "buyer's premium"—an extra fee—of 10%. How much did the bidder pay in all?

58. According to the news report, 80 tons of food met only 20% of the food needs in the refugee camp. How much additional food was needed?

59. A traveler needs 14,000 more frequent-flier miles to earn a free trip to Hawaii, which is 20% of the total number needed. How many frequent-flier miles in all does this award require?

60. The sales tax rate on a flat panel TV bought in New Orleans is 9%. What was the selling price (not including the sales tax) if the sales tax amounted to \$53.91? (*Source:* forbes.com)

61. A sales representative for wholesale products earned \$49,000 per year plus 10% commission on sales totaling \$25,000. What was his total income for the year?

62. At the end of the year, the receipts of a retail store amounted to \$200,000. Of these receipts, 85% went for expenses; the rest was profit. How much profit did the store make?

63. If a bank customer deposits \$7,000 in a bank account that pays a 5.5% rate of interest compounded annually, what will be the balance after 2 years?

64. Suppose that a country's economy expands by 2% per year. By what percent will it expand in 10 years, to the nearest whole percent?

65. Complete the following table which shows the net income for Texas Instruments Inc. in four consecutive quarters.

Quarter Ending	Income (in millions of dollars)	Percent of Total Income (rounded to the nearest tenth of a percent)
Jun 30, 2009	260	
Sep 30, 2009	538	
Dec 31, 2009	655	
Mar 31, 2010	658	
Total		

(*Source:* finance.yahoo.com)

66. The following graph shows the sources from which the federal government received income in a recent year:

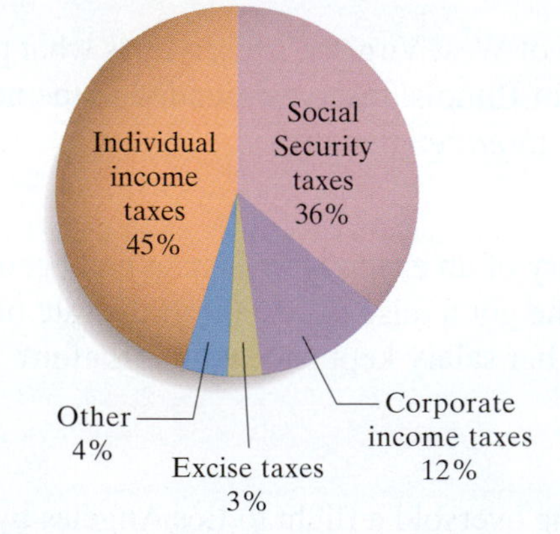

If the total amount of money taken in was \$2,500 billion, compute how much money was received from each source, to the nearest billion dollars. (*Source:* taxpolicycenter.org)

• Check your answers on page A-9.

CHAPTER 6 Posttest

FOR EXTRA HELP

Chapter Test Prep Videos

The Chapter Test Prep Videos with test solutions are available on DVD, in MyMathLab, and on YouTube (search "AkstBasicMath" and click on "Channels").

To see if you have mastered the topics in this chapter, take this test.

Rewrite.

1. 4% as a fraction

2. $27\frac{1}{2}\%$ as a fraction

3. 174% as a decimal

4. 8% as a decimal

5. 0.009 as a percent

6. 10 as a percent

7. $\frac{5}{6}$ as a percent, rounded to the nearest whole percent

8. $2\frac{1}{5}$ as a percent

Solve.

9. What is 25% of 30 miles?

10. Find 120% of 40.

11. 3% of what number is 9?

12. 8% of what number is 16?

13. What percent of 10 is 6?

14. What percent of 4 is 10?

15. To pay for tuition, a college student borrows $2,000 from a relative for 2 years at 5% simple interest. Find the amount of simple interest that is due.

16. In a parking lot that has 150 spaces, 4% are for handicap parking. How many handicap spaces are in the lot?

17. A customer paid $14.95 in sales tax on an iPhone that sells for $299 (before tax). What was the sales tax rate?

18. Milk is approximately 50% cream. How much milk is needed to produce 2 pints of cream?

19. For some Broadway and off-Broadway performances, TKTS discount booths sell tickets at 30% off the full price plus a $4 per ticket service charge. What is the total cost of three tickets that sell for $98 at full price? (*Source*: tdf.org)

20. A college ended six straight years of tuition increases by raising its tuition from $3,000 to $3,100. Find the percent increase.

• Check your answers on page A-10.

Cumulative Review Exercises

To help you review, solve the following.

1. Express 10,000,000 as a power of 10.

2. Divide: 1,962 ÷ 18

3. Find the sum of $3\frac{4}{5}$ and $1\frac{9}{10}$.

4. Find the difference: 32.25 − 4.68

5. Multiply: 0.2 × 3.5

6. Express $\frac{5}{6}$ as a decimal, rounded to the nearest hundredth.

7. Divide, rounding to the nearest hundredth: 5.122 ÷ 0.7

8. Translate the phrase "the difference between a number and 6.7" into an algebraic expression.

9. Solve and check: $w + 17\frac{2}{5} = 41$

10. Solve for x: $\frac{x}{3} = 2.5$

11. Write as a unit rate: $327.60 for 40 hours.

12. Solve and check: $\frac{1.2}{x} = \frac{1.8}{21}$

13. Change $18\frac{2}{11}\%$ to a fraction.

14. 20% of what amount is $200?

Solve.

15. The government withdrew $\frac{1}{4}$ million of its 2 million troops. What fraction of the total is this?

16. Three FM stations are highlighted on the radio dial shown. These stations have frequencies 99.5 (WBAI), 104.3 (WAXQ), and 105.9 (WQXR). Label the three stations on the dial.

88 108

17. At the Westminster Dog Show, the Best in Show prize has been won by terriers three times as often as working group breeds (such as boxers and Great Danes). If terriers have won 45 out of the 103 times that the prize has been awarded, determine how many times working group dogs have won. (*Source*: wikipedia.org)

18. Twitter is a popular social networking service used for tweeting, that is, for sending brief messages. According to a recent survey, 11 of 100 adults who use the internet have tweeted. At this rate among 6,500 such adults, how many would be expected to have tweeted? (*Sources*: wikipedia.org and socialmediatoday.com)

19. In a recent year, about 28% of the 992 thousand U.S. doctors were female. How many female doctors were there, to the nearest thousand? (*Source:* ama.assn.org)

20. Between the years 2010 and 2050, the U.S. population is projected to double. The portion of this population 85 and over is projected to grow more rapidly, increasing from 4 million to 21 million. What percent increase is this? (*Source*: census.gov)

• Check your answers on page A-10.

CHAPTER 7

Signed Numbers

Signed Numbers and Chemistry

In chemistry, a valence is assigned to each element in a compound. Valences help us study the ways in which the elements combine to form the compound.

The valence is a positive or negative whole number that expresses the combining capacity of the element. For example in the compound $CaCl_2$ (calcium chloride), the element calcium (Ca) has a valence of +2, whereas the element chlorine (Cl) has a valence of −1.

The valences in any chemical compound add up to 0. So if you know how to perform signed number computations, you can predict the chemical formula of any compound.

(*Source:* Karen C. Timberlake, *Basic Chemistry*, Prentice Hall, 2011)

Comparing Signed Numbers

The number line helps us compare two signed numbers, that is, to decide which number is larger and which is smaller. On the number line, a number to the right is the larger number.

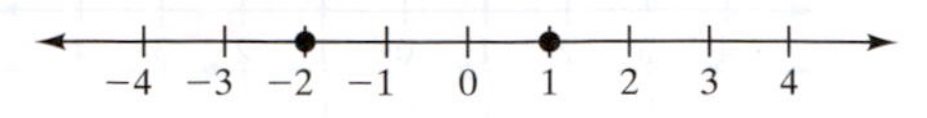

So $1 > -2$.

To Compare Signed Numbers

- Locate the points being compared on the number line; a number to the right is larger than a number to the left.

When comparing signed numbers, remember the following:

- Zero is greater than any negative number because all negative numbers lie to the left of 0.
- Zero is less than any positive number because all positive numbers lie to the right of 0.
- Any positive number is greater than any negative number because all positive numbers lie to the right of all negative numbers.

EXAMPLE 5

Which is larger?

a. $\frac{1}{2}$ or 0 **b.** -1 or -3 **c.** 1.4 or -3

Solution

a. Because $\frac{1}{2}$ (or $+\frac{1}{2}$) is to the right of 0 on the number line, $\frac{1}{2}$ is greater than 0.

−4 −3 −2 −1 0 1 2 3 4

b. Because -1 is to the right of -3, $-1 > -3$, that is, -1 is larger.

−4 −3 −2 −1 0 1 2 3 4

c. Because 1.4 is to the right of -3, $1.4 > -3$, that is, 1.4 is the larger of the two numbers.

−4 −3 −2 −1 0 1 2 3 4

PRACTICE 5

Which is smaller?

a. 0 or 2

b. -5 or -2

c. 2.3 or -4

Now, let's try some practical applications of comparing signed numbers. The key is to be able to determine if a number is negative or positive. You should become familiar with the following words that indicate the sign of a number.

Negative	Positive
Loss	Gain
Below	Above
Decrease	Increase
Down	Up
Withdrawal	Deposit
Past	Future
Before	After

EXAMPLE 6

Express as a signed number: Badwater Basin in Death Valley is the lowest elevation in the Western Hemisphere at 282 feet below sea level. (*Source:* National Park Service)

Solution The number in question represents an elevation below sea level, so we write it as a negative number: -282 feet.

PRACTICE 6

Represent as a signed number: The New York Giants gained 2 yards on a play.

EXAMPLE 7

The following table shows the temperature below which various plants freeze and die.

Plant	Asters	Carnations	Mums
Hardy to	-20°F	-5°F	-30°F

In a very cold climate, which would be planted? (*Source: The American Horticultural Society A–Z Encyclopedia of Garden Plants*)

Solution First, we compare the temperatures of the asters and the carnations. Because $-20° < -5°$, the asters are hardier than the carnations. Next, we compare the temperatures of the asters and the mums. Because $-20° > -30°$, the mums are hardier. So the mums would be the best of the three to plant.

PRACTICE 7

A student studying astronomy learns that *apparent magnitude* is how bright a star appears when viewed from Earth. The lower the apparent magnitude, the brighter the star appears. The table shows the apparent magnitude for various stars. (*Source: Encyclopedia Britannica Almanac*)

Star	Canopus	Sirius	Alpha Centauri
Apparent Magnitude	-0.72	-1.46	-0.01

Which of these stars is the brightest?

87. Write as a signed number:
a. a withdrawal of $10.98 from a bank account
b. a deposit of $100 into the account

88. What are the sign and absolute value of the following numbers?
a. 4 **b.** $-\frac{2}{3}$

	Sign	**Absolute Value**
a.		
b.		

89. Evaluate: **a.** $|0.5|$ **b.** $|-11|$

90. Which number is larger, -4.95 or -4?

91. Complete using the symbol < or >.
a. -9 ▢ -6 **b.** 0 ▢ $-8\frac{2}{3}$

92. Rewrite -1.7, -2, and $-\frac{3}{4}$ from largest to smallest.

Applications

E *Solve.*

93. The Mariana Trench, the deepest point in the Pacific Ocean, is 11,033 meters below sea level, and the Puerto Rico Trench, the deepest point in the Atlantic Ocean, is 8,648 meters below sea level. Which trench is deeper? (*Source:* marianatrench.com)

94. A small toy company shows a loss of $0.3 million for the second quarter of its business and a loss of $0.9 million for the third quarter. In which quarter did the company show the greater loss?

95. Would a patient be receiving more medication if his dosage is decreased by 50 milligrams or if it is decreased by 25 milligrams?

96. Would a group of passengers be higher if they took the elevator down 2 floors or if they took it down 5 floors?

97. A bone density test is used to determine whether a person has osteoporosis (brittle bone disease). If the result of a bone density test, called the T-score, is below -2.5, then a person has osteoporosis. Does a patient whose T-score is -1.8 have osteoporosis? (*Source:* mayoclinic.com)

98. A bank customer has a checking account with overdraft privileges. The account is currently overdrawn by $109.45. If the customer pays off $100 of the overdraft, will his account still be overdrawn?

99. The following table shows the average surface temperature on several planets.

Planet	Temperature (in degrees Fahrenheit)
Mars	−81
Saturn	−218
Uranus	−323

Which planet is the warmest? (*Source:* nasa.gov)

100. The following graph gives the boiling point (in degrees Celsius) of three liquids.

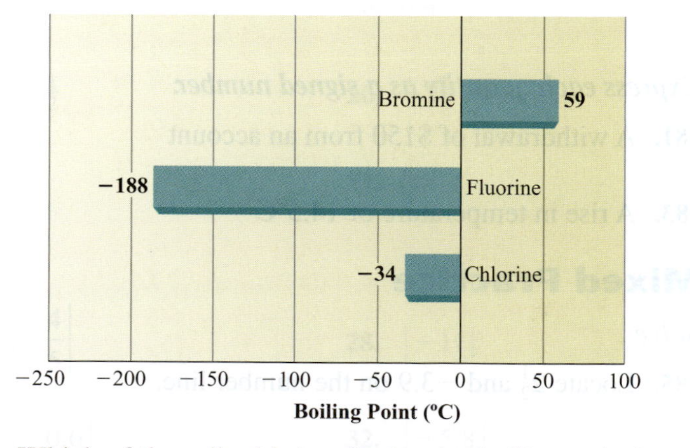

Which of these liquids has the lowest boiling point? (*Source: CRC Handbook of Chemistry and Physics*)

101. In ice hockey, the *plus/minus* statistic is used to rate individual players. If a player is on the ice when *his* team scores a goal, then he gets a *plus* point. If he is on the ice when the *other* team scores a goal, he gets a *minus* point. In theory, the higher a player's plus/minus rating, the better the player. The following table shows an individual hockey player's plus/minus rating for each of the three periods of a certain game. (*Source:* wiki.answers.com)

Period	Plus/Minus Rating
First	−3
Second	1
Third	−2

a. Locate the scores on the number line below. What does 0 on the number line represent?

−4 −3 −2 −1 0 1 2 3 4

b. In which period were the most goals scored by the other team while the player was on the ice?

102. The following graph shows the estimated change in population of four Midwestern cities during the first decade of the 21st century.

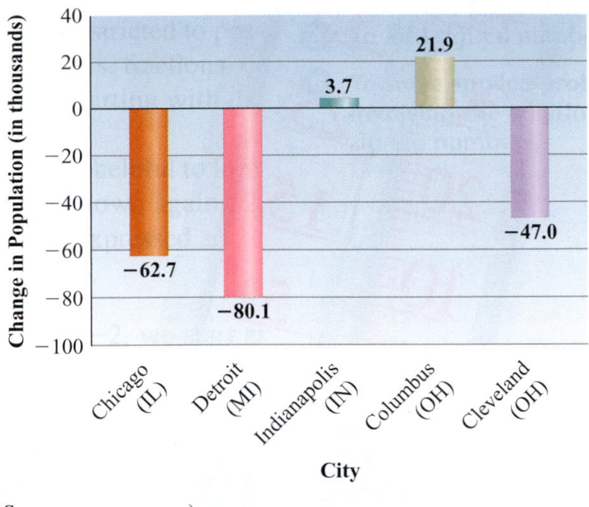

(*Source:* census.gov)

a. Which cities grew in population?

b. Which of the cities had the largest decline in population?

• Check your answers on page A-10.

MINDStretchers

Groupwork

1. a. List several numbers between −2 and −3.

b. How many numbers are there between −2 and −3?

Mathematical Reasoning

2. On the thermometer at the right, highlight all temperatures within 4 degrees of −1°.

20
10
0
−10
−20

4
2
0
−2
−4
−6

Technology

3. Using a computer spreadsheet application such as Microsoft Excel®, enter the numbers

$$-4, 9, 2, 0, -5, -1, 7, 2, 0, 9, 9, 7, -3, -4, 6, 1, 4, 3, 0$$

and sort these numbers in ascending order. Then, sort them in descending order. Is the same number in the middle with both sorts? Explain how you could have predicted this result.

EXAMPLE 5

Simplify: $4 - [2 - (-3)]$

Solution This problem involves an operation within brackets. According to the order of operations rule, first we work within the brackets.

$$4 - [2 - (-3)] = 4 - [2 + 3] \quad \text{Subtract } -3 \text{ from 2.}$$
$$= 4 - 5 \quad \text{Add 2 and 3.}$$
$$= -1 \quad \text{Subtract 5 from 4.}$$

PRACTICE 5

Calculate: $7 - [3 + (-2)]$

EXAMPLE 6

Normally we think of oxygen as a gas. However, when cooled to $-183°C$ (its boiling point), oxygen becomes a liquid. If it is cooled further to $-218°C$ (its melting point), oxygen becomes a solid. How much higher is the boiling point of oxygen than its melting point? (*Source: Handbook of Chemistry & Physics*)

Solution We need to compute how much greater is -183 than -218.

$$(-183) - (-218) = (-183) + (+218)$$
$$= +35$$

The boiling point of oxygen is 35°C higher than its melting point.

PRACTICE 6

Paper was invented in China in about 100 B.C. How many years ago was that, to the nearest hundred years? (*Source: World of Invention*)

7.3 Exercises

FOR EXTRA HELP MyMathLab MathXL PRACTICE WATCH READ REVIEW

Mathematically Speaking

Fill in each blank with the most appropriate term or phrase from the given list.

absolute value	order of operations	addition
sum	multiplication	signed numbers
difference	opposite	

1. To subtract two signed numbers, change the operation of subtraction to addition, and change the number being subtracted to its __________. Then, follow the rule for adding signed numbers.

2. Every subtraction problem has a related __________ problem.

3. When a signed number problem involves addition and subtraction, work from left to right according to the __________ rule.

4. When subtracting a negative number, the __________ is greater than the original number.

A *Find the difference.*

5. $5 - (-2)$

6. $7 - (-3)$

7. $4 - 8$

8. $5 - 9$

9. $-9 - 5$

10. $-44 - 2$

11. $42 - (-2)$

12. $36 - (-4)$

13. $50 - 75$

14. $44 - 83$

15. $-20 - (-1)$

16. $-18 - (-3)$

17. $3 - (-3)$

18. $4 - (-4)$

19. $0 - 38$

20. $0 - 56$

21. $-13 - 13$

22. $-15 - 15$

23. $13 - (-13)$

24. $14 - (-14)$

25. $8 - 23$

26. $7 - 34$

27. $800 - (-200)$

28. $300 - (-100)$

29. $7 - 8.52$

30. $9.1 - 10.84$

31. $9.2 - (-0.5)$

32. $8.6 - (-0.7)$

33. $-5.2 - (-5.2)$

34. $-0.5 - (-0.5)$

35. $8.6 - (-1.9)$

36. $7.4 - (-3.1)$

37. $-10 - (-9.5)$

38. $-6 - (-8.7)$

39. $4\frac{1}{2} - 9\frac{1}{2}$

40. $6\frac{1}{5} - 8\frac{1}{5}$

41. $10 - 2\frac{1}{4}$

42. $12 - 5\frac{2}{3}$

43. $-7 - \frac{1}{4}$

44. $-9 - \frac{1}{8}$

45. $5\frac{3}{4} - \left(-1\frac{1}{2}\right)$

46. $6\frac{1}{2} - \left(-1\frac{1}{3}\right)$

Combine.

47. $4 + (-6) - (-9)$

48. $10 + (-6) - (-8)$

49. $7 - 7 + (-5)$

50. $8 - 8 + (-9)$

51. $-12 + 3.6 - (-6.5)$

52. $4.6 - (-5) + (-3.6)$

53. $6 + \left(-4\frac{1}{5}\right) + \left(-2\frac{3}{10}\right)$

54. $-2\frac{1}{2} - (-3) + 5\frac{1}{4}$

55. $-8 + (-4) - 9 + 7 + (-1)$

56. $-5 - (-1) + 6 + (-3) - 4$

57. $6 - [5 - (-4)]$

58. $2 - [3 + (-5)]$

59. $7.043 - 9.002 - 1.883$

60. $-6.192 - 0.337 - (-23.94)$

61. $-8.722 + (-3.913) - 3.86$

62. $2.884 - 0.883 + (-6.125)$

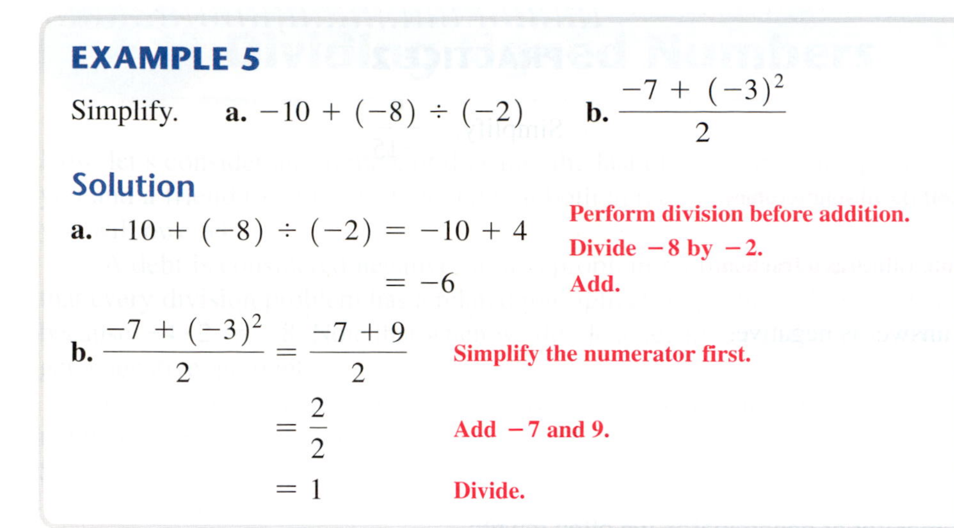

EXAMPLE 5

Simplify. **a.** $-10 + (-8) \div (-2)$ **b.** $\dfrac{-7 + (-3)^2}{2}$

Solution

a. $-10 + (-8) \div (-2) = -10 + 4$ Perform division before addition. Divide -8 by -2.

$= -6$ Add.

b. $\dfrac{-7 + (-3)^2}{2} = \dfrac{-7 + 9}{2}$ Simplify the numerator first.

$= \dfrac{2}{2}$ Add -7 and 9.

$= 1$ Divide.

PRACTICE 5

Simplify.

a. $6 - (-12) \div (-2)$

b. $\dfrac{5 - (-1)^2}{-4}$

EXAMPLE 6

The federal deficit in 1910 was about \$20 million. Five years later, it was \$60 million. How many times greater was the deficit of 1915 than that of 1910? (*Source:* infoplease.com)

Solution The problem asks us to compute $-60 \div (-20)$. The quotient of numbers with the same sign is positive, so the answer is 3. That is, the 1915 deficit was 3 times as great as the deficit of 1910.

PRACTICE 6

About $\frac{3}{4}$ of Earth's coral reefs are located in the Indian and Pacific Oceans. These reefs are declining at a rate of approximately 600 square miles per year. By how many square miles per month are the reefs declining, expressed as a signed number? (*Source:* news.bbc.co.uk)

EXAMPLE 7

The table shows the change in the closing price of a share of a software company's stock each day over a 5-day period.

Day	Change in Closing Price (in cents)
Monday	+32
Tuesday	−18
Wednesday	−21
Thursday	+16
Friday	−54

What was the average daily change in the closing price of a share of the stock?

Solution To compute the average change, we add the five changes and divide the sum by 5.

$$\frac{+32 + (-18) + (-21) + (+16) + (-54)}{5}$$

Recall from the order of operations rule that we must find the sum in the numerator before dividing by the denominator.

$$\frac{+48 + (-93)}{5} = \frac{-45}{5} = -9$$

So the average daily change over the 5-day period was down 9 cents per share.

PRACTICE 7

A young girl has a fever. The following chart shows how her temperature changed each day this week.

Monday	Up 2°
Tuesday	Up 1°
Wednesday	Down 1°
Thursday	Up 1°
Friday	Down 3°

What was the average daily change in her temperature?

7.5 Exercises

FOR EXTRA HELP MyMathLab MathXL PRACTICE WATCH READ REVIEW

Mathematically Speaking

Fill in each blank with the most appropriate term or phrase from the given list.

addition	positive	negative
unequal	equal	multiplication

1. The quotient of two numbers with the same signs is __________.

2. The quotient of two numbers with different signs is __________.

3. The fractions $\frac{-2}{3}, \frac{2}{-3}$ and $-\frac{2}{3}$, are __________ in value.

4. Every division problem has a related __________ problem.

A *Find the quotient. Simplify.*

5. $-20 \div (-4)$
6. $-7 \div (-1)$
7. $0 \div 5$
8. $0 \div 3$
9. $10 \div (-2)$
10. $-9 \div 3$
11. $16 \div (-8)$
12. $-12 \div 4$
13. $-250 \div (-10)$
14. $-300 \div (-3)$
15. $-200 \div 8$
16. $-20 \div 10$
17. $-35 \div (-5)$
18. $-8 \div (-4)$
19. $6 \div (-3)$
20. $-8 \div 2$
21. $-17 \div (-1)$
22. $-20 \div (-2)$
23. $-72 \div (-12)$
24. $-440 \div (-10)$
25. $-2.4 \div 8$
26. $-0.26 \div 2$
27. $-4 \div 0.2$
28. $9 \div (-0.6)$
29. $-4.8 \div (-0.3)$
30. $-2.6 \div (-0.2)$
31. $\left(-\frac{2}{3}\right) \div \frac{4}{5}$
32. $\left(-\frac{5}{6}\right) \div \left(-\frac{5}{6}\right)$
33. $7 \div \left(-\frac{1}{3}\right)$
34. $9 \div \left(-\frac{1}{3}\right)$
35. $-40 \div 2\frac{1}{2}$
36. $-30 \div 1\frac{1}{2}$
37. $(-15.1214) \div (-2.45)$
38. $-0.749 \div -0.214$
39. $-12.25 \div 3.5$
40. $50.8369 \div (-7.13)$

Simplify.

41. $\frac{-1}{5}$
42. $\frac{-1}{7}$
43. $\frac{-11}{-11}$
44. $\frac{-3}{-11}$
45. $\frac{4}{-10}$
46. $\frac{5}{-10}$
47. $\frac{-11}{-2}$
48. $\frac{-2}{-11}$
49. $\frac{-17}{-4}$
50. $\frac{-26}{-5}$
51. $\frac{-9}{-12}$
52. $\frac{-14}{-16}$
53. $-8 \div (-2)(-2)$
54. $-3(-4) \div (-2)$
55. $(3 - 7)^2 \div (-4)$
56. $(4 - 6)^2 \div (1 - 5)^2$
57. $\frac{2^2 - (-6)}{2}$
58. $\frac{3^2 - (-7)}{2}$
59. $\frac{2^2 + (-6)}{-2}$
60. $\frac{3^2 + (-7)}{-1}$
61. $\left(\frac{-8}{-2}\right)\left(\frac{8}{-2}\right)$
62. $\frac{-10}{2} \cdot \frac{-6}{5}$
63. $\frac{-9 - (-3)}{2}$
64. $\frac{-5 - (-7)}{2}$

Basic St

Statistics and the

Lawyers make frequent use of

Statistics on the distribu
lation are commonly used as
For those plaintiffs who claim
agent, their lawyers often pr
dence of their illness. And ca
on such statistics as the propo
hired by a company or the av
in positions before being pron

This use of statistics in
nineteenth-century trial, wh
Sylvia Howland on her will w
testimony of an expert witnes
system of statistically analyzin
of the deceased. On the basis
on the will was unreasonably s
had probably been traced.

(*Source:* Jack B. Weinstein, "Litigation
pp. 286–297)

opics in this

n

arious

ks its emergency response
onse times were: 12 minutes
minutes, 6 minutes, 15 min-
, and 9 minutes. What was

ur grades were: Spanish I
edits)—A; Social Science
al Education (1 credit)—B.
e following points: A = 4,
d F = 0. Calculate your GPA.

10
5.6
1.4

8. The following pictograph shows the average circulation of some major newspapers across the United States in a recent year.

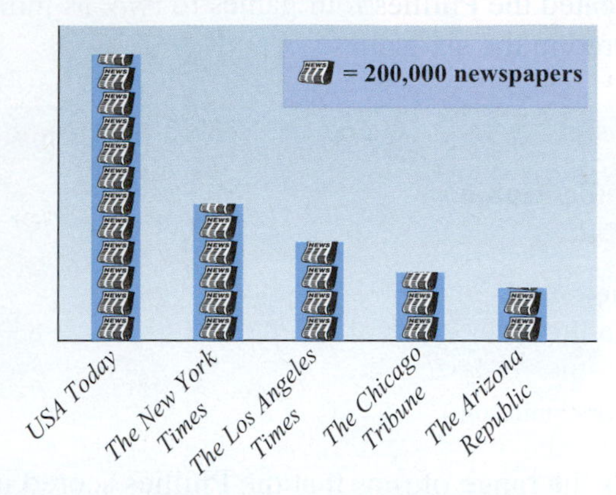

(*Source: Top Ten of Everything, 2010*)

What was the approximate daily circulation of *The New York Times*?

10. The graph shows the breakdown of days of school missed in a recent 12-month period due to illness or injury for U.S. children 5–17 years of age.

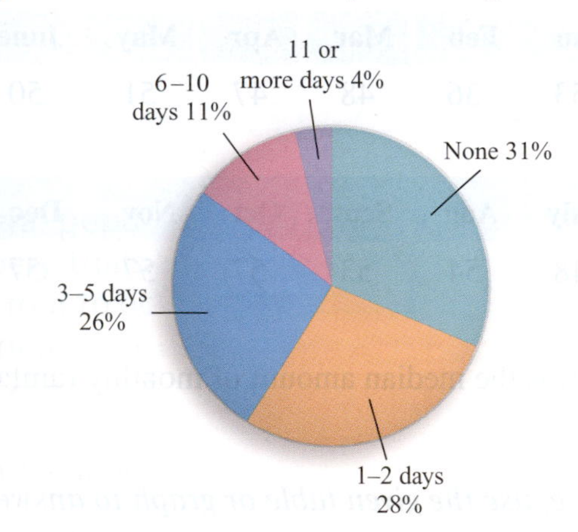

(*Source:* cdc.gov)

What percent of children missed 3 or more days of school?

• Check your answers on page A-11.

Cultural Note

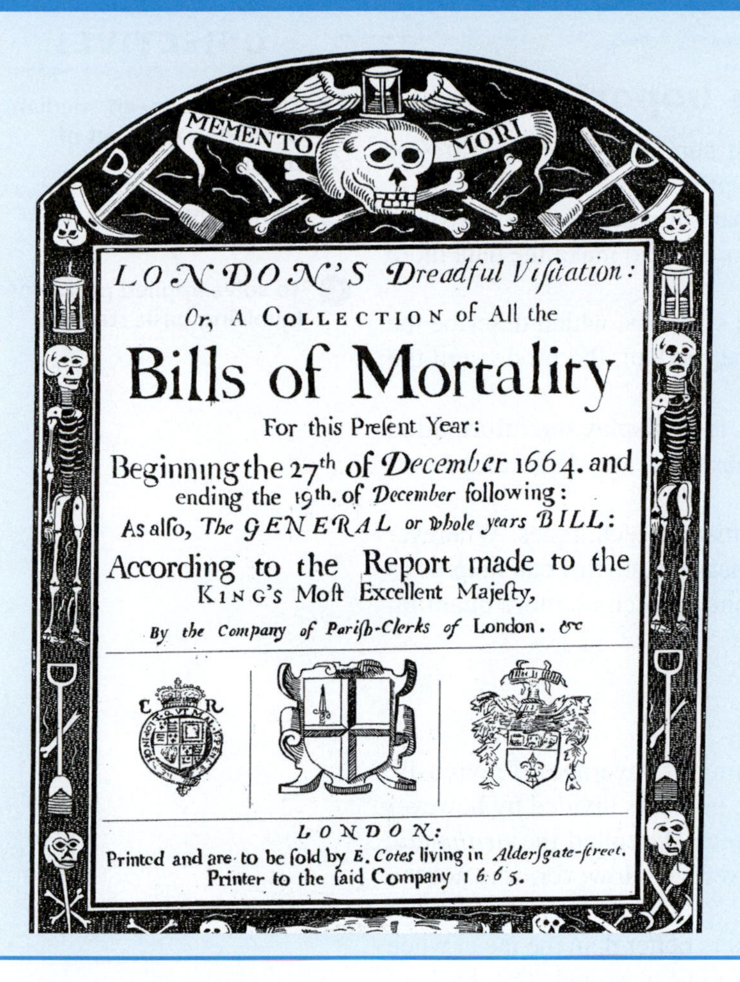
MEMENTO MORI

LONDON'S Dreadful Viſitation:
Or, A COLLECTION of All the
Bills of Mortality
For this Preſent Year:
Beginning the 27th of December 1664. and
ending the 19th. of December following:
As alſo, The GENERAL or whole years BILL:
According to the Report made to the
KING's Moſt Excellent Majeſty,
By the Company of Pariſh-Clerks of London. &c

LONDON:
Printed and are to be ſold by E. Cotes living in Alderſgate-ſtreet.
Printer to the ſaid Company 1665.

A seventeenth-century English clothing salesman named John Graunt had the insight to apply a numerical approach to major social problems. In 1662, he published a book entitled *Natural and Political Observations upon the Bills of Mortality*, and thus founded the science of statistics.

Graunt was curious about the periodic outbreaks of the bubonic plague in London, and his book analyzed the number of deaths in London each week due to various causes. He was the first to discover that, at least in London, the number of male births exceeded the number of female births. He also found that there was a higher death rate in urban areas than in rural areas and that more men than women died violent deaths. Graunt summarized large amounts of information to make it understandable and made conjectures about large populations based on small samples. Graunt was also a pioneer in examining expected life span—a statistic that became vital to the insurance companies formed at the end of the seventeenth century.

Sources: Morris Kline, *Mathematics, a Cultural Approach* (Reading, Mass.: Addison-Wesley Publishing Company, 1962), p. 614.

F. N. David, *Games, Gods and Gambling* (New York: Hafner Publishing Company, 1962).

8.1 Introduction to Basic Statistics

OBJECTIVES

- **A** To find the mean, median, or mode(s) of a set of numbers
- **B** To find the range of a set of numbers
- **C** To solve applied problems involving basic statistics

What Basic Statistics Is and Why It Is Important

Statistics is the branch of mathematics that deals with ways of handling large quantities of information. The goal is to make this information easier to interpret.

With unorganized data, spotting trends and making comparisons is difficult. The study of statistics teaches you how to organize data in various ways in order to make the data more understandable.

One approach is to calculate special numbers, also called statistics, which describe the data. In this section, we consider four statistics: the mean, the median, the mode, and the range.

You have already seen that another way to organize data is to display the information in the form of a table or graph. We will discuss tables and graphs in greater detail in the next section of this chapter.

Many situations lend themselves to the application of statistical techniques. Wherever there are large quantities of information—from sports to business—statistics can help us to find meaning where, at first glance, there seems to be none, and to become more quantitatively literate.

Averages

We begin our introduction to statistics by revisiting the meaning of "average." Previously, we defined the average of a set of numbers to be the sum of the numbers divided by however many numbers are in the set. This statistic, which is more precisely called the *arithmetic mean*, or just the **mean**, is what most people think of as the average. However, it is not the only kind of average used to represent the numbers in a set.

A second average, the *median*, may describe a set of numbers better than the mean when there is an unusually large or unusually small number in the set to be averaged. The third average, the *mode*, has a special property—unlike the mean and the median, it is always in the set of numbers being averaged.

Mean

Let's look at an example of the mean.

EXAMPLE 1

The area of the United States is about 3,800,000 square miles.

a. Approximately what is the average area of each of the 50 states?

b. Is Michigan above or below average in area?

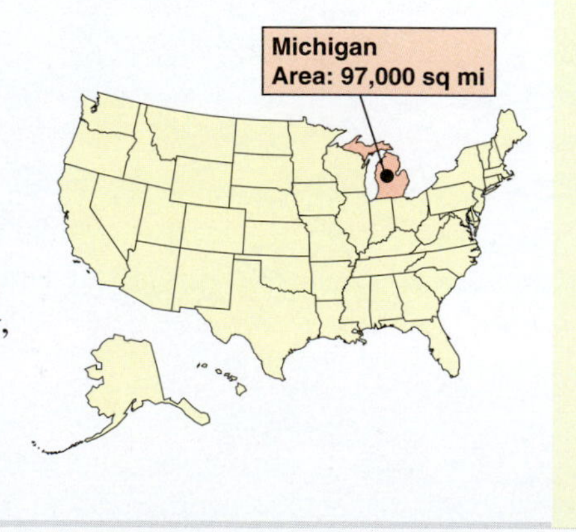

Solution

a. The mean area of a state is $\frac{3,800,000}{50}$, or 76,000 square miles.

b. Since 97,000 is greater than 76,000, Michigan is above average in area.

PRACTICE 1

Reggie Jackson hit five home runs in the 1977 World Series, which lasted six games. By contrast, Lou Gehrig hit four home runs in the 1928 World Series, a four-game series. On the average, which baseball player hit fewer home runs per game?

Note that a property of the mean is that it is changed substantially if even a single number in a set of numbers is replaced by one much larger or much smaller. For instance, if five people each make \$10 per hour, then their mean hourly wage is \$10. However, if the hourly wage of one of these individuals jumps to \$500, then the mean wage skyrockets to \$108, more than 10 times the previous mean.

Another kind of mean, called the *weighted average*, is used when some numbers in a set count more heavily than others. Weighted average comes into play, for instance, if you want to compute the average of your test scores in a class and the final exam counts twice as much as any of the other tests. Or, if you are computing your grade point average (GPA) and some courses carry more credits than others.

EXAMPLE 2

Last term, a student's grades were as follows.

Course	Credits	Grade	Grade Equivalent
Psychology	4	A	4
English	4	C	2
Art	3	B	3
Physical Education	1	B	3

If the student's GPA is 3.5 this term, did she have a higher or lower GPA last term?

Solution To calculate the GPA for last term, we first multiply the number of credits each course carries by the numerical grade equivalent received. We then add these products to find the total number of grade points. Finally, we divide this sum by the total number of credits.

Number of credits for the first course

Grade equivalent of the first course

$$\text{GPA} = \frac{4 \cdot 4 + 4 \cdot 2 + 3 \cdot 3 + 1 \cdot 3}{12}$$

Total number of credits

$$= \frac{16 + 8 + 9 + 3}{12} = \frac{36}{12} = 3$$

A GPA of 3 is less than a GPA of 3.5. So the student had a lower GPA last term.

PRACTICE 2

The following table shows the test scores that a classmate earned.

Exam	Score
1	95
2	80
3	80
Final	90

If the final exam is equivalent to two other exams, did the classmate earn an exam average above or below 85?

Median

As we have seen, a very large number can affect the mean of a set of numbers to such an extent that it is not representative of the set. Another kind of average, the *median*, is used when we wish to reduce the impact of an extreme number in the set, for instance, in computing average salary.

DEFINITION

In a set of numbers arranged in numerical order, the **median** of the numbers is the number in the middle. If there are two numbers in the middle, the median is the mean of the two middle numbers.

Applications

C *Solve and check.*

13. Here are a student's grades last term: A in College Skills (2 credits), B in World History (4 credits), C in Music (2 credits), A in Spanish (3 credits), and B in Physical Education (1 credit). Did the student make the Dean's List, which requires a GPA of 3.5? Explain. (*Reminder:* A = 4, B = 3, C = 2, and D = 1.)

14. On a test, 9 students earned 80, 10 students earned 70, and 1 student earned 75. Was the grade of 75 below the class average (mean), exactly average, or above the class average? Explain.

15. A grandmother leaves a total of $1,000,000 to her 10 grandchildren. What is the mean amount left to each grandchild? Can you compute the median amount with the given information? Explain.

16. A woman and four men are riding in an elevator. Two men are taller than the woman, and two are shorter. Who has the median height of the people in the elevator?

17. In the U.S. House of Representatives, 435 members of Congress represent the 50 states. The table below shows the number of representatives of 8 states.

State	Number of Representatives
Maine	2
Indiana	9
Wisconsin	8
Hawaii	2
Colorado	7
North Carolina	13
Tennessee	9
Nebraska	3

(*Source:* 2010.census.gov)

Which of these 8 states has representation that is above the average for all 50 states?

18. The table shows the quarterly revenues for Ford Motor Company in a recent year.

Quarter	Revenue (in billions)
1	$31.7
2	$35.4
3	$31.6
4	$35.1

(*Source:* finance.yahoo.com)

What was the median quarterly revenue?

19. The table shows the salary of six teachers based on the number of years of service at a local school.

Years of Service	Salary
6	$44,424
10	$57,418
1	$37,925
4	$42,656
13	$58,358
18	$70,852

a. Find the median salary.

b. What is the range?

20. The table shows the prime interest rate on June 1st of the years 2000 through 2010.

2000	2001	2002	2003	2004	2005
9.50%	7.00%	4.75%	4.25%	4.00%	6.00%

2006	2007	2008	2009	2010
8.00%	8.25%	5.00%	3.25%	3.25%

(*Source:* moneycafe.com)

a. What is the mode of the interest rates for the given years?

b. Find the range.

21. The diameters for the eight planets of the solar system, rounded to the nearest 1,000 miles, are as follows:

Planet	Diameter (in thousands of miles)
Mercury	3
Venus	8
Earth	8
Mars	4
Jupiter	89
Saturn	75
Uranus	32
Neptune	31

(*Source: Encyclopedia Americana*)

Find each of the following distances, rounded to the nearest 1,000 miles:

a. mean diameter

b. median diameter

c. mode(s) of the diameters

d. range of the diameters

22. Consider the following utility bills for the past 10 months:

Month	Utility Bill
January	$90
February	$80
March	$90
April	$70
May	$100
June	$110
July	$140
August	$140
September	$100
October	$90

Find each of the following:

a. the mean bill

b. the median bill

c. the mode(s) of the bills

d. the range of the bills

23. In the year 1990, when the number of U.S. residents was about 249 million, the U.S. Postal Service delivered some 166 billion pieces of mail. By 2003, when the population had grown to 292 million, the Service delivered approximately 202 billion pieces of mail. On the average, how many more pieces of mail, to the nearest whole number, did a resident receive in 2003 than in 1990? (*Source:* U.S. Bureau of the Census)

24. Students earned the following grades on a college math test:

85 90 60 45 95 70 60 90 100 25 85 70 80
75 55 85 100 40 95 50 75 65 90 75 60 50

Using the mean as the average, how far above or below the class average, to the nearest whole number, was the test score of 75?

• Check your answers on page A-11.

MINDStretchers

Groupwork

1. Working with a partner, construct an example of a set of 10 numbers

a. whose mean, median, and mode are equal.

b. whose mean is less than its median.

c. that has two modes.

Mathematical Reasoning

2. Can the range of a set of numbers be equal to a negative number? Explain.

Investigation

3. In your college library or on the web, research the legal drinking age in 10 countries of your choosing. Then, determine the mean, median, mode, and range of these ages.

Graphs

Now, let's discuss displaying data in the form of graphs. We deal with five kinds of graphs: pictographs, bar graphs, histograms, line graphs, and circle graphs.

Pictographs

A **pictograph** is a kind of graph in which images such as people, books, or coins are used to represent and to compare quantities. A *key* is given to explain what each image represents.

Pictographs are visually appealing. However, they make it difficult to distinguish between small differences—say, between a half and a third of an image.

EXAMPLE 2

The following graph shows the number of degrees awarded in the United States in a recent year:

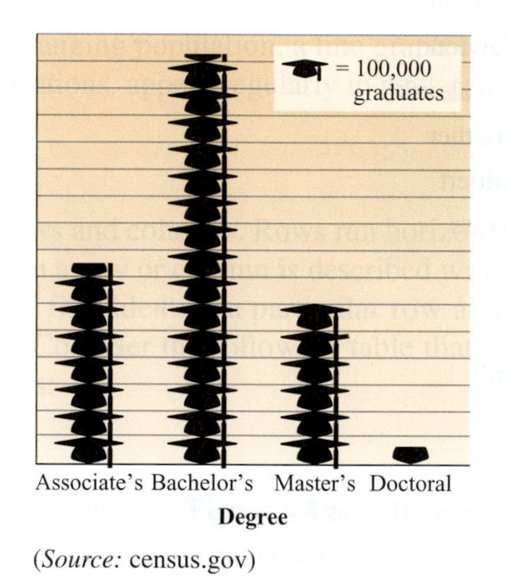

(*Source:* census.gov)

a. What does the symbol 🎓 mean in the key at the top of the graph?

b. About how many master's degrees were awarded?

c. About how many more bachelor's degrees than associate's degrees were awarded?

Solution

a. According to the key, the symbol 🎓 represents 100,000 graduates.

b. The number of master's degrees awarded was about 6(100,000), or 600,000.

c. About 15(100,000), or 1,500,000, bachelor's degrees were awarded, in contrast to about $7\frac{1}{2}$(100,000), or 750,000, associate's degrees.
So there were approximately 750,000 more bachelor's degrees awarded.

PRACTICE 2

The following pictograph shows the number of passengers who departed from or arrived at four busy U.S. airports in a recent year.

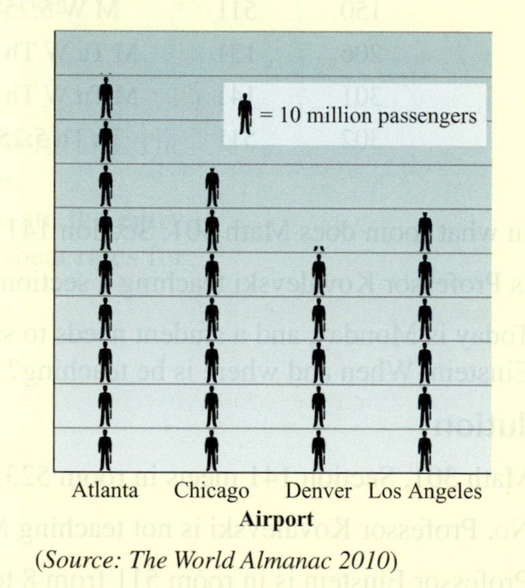

(*Source: The World Almanac 2010*)

a. What does the symbol 🚶 represent?

b. Which of the four airports was the busiest in terms of passengers?

c. Approximately how many passengers did the Los Angeles airport serve?

Bar Graphs

On a **bar graph**, quantities are represented by thin, parallel rectangles called bars. The length of each bar is proportional to the quantity that it represents.

On some graphs, the bars extend to the right. On other graphs, they extend upward or downward. Sometimes, bar lengths are labeled. Other times, they are read against an *axis*—a straight line parallel to the bars and similar to a number line.

Bar graphs are especially useful for making comparisons or contrasts among a few quantities, as the following example illustrates.

EXAMPLE 3

The graph shows the net income of Delta Air Lines Inc. in recent years.

a. What was the approximate net income of the company in fiscal year 2009?

b. About how much greater was the net income in fiscal year 2007 than in fiscal year 2006?

c. Describe the information shown by the graph.

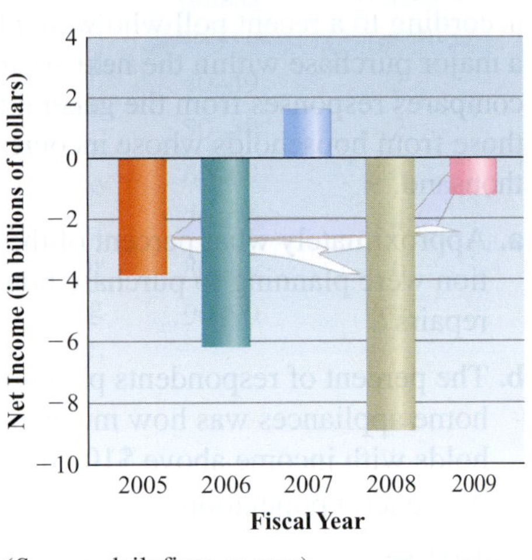

(*Source:* dailyfinance.com)

Solution

a. In 2009, the net income of the company was about −\$1.2 billion, that is, a loss of about \$1.2 billion.

b. In 2006, the net income was about −\$6.2 billion. The next year, it was about \$1.6 billion. So the net income for 2007 was approximately \$7.8 billion greater than in 2006.

c. The company operated at a profit in 2007. In other years, however, it operated at a loss, especially in 2008.

PRACTICE 3

The following graph shows the value of five leading farm commodities in California for a recent year.

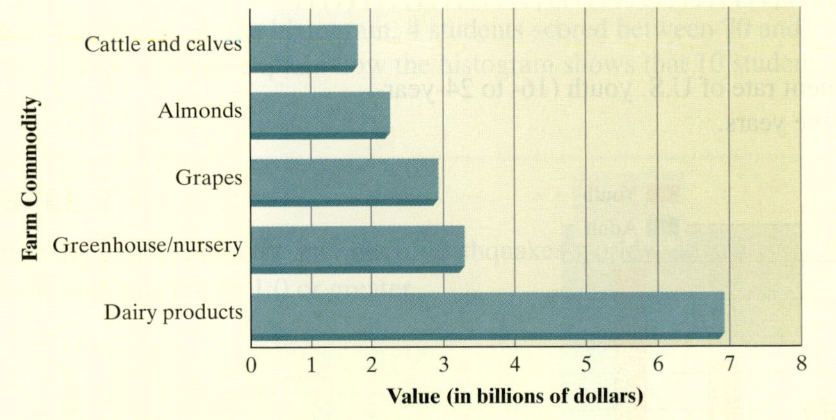

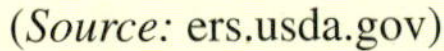

(*Source:* ers.usda.gov)

a. Which commodity had the greatest value?

b. What was the approximate value of grapes?

c. About how much greater was the value of greenhouse/nursery than almonds?

The next graph is an example of a *double-bar graph*. This kind of graph is used to compare two sets of data in various ways, as the following example illustrates.

EXAMPLE 5 (continued)

a. Approximately how many earthquakes were there with magnitude between 5.0 and 5.9?

b. To the nearest thousand, how many earthquakes were there with magnitude 4.9 or below?

c. What is the approximate ratio in simplified form of the number of earthquakes with magnitude 4.0–4.9 to those with magnitude 3.0–3.9?

Solution

a. The height of the bar for the class interval 5.0–5.9 is approximately 1,800. So there were about 1,800 earthquakes with magnitude between 5.0 and 5.9.

b. The number of earthquakes with magnitude 4.9 or below is the sum of the height of the bar for the class interval 4.0–4.9, as well as the height of all bars to the left. To the nearest thousand, this sum is 13,000.

c. The ratio of the number of earthquakes with magnitude 4.0–4.9 to those with magnitude 3.0–3.9 is about 7 to 3.

a. Approximately how many presidents were in their fifties when they were first inaugurated?

b. Were any presidents younger than 40 at their initial inauguration?

c. About how many presidents were 59 or younger when they were first inaugurated?

Line Graphs

On a **line graph**, points are connected by straight-line segments. The position of any point on a line graph is read against the vertical axis and the horizontal axis.

A line graph, also called a **broken-line graph**, is commonly used to highlight changes and trends over a period of time. Especially when we have data for many points in time, we are more likely to use a line graph than a bar graph.

EXAMPLE 6

The following graph shows the number of Americans 65 years of age and older from 1900 to 2010.

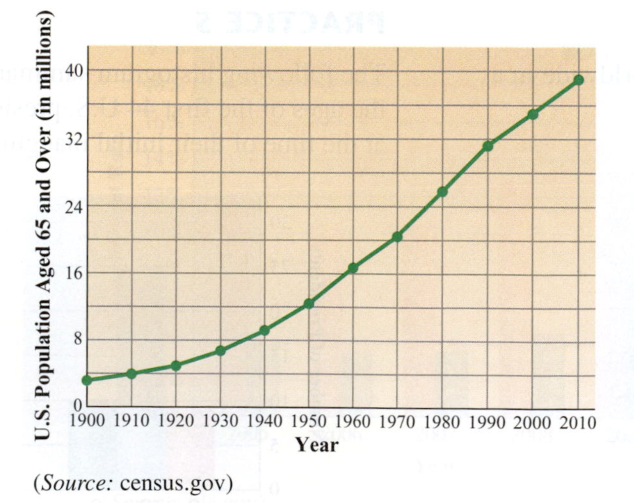

(*Source:* census.gov)

a. Approximately how big was this population in the year 2000?

b. In what year did this population number about 21 million?

c. In the year 2000, the U.S. population overall was approximately 4 times as large as it had been in the year 1900. Did the population shown in the graph grow more quickly?

PRACTICE 6

The following graph shows the mean temperatures in Chicago over a 30-year period for each month of the year.

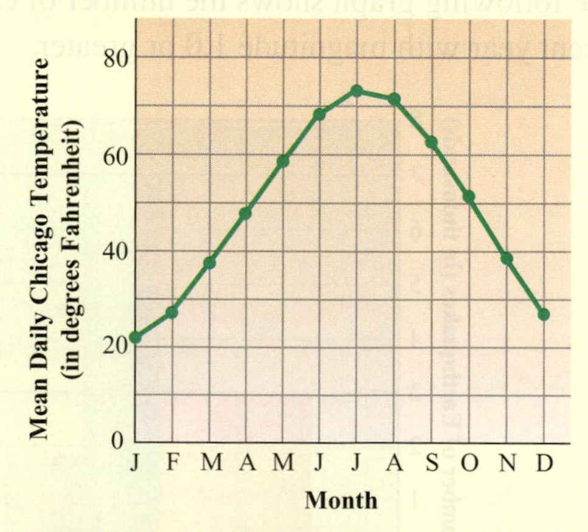

(*Source:* The U.S. National Climatic Data Center)

a. Which month in Chicago has the highest mean temperature?

Solution

a. In 2000, there were about 35 million Americans aged 65 and above.

b. There were approximately 21 million Americans aged 65 and above in the year 1970.

c. In 2000, the overall U.S. population was 4 times what it had been in 1900. But the population shown in the graph grew by a factor of about 10 and so grew more quickly.

b. Approximately what is the mean temperature in February?

c. What trend does the graph show?

Comparison line graphs show two or more changing quantities, as Example 7 illustrates.

EXAMPLE 7

The graph gives the number of male and female participants in high school athletic programs in the U.S. for selected years.

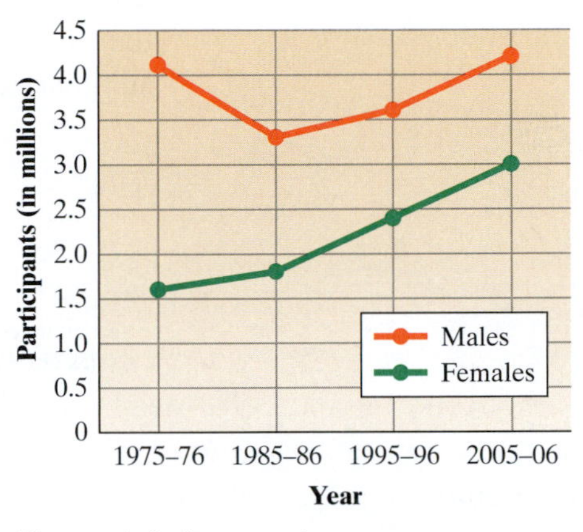

(*Source:* infoplease.com)

a. Estimate the number of females who participated in high school athletic programs in 2005–06.

b. Was the difference between the number of male and the number of female participants greater in 1985–86 or 1995–96?

c. Describe the trend that this graph shows.

Solution

a. In 2005–06, the number of female participants in high school athletic programs was approximately 3 million.

b. In 1985–86, the number of male participants was about 3.3 million and the number of female participants about 1.8 million. So the difference was about 1.5 million. By contrast in 1995–96, there were about 3.6 million males and 2.4 million females, with a difference of approximately 1.2 million. Since 1.5 million is larger than 1.2 million, the difference was greater in 1985–86.

c. Each year, there were more male participants than female participants. The number of female participants increased each year. The number of male participants decreased in 1985–86 and then increased in subsequent years.

PRACTICE 7

The following graph shows the mean precipitation for the cities of Seattle, Washington and Orlando, Florida in selected months.

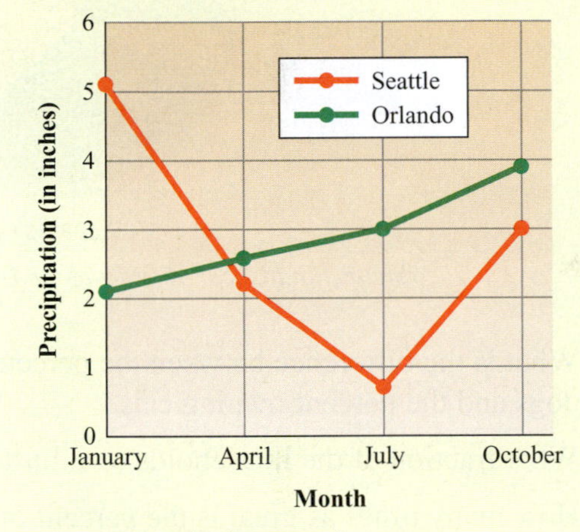

(*Source:* weatherbase.com)

a. On the average, does it rain more in Seattle or in Orlando during the month of October?

b. Approximately what is the average precipitation in Seattle for the month of January?

c. What trend does this graph show?

Circle Graphs

Circle graphs are commonly used to show how a whole amount—say, an entire budget or population—is broken into its parts. The graph resembles a pie (the whole amount) that has been cut into slices (the parts).

Each slice (or *sector*) is proportional in size to the part of the whole that it represents. Each slice is appropriately labeled with either its actual count or the percent of the whole that it represents.

The following example illustrates how to read and interpret the information given by a circle graph.

EXAMPLE 8

The following graph shows the percents of American households that own a single kind of pet:

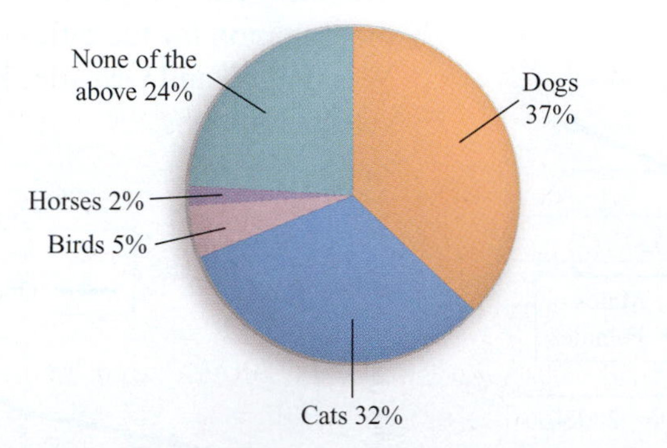

(*Source: Statistical Abstract of the United States, 2010*)

a. What is the difference between the percent of households owning dogs and the percent owning cats?

b. What fraction of the households own birds?

c. How many times as great is the percent of households owning cats as the percent owning horses?

Solution

a. Dog owners comprise 37% of the households, in contrast to 32% for cats. So the difference is 5%.

b. 5%, or 5 out of every 100, of the households own birds, which is equivalent to $\frac{5}{100}$, or $\frac{1}{20}$.

c. 32% of the households own cats, and 2% own horses. So the percent of households that owns cats is 16 times the percent that owns horses.

PRACTICE 8

A sample of U.S. e-mail users were asked how frequently they check their e-mail. The following graph summarizes their responses.

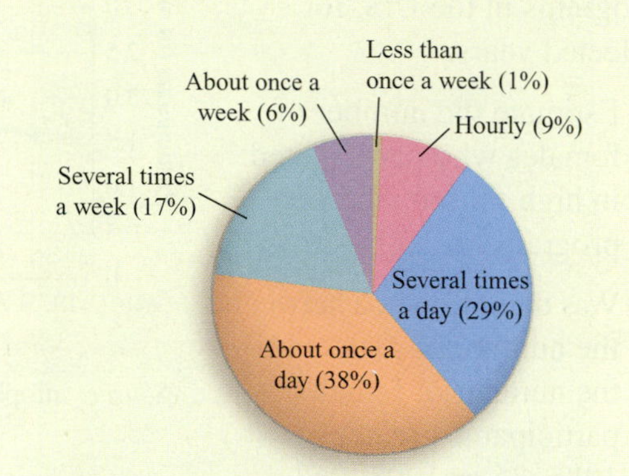

(*Source:* digitalcenter.org)

a. What fraction of the respondents check their e-mail about once a week?

b. What percent check their e-mail at least several times a day?

c. How many times as great is the percent of users who check their e-mail hourly than the percent who check their e-mail less than once a week?

8.2 Exercises

FOR EXTRA HELP MyMathLab Math XL PRACTICE WATCH READ REVIEW

Mathematically Speaking

Fill in each blank with the most appropriate term or phrase from the given list.

line graph	circle graph	heading
rows	columns	histogram
bar graph	graph	
pictograph	table	

1. A ________ is a rectangular display of data.

2. A ________ is a picture or diagram of data.

3. In a table, ________ run horizontally.

4. On a ____________, images such as people, books or coins are used to represent quantities.

5. On a ________, quantities are represented by thin, parallel rectangles.

6. A ___________ is a graph of a frequency table.

7. A ___________ is commonly used to highlight changes and trends over a period of time.

8. A ____________ resembles a pie (the whole) that has been cut into slices (the parts).

Applications

A *Read each table and solve.*

9. The following table shows how to determine a stockbroker's commission in a stock transaction. The commission depends on both the number of shares sold and the price per share.

	Number of Shares				
Price per Share	**100**	**200**	**300**	**400**	**500**
$1–$20	$40	$50	$60	$70	$80
>$20	$40	$60	$80	$90	$100

a. What is the broker's commission on a sale of 300 shares of stock at $15.75 a share?

b. What is the commission on a sale of 500 shares of stock at $30 a share?

c. Will an investor pay her broker a lower commission if she sells 400 shares of stock in a single deal or 200 shares of stock in each of two deals?

10. The table shows the 2009 federal income tax schedule for single filers.

If taxable income is over	But not over	The tax is
$0	$8,350	10% of the amount over $0
$8,350	$33,950	$835 plus 15% of the amount over $8,350
$33,950	$82,250	$4,675 plus 25% of the amount over $33,950
$82,250	$171,550	$16,750 plus 28% of the amount over $82,250
$171,550	$372,950	$41,754 plus 33% of the amount over $171,550
$372,950	No limit	$108,216 plus 33% of the amount over $372,950

(*Source:* irs.gov)

a. What is the tax for a person whose taxable income was $33,950?

b. Compute the tax for a person whose taxable income is $27,000.

Key Concepts and Skills

CONCEPT SKILL

Concept/Skill	Description	Example
[8.1] Mean	Given a set of numbers, the sum of the numbers divided by however many numbers are in the set.	For 0, 0, 1, 3, and 5, the mean is: $\frac{0+0+1+3+5}{5} = \frac{9}{5} = 1.8$
[8.1] Median	Given a set of numbers arranged in numerical order, the number in the middle. If there are two numbers in the middle, the mean of the two middle numbers.	For 0, 0, 1, 3, and 5, the median is 1.
[8.1] Mode	Given a set of numbers, the number (or numbers) occurring most frequently in the set.	For 0, 0, 1, 3, and 5, the mode is 0.
[8.1] Range	Given a set of numbers, the difference between the largest and the smallest number in the set of numbers.	For 0, 0, 1, 3, and 5, the range is $5 - 0$, or 5.
[8.2] Table	A rectangular display of data.	Column ↓; Row → Men: Dems. 37, Reps. 45; Women: Dems. 14, Reps. 15
[8.2] Pictograph	A graph in which images such as people, books, or coins are used to represent the quantities.	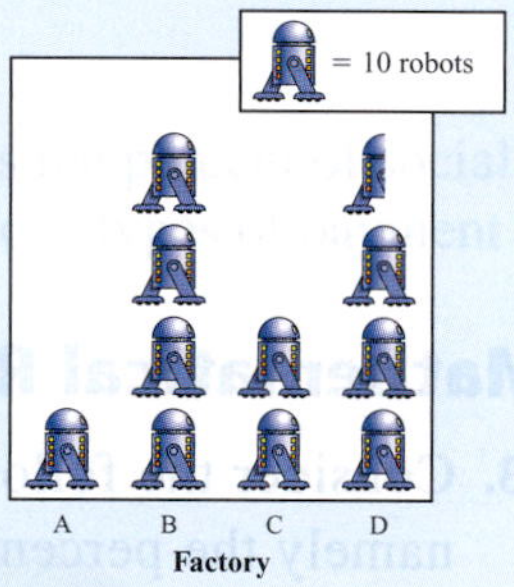
[8.2] Bar graph	A graph in which quantities are represented by thin, parallel rectangles called bars. The length of each bar is proportional to the quantity that it represents.	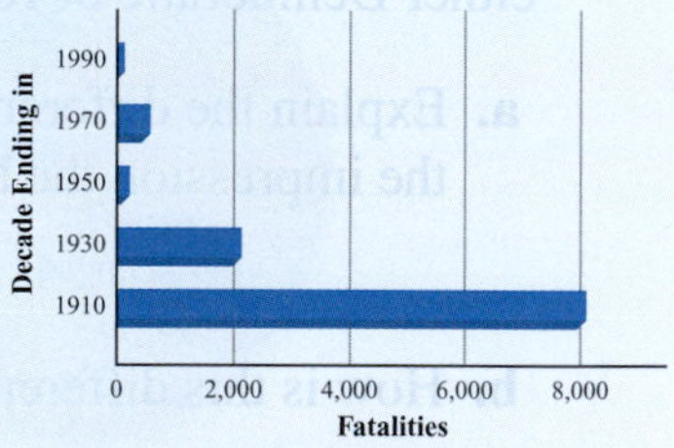
[8.2] Histogram	A graph of a frequency table.	

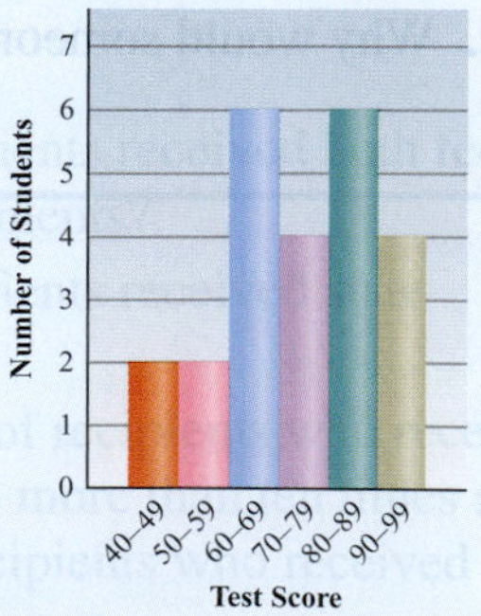

	Dems.	Reps.
Men	37	45
Women	14	15

CONCEPT SKILL

Concept/Skill	Description	Example
[8.2] **Line graph**	A graph in which points are connected by straight-line segments. The position of any point on a line graph is read against the vertical axis and the horizontal axis.	Number of Cases of Measles (in thousands); 0, 5, 10, 15, 20, 25, 30; 1986 1987 1988 1989 1990 1991 1992 1993; Year
[8.2] **Circle graph**	A graph that resembles a pie (a whole amount) that has been cut into slices (the parts).	Other $197; Europe $364; Canada $132; Japan $92; Western Hemisphere $283

9.1 Solving Equations

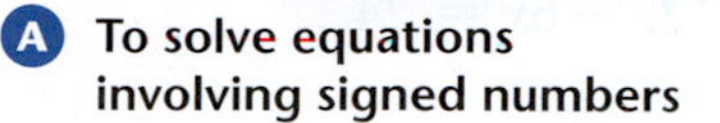

OBJECTIVES

- **A** To solve equations involving signed numbers
- **B** To solve equations with more than one operation
- **C** To solve applied problems involving equations with signed numbers or more than one operation

In Chapter 4, we solved equations with one operation such as the following:

$$x + 3 = 5, \quad x - 3 = 7, \quad 3x = 6, \quad \text{and} \quad \frac{x}{4} = 3$$

In this chapter, we extend the discussion to include equations with

- both positive and negative numbers or
- more than one operation.

Solving Equations with Signed Numbers

Recall that in solving equations with one operation, the key is to isolate the variable, that is, to get the variable alone on one side of the equation. We do this by performing the appropriate opposite operation.

DEFINITION

A **solution** to an equation is a value of the variable that makes the equation a true statement. To **solve** an equation means to find all solutions of the equation.

We have already discussed rules for solving addition, subtraction, multiplication, and division equations involving positive numbers. Let's now apply these rules to simple equations involving both positive and negative numbers.

EXAMPLE 1

Solve and check: $y + 5 = 3$

Solution $y + 5 = 3$

$y + 5 - 5 = 3 - 5$ Subtract 5 from each side of the equation.

$y + 0 = -2$ $5 - 5 = 0$ and $3 - 5 = 3 + (-5) = -2$

$y = -2$ $y + 0 = y$

Check $y + 5 = 3$

$-2 + 5 \stackrel{?}{=} 3$ Substitute -2 for y in the original equation.

$3 \stackrel{\checkmark}{=} 3$

The solution is -2.

PRACTICE 1

Solve and check: $x + 7 = 4$

EXAMPLE 2

Solve and check: $n - 5 = -11$

Solution $n - 5 = -11$

$$n - 5 + 5 = -11 + 5$$ Add 5 to each side of the equation.

$$n + 0 = -6$$

$$n = -6$$

Check $n - 5 = -11$

$$-6 - 5 \stackrel{?}{=} -11$$ Substitute -6 for n in the original equation.

$$-11 \stackrel{\checkmark}{=} -11$$

The solution is -6.

PRACTICE 2

Solve and check: $m - 7 = -19$

EXAMPLE 3

Solve and check: $-7x = 21$

Solution $-7x = 21$

$$\frac{-7x}{-7} = \frac{21}{-7}$$ Divide each side of the equation by -7.

$$x = -3$$

Check $-7x = 21$

$$-7(-3) \stackrel{?}{=} 21$$ Substitute -3 for x in the original equation.

$$21 \stackrel{\checkmark}{=} 21$$

The solution is -3.

PRACTICE 3

Solve and check: $9y = -18$

EXAMPLE 4

Solve and check: $-1 = \frac{x}{6}$

Solution $-1 = \frac{x}{6}$

$$-1 \cdot 6 = \frac{x}{\cancel{6}} \cdot \cancel{6}$$ Multiply each side of the equation by 6.

$$-6 = x$$

$$x = -6$$

Check $-1 = \frac{x}{6}$

$$-1 \stackrel{?}{=} \frac{-6}{6}$$ Substitute -6 for x in the original equation.

$$-1 \stackrel{\checkmark}{=} -1$$

The solution is -6.

PRACTICE 4

Solve and check: $\frac{y}{-3} = -2$

EXAMPLE 3

Solve and check: $\frac{4}{5}z - \frac{3}{5}z = 2$

Solution

$$\frac{4}{5}z - \frac{3}{5}z = 2$$

$$\frac{1}{5}z = 2$$

Combine like terms: $\frac{4}{5}z - \frac{3}{5}z = \left(\frac{4}{5} - \frac{3}{5}\right)z = \frac{1}{5}z$

$$5 \cdot \frac{1}{5}z = 5 \cdot 2$$

Multiply each side of the equation by 5.

$$z = 10$$

Check

$$\frac{4}{5}z - \frac{3}{5}z = 2$$

$$\frac{4}{5}(10) - \frac{3}{5}(10) \stackrel{?}{=} 2$$

Substitute 10 for z in the original equation.

$$8 - 6 \stackrel{?}{=} 2$$

$$2 \stackrel{\checkmark}{=} 2$$

The solution is 10.

PRACTICE 3

Solve and check: $-\frac{2}{7}x + \frac{1}{7}x = 3$

Some equations have like terms on both sides. To solve this type of equation, we use the addition or subtraction rule to get like terms on the same side so that they can be combined.

EXAMPLE 4

Solve and check: $8y - 11 = 5y - 2$

Solution

$$8y - 11 = 5y - 2$$

$$8y - 5y - 11 = 5y - 5y - 2$$

Subtract $5y$ from each side of the equation.

$$3y - 11 = -2$$

Combine like terms.

$$3y - 11 + 11 = -2 + 11$$

Add 11 to each side of the equation.

$$3y = 9$$

$$y = 3$$

Check

$$8y - 11 = 5y - 2$$

$$8(3) - 11 \stackrel{?}{=} 5(3) - 2$$

Substitute 3 for y in the original equation.

$$24 - 11 \stackrel{?}{=} 15 - 2$$

$$13 \stackrel{\checkmark}{=} 13$$

The solution is 3.

PRACTICE 4

Solve and check: $10x - 1 = 2x + 7$

Recall that in Chapter 4 we translated to algebraic expressions phrases, such as *four more than n* and *twice n*. Now, we look at sentences that involve more than one operation, such as *four more than twice n equals ten*, which translates to $2n + 4 = 10$. Such translations lead to equations like the one shown in Example 5.

EXAMPLE 2

Solve and check: $n - 5 = -11$

Solution $n - 5 = -11$

$n - 5 + 5 = -11 + 5$ Add 5 to each side of the equation.

$n + 0 = -6$

$n = -6$

Check $n - 5 = -11$

$-6 - 5 \stackrel{?}{=} -11$ Substitute −6 for *n* in the original equation.

$-11 \stackrel{\checkmark}{=} -11$

The solution is −6.

PRACTICE 2

Solve and check: $m - 7 = -19$

EXAMPLE 3

Solve and check: $-7x = 21$

Solution $-7x = 21$

$\frac{-7x}{-7} = \frac{21}{-7}$ Divide each side of the equation by −7.

$x = -3$

Check $-7x = 21$

$-7(-3) \stackrel{?}{=} 21$ Substitute −3 for *x* in the original equation.

$21 \stackrel{\checkmark}{=} 21$

The solution is −3.

PRACTICE 3

Solve and check: $9y = -18$

EXAMPLE 4

Solve and check: $-1 = \frac{x}{6}$

Solution $-1 = \frac{x}{6}$

$-1 \cdot 6 = \frac{x}{\cancel{6}} \cdot \cancel{6}$ Multiply each side of the equation by 6.

$-6 = x$

$x = -6$

Check $-1 = \frac{x}{6}$

$-1 \stackrel{?}{=} \frac{-6}{6}$ Substitute −6 for *x* in the original equation.

$-1 \stackrel{\checkmark}{=} -1$

The solution is −6.

PRACTICE 4

Solve and check: $\frac{y}{-3} = -2$

EXAMPLE 5

A student lost 8 points for each incorrect answer on an exam. How many incorrect answers did he get if he lost a total of 32 points?

Solution We can represent a loss of 8 points as -8 and a loss of 32 points as -32. To find the number of incorrect answers, we write the following equation, letting x represent the number of incorrect answers:

$$-8x = -32$$

We solve this equation for x.

$$-8x = -32$$
$$\frac{-8x}{-8} = \frac{-32}{-8}$$
$$x = 4$$

Check

$$-8x = -32$$
$$-8(4) \stackrel{?}{=} -32$$
$$-32 \stackrel{\checkmark}{=} -32$$

So the student got 4 incorrect answers.

PRACTICE 5

A patient's daily dosage of a medication was decreased by 15 milligrams per week. After how many weeks had the dosage decreased by 105 milligrams?

Solving Equations with More Than One Operation

We now turn our attention to solving equations such as:

$$3y + 5 = 26 \quad \text{and} \quad \frac{c}{3} - 4 = 7$$

To Solve an Equation with More than One Operation

- First, use the rule for solving addition or subtraction equations.
- Then, use the rule for solving multiplication or division equations.

EXAMPLE 6

Solve and check: $3y + 11 = 26$

Solution

$$3y + 11 = 26$$
$$3y + 11 - 11 = 26 - 11 \quad \text{Subtract 11 from each side of the equation.}$$
$$3y = 15$$
$$\frac{3y}{3} = \frac{15}{3} \quad \text{Divide each side of the equation by 3.}$$
$$y = 5$$

Check

$$3y + 11 = 26$$
$$3(5) + 11 \stackrel{?}{=} 26 \quad \text{Substitute 5 for } y \text{ in the original equation.}$$
$$15 + 11 \stackrel{?}{=} 26$$
$$26 \stackrel{\checkmark}{=} 26$$

The solution is 5.

PRACTICE 6

Solve and check: $2x + 8 = -6$

EXAMPLE 7

Solve and check: $\frac{c}{3} - 4 = 7$

Solution $\frac{c}{3} - 4 = 7$

$\frac{c}{3} - 4 + 4 = 7 + 4$ Add 4 to each side of the equation.

$\frac{c}{3} = 11$

$3 \cdot \frac{c}{3} = 3 \cdot 11$ Multiply each side of the equation by 3.

$c = 33$

Check $\frac{c}{3} - 4 = 7$

$\frac{33}{3} - 4 \stackrel{?}{=} 7$ Substitute 33 for c in the original equation.

$11 - 4 \stackrel{?}{=} 7$

$7 \stackrel{\checkmark}{=} 7$

The solution is 33.

PRACTICE 7

Solve and check: $\frac{k}{5} - 6 = -3$

EXAMPLE 8

Solve and check: $-2x + 1 = 5$

Solution $-2x + 1 = 5$

$-2x + 1 - 1 = 5 - 1$ Subtract 1 from each side of the equation.

$-2x = 4$

$\frac{-2x}{-2} = \frac{4}{-2}$ Divide each side of the equation by -2.

$x = -2$

Check $-2x + 1 = 5$

$-2(-2) + 1 \stackrel{?}{=} 5$ Substitute -2 for x in the original equation.

$4 + 1 \stackrel{?}{=} 5$

$5 \stackrel{\checkmark}{=} 5$

The solution is -2.

PRACTICE 8

Solve and check: $-3d + 8 = -4$

Note that in Example 8 we solved by subtracting before dividing. Can you think of any other way to solve this equation?

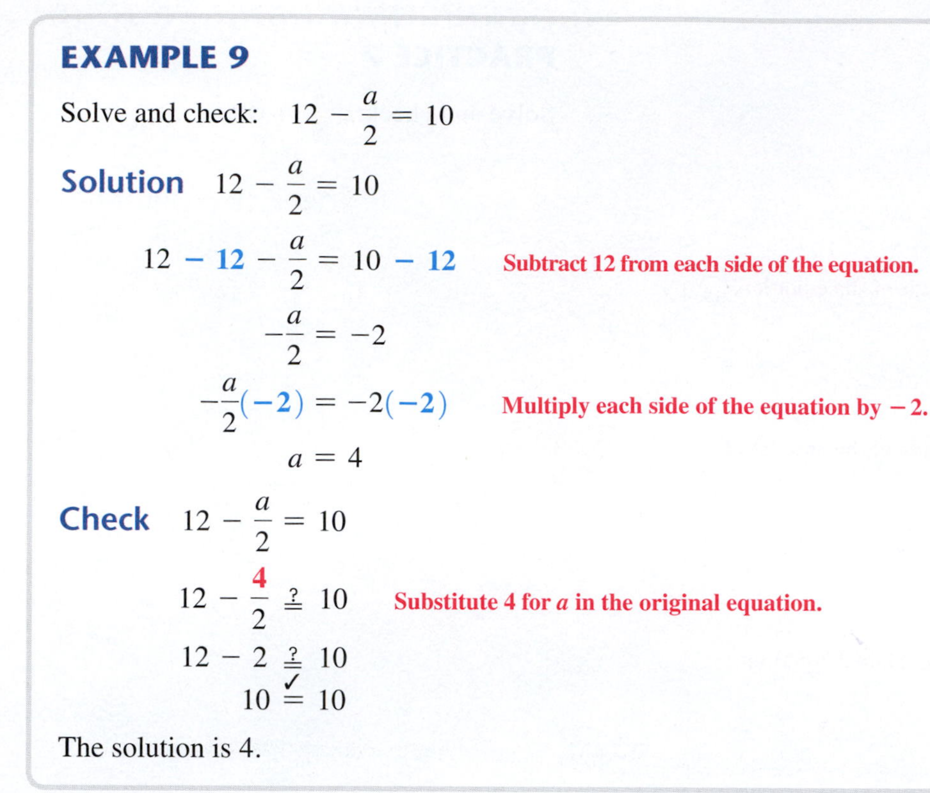

EXAMPLE 9

Solve and check: $12 - \frac{a}{2} = 10$

Solution $12 - \frac{a}{2} = 10$

$$12 - 12 - \frac{a}{2} = 10 - 12$$ Subtract 12 from each side of the equation.

$$-\frac{a}{2} = -2$$

$$-\frac{a}{2}(-2) = -2(-2)$$ Multiply each side of the equation by -2.

$$a = 4$$

Check $12 - \frac{a}{2} = 10$

$$12 - \frac{4}{2} \stackrel{?}{=} 10$$ Substitute 4 for a in the original equation.

$$12 - 2 \stackrel{?}{=} 10$$

$$10 \stackrel{\checkmark}{=} 10$$

The solution is 4.

PRACTICE 9

Solve and check: $1 - \frac{x}{8} = -9$

EXAMPLE 10

A customer paid $50 for monthly cable service and $9.50 for each pay-per-view movie he ordered.

a. Write an equation to determine how many pay-per-view movies the customer ordered if his monthly bill was $107.

b. Solve this equation.

Solution

a. If we let x represent the number of pay-per-view movies ordered, then $9.5x$ represents the total amount paid (in dollars) for the pay-per-view movies.

Next, we write a sentence for the problem and then translate it to an algebraic equation.

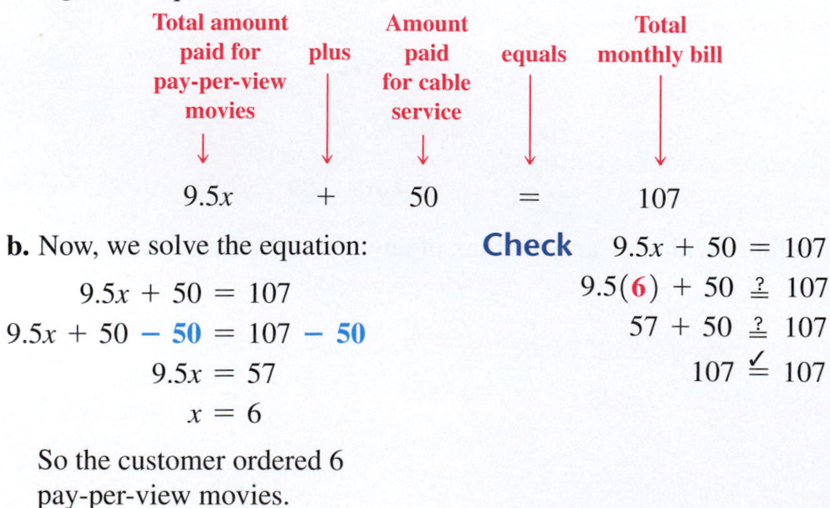

$$9.5x + 50 = 107$$

b. Now, we solve the equation:

$$9.5x + 50 = 107$$

$$9.5x + 50 - 50 = 107 - 50$$

$$9.5x = 57$$

$$x = 6$$

Check

$$9.5x + 50 = 107$$

$$9.5(6) + 50 \stackrel{?}{=} 107$$

$$57 + 50 \stackrel{?}{=} 107$$

$$107 \stackrel{\checkmark}{=} 107$$

So the customer ordered 6 pay-per-view movies.

PRACTICE 10

Suppose that an empty crate for oranges weighs 2 kilograms. A typical orange weighs about 0.2 kilogram and the total weight of the crate with oranges is 10 kilograms.

a. Write an equation to determine how many oranges the crate contains.

b. Solve this equation.

9.1 Exercises

FOR EXTRA HELP MyMathLab MathXL PRACTICE WATCH READ REVIEW

A *Solve and check.*

1. $a - 7 = -21$

2. $x - 6 = -9$

3. $b + 4 = -7$

4. $y + 12 = -12$

5. $-11 = z - 4$

6. $-15 = m - 20$

7. $x + 21 = 19$

8. $a + 12 = 10$

9. $c + 33 = 14$

10. $d + 27 = 15$

11. $39 = z + 51$

12. $33 = c + 49$

13. $z + 2.4 = -5.3$

14. $t + 2.3 = -6.7$

15. $2.3 = x - 5.9$

16. $4.1 = d - 6.9$

17. $y - 2\frac{1}{3} = -3$

18. $s - 4\frac{1}{2} = -8$

19. $n + \frac{1}{3} = \frac{1}{2}$

20. $\frac{1}{4} + t = -\frac{1}{6}$

21. $-5 = t + 1\frac{1}{4}$

22. $-3 = 1\frac{2}{3} + c$

23. $-5x = 30$

24. $-8y = 8$

25. $-36 = -9n$

26. $-125 = -25x$

27. $\frac{w}{10} = -24$

28. $\frac{a}{5} = -40$

29. $-6 = \frac{x}{-2}$

30. $-8 = \frac{y}{-4}$

31. $1.7t = -51$

32. $-1.5x = 45$

33. $-10y = 4$

34. $-15a = 3$

35. $\frac{m}{-1.5} = 1$

36. $\frac{x}{-7} = 1.3$

37. $\frac{y}{9} = -\frac{5}{3}$

38. $\frac{z}{3} = \frac{-4}{3}$

B *Solve and check.*

39. $4n - 20 = 36$

40. $3a - 13 = 11$

41. $3x + 1 = 7$

42. $7m + 1 = 22$

43. $6k + 23 = 5$

44. $2x + 21 = 7$

45. $3x + 20 = 20$

46. $4y + 28 = 28$

47. $31 = 3 - 4h$

48. $-68 = 10 - 3x$

49. $34 = 13 - 4p$

50. $36 = 25 - 3c$

51. $-7b + 8 = -6$

52. $-2x + 15 = -9$

53. $21 + \frac{a}{3} = 10$

54. $25 + \frac{w}{5} = 15$

55. $\frac{1}{2}y + 5 = -13$

56. $\frac{x}{5} + 15 = 0$

57. $5 - \frac{x}{12} = 1$

58. $16 - \frac{a}{2} = 15$

59. $\frac{c}{3} + 3 = -4$

60. $\frac{m}{4} + 1 = -5$

61. $\frac{4}{9}x - 13 = -5$

62. $\frac{5}{4}y - 19 = 26$

63. $-8 - x = 11$

64. $-2 - x = 24$

65. $8 = -4 - x$

66. $10 = -3 - x$

Solve. Round each solution to the nearest tenth. Check.

67. $8{,}950 = -6.24n$

68. $-1{,}458 = 20.9p$

69. $-2.57 = \dfrac{x}{5.91}$

70. $-4.6 = \dfrac{z}{-2.78}$

71. $58.3r + 23.58 = 2.79$

72. $-51.5 = 29m - 4.06$

73. $\dfrac{x}{2.4} - 0.03 = -0.14$

74. $\dfrac{a}{2.7} + 11.9 = 0.02$

Mixed Practice

Solve and check.

75. $\dfrac{t}{-8} = -1.2$

76. $-10 = m + 6$

77. $6n + 21 = 15$

78. $1.9 + \dfrac{x}{4} = 2.1$

79. $32 = 27 - 2c$

80. $-1.4a = 42$

81. $-3 = y - 1\dfrac{1}{5}$

82. $\dfrac{2}{3}n - 5 = 4$

C Applications

Write an equation. Solve and check.

83. On a particular day, the price of one share of a video gaming company's stock dropped by \$1.50. If a stockholder lost \$750 that day, how many shares of stock did he own?

84. The value of a networked color printer purchased new decreased by the same amount each year. After 4 years, the value of the printer dropped by \$600. By how much did the value decrease each year?

85. A small company lost a total of \$15,671 in the first two quarters of last year. If the company lost \$9,046 in the first quarter, how much did it lose in the second quarter?

86. If the holder of a checking account were to write a check for \$350, the account would be overdrawn by \$200. What is the present balance of the account?

87. The calendar used by the Mayas in the third century A.D. consisted of 365 days divided into 18 months. Every month contained the same number of days, with the five left-over days considered unlucky. How many days were in each of the Mayan months? (*Source:* exploratorium.edu)

88. A car rents for \$85 per day plus \$0.35 for every mile driven more than 200 miles. If the car was driven 300 miles and the rental paid was \$205, for how many days was the car rented?

89. At a graduation dinner, an equal number of guests were seated at each of 8 large tables, and 3 late-arriving guests were seated at a small table. If there were 43 guests in all, how many guests were seated at each of the large tables?

90. A toy maker has daily manufacturing costs of \$890, plus \$3 for each action figure it produces. How many action figures does the toy maker produce each day if the total daily manufacturing cost is \$5,390?

91. A homeowner's electric bill was \$94.98 last month. How many kilowatt hours (to the nearest kilowatt hour) were used if the electric company charges \$0.1374 per kilowatt hour plus a basic customer fee of \$6.29 per month?

92. A long-distance service provider charges a flat monthly rate of \$4.99 plus \$0.11 per minute for long-distance phone calls. If a customer's monthly bill (excluding taxes and other fees) for long-distance service was \$41.09, how many minutes (to the nearest minute) was she charged for long-distance calls this month?

• Check your answers on page A-12.

MINDStretchers

Groupwork

1. In the following magic square, the sum of every row, column, and diagonal is 15. Working with a partner, solve for a, b, and c.

6	1	$\frac{c}{4} + 2$
7	$2b + 1$	3
$3a - 1$	9	4

Mathematical Reasoning

2. You are given three objects: **a**, **b**, and **c**. Suppose that two of the three objects are equal in weight but that the third object weighs more than either of the other two. On the balance scale shown, indicate how to identify which object is the heaviest one by *only one weighing*.

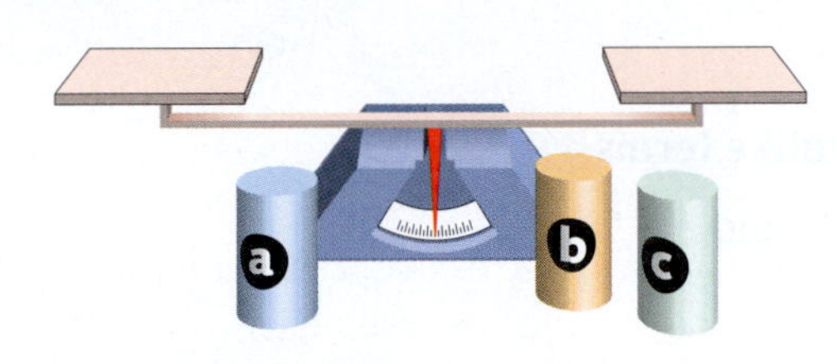

Writing

3. For each of the following equations, write two different situations that the equation models.

a. $12x + 500 = 24{,}500$

b. $\frac{x}{2} - 40 = 60$

9.2 More on Solving Equations

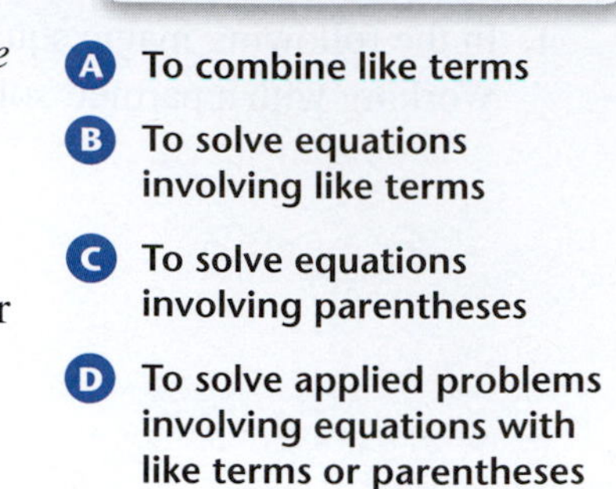

In this section, we extend the discussion of solving equations to include those that involve *like terms* or contain parentheses.

Solving Equations with Like Terms

In Chapter 4, we considered algebraic expressions such as $2x$ or $x - 1$ that consist of one or two terms. Some algebraic expressions, such as $2x + x + 1$, have three terms.

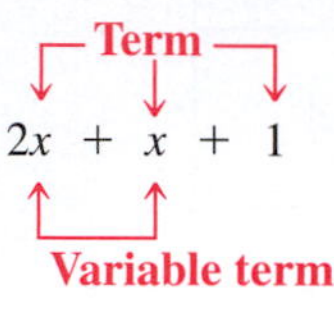

The terms $2x$ and x are like terms.

DEFINITION

Like terms are terms that have the same variables with the same exponents. Terms that are not like are called **unlike terms.**

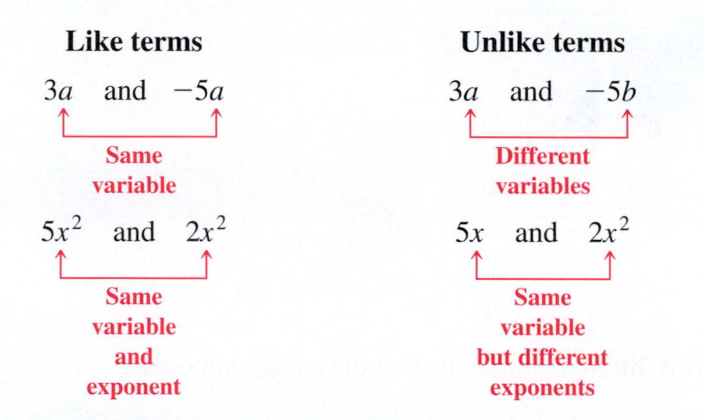

We cannot combine unlike terms, such as $3a$ and $-5b$. However, like terms can be **combined** (that is, added or subtracted) by using the distributive property.

Recall that this property states that multiplying a factor by the sum of two numbers gives the same result as multiplying the factor by each of the two numbers and then adding, for example,

$$3 \cdot (7 + 2) = 3 \cdot 7 + 3 \cdot 2$$

The distributive property also implies

$$7 \cdot 3 + 2 \cdot 3 = (7 + 2) \cdot 3 = 9 \cdot 3,$$

where we add 7 threes and 2 threes to get 9 threes

and

$$7 \cdot 3 - 2 \cdot 3 = (7 - 2) \cdot 3 = 5 \cdot 3,$$

where we subtract 2 threes from 7 threes to get 5 threes.

Now, let's apply the distributive property to combine like terms:

$$7x + 2x = (7 + 2)x = 9x \qquad 7x - 2x = (7 - 2)x = 5x$$

More generally, we can use the following rule to combine like terms and to simplify algebraic expressions:

To Combine Like Terms

- Use the distributive property.
- Add or subtract.

EXAMPLE 1

Combine like terms.

a. $7x + 3x$ **b.** $\frac{1}{2}y - \frac{1}{4}y$ **c.** $a - 3a + 9$

Solution

a. $7x + 3x = (7 + 3)x$ Use the distributive property.

$= 10x$ Add 7 and 3.

b. $\frac{1}{2}y - \frac{1}{4}y = \left(\frac{1}{2} - \frac{1}{4}\right)y$ Use the distributive property.

$= \frac{1}{4}y$ Subtract $\frac{1}{4}$ from $\frac{1}{2}$.

c. $a - 3a + 9 = (1 - 3)a + 9$ Use the distributive property. Recall that $a = 1a$.

$= -2a + 9$ Subtract 3 from 1.

PRACTICE 1

Simplify.

a. $5x + 7x$

b. $\frac{5}{8}n - \frac{1}{2}n$

c. $4z - 5z + 6$

Some equations have like terms on one side. To solve this type of equation, we begin by combining all like terms.

EXAMPLE 2

Solve and check: $3x - 5x = 10$

Solution

$3x - 5x = 10$

$-2x = 10$ Combine like terms: $3x - 5x = (3 - 5)x = -2x$

$\frac{-2x}{-2} = \frac{10}{-2}$ Divide each side of the equation by -2.

$x = -5$

Check

$3x - 5x = 10$

$3(-5) - 5(-5) \stackrel{?}{=} 10$ Substitute -5 for x in the original equation.

$-15 - (-25) \stackrel{?}{=} 10$

$10 \stackrel{\checkmark}{=} 10$

The solution is -5.

PRACTICE 2

Solve and check: $2y - 3y = 8$

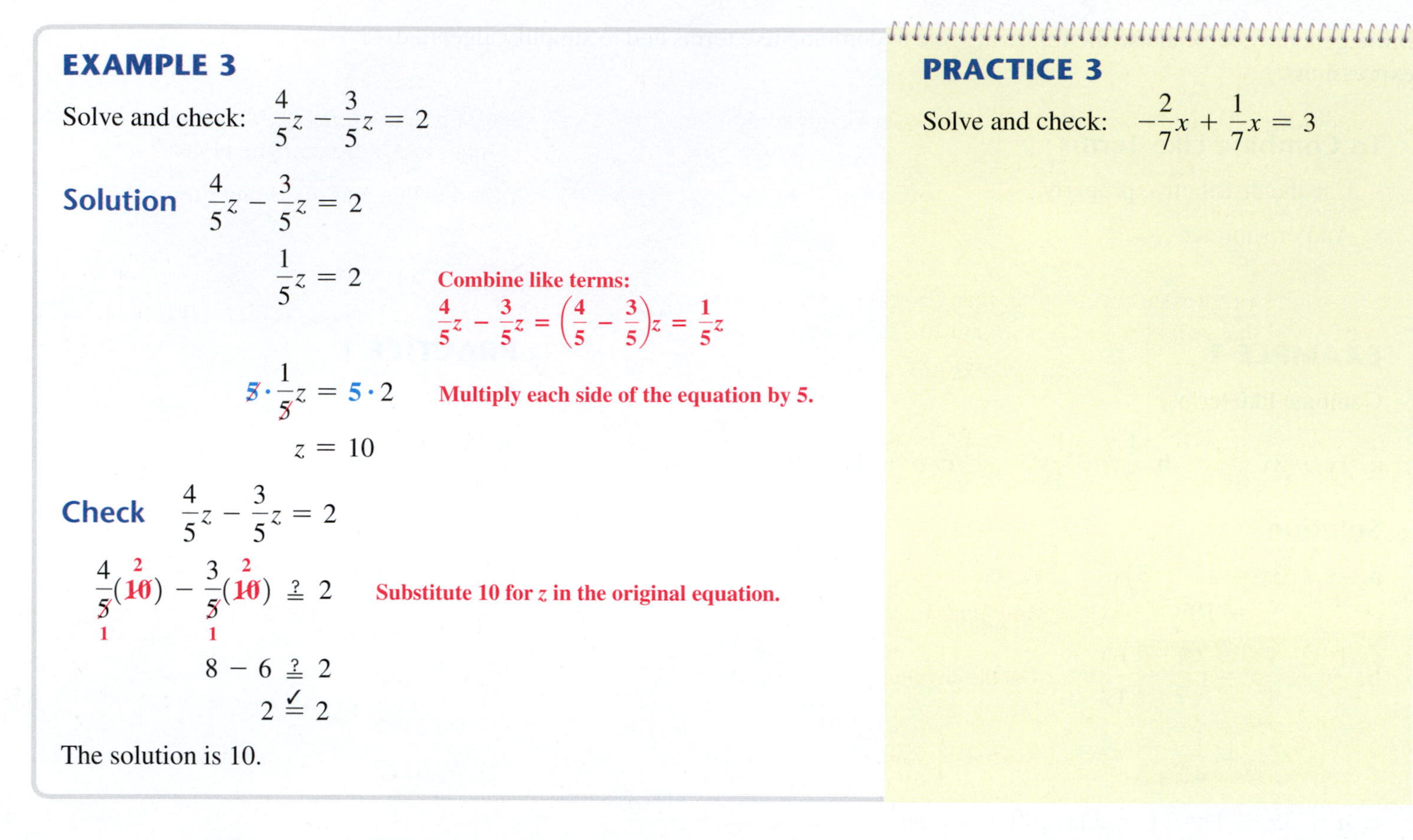

EXAMPLE 3

Solve and check: $\frac{4}{5}z - \frac{3}{5}z = 2$

Solution $\frac{4}{5}z - \frac{3}{5}z = 2$

$\frac{1}{5}z = 2$ Combine like terms: $\frac{4}{5}z - \frac{3}{5}z = \left(\frac{4}{5} - \frac{3}{5}\right)z = \frac{1}{5}z$

$5 \cdot \frac{1}{5}z = 5 \cdot 2$ Multiply each side of the equation by 5.

$z = 10$

Check $\frac{4}{5}z - \frac{3}{5}z = 2$

$\frac{4}{5}(10) - \frac{3}{5}(10) \stackrel{?}{=} 2$ Substitute 10 for z in the original equation.

$8 - 6 \stackrel{?}{=} 2$

$2 \stackrel{\checkmark}{=} 2$

The solution is 10.

PRACTICE 3

Solve and check: $-\frac{2}{7}x + \frac{1}{7}x = 3$

Some equations have like terms on both sides. To solve this type of equation, we use the addition or subtraction rule to get like terms on the same side so that they can be combined.

EXAMPLE 4

Solve and check: $8y - 11 = 5y - 2$

Solution $8y - 11 = 5y - 2$

$8y - 5y - 11 = 5y - 5y - 2$ Subtract $5y$ from each side of the equation.

$3y - 11 = -2$ Combine like terms.

$3y - 11 + 11 = -2 + 11$ Add 11 to each side of the equation.

$3y = 9$

$y = 3$

Check $8y - 11 = 5y - 2$

$8(3) - 11 \stackrel{?}{=} 5(3) - 2$ Substitute 3 for y in the original equation.

$24 - 11 \stackrel{?}{=} 15 - 2$

$13 \stackrel{\checkmark}{=} 13$

The solution is 3.

PRACTICE 4

Solve and check: $10x - 1 = 2x + 7$

Recall that in Chapter 4 we translated to algebraic expressions phrases, such as *four more than n* and *twice n*. Now, we look at sentences that involve more than one operation, such as *four more than twice n equals ten*, which translates to $2n + 4 = 10$. Such translations lead to equations like the one shown in Example 5.

EXAMPLE 5

The product of 5 and a number is 24 less than twice that number.

a. Write an equation to find the number.

b. Solve this equation.

Solution

a. If we let x represent the number, then $5x$ represents the product of 5 and that number. Next, we represent 24 less than twice that number by $2x - 24$. We can now translate the given problem to an equation: $5x = 2x - 24$.

b. Now, we solve the equation $5x = 2x - 24$.

$$5x = 2x - 24$$
$$5x - 2x = 2x - 2x - 24$$
$$3x = -24$$
$$x = -8$$

Check $5x = 2x - 24$

$$5(-8) \stackrel{?}{=} 2(-8) - 24$$
$$-40 \stackrel{?}{=} -16 - 24$$
$$-40 \stackrel{\checkmark}{=} -40$$

The solution is -8.

PRACTICE 5

The sum of a number and 3 times that number is 9 more than the number.

a. Write an equation to find the number.

b. Solve this equation.

Solving Equations Containing Parentheses

Some equations contain parentheses. To solve this type of equation, we first remove the parentheses, using the distributive property.

EXAMPLE 6

Solve and check: $2(6 - x) = -8$

Solution

$$2(6 - x) = -8$$
$$12 - 2x = -8 \quad \text{Use the distributive property.}$$
$$12 - 12 - 2x = -8 - 12 \quad \text{Subtract 12 from each side of the equation.}$$
$$-2x = -20$$
$$x = 10$$

Check $2(6 - x) = -8$

$$2(6 - 10) \stackrel{?}{=} -8 \quad \text{Substitute 10 for } x \text{ in the original equation.}$$
$$2(-4) \stackrel{?}{=} -8$$
$$-8 \stackrel{\checkmark}{=} -8$$

The solution is 10.

PRACTICE 6

Solve and check: $3(1 - 2x) = 9$

EXAMPLE 7

Solve and check: $6x = -4(x + 5)$

Solution

$$
\begin{aligned}
6x &= -4(x + 5) \\
6x &= (-4)(x) + (-4)(5) && \text{Use the distributive property.} \\
6x &= -4x - 20 \\
6x + 4x &= -4x + 4x - 20 && \text{Add } 4x \text{ to each side of the equation.} \\
10x &= -20 && \text{Combine like terms.} \\
x &= -2
\end{aligned}
$$

Check

$$
\begin{aligned}
6x &= -4(x + 5) \\
6(-2) &\stackrel{?}{=} -4(-2 + 5) && \text{Substitute } -2 \text{ for } x \text{ in the original equation.} \\
-12 &\stackrel{?}{=} -4(3) \\
-12 &\stackrel{\checkmark}{=} -12
\end{aligned}
$$

The solution is -2.

PRACTICE 7

Solve and check: $-3(x - 4) = x$

EXAMPLE 8

A parking garage charges $15 for the first hour and $2.50 for each additional hour of parking. If a driver paid $27.50 to park in the garage, how many hours did she park?

Solution Let h represent the number of hours she parked. Then, $h - 1$ represents the number of hours for which she paid the hourly rate of $2.50.

Amount for first hour	plus	Amount for each additional hour	equals	Total cost of parking
↓	↓	↓	↓	↓
15	$+$	$2.5(h - 1)$	$=$	27.5

$$
\begin{aligned}
15 + 2.5(h - 1) &= 27.5 \\
15 + 2.5h - 2.5 &= 27.5 \\
12.5 + 2.5h &= 27.5 \\
2.5h &= 15 \\
h &= 6
\end{aligned}
$$

So the driver parked in the garage for 6 hours.

PRACTICE 8

A pizza parlor charges $9.70 for a 14-inch cheese pizza with one topping and $1.10 for each additional topping. If a customer's total charge was $13.00, how many toppings did he have on his pizza?

9.2 Exercises

FOR EXTRA HELP MyMathLab MathXL PRACTICE WATCH READ REVIEW

A *Simplify.*

1. $4x + 3x$
2. $y + 5y$
3. $4a - a$
4. $3d - 2d$
5. $6y - 9y$
6. $2x - 7x$
7. $2n - 3n$
8. $4z - 7z$
9. $\frac{1}{2}x + \frac{1}{3}x$
10. $\frac{5}{8}n + \frac{3}{4}n$
11. $2c + c + 12$
12. $5a + a - 1$
13. $8 + x - 7x$
14. $5 - a + 4a$
15. $-y + 5 + 3y$
16. $2x - 4 - x$

B *Solve and check.*

17. $5m + 4m = 36$
18. $3y + 8y = 22$
19. $18 = 4y - 2y$
20. $24 = 3x - x$
21. $2a - 3a = 0$
22. $x - 7x = 12$
23. $7 = -5b + b$
24. $5 = -y + 3y$
25. $\frac{2}{5}b + \frac{3}{10}b = 14$
26. $\frac{5}{6}t + \frac{2}{3}t = 3$
27. $n + n - 13 = 13$
28. $y + 2y - 3 = 18$
29. $6 = 7x - 3x - 6$
30. $11 = s - 4s + 5$
31. $n + 3n - 7 = 29$
32. $6m + m - 9 = 30$
33. $6a - a + 4a = -6$
34. $2y + 3y - y = -8$
35. $5x = 2x + 12$
36. $3t + 8 = t$
37. $4p + 1 = 3p - 1$
38. $10w + 1 = 3w + 1$
39. $8x + 1 = x - 6$
40. $5x - 7 = 2x + 2$
41. $3p - 2 = -p + 4$
42. $2y - 3 = y + 1$
43. $4n - 6 = 3n + 6$
44. $2x + 1 = -x - 4$
45. $3 - 6t = 5t - 19$
46. $5 - 8r = 2r - 25$
47. $-7y + 2 = 3y - 8$
48. $-6s - 2 = -8s - 4$
49. $1.5r + 6 = 1.2r + 15$
50. $0.7x + 1 = 0.3x + 9$
51. $3x - 2\frac{1}{2} = 3\frac{1}{3} - 3x - x$
52. $4y - 7 = 4\frac{2}{3} - y - 2y$

C *Solve and check.*

53. $2(n + 3) = 12$
54. $3(b + 4) = 24$
55. $8(x - 1) = -24$
56. $3(r - 5) = -18$
57. $\frac{1}{2}(x + 12) = 7$
58. $0.4(10 - x) = 8$
59. $6n = 5(n + 7)$
60. $3x = 2(x + 4)$
61. $3(y - 5) = 2y$
62. $5(m - 3) = 2m$
63. $5y = -3(y + 1)$
64. $6x = -4(x - 1)$
65. $-4n = 7(9 - n)$
66. $2(3 - x) = -x$
67. $6r + 2(r - 1) = 14$
68. $3t + 2(t + 4) = 13$
69. $5y - 3(y + 2) = 4$
70. $7n - 2(n - 3) = 16$

Write an equation to find the number. Solve and check.

71. Nine less than 6 times a number is equal to 18 more than -3 times the number.

72. Twenty-one less than 5 times a number is equal to 28 more than -2 times the number.

73. Three times the sum of a number and 5 equals twice the number.

74. Four times the sum of a number and 2 equals 3 times the number.

Solve. Round each solution to the nearest tenth. Check.

75. $60r - 17r + 23.58 = 2.79$

76. $1.02m + 3.007m = 50.1$

77. $3.61n = 4 + 2.135n$

78. $0.138a = -4.667a + 2.931a - 4.625$

79. $1.72y = 3.16(y - 8.72)$

80. $6.19t + 3.81(1 - t) = 2.72$

Mixed Practice

Simplify.

81. $-x + 6 + 5x$

82. $-1 + a - 9a$

Solve and check.

83. $6 = -5y + y$

84. $7n - 5 = 2n - 5$

85. $2x + 3(x + 1) = -x$

86. $-8 = 8 - t + 5t$

87. $-5m - 1 = 1 + 5m$

88. $-2(y - 3) = 5.4$

Applications

D *Write an equation. Solve and check.*

89. In a national telephone survey, 83% of the interviewees indicated that strengthening the country's economy should be the top domestic priority for the President and the Congress. The remaining interviewees, who numbered 255, disagreed. How many interviewees were there in all? (*Source:* people-press.org)

90. A student purchased 8 tickets for a monster truck rally through an online ticket agency. A service charge of $5 was added to the price of each ticket. If the total cost of the tickets was $528, what was the price of one ticket?

91. The perimeter of a rectangle 70 feet long is 200 feet. Find the width, w.

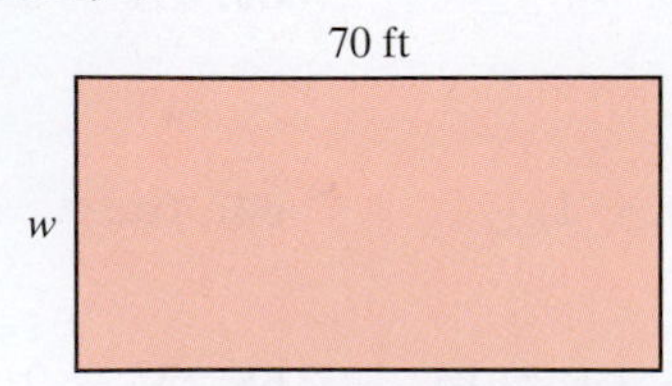

92. The sum of the measures of the angles of a triangle is 180°. In the triangle shown, what is the measure of the two equal angles, x?

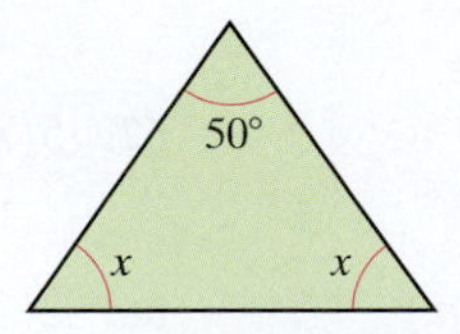

93. On a large wide-body plane, one-sixth of the passenger seats are in business class and three-fourths are in economy class. If the airline advertises that the passenger seats in these two cabins number 440, how many passenger seats in all are there on the plane?

94. A family budgets one-third of their monthly income for housing and one-fourth for food. If the difference between the amount budgeted for housing and that for food is $150, what is their monthly income?

95. Two car dealerships offer lease options on comparable midsize sedans. One dealership requires \$3,000 down and \$120 per month. The other dealership requires \$2,400 down and \$145 per month. In how many months will a customer spend the same amount for each lease?

96. In any polygon with n sides, the sum of the measures of the interior angles is $180(n - 2)$ degrees. If the sum of the measures of the angles of a polygon is 540 degrees, how many sides does the polygon have?

97. In the spreadsheet shown, $C1 = A1 \cdot B1$, $B3 = B1 + B2$, $C2 = A2 \cdot B2$, and $C3 = C1 + C2$. Find B1.

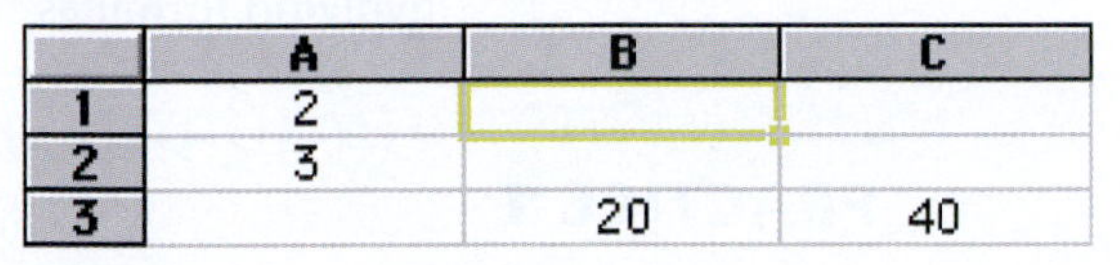

	A	B	C
1	2		
2	3		
3		20	40

98. A company bought a copier for \$10,000. After n years of depreciation, the copier is valued at $10{,}000\left(1 - \dfrac{n}{20}\right)$ dollars. After how many years will the copier be valued at \$5,500?

99. In a family cell phone plan, the monthly cost for the first two lines is \$119.99, with \$49.99 for each additional line. If such a plan costs \$269.96, how many lines does the plan cover? (*Source:* wireless.att.com)

100. Alaska and Texas are the two largest states in the United States. Alaska has 17.5% of the country's area, and Texas has 7.1%. If the combined area of these two states is 931,848 square miles, find the area of the United States to the nearest hundred thousand square miles. (*Source:* wikipedia.org)

• Check your answers on page A-12.

MINDStretchers

Mathematical Reasoning

1. The *algebra tiles* pictured represent $3x + 1$ and $2x - 3$, respectively.

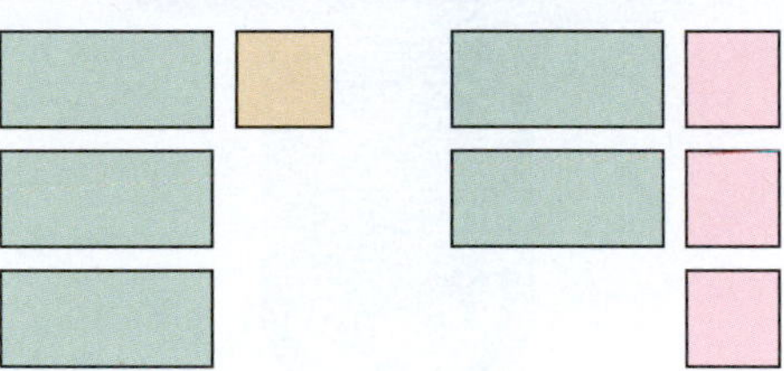

Represent each of the following by algebra tiles:

a. $2x - 1$

b. $x + 4$

c. The sum of the expressions in parts (a) and (b).

Writing

2. Write a list of steps for solving equations with variables on both sides.

Critical Thinking

3. Suppose that you work on an equation with more than one operation and obtain $6x = 5x$. Can you solve this equation by dividing both sides by x? Explain.

9.3 Using Formulas

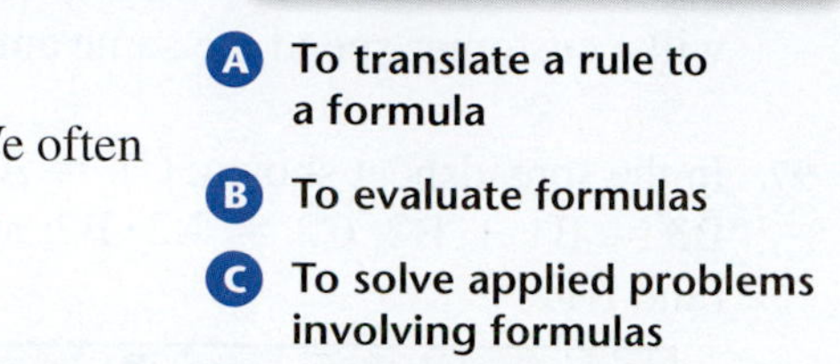

OBJECTIVES

- **A** To translate a rule to a formula
- **B** To evaluate formulas
- **C** To solve applied problems involving formulas

Translating Rules to Formulas

A **formula** is an equation that indicates how the variables are related to one another. We often use formulas as shorthand for expressing a rule or relationship involving the variables.

EXAMPLE 1

The Brannock device, found in many men's shoe stores, is used to determine a customer's shoe size s based on his foot length f. To find his shoe size, the customer inserts his foot in the measurement device, which triples the length of the foot (in inches) and then subtracts 22. Write a formula for this relationship.

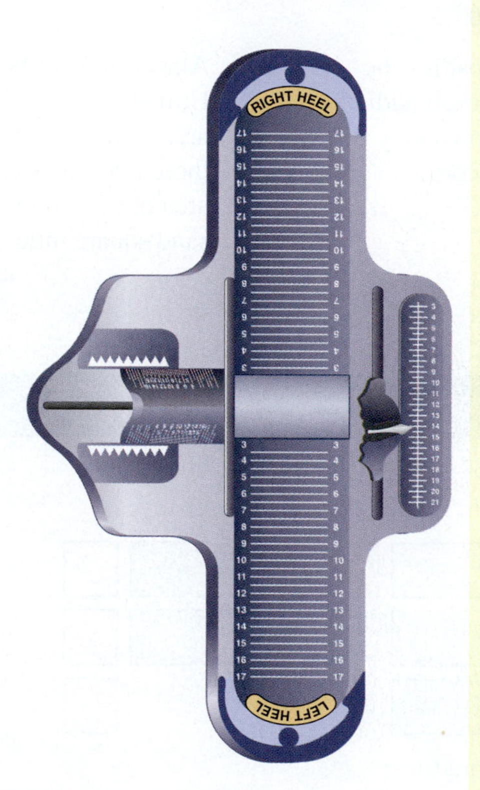

Solution

Stating the rule briefly in words, we get:

Shoe size equals three times foot length subtract 22.

Now, we can easily translate the rule to mathematical symbols:

$$s = 3f - 22$$

PRACTICE 1

To predict the temperature t at a particular altitude, a scientists who study weather conditions subtract $\frac{1}{200}$ of the altitude from the temperature on the ground g. Here t and g are in degrees Fahrenheit and a is in feet. Write this relationship as a formula.

Evaluating Formulas

In business and the health and physical sciences, as well as in many other areas of life, we often must evaluate a formula.

EXAMPLE 2

The formula $r = \frac{72}{y}$, known as the Rule of 72, gives the approximate rate of compound interest r (expressed as a percent) on an investment that doubles in y years. Find the approximate rate on an investment that doubles in 5 years.

PRACTICE 2

The formula for finding simple interest is $I = Prt$, where I is the interest, P is the principal, r is the rate of interest, and t is the time in years that the principal has been on deposit. Evaluate $I = Prt$ when $P = \$3{,}000$, $r = 0.06$, and $t = 2$ years.

Solution

$$r = \frac{72}{y}$$

$$= \frac{72}{5} \quad \text{Replace } y \text{ with 5.}$$

$$= 14.4$$

So r is 14.4%.

EXAMPLE 3

A state provides building owners with a tax credit for planting a "green" space—a portion of a building's roof covered with vegetation. A formula for computing the amount of tax credit T in dollars for a rectangular green space is $T = 4.5lw$, where l is the length and w is the width of the space. Find the tax credit for a green space that measures 30 feet by 20 feet.
(*Source*: capitolgreenroofs.groupsite.com)

Solution We use the formula $T = 4.5lw$ and substitute for l and w.

$$T = 4.5lw$$

$$= 4.5(30)(20)$$

$$= 2{,}700$$

So the tax credit for the green space was $2,700.

PRACTICE 3

When a cricket chirps n times per minute, the temperature outside in degrees Fahrenheit F can be found by using the following formula:

$$F = \frac{n}{4} + 37$$

What is the temperature outside when a cricket chirps 10 times per minute?

EXAMPLE 4

The top of a can of soda is a circle with radius 1 inch. Find the area of the top of the can. Use 3.14 for π.

Solution We use the formula $A = \pi r^2$ and substitute for r:

$$A = \pi r^2$$

$$\approx 3.14 \times (1)^2$$

$$\approx 3.14 \times 1$$

$$\approx 3.14$$

So the top of the can has area 3.14 square inches. Note that since the radius of the circle is measured in inches, the area of the circle is measured in square inches.

PRACTICE 4

The body mass index (BMI), a ratio used by doctors who study obesity, can be found using the following formula:

$$BMI = \frac{w}{h^2}$$

where w is the weight in kilograms and h is the height in meters. Find the BMI, measured in kilograms per square meter, for a person 2 meters tall who weighs 83 kilograms. Round to the nearest whole number.

10.1 U.S. Customary Units

OBJECTIVES

- **A** To change a measurement from one U.S. customary unit to another
- **B** To add or subtract measurements expressed in U.S. customary units
- **C** To solve applied problems involving U.S. customary units

What Measurement and Units Are and Why They Are Important

When measuring, we express quantities in standardized units such as pounds and yards, which enable us to compare characteristics of physical objects. With standardized units, we can decide which of two packages is heavier or how many times longer one room is than another.

Throughout this chapter, we focus on four kinds of measure: length, weight, capacity, and time. The units that we use to express these measurements come from two systems of measurement: the U.S. customary system and the metric system.

U.S. Customary Units of Length, Weight, Capacity, and Time

We begin with U.S. customary units (also called U.S. units). These are the units that we use most often in everyday situations.

Sometimes, we need to change the unit in which a measurement is expressed. To convert units, we must know how different units are related.

In measuring *length*, the main U.S. units are miles (mi), yards (yd), feet (ft), and inches (in.). The key conversion relationships among these units are as shown in the table to the right.

Length
5,280 ft = 1 mi
3 ft = 1 yd
12 in. = 1 ft

The main U.S. units of *weight* are tons, pounds (lb), and ounces (oz). These relationships are shown in the table to the right.

Weight
2,000 lb = 1 ton
16 oz = 1 lb

Next, let's consider the main U.S. units of *capacity* (*liquid volume*): gallons (gal), quarts (qt), pints (pt), cups (c), and fluid ounces (fl oz). Here, the table to the right shows is how these units are related.

Capacity
4 qt = 1 gal
2 pt = 1 qt
2 c = 1 pt
8 fl oz = 1 c

Finally, there are the main U.S. units of *time*: years (yr), months (mo), weeks (wk), days, hours (hr), minutes (min), and seconds (sec). Some of the key relationships among these units of time are as shown in the table to the right.

Time
365 days = 1 yr
12 mo = 1 yr
52 wk = 1 yr
7 days = 1 wk
24 hr = 1 day
60 min = 1 hr
60 sec = 1 min

Study these relationships so that you can use them to solve problems involving units.

Note that the abbreviations of the units are the same regardless of whether they are singular or plural. For example, the abbreviation for foot (ft) is the same as the abbreviation for feet (ft).

Changing Units

Suppose that you are ordering from a catalog the placemat pictured below. If you know that the length of the space to be covered is 2 ft, how long is it in inches? To change a length given in feet to inches, we need to know that there are 12 in. in 1 ft.

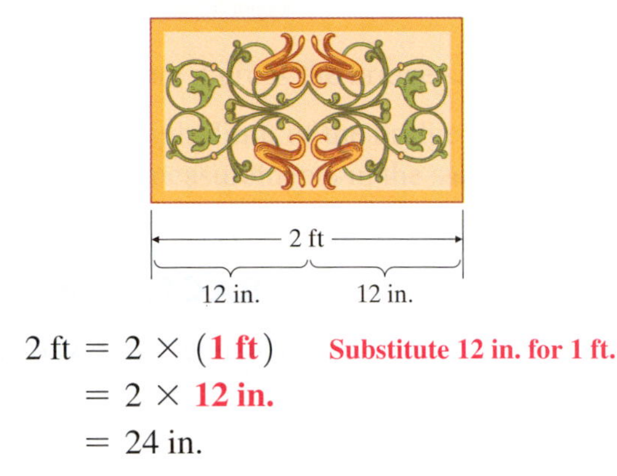

$$\begin{aligned} 2 \text{ ft} &= 2 \times (\mathbf{1\ ft}) \quad \textbf{Substitute 12 in. for 1 ft.} \\ &= 2 \times \mathbf{12\ in.} \\ &= 24 \text{ in.} \end{aligned}$$

So the rug is 24 in. long.

Another way of solving this problem is to multiply the original measurement by the *unit factor* $\frac{12 \text{ in.}}{1 \text{ ft}}$. The unit factor method, commonly used in the physical and health sciences, is particularly helpful in solving complex conversion problems. Because the numerator 12 in. and the denominator 1 ft represent the same length, the unit factor $\frac{12 \text{ in.}}{1 \text{ ft}}$ is equivalent to 1.

$$\begin{aligned} 2 \text{ ft} &= 2 \cancel{\text{ft}} \times \frac{12 \text{ in.}}{1 \cancel{\text{ft}}} \\ &= \frac{2 \times 12 \text{ in.}}{1} = 24 \text{ in.} \end{aligned}$$

Note how we simplified the answer by canceling common units, as if the units were numbers.

Or suppose that we wanted to change 24 in. to feet. We can solve this problem by multiplying the original measurement by the unit factor $\frac{1 \text{ ft}}{12 \text{ in.}}$.

$$\begin{aligned} 24 \text{ in.} &= \overset{2}{\cancel{24}} \cancel{\text{in.}} \times \frac{1 \text{ ft}}{\underset{1}{\cancel{12}} \cancel{\text{in.}}} \\ &= 2 \text{ ft} \end{aligned}$$

Let's look back at the two problems that we just solved. Both involved inches and feet. In the first problem, we multiplied by the unit factor $\frac{12 \text{ in.}}{1 \text{ ft}}$; in the second problem, we multiplied by its reciprocal, $\frac{1 \text{ ft}}{12 \text{ in.}}$. (Note that both unit factors are equivalent to 1.)

TIP When converting from one unit to another unit, multiply the original measurement by the unit factor that has the *desired unit in its numerator* and the *original unit in its denominator.*

There are many practical situations in which we need to add or subtract measurements written in mixed units—for instance, when we want to find out how much longer one car is than another or how much paint there will be if we combine the contents of several cans.

In addition or subtraction problems, only quantities having the same unit can be added or subtracted.

	1 ft 4 in.	2 hr 10 min
	+1 ft 2 in.	−1 hr 3 min
	2 ft 6 in.	1 hr 7 min

Often addition or subtraction problems involve changing units.

EXAMPLE 7

Find the sum:

7 ft 9 in.
+2 ft 5 in.

Solution

7 ft 9 in.
+2 ft 5 in.
14 in.

Start with the inches column. (The smaller unit is always in the column on the right.) Add the numbers in the inches column.

1 ft
7 ft 9 in.
+2 ft 5 in.
~~14 in.~~ 2 in.

Because 14 in. = 1 ft 2 in., replace the 14 in. by 2 in. and carry the 1 ft to the feet column.

1 ft
7 ft 9 in.
+2 ft 5 in.
10 ft 2 in.

Add the numbers in the feet column.

The sum is 10 ft 2 in.

PRACTICE 7

Add 3 lb 10 oz and 1 lb 14 oz.

EXAMPLE 8

A local theater was showing a double feature of two films—*Pirates of the Caribbean: At World's End* and *Pirates of the Caribbean: Dead Man's Chest*. The first film ran 2 hr 49 min and the second, 2 hr 30 min. How long was the double feature? (*Source:* wikipedia.org)

Solution

1 hr
2 hr 49 min
2 hr 30 min
5 hr ~~79~~ min
19 min

Start with the minutes column. Because 79 min = 1 hr 19 min, replace the 79 min with 19 min and carry the 1 hr to the hours column. Add the numbers in the hours column.

So the double feature ran 5 hr 19 min.

PRACTICE 8

An online camping gear company ships two camping tents to a customer. If each tent weighs 6 lb 10 oz, what is their combined weight?

Now, let's look at some examples of subtraction.

EXAMPLE 9

Subtract 1 yd 2 ft from 3 yd 1 ft.

Solution

$$\begin{array}{r} \overset{2}{\cancel{3}}\text{ yd } \overset{4}{\cancel{1}}\text{ ft} \\ -1\text{ yd } 2\text{ ft} \\ \hline \end{array}$$

Start with the feet column. Because 2 ft is larger than 1 ft, replace the 3 yd with 2 yd, and the borrowed yard with 3 ft, which when added to 1 ft gives 4 ft.

$$\begin{array}{r} 2\text{ yd } 4\text{ ft} \\ -1\text{ yd } 2\text{ ft} \\ \hline 1\text{ yd } 2\text{ ft} \end{array}$$

Subtract the numbers within each column.

Check Remember that we can check a subtraction by using addition.

$$\begin{array}{r} \overset{1\text{ yd}}{1\text{ yd}}\ 2\text{ ft} \\ +1\text{ yd } 2\text{ ft} \\ \hline 3\text{ yd } \underset{1\text{ ft}}{\cancel{4}}\text{ ft} \end{array}$$

Adding, we get 3 yd 1 ft, so our answer checks.

PRACTICE 9

How much greater is 6 gal than 5 gal 1 qt?

EXAMPLE 10

The *Mona Lisa* is one of the best known paintings in the world.

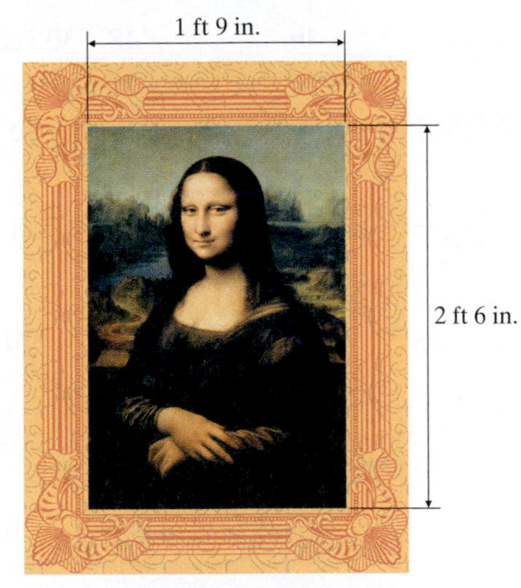

What is the difference between the height and width of the painting?

Solution

$$\begin{array}{r} \overset{1}{\cancel{2}}\text{ ft } \overset{18}{\cancel{6}}\text{ in.} \\ -1\text{ ft } 9\text{ in.} \\ \hline 0\text{ ft } 9\text{ in.} \end{array}$$

Because 9 in. is larger than 6 in., replace the 2 ft with 1 ft, and the borrowed foot with 12 in., which when added to 6 in. gives 18 in.

So the difference between the height and the width is 9 in.

PRACTICE 10

Two of the most famous horses in history were Citation and Secretariat. In 1948, Citation won the Kentucky Derby with a time of 2 min 5 sec. Twenty-five years later, Secretariat won in 1 min 59 sec. What was the difference between the two times? (*Source: Facts and Dates of American Sports*)

10.1 Exercises

FOR EXTRA HELP MyMathLab MathXL PRACTICE WATCH READ REVIEW

Mathematically Speaking

Fill in each blank with the most appropriate term or phrase from the given list.

smaller	pound	gallon	weight
larger	numerator	denominator	unit factor
sign	length	unit	

1. In the U.S. customary system, the yard is a unit of ______________.

2. In the U.S. customary system, the ______________ is a unit of capacity.

3. When we change a large unit to a small unit, the numerical part of the answer is ______________ than the original number.

4. When changing from miles to feet, we multiply the original measurement by $\frac{5{,}280 \text{ ft}}{1 \text{ mi}}$, which is called a ______________.

5. In addition or subtraction problems, only quantities having the same ______________ can be added or subtracted.

6. When converting from one unit to another unit, multiply the original measurement by the unit factor that has the desired unit in its ______________.

A *Change each quantity to the indicated unit.*

7. 48 in. = ____ ft
8. 6 pt = ____ qt
9. 9 ft = ____ yd
10. 48 mo = ____ yr
11. 60 ft = ____ in.
12. 8 qt = ____ pt
13. 7 min = ____ sec
14. 4 lb = ____ oz
15. 10 yd = ____ ft
16. 2 yr = ____ mo
17. 32 oz = ____ lb
18. 30 sec = ____ min
19. 2 mi = ____ yd
20. 5 hr = ____ min
21. 32 pt = ____ fl oz
22. $\frac{1}{4}$ day = ____ hr
23. $\frac{1}{2}$ gal = ____ qt
24. 32 fl oz = ____ pt
25. $2\frac{1}{2}$ qt = ____ pt
26. $1\frac{1}{2}$ qt = ____ gal
27. 7 pt = ____ qt
28. 36 hr = ____ days
29. $1\frac{1}{2}$ tons = ____ lb
30. 7,000 lb = ____ tons
31. 45 min = ____ hr
32. 7,920 yd = ____ mi
33. $\frac{1}{2}$ day = ____ hr
34. $\frac{1}{2}$ hr = ____ day
35. 5 min 10 sec = ____ sec
36. 1 lb 2 oz = ____ oz
37. 90 in. = ____ ft ____ in.
38. 50 hr = ____ days ____ hr

Complete each table.

	Length	Inches	Feet	Yards
39.	Giraffe (height)		16	
40.	Baseball (diameter)	3		
41.	Dog pen (width)			5
42.	Pond (depth)		12	

	Weight	Ounces	Pounds	Tons
43.	Ostrich egg		3	
44.	Baseball bat	29		
45.	Tongue of a blue whale			4
46.	Pony		200	

	Capacity	Fluid Ounces	Pints	Quarts
47.	Can of paint			1
48.	Case of cream		24	
49.	Bottle of mouthwash	16		
50.	Container of milk			2

	Time	Seconds	Minutes	Hours
51.	Rocket blast	50		
52.	Baseball game			3
53.	News report		22	
54.	Standing ovation		8	

B *Compute.*

55. $\begin{array}{r} 4 \text{ lb } 7 \text{ oz} \\ -2 \text{ lb } 9 \text{ oz} \\ \hline \end{array}$

56. $\begin{array}{r} 2 \text{ hr} \\ -1 \text{ hr } 2 \text{ min} \\ \hline \end{array}$

57. $\begin{array}{r} 20 \text{ lb } 5 \text{ oz} \\ +\ 9 \text{ lb } 10 \text{ oz} \\ \hline \end{array}$

58. $\begin{array}{r} 5 \text{ lb } 10 \text{ oz} \\ +1 \text{ lb } 8 \text{ oz} \\ \hline \end{array}$

59. 5 yr 7 mo + 3 yr 11 mo

60. 4 yr − 2 yr 3 mo

61. 5 gal 1 qt − 2 gal 2 qt

62. 1 pt 10 fl oz + 3 pt 8 fl oz

63. 2 qt 1 pt + 1 qt 1 pt

64. 2 ft − 5 in.

65. $\begin{array}{r} 6 \text{ min } 2 \text{ sec} \\ 1 \text{ min } 10 \text{ sec} \\ +1 \text{ min } 3 \text{ sec} \\ \hline \end{array}$

66. $\begin{array}{r} 20 \text{ ft } 5 \text{ in.} \\ 5 \text{ ft } 7 \text{ in.} \\ +\ 9 \text{ ft } 10 \text{ in.} \\ \hline \end{array}$

Mixed Practice *Solve.*

67. Find the sum of 10 lb 12 oz and 3 lb 5 oz.

68. 2 gal = ________ pt

69. Express 4 min 7 sec in seconds.

70. Compute: 4 yr − 1 yr 3 mo

71. Convert 15 ft to yards.

72. How many pints are equivalent to 20 fl oz?

Applications

C *Solve.*

73. The longest field goal made by Tom Dempsey while playing football for the New Orleans Saints was 189 ft. The longest field goal made by Jason Elam while playing for the Denver Broncos was 63 yd. Determine if Elam's field goal was less than, greater than, or equal to that of Demsey. (*Source:* mmbolding.com)

74. The Tyrannosaurus rex (T. rex) was a dinosaur that lived in what is now western North America some 65 million years ago. Fossils show that the T. rex measured up to 42 ft in length. Express this length in yards. (*Source:* wikipedia.org)

75. In 1940, Cornelius Warmerdam used a bamboo pole to vault 15 ft 8 in., setting a record. In 1962, Dave Tork, using a fiberglass pole, vaulted 16 ft 2 in. How much higher was Tork's vault than Warmerdam's? (*Source: Facts and Dates of American Sports*)

76. One of the tallest women who ever lived was an American named Sandy Allen, who, at age 22, was 91 in. tall. What was her height in feet and inches? (*Source: Guinness Book of World Records*)

77. The record for a person holding his or her breath under water without special equipment is 823 sec. Express this time in minutes and seconds. (*Source: Atlantic Monthly*)

78. In an Olympic marathon, an athlete runs 26 mi plus an additional 385 yd. How many total yards does an athlete run in a marathon?

79. Born in 1934, the Dionne sisters were Canadians who became world famous as the first quintuplets to survive beyond infancy. At birth, the tiniest of these babies weighed 1 lb 15 oz, and the largest weighed 3 lb 4 oz. What was the difference between their weights? (*Source: Encyclopaedia Brittanica*)

80. Abraham Lincoln spoke of "four score and seven years ago." If a score is 20, how many months are there in four score and seven years?

81. The lease on a tenant's apartment runs for 3 yr. If he has lived in the apartment already for 1 yr 2 mo, does he have more or less than $1\frac{1}{2}$ yr left on the lease?

82. A United Airlines nonstop flight from New Orleans to Los Angeles is scheduled to last 4 hr 20 min. According to the schedule, how much time remains $1\frac{1}{2}$ hr into the flight? (*Source:* expedia.com)

83. In the Spring of 2009, a first annual fishing competition was held in Southern California in which the anglers who caught the largest halibut (by weight) shared prize money. Two of the heaviest halibut caught in the contest weighed 31 lb and 17 lb 2 oz. What was their difference in weight? (*Source:* danawharf.com)

84. Evergreen trees come in different sizes. In one year, a fast-growing evergreen that had been 5 ft 10 in. tall grew by 4 ft 7 in. At the end of the year, what was the height of the tree?

85. As part of a kitchen remodel, a homeowner bought a new refrigerator.

a. The refrigerator is 36 in. wide and 30 in. deep. Express the width and depth in feet.

b. Using the answer from part (a), calculate the area of floor space the refrigerator will occupy.

86. A recipe calls for 8 fl oz of chicken broth.

a. How many pints of chicken broth are needed for the recipe?

b. If 1 qt of chicken broth is available, will there be enough to triple the recipe?

87. The highest mountain in the world is Mt. Everest. If its peak is 29,035 ft above sea level, find the height of the mountain to the nearest tenth of a mile.

88. Earth is made up of three main layers: the crust, the mantle, and the core. The core, which is composed of a liquid outer core and a solid inner core, is about 2,156 mi thick. How many feet thick is the core, rounded to the nearest million? (*Source:* U.S. Geological Survey)

• Check your answers on page A-13.

MINDStretchers

Groupwork

1. The U.S. customary system includes units that measure not only length, weight, capacity or time but also energy. For example, the British Thermal Unit (BTU) is commonly used to measure the amount of energy put out by a furnace or the cooling capacity of an air conditioner. Working with a partner, identify some other units in the U.S. customary system not discussed in this section. What do these units measure?

Mathematical Reasoning

2. In measuring, we often introduce errors. Suppose that each of two measurements could be as much as an inch off. If we then add the two measurements, how far from the truth could our sum be? Explain.

History

3. The foot is not the only body part used as a measure. For instance, the ancient Egyptians used the *mouthful* as a unit of measure of volume. In your college library or on the Web, investigate other examples.

Cultural Note

Joseph Louis Lagrange (1736–1813) was a French mathematician who was influential during the years following the French Revolution of 1789 in developing the metric system of measures based on decimals and powers of 10. Since then, the United States has been resistant to adopting the metric system, although in 1790, Thomas Jefferson argued that the country should adopt a decimal system of weights and measures.

Source: Gullberg, *Mathematics: From the Birth of Numbers* (New York: W. W. Norton, 1997), p. 52.

10.2 Metric Units and Metric/U.S. Customary Unit Conversions

OBJECTIVES

- **A** To identify units in the metric system
- **B** To change a measurement from one metric unit to another
- **C** To change a measurement from a metric unit to a U.S. customary unit, and vice versa
- **D** To solve applied problems involving metric units or U.S. customary units

Metric Units of Length, Weight, and Capacity

Now, we turn to metric units. Developed by French scientists over 200 years ago, the metric system (formally known as the International System of Units, or SI) has become standard in most countries of the world. Even in the United States, metric units predominate in many important fields, including scientific research, medicine, the film industry, food and drink packaging, sports, and the import–export industry.

As in Section 10.1, which dealt with the U.S. customary system, here we consider measurements of length, weight, and capacity. Time units are identical in both systems, so we do not discuss them in this section. Again, abbreviations of units in the singular and plural are the same.

We begin this discussion of the metric system by considering the basic metric units:

- the **meter (m)**, a unit of length, which gives the metric system its name;
- the **gram (g)**, a unit of weight, or more precisely of mass; and
- the **liter (L)**, a unit of capacity, that is, of liquid volume.

There are quite a few other metric units as well. The names for many of the other units are formed by combining a basic unit with one of the metric prefixes. The following table contains a list of metric prefixes, with those most commonly used in bold:

METRIC PREFIXES

Prefix	Symbol	Meaning
Kilo-	k	One thousand (1,000)
Hecto-	h	Hundred (100)
Deka-	da	Ten (10)
Deci-	d	One ten**th** $\left(\frac{1}{10}\right)$
Centi-	c	One hundred**th** $\left(\frac{1}{100}\right)$
Milli-	m	One thousand**th** $\left(\frac{1}{1,000}\right)$

Next, let's see how the three basic metric units combine with the metric prefixes to form new units. We begin with units of length.

Length

The table shows the four most commonly used metric units of length: kilometers (km), meters (m), centimeters (cm), and millimeters (mm). Memorize the table to the right, noting what each unit means as well as its symbol.

METRIC UNITS OF LENGTH

Unit	Symbol	Meaning
Kilometer	km	1,000 meters
Meter	m	1 meter
Centimeter	cm	$\frac{1}{100}$ meter
Millimeter	mm	$\frac{1}{1,000}$ meter

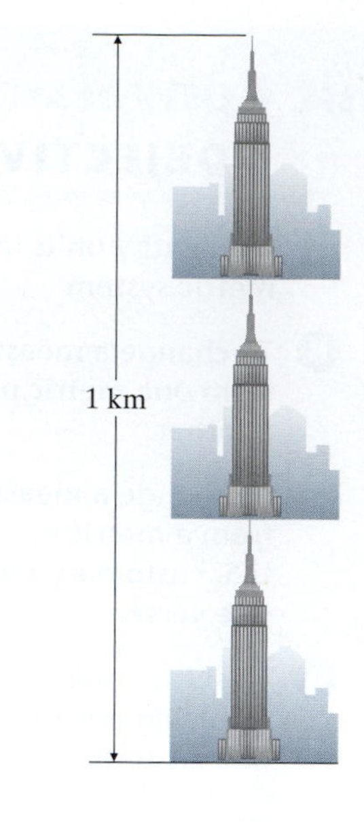

In the preceding table, the largest unit of length is the *kilometer*. A kilometer is a little more than half a mile, or about 3 times the height of the Empire State Building, as shown to the left. Great lengths, such as the distance between two cities, are expressed in kilometers.

The *meter*, the basic metric unit of length, is a little longer than a yard, or about the width of a twin bed. In the metric system, we use meters to measure medium-size lengths—say, the length of a room.

1 m

The next unit of length, the *centimeter*, is approximately the width of your little finger, or somewhat less than half an inch. In the metric system, the width of an envelope is expressed in centimeters.

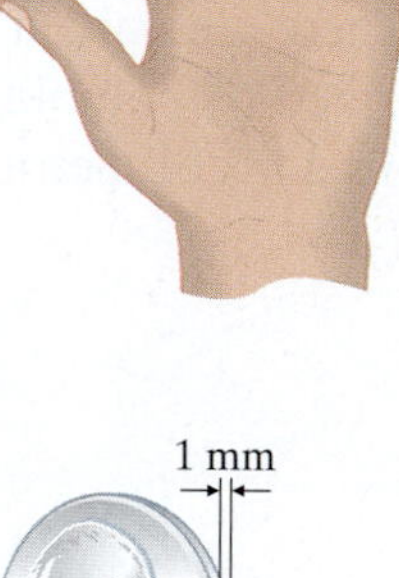

The smallest unit of length in the table is the *millimeter*. A millimeter is about the thickness of a dime. We use the millimeter to measure short lengths—say, the dimensions of an insect.

1 mm

Weight

Now, we turn to the metric units of weight, shown in the following table: kilograms (kg), grams (g), and milligrams (mg). Memorize this table, which deals with the three most commonly used metric units of weight.

METRIC UNITS OF WEIGHT

Unit	Symbol	Meaning
Kilogram	kg	1,000 grams
Gram	g	1 gram
Milligram	mg	$\frac{1}{1{,}000}$ gram

The largest unit of weight in the table is the *kilogram*. A kilogram is approximately 2 lb, or about the weight of this textbook. Large weights—say, that of a car or of a person—are expressed in kilograms.

This textbook

10.2 Metric Units and Metric/U.S. Customary Unit Conversions

OBJECTIVES

- **A** To identify units in the metric system
- **B** To change a measurement from one metric unit to another
- **C** To change a measurement from a metric unit to a U.S. customary unit, and vice versa
- **D** To solve applied problems involving metric units or U.S. customary units

Metric Units of Length, Weight,and Capacity

Now, we turn to metric units. Developed by French scientists over 200 years ago, the metric system (formally known as the International System of Units, or SI) has become standard in most countries of the world. Even in the United States, metric units predominate in many important fields, including scientific research, medicine, the film industry, food and drink packaging, sports, and the import–export industry.

As in Section 10.1, which dealt with the U.S. customary system, here we consider measurements of length, weight, and capacity. Time units are identical in both systems, so we do not discuss them in this section. Again, abbreviations of units in the singular and plural are the same.

We begin this discussion of the metric system by considering the basic metric units:

- the **meter (m)**, a unit of length, which gives the metric system its name;
- the **gram (g)**, a unit of weight, or more precisely of mass; and
- the **liter (L)**, a unit of capacity, that is, of liquid volume.

There are quite a few other metric units as well. The names for many of the other units are formed by combining a basic unit with one of the metric prefixes. The following table contains a list of metric prefixes, with those most commonly used in bold:

METRIC PREFIXES

Prefix	Symbol	Meaning
Kilo-	k	One thousand (1,000)
Hecto-	h	Hundred (100)
Deka-	da	Ten (10)
Deci-	d	One ten**th** $\left(\frac{1}{10}\right)$
Centi-	c	One hundred**th** $\left(\frac{1}{100}\right)$
Milli-	m	One thousand**th** $\left(\frac{1}{1,000}\right)$

Next, let's see how the three basic metric units combine with the metric prefixes to form new units. We begin with units of length.

Length

The table shows the four most commonly used metric units of length: kilometers (km), meters (m), centimeters (cm), and millimeters (mm). Memorize the table to the right, noting what each unit means as well as its symbol.

METRIC UNITS OF LENGTH

Unit	Symbol	Meaning
Kilometer	km	1,000 meters
Meter	m	1 meter
Centimeter	cm	$\frac{1}{100}$ meter
Millimeter	mm	$\frac{1}{1,000}$ meter

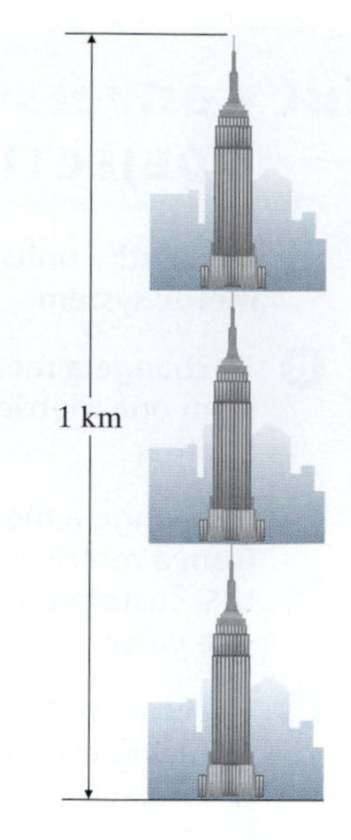

In the preceding table, the largest unit of length is the *kilometer*. A kilometer is a little more than half a mile, or about 3 times the height of the Empire State Building, as shown to the left. Great lengths, such as the distance between two cities, are expressed in kilometers.

The *meter*, the basic metric unit of length, is a little longer than a yard, or about the width of a twin bed. In the metric system, we use meters to measure medium-size lengths—say, the length of a room.

The next unit of length, the *centimeter*, is approximately the width of your little finger, or somewhat less than half an inch. In the metric system, the width of an envelope is expressed in centimeters.

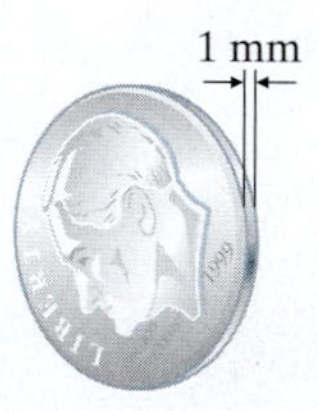

The smallest unit of length in the table is the *millimeter*. A millimeter is about the thickness of a dime. We use the millimeter to measure short lengths—say, the dimensions of an insect.

Weight

Now, we turn to the metric units of weight, shown in the following table: kilograms (kg), grams (g), and milligrams (mg). Memorize this table, which deals with the three most commonly used metric units of weight.

METRIC UNITS OF WEIGHT

Unit	Symbol	Meaning
Kilogram	kg	1,000 grams
Gram	g	1 gram
Milligram	mg	$\frac{1}{1,000}$ gram

This textbook

The largest unit of weight in the table is the *kilogram*. A kilogram is approximately 2 lb, or about the weight of this textbook. Large weights—say, that of a car or of a person—are expressed in kilograms.

The next unit, the *gram*, is smaller, only about $\frac{1}{30}$ oz, or about the weight of a raisin.

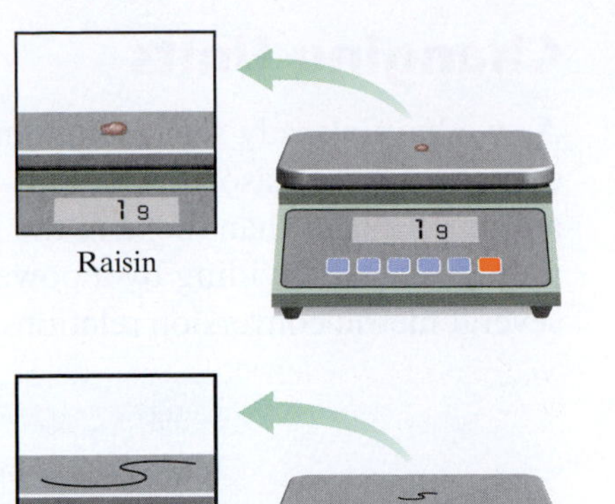

Raisin

The smallest unit, the *milligram*, is tiny—about the weight of a strand of hair. It is therefore used to measure small weights—say, that of a pill.

Strand of hair

Capacity (Liquid Volume)

Amounts of liquid are commonly measured in terms of liquid volume, or equivalently, the capacity of containers that hold the liquid. The following table deals with three metric units of capacity: kiloliters (kL), liters (L), and milliliters (mL). Memorize this table, which describes the three primary metric units of capacity.

METRIC UNITS OF CAPACITY

Unit	Symbol	Meaning
Kiloliter	kL	1,000 liters
Liter	L	1 liter
Milliliter	mL	$\frac{1}{1,000}$ liter

In this table, the largest unit of capacity is the *kiloliter*. The amount of water that a typical collapsible swimming pool holds is about 1 kiloliter. Kiloliters are used to measure large volumes of liquid, for instance, the capacity of an oil barge or the amount of soda that a factory produces annually.

The second unit of liquid volume, the *liter*, is slightly more than a quart, a typical size for a bottle of mouthwash. Liters are used in measuring larger quantities of liquid, such as the amount of water that a sink will hold or, in some countries, the amount of gasoline purchased at the pump.

The smallest unit in this table is the *milliliter*, which represents a very small amount of liquid—about as much as an eyedropper contains. Milliliters are used in measuring small volumes of liquid—say, the amount of perfume in a tiny bottle.

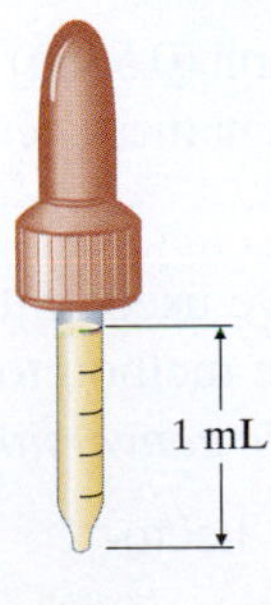

Changing Units

As we have already seen, sometimes we need to change the unit in which a measurement is expressed. One reason the metric system is widely used is that unit conversions in this system are much easier than those in the U.S. customary system. Such conversions simply involve multiplying or dividing by a power of 10, such as 100 or 1,000. The following table shows several metric conversion relationships.

Length	Weight	Capacity
1,000 m = 1 km	1,000 g = 1 kg	1,000 L = 1 kL
100 cm = 1 m	1,000 mg = 1 g	1,000 ml = 1 L
1,000 mm = 1 m		

From these relationships, we can set up unit factors to carry out unit conversions.

EXAMPLE 1

1.5 g = ________ mg

Solution To convert to *milligrams*, we use the unit factor $\frac{1{,}000 \text{ mg}}{1 \text{ g}}$ because 1,000 mg = 1 g.

$$1.5 \text{ g} = 1.5 \cancel{\text{g}} \times \frac{1{,}000 \text{ mg}}{1 \cancel{\text{g}}}$$

$$= 1.5 \times 1{,}000 \text{ mg}$$

To multiply 1.5 by 1,000, move the decimal point in 1.5 three places to the right.

$$= 1{,}500 \text{ mg}$$

PRACTICE 1

3,100 mg = ____ g

As illustrated in Example 1, a part of a metric unit is usually expressed as a decimal, not as a fraction. That is, we write 1.5 g, not $1\frac{1}{2}$ g.

EXAMPLE 2

500 m = ________ km

Solution To convert meters to kilometers, we multiply the original measurement by the unit factor $\frac{1 \text{ km}}{1{,}000 \text{ m}}$.

$$500 \text{ m} = 500 \cancel{\text{m}} \times \frac{1 \text{ km}}{1000 \cancel{\text{m}}}$$

$$= \frac{500}{1{,}000} \text{ km}$$

$$= 0.5 \text{ km}$$

To divide 500 by 1,000, move the decimal point in 500 three places to the left.

As a quick check, note that the number of the larger unit (0.5 km) is less than the number of the smaller unit (500 m) for an equivalent measurement.

PRACTICE 2

2,500 cm = ____ m

So far we have used unit factors to convert between units of measure in the metric system. An alternative method to unit factors is moving the decimal point to the right or to the left using the *metric conversion line*.

kilo- hecto- deka- *unit* deci- centi- milli-

For example, let's revisit Example 1, where we changed 1.5 g to milligrams. Here we consider the metric conversion line for weight.

kg hg dag g dg cg mg

Note that each unit is equivalent to 10 times the unit to its right. We locate the given unit grams (g) on the line. To change to milligrams (mg), we need to move along the line 3 units to the right.

kg hg dag g dg cg mg

Similarly, in 1.5 we move the decimal point 3 places to the right.

$$1.5 \text{ g} = 1\,500. \text{ mg} = 1{,}500 \text{ mg}$$

So, as before, we find that 1.5 g is equivalent to 1,500 mg.

In the next example, we revisit Example 2, where we used the unit factor method to convert 500 m to kilometers. This time, however, we use the metric conversion line.

EXAMPLE 3

Change 500 m to kilometers.

Solution First, we list the metric units of length from largest to smallest. Then, we locate meters (m) on the metric conversion line and move to kilometers (km).

km hm dam m dm cm mm

Similarly, we move the decimal point 3 places to the left in 500.

$$500. \text{ m} = 0.500 \text{ km} = 0.5 \text{ km}$$

As with the unit factor method, we conclude that 500 m = 0.5 km.

PRACTICE 3

Use the metric conversion line to change 2,500 cm to meters.

TIP When using the metric conversion line to change one metric unit to another, locate the original unit on the line and move to the desired unit. Then, in the numerical part of the original measurement, move the decimal point the same number of places in the same direction as you moved on the metric conversion line.

EXAMPLE 4

Express 3 kL in milliliters.

Solution First, we consider the unit factor method. Because the table of capacity on page 443 does not indicate how many milliliters are equivalent to a kiloliter, we need to solve the problem in steps.

Step 1. Change kiloliters to liters.

Step 2. Change liters to milliliters.

To combine these two steps, we multiply the original measurement by a chain of appropriate unit factors to get *milliliters* for the final answer.

$$\begin{aligned} 3 \text{ kL} &= 3 \text{ kL} \times \frac{1{,}000 \text{ L}}{1 \text{ kL}} \times \frac{1{,}000 \text{ mL}}{1 \text{ L}} \\ &= 3 \times 1{,}000 \times 1{,}000 \text{ mL} \\ &= 3{,}000{,}000 \text{ mL} \end{aligned}$$

Alternatively, let's use the metric conversion line. We list the units of capacity in decreasing order from left to right, and then move from kiloliters to milliliters on the metric conversion line.

kL hL daL L dL cL mL

6 units to the right

$$3 \text{ kL} = 3. \text{ kL} = 3\,000\,000. \text{ mL} = 3{,}000{,}000 \text{ mL}$$

Note that with either method, we get 3,000,000 mL.

PRACTICE 4

Change 5,000,000 mL to kiloliters.

EXAMPLE 5

For a healthy 20-year-old female, the U.S. recommended dietary allowances (RDAs) for calcium and iron are 1 g and 18 mg, respectively. Which RDA is higher? (*Source:* iron.edu)

Solution When comparing quantities expressed in different units, we convert them to the same unit, usually the smaller unit. Here, we change 1 g to milligrams.

$$1\text{ g} = 1\text{ g} \times \frac{1{,}000\text{ mg}}{1\text{ g}}$$
$$= 1{,}000\text{ mg}$$

Using the metric conversion line gives us:

kg hg dag g dg cg mg

3 units to the right

1. g = 1 000. mg = 1,000 mg

So the RDA for calcium is 1,000 mg, which is higher than the 18-mg RDA for iron.

PRACTICE 5

The small intestine is 6 m long, and the large intestine is 150 cm long. Which is shorter? (*Source: Webster's New World Book of Facts*)

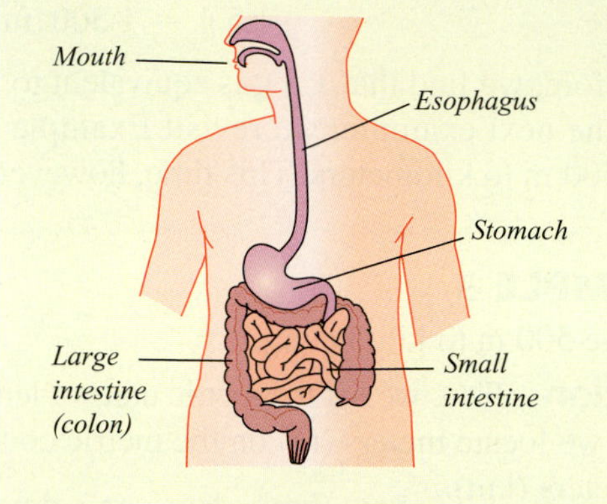

Changing Extreme Units

In addition to the metric prefixes already discussed, there are others that are in common use. With these prefixes, we can create very large or very small units. Often the need for such units has arisen because of advances in technology, science, or medicine. For example, the prefix "mega-" meaning 1,000,000 makes it convenient to express the large memories and high speeds that today's computers possess. At the other end of the spectrum, the prefix "micro-", meaning $\frac{1}{1{,}000{,}000}$, is helpful in measuring tiny doses of medicines or vaccines. The following table lists some of these metric prefixes. Commit these to memory.

OTHER METRIC PREFIXES

Prefix	Symbol	Meaning
Giga-	G	One billion (1,000,000,000)
Mega-	M	One million (1,000,000)
Micro-	mc	One million**th** $\left(\frac{1}{1{,}000{,}000}\right)$

We can combine these prefixes with the metric units. For instance, the microgram is a tiny unit of weight and the microsecond is a tiny unit of time.

These prefixes are also sometimes combined with units that, while in common use, are technically not part of the metric system. For example, the megabyte and the gigabyte are large units of computer memory based on the byte (B).

Consider the following table.

OTHER METRIC UNITS

Unit	Symbol	Meaning	Relationship
Gigabyte	GB	1,000,000,000 bytes	1,000,000,000 B = 1 GB
Megabyte	MB	1,000,000 bytes	1,000,000 B = 1 MB
Microsecond	mcsec	$\frac{1}{1{,}000{,}000}$ second	1,000,000 mcsec = 1 sec
Microgram	mcg	$\frac{1}{1{,}000{,}000}$ gram	1,000,000 mcg = 1 g

We can use the unit factor method to convert the units in this table to other units, as the following examples illustrate.

EXAMPLE 6

2,000 mcg = ________ g

Solution Because 1 g = 1,000,000 mcg, to convert micrograms to grams we use the unit factor $\frac{1 \text{ g}}{1{,}000{,}000 \text{ mcg}}$.

$$2{,}000 \text{ mcg} = 2{,}000 \cancel{\text{mcg}} \times \frac{1 \text{ g}}{1{,}000{,}000 \cancel{\text{mcg}}}$$

$$= \frac{2{,}000}{1{,}000{,}000} \text{ g}$$

$$= 0.002 \text{ g}$$

Move the decimal point in 2,000 six places to the left.

PRACTICE 6

5.3 sec = ________ mcsec

EXAMPLE 7

Many computers come equipped with 4 GB of RAM (random access memory) to run today's demanding operating systems. Express this amount of RAM in terms of megabytes.

Solution

$$4 \text{ GB} = 4 \cancel{\text{GB}} \times \frac{1{,}000{,}000{,}000 \cancel{\text{B}}}{1 \cancel{\text{GB}}} \times \frac{1 \text{ MB}}{1{,}000{,}000 \cancel{\text{B}}}$$

$$= 4 \times 1{,}000 \text{ MB}$$

$$= 4{,}000 \text{ MB}$$

So the amount of RAM is 4,000 MB.

PRACTICE 7

Vitamin D is important for the proper absorption of calcium from food. The U.S. Institute of Medicine recommends that all individuals under the age of 50 take 5 mcg of Vitamin D daily. Express this amount in milligrams. (*Source:* mayoclinic.com)

Metric/U.S. Customary Unit Conversions

In some situations, we need to change a measurement expressed in a U.S. unit to a metric unit, or vice versa. For example, if Americans were driving in Ireland and saw a road sign giving the distance to the next town in kilometers, they might want to express that distance in miles.

Or suppose that shoppers wanted to buy mouthwash and wondered how many pint bottles are equal in capacity to a 750-mL bottle. Here, they might want to change pints to milliliters.

To convert, we must either have memorized or have access to metric/U.S. unit conversion relationships. The following table shows some of these key relationships:

METRIC/U.S. UNIT CONVERSION RELATIONSHIPS

Length	Weight	Capacity
1.6 km ≈ 1 mi	910 kg ≈ 1 ton	260 gal ≈ 1 kL
1,600 m ≈ 1 mi	2.2 lb ≈ 1 kg	3.8 L ≈ 1 gal
3,300 ft ≈ 1 km	450 g ≈ 1 lb	1.1 qt ≈ 1 L
3.3 ft ≈ 1 m	28 g ≈ 1 oz	2.1 pt ≈ 1 L
39 in. ≈ 1 m		470 mL ≈ 1 pt
30 cm ≈ 1 ft		
2.5 cm ≈ 1 in.		

EXAMPLE 8

Express 2 oz in grams.

Solution According to the conversion table, 1 oz ≈ 28 g.

To express in *grams*, we multiply 2 oz by the unit factor $\frac{28\text{ g}}{1\text{ oz}}$.

$$2\text{ oz} \approx 2\text{ oz} \times \frac{28\text{ g}}{1\text{ oz}}$$
$$\approx 56\text{ g}$$

So 2 ounces is about 56 grams.

PRACTICE 8

Express 10 gal in terms of liters.

Note that our answer in Example 8 is only an approximation because the unit factor is not exact. Also note that the number of grams is more than the number of ounces because an ounce is larger than a gram.

EXAMPLE 9

Since 1866, the capacity of an oil barrel in the United States has been standardized at 42 gal. How many liters of oil does an oil barrel hold, rounded to the nearest liter? (*Source:* Daniel Yergin, *The Prize*)

Solution The conversion table indicates that 3.8 L ≈ 1 gal.

To convert to liters, we use the conversion factor $\frac{3.8\text{ L}}{1\text{ gal}}$.

$$42\text{ gal} \approx 42\text{ gal} \times \frac{3.8\text{ L}}{1\text{ gal}}$$
$$\approx 159.6\text{ L}$$

So an oil barrel holds approximately 160 L of oil.

PRACTICE 9

An iron-nickel meteorite from Argentina was for sale on an auction website. If the meteorite weighs 1,760 grams, find its weight in pounds to the nearest whole number. (*Source:* astromart.com)

10.2 Exercises

FOR EXTRA HELP MyMathLab MathXL PRACTICE WATCH READ REVIEW

Mathematically Speaking

Fill in each blank with the most appropriate term or phrase from the given list.

kilo-	centi-	milli-
weight	deci-	quart
liter	length	

1. In the metric system, the meter is a unit of _______.

2. In the metric system, the _______ is a unit of capacity.

3. The prefix _______ means one thousand.

4. The prefix _______ means one-thousandth.

5. The prefix _______ means one-hundredth.

6. The prefix _______ means one-tenth.

A *Choose the unit that would most likely be used to measure each quantity.*

7. The volume of liquid in a test tube
a. millimeter **b.** milligram **c.** milliliter

8. The weight of a television set
a. milligram **b.** gram **c.** kilogram

9. The width of a street
a. millimeter **b.** meter **c.** kilometer

10. The length of a river
a. kilometer **b.** kilogram **c.** millimeter

Choose the best estimate in each case.

11. The capacity of a large bottle of soda
a. 1 mL **b.** 1 L **c.** 1 kg

12. The width of film for slides
a. 35 mm **b.** 35 cm **c.** 35 m

13. The height of the Washington Monument
a. 170 cm **b.** 170 m **c.** 170 km

14. The length of a pencil
a. 20 mm **b.** 20 cm **c.** 20 m

15. The capacity of a bottle of hydrogen peroxide
a. 400 mL **b.** 400 L **c.** 400 g

16. The weight of an adult
a. 70 mg **b.** 70 g **c.** 70 kg

B *Change each quantity to the indicated unit.*

17. 1,000 mg = _____ g
18. 253 mm = ____ m
19. 750 g = _____ kg
20. 2 L = ____ mL
21. 0.08 kL = ____ L
22. 4.3 kg = ____ g
23. 3.5 m = ____ mm
24. 900 m = _____ km
25. 5 mL = ____ L
26. 250 mg = _____ g
27. 4,000 mm = ____ m
28. 5 L = ____ mL
29. 7,000 L = ____ kL
30. 2,500 mg = ____ g
31. 413 cm = _____ m
32. 2.8 m = ____ cm
33. 0.002 kg = ____ mg
34. 3,000 mm = ____ cm
35. 7,500 mL = _____ kL
36. 2.1 km = ______ cm
37. 0.03 g = _______ mcg
38. 5,000 mcsec = ____ sec
39. 4 MB = ________ B
40. 8,000,000 B = ____ GB
41. 7,000,000 mcg = ____ g
42. 0.09 sec = _____ mcsec
43. 1,280 MB = _____ GB
44. 0.04 mg = ____ mcg

Complete each table.

	Length	Millimeters	Centimeters	Meters
45.	Swan's wingspan		238	
46.	Mouse pad		5	
47.	Jumping spider	10		
48.	Power cord			3

	Weight	Milligrams	Grams	Kilograms
49.	Capsule	300		
50.	Human liver		1,560	
51.	Pastry		450	
52.	Kangaroo cub			1

	Capacity	Milliliters	Liters	Kiloliters
53.	Container of tile cleaner	709		
54.	Bottle of spring water		3	
55.	Gas station fuel tank			17
56.	Aquarium		110	

Compute.

57. 3 km + 250 m **58.** 5 L − 600 mL **59.** 98 kg + 25.6 g **60.** 30 cm + 2 m

C *Change each quantity to the indicated unit. If needed, round the answer to the nearest tenth of the unit.*

61. 30 oz ≈ ___ g **62.** 4 mi ≈ ____ km **63.** 10 cm ≈ ___ in. **64.** 900 g ≈ ___lb

65. 48 in. ≈ ____ m **66.** 6 qt ≈ ____ L **67.** 5 pt ≈ ___ L **68.** 6 ft ≈ _____cm

Mixed Practice

Solve.

69. Combine 3 m and 50 cm.

70. Change 500 g to kilograms.

71. 2.5 m = ____ mm

72. Express 2,000 mL in liters.

73. The distance between a student's home and her college would most likely be measured in

a. kilometers **b.** kilograms **c.** millimeters

74. Express 60 cm in feet.

75. Change 3 L to quarts.

76. The best estimate for the capacity of a bottle of olive oil is

a. 500 L **b.** 500 kg **c.** 500 mL

Applications

D *Solve. If needed, round the answer to the nearest tenth of the unit.*

77. According to a medical journal, the average daily U.S. diet contains 6,000 mg of sodium. How many grams is this? (*Source: Journal of the American Medical Association*)

78. Some road signs in the United States give the speed limit in both miles per hour (mph) and kilometers per hour (kph). Confirm that the speed limits in the South Dakota sign to the right are approximately the same. (*Source:* worldofstock.com)

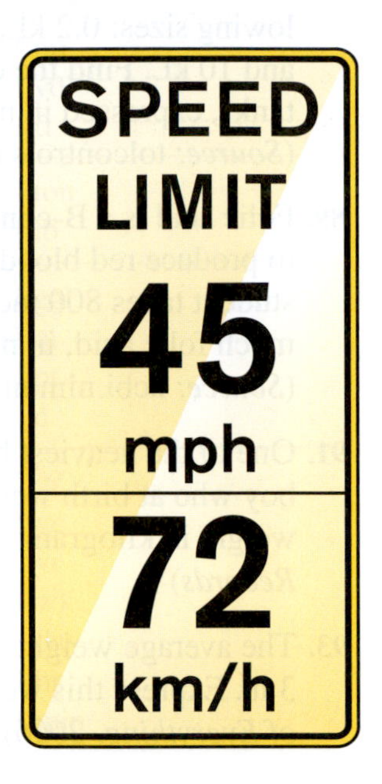

79. A nurse must administer a 4-mL dose of a drug daily to a patient. If there is 1 L of this drug on hand, will it last the patient 120 days?

80. Vitamin C commonly comes in pills with a strength of 500 mg. How many of these pills will an adult need to take if she wants a dosage of half a gram?

81. A student in a physics lab measured the length of a pendulum string as 7.5 cm. Express this length in inches.

82. In the Summer Olympics, a major track-and-field event is the 100-m dash. How long is this race in kilometers?

83. The diameter of the primary mirror of the Hubble Space Telescope is 2.4 m. Express this diameter in millimeters. (*Source:* http://hubblesite.org)

84. The Willis Tower, formerly the Sears Tower, in Chicago is 442 m tall. Express this height in kilometers. (*Source:* Emporis Buildings)

85. The side of a square tile is 75 mm long. If 100 of these tiles are placed on the floor side by side, what is their total length in centimeters?

86. In 1948, a medical researcher studied the possible use of large doses of ascorbic acid to cure tuberculosis (TB). He reported administering about 3,000,000 mg of ascorbic acid to patients with TB. Express this amount in kilograms. (*Source:* vitamincfoundation.org)

OTHER METRIC PREFIXES

Prefix	Symbol	Meaning
Giga-	G	One billion (1,000,000,000)
Mega-	M	One million (1,000,000)
Micro-	mc	One millionth $\left(\frac{1}{1{,}000{,}000}\right)$

OTHER METRIC UNITS

Unit	Symbol	Meaning	Relationship
Gigabyte	GB	1,000,000,000 bytes	1,000,000,000 B = 1 GB
Megabyte	MB	1,000,000 bytes	1,000,000 B = 1 MB
Microsecond	mcsec	$\frac{1}{1{,}000{,}000}$ second	1,000,000 mcsec = 1 sec
Microgram	mcg	$\frac{1}{1{,}000{,}000}$ gram	1,000,000 mcg = 1 g

Key Metric/U.S. Unit Conversion Relationships

[10.2] **Length**	1.6 km ≈ 1 mi 1,600 m ≈ 1 mi 3,300 ft ≈ 1 km 3.3 ft ≈ 1 m 39 in. ≈ 1 m 30 cm ≈ 1 ft 2.5 cm ≈ 1 in.
[10.2] **Weight**	910 kg ≈ 1 ton 2.2 lb ≈ 1 kg 450 g ≈ 1 lb 28 g ≈ 1 oz
[10.2] **Capacity (liquid volume)**	260 gal ≈ 1 kL 3.8 L ≈ 1 gal 1.1 qt ≈ 1 L 2.1 pt ≈ 1 L 470 mL ≈ 1 pt

CHAPTER 10 Review Exercises

Say Why

Fill in each blank.

1. To change 6 days to hours, we _______ (do/do not) multiply by $\frac{1 \text{ day}}{24 \text{ hr}}$ because ____.

2. A container _______ (can/cannot) weigh 1 quart because ____.

3. A centimeter _______ (is/is not) longer than a millimeter because ____.

4. 7.3 km _______ (is/is not) equivalent to 7,300 m because ____.

5. For a digital camera, the unit *megapixel* _______ (is/is not) equivalent to one million pixels because ____.

6. In chemistry, the unit *microliter* _______ (is/is not) equivalent to $\frac{1}{1{,}000{,}000{,}000}$ of a liter because ____.

[10.1] ***Change each quantity to the indicated unit.***

7. 5 yd = ____ ft
8. 20 mo = ____ yr
9. 32 oz = ____ lb
10. 10 ft = ____ yd
11. $1\frac{1}{2}$ tons = ____ lb
12. $8\frac{1}{2}$ lb = ____ oz
13. 3 pt = ____ fl oz
14. 150 sec = ____ min
15. 7 hr 15 min = ____ min
16. 50 in. = ____ ft ____ in.
17. 10,560 ft = ____ mi
18. 2,000 oz = ____ lb

Compute the given sum or difference.

19. 4 hr 20 min
+3 hr 50 min
20. 20 ft
− 1 ft 3 in.
21. 3 gal 2 qt − 1 gal 3 qt
22. 3 lb 6 oz + 2 lb 9 oz + 1 lb 3 oz

[10.2] ***Choose the unit that you would most likely use to measure each quantity.***

23. The weight of a car
    a. milligrams **b.** grams **c.** kilograms
24. The width of a pencil's point
    a. millimeters **b.** centimeters **c.** meters
25. The capacity of an oil barrel
    a. milliliters **b.** liters **c.** meters
26. The distance a commuter drives
    a. millimeters **b.** centimeters **c.** kilometers

Choose the best estimate in each case.

27. The width of a piece of typing paper
    a. 16 mm **b.** 16 cm **c.** 16 km
28. The capacity of a bottle of mouthwash
    a. 100 mL **b.** 100 L **c.** 100 g
29. The weight of an aspirin pill
    a. 200 mg **b.** 200 g **c.** 200 kg
30. The length of an athlete's long jump
    a. 6.72 mm **b.** 6.72 cm **c.** 6.72 m

Change each quantity to the indicated unit.

31. 37 mg = ____ g

32. 4 kL = ____ L

33. 8 m = ____ cm

34. 2.1 km = ____ m

35. 600 mm = ____ m

36. 5,100 g = ____ kg

37. 5,000 mcsec = ____ sec

38. 4 GB = ____ MB

Change each quantity to the indicated unit, rounding to the nearest unit.

39. 4 oz ≈ ____ g

40. 5 cm ≈ ____ in.

41. 32 km ≈ ____ mi

42. 4 gal ≈ ____ L

Mixed Applications

Solve.

43. A DVD plays for 72 min. Express this playing time in hours.

44. In a recent year, a typical U.S. resident used about 1,430 gal of water a day for residential, agricultural, and industrial purposes. How many pints is this? (*Source:* usgs.gov)

45. *Frankenstein* (130 minutes) and *Dracula* (1 hr 15 min) are two classic horror films made in 1931. Which film is longer?

46. Some doctors recommend that athletes drink about 600 mL of fluid each hour. Express this amount in liters.

47. A teaspoon of common table salt contains about 2,000 mg of sodium. How many grams of sodium is this?

48. In a factory, a chemical process produced 3 mg of a special compound each hour. How many grams were produced in 24 hr?

49. A high blood level of cholesterol is a risk factor for heart disease. Having a low level of cholesterol, less than 200 mg per deciliter of blood, is considered desirable. Express this amount of cholesterol in grams. (*Source:* ext.colostate.edu)

50. It is estimated that 750 kg of pesticide is sprayed on a typical U.S. golf course each year. How many grams is this? (*Source: Journal of Pesticide Reform*)

51. A computer virus checker took 349 min to scan each file on a 320 GB hard drive. Express this length of time to the nearest hour.

52. A daily reference value (DRV) is a reference point that serves as a general guideline for a healthy diet. For a 2,000-calorie diet, the DRV for fiber is 25 g. Express this DRV in milligrams. (*Source:* U.S. Food and Drug Administration)

53. In Olympic gymnastics, the floor exercise is performed on a square mat measuring 10 m on a side. What is the length of one side to the nearest foot?

54. The weight of a precious stone is given in carats, where 1 carat is equal to 200 mg. The Hope Diamond weighs 45.52 carats. Express this weight in grams. (*Source:* Smithsonian Institution)

55. A website for bodybuilders recommends taking the dietary supplement melatonin to promote sleep without the hazards of prescription sleeping pills. The site notes that successful results can be achieved with dosages ranging from 100 mcg to 200 mg. What is the difference between these dosages in micrograms? (*Source:* bodybuilding.com)

56. A computer has a hard drive with capacity 50 GB. Express this capacity in megabytes.

57. In pairs figure skating, the free skate is 1 min 40 sec longer than the short program. If the short program is 2 min 50 sec, how long is the free skate? (*Source:* usfsa.org)

58. The following diagram shows the heights of an average U.S. woman and an average 10-year-old U.S. girl.

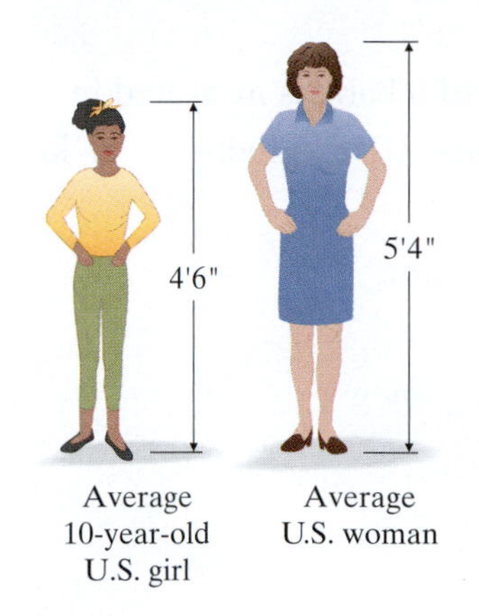

What is the difference in their heights? (*Source: Archives of Pediatrics and Adolescent Medicine*)

59. Thimerosal is a compound widely used since the 1930s as a preservative in a variety of drugs, including vaccines. Because of concerns about this use of thimerosal, a study was conducted in which individuals received up to 26,000 micrograms of thimerosal. No toxic effects were found. What was the largest quantity of thimerosal in milligrams that these individuals received? (*Source:* fda.gov)

60. In 1956, IBM shipped the first hard drive, about the size of two refrigerators. This hard drive held 5,000,000 bytes of data. Express this quantity in megabytes. (*Source:* pcworld.com)

61. One of the shortest dinosaurs that ever lived was only 60 cm long when fully grown. What was the length of this dinosaur, rounded to the nearest inch? (*Source: Encarta Learning Zone Encyclopedia*)

62. About 43,000 pints of donated blood are used each day in the United States and Canada. How many liters is this, to the nearest 10,000 L? (*Source:* americasblood.org)

63. The table shows the average gestation period, in days, for various mammals. (*Source: The World Almanac and Book of Facts, 2010*)

Mammal	Gestation Period
Polar bear	240
Cow	284
Hippopotamus	238
Gorilla	258
Sea lion	350

How many more weeks is the gestation period for a sea lion than a hippopotamus?

64. The table shows the Saffir-Simpson Hurricane Scale, which is used to rate a hurricane based on its intensity. (*Source:* National Weather Service, NOAA)

Category	Wind Speed (mph)
1	74–95
2	96–110
3	111–130
4	131–155
5	156 and above

Express the range of wind speeds for a category 3 hurricane to the nearest kilometer per hour.

• Check your answers on page A-13.

CHAPTER 10 Posttest

To see if you have mastered the topics in this chapter, take this test.

1. 120 sec = ____ min

2. 7 yd = ____ ft

3. 3 gal = ____ pt

4. Add: 1 ft 9 in. and 1 ft 6 in.

5. Subtract 1 hr 25 min from 3 hr 10 min.

6. Which of the following units is a measure of capacity?
   a. a gram **b.** a meter **c.** a liter **d.** an hour

7. The weight of a baby is measured in
   a. milligrams **b.** grams **c.** kilograms

8. 400 cm = ____ m

9. 6 m = ____ mm

10. 4.6 sec = ________ mcsec

11. 2 mg = ____ mcg

12. 5 GB = ____ MB

13. 3 oz ≈ ____ g

14. 3 L ≈ ____ qt

15. A scientist dissecting a snake found in its stomach a centipede 140 mm long. Express the length of the centipede in centimeters. (*Source:* scienceblogs.com)

16. The cruising altitude of a passenger jet is 33,000 ft. Express this altitude in miles, rounded to the nearest whole number.

17. An adult humpback whale weighs about 60,000 lb and an adult finback whale weights about 40 tons. Which whale weighs more? (*Source:* Whale Center of New England)

18. On December 26, 2004, a powerful earthquake created huge waves that raced across the Indian Ocean. By the end of the day, more than 150,000 people in 11 countries were dead or missing, making this perhaps the most destructive tsunami in history and resulting in Earth spinning 3 mcsec faster. Express in seconds the change in spin time. (*Source:* nytimes.com)

19. A seven-year-old computer came with 960 MB of memory, in contrast to a more recent computer with 2 GB of memory. The new computer had how many times as much memory as the old, to the nearest tenth?

20. In order for an orbiting satellite to remain in the same spot over Earth, it must be approximately 35,786 km above the surface of Earth. Express this distance to the nearest thousand miles. (*Source:* Marshall Space Flight Center, NASA)

• Check your answers on page A-13.

Cumulative Review Exercises

To help you review, solve the following:

1. Compute: 9^3

2. Express as a mixed number: $\dfrac{11}{2}$

3. Combine and simplify: $\dfrac{5}{6} + \dfrac{1}{3} - \dfrac{11}{12}$

4. Divide: $\dfrac{2.8}{0.2}$

5. Solve and check: $\dfrac{20}{27} = \dfrac{5}{9}k$

6. Determine the unit rate: 120 mg of sodium in 5 crackers

7. \$36,000 is what percent of \$54,000?

8. Find the price of a book that normally sells for \$24 but is on sale at a 25% discount.

9. Find the difference: $-8 - (-3\frac{2}{5})$

10. Find the product: $\left(-\dfrac{2}{5}\right)\left(-1\dfrac{2}{3}\right)$

11. Find the range: 8, 6, 2, 9, 1, and 6

12. Solve: $3(x - 2) = -1$

13. Fill in the blank: 7 ft = ____ in.

14. Change 0.04 kg to grams.

15. On an Indiana farm, 160 bushels of wheat were raised on $5\frac{1}{3}$ acres. How many bushels per acre of wheat were raised?

16. The closing price of a share of technology stock on Monday was \$23.86. On Tuesday, the closing price was \$22.39. What was the change in the price of the stock?

17. In Finland, police levied traffic fines proportional to an offender's income. For speeding, a wealthy driver was fined about \$70,000, and a student, with a monthly income of approximately \$700, was fined about \$100. Estimate, to the nearest hundred thousand dollars, the monthly income of the wealthy driver. (*Source:* Steve Stecklow, "Helsinki on Wheels: Fast Finns Find Fines Fit Their Finances," *Wall Street Journal*)

18. The bar graph shows the number of American Kennel Club registrations for various breeds in a recent year.

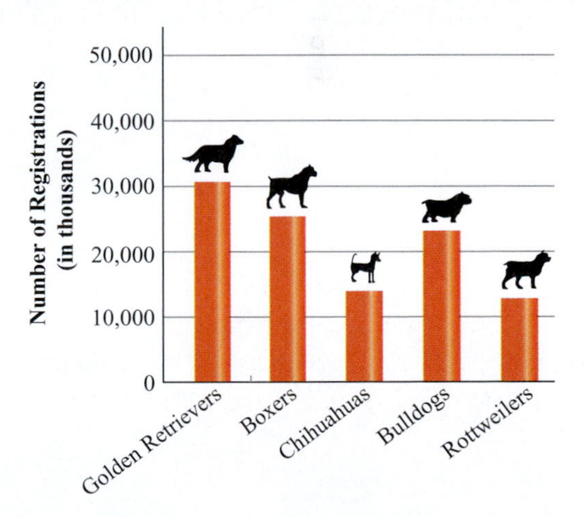

About how many more golden retrievers were registered than bulldogs? (*Source:* akc.org)

19. Fishermen commonly want to find the weight of a fish that they have caught, even when they are going to throw it back into the water. They can approximate this weight W (in pounds) by measuring the length of the fish L and the girth of the fish G (both in inches). Here girth means the distance around the body of the fish at its largest point.

The formula $W = \dfrac{L \cdot G^2}{800}$ approximates the weight of a trout. Find, to the nearest pound, the weight of a trout 30 inches long with girth 15 inches. (*Source:* dnr.wi.gov)

20. The distance from Memphis to Atlanta is given on the map shown below. Express this distance to the nearest hundred miles. (*Source: The World Almanac 2010*)

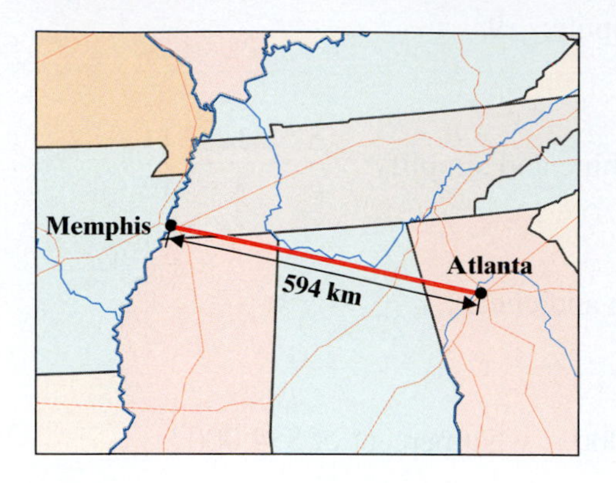

• Check your answers on page A-13.

CHAPTER 11

Basic Geometry

Geometry and Architecture

Students of geometry study abstract figures in space, whereas architects design real structures in space. The two fields, geometry and architecture, are, therefore, closely related.

The simplest architectural structures have basic geometric shapes. An igloo in the far north and a dome that graces a state capitol are shaped like hemispheres. A tepee is in the shape of a cone, and the peak of a roof is a pyramid.

The rectangle plays an especially important role in architectural design. Bricks, windows, doors, rooms, buildings, lots, city blocks, and street grids are all based on the rectangle—one of the most adaptable shapes for human needs.

Of all the rectangles with a given area, the square has the smallest perimeter. As a result, warehouses are often built in the form of squares. On the other hand, houses, hotels, and hospitals—for which daylight and a long perimeter are more important—are seldom square in shape.

(*Source:* William Blackwell, *Geometry in Architecture,* John Wiley and Sons, 1984)

CHAPTER 11 PRETEST

To see if you have already mastered the topics in this chapter, take this test.

1. Sketch and label an example of an obtuse $\angle PQR$

2. Find: $\sqrt{121}$

3. Find the supplement of 100°.

4. Find the complement of 36°.

Find each perimeter or circumference. Use $\pi \approx 3.14$, when needed.

5.

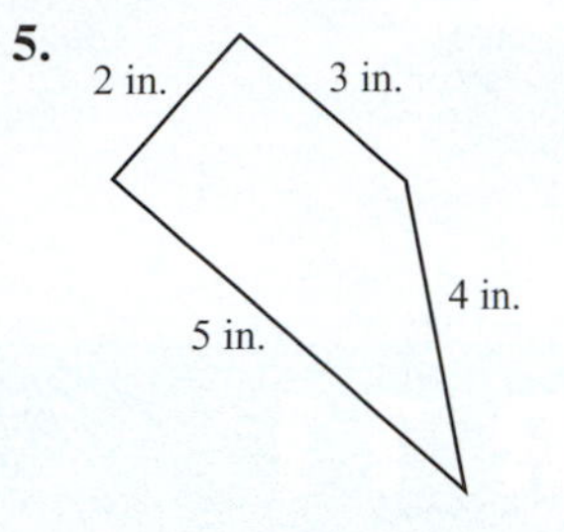

6.

8 ft

2 ft

7. A circle with a diameter of 4 inches.

8. A square with side 2.6 meters.

Find each area. Use $\pi \approx 3.14$, when needed.

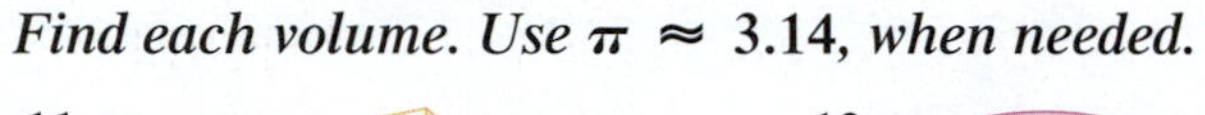

Find each volume. Use $\pi \approx 3.14$, when needed.

9. **10.**

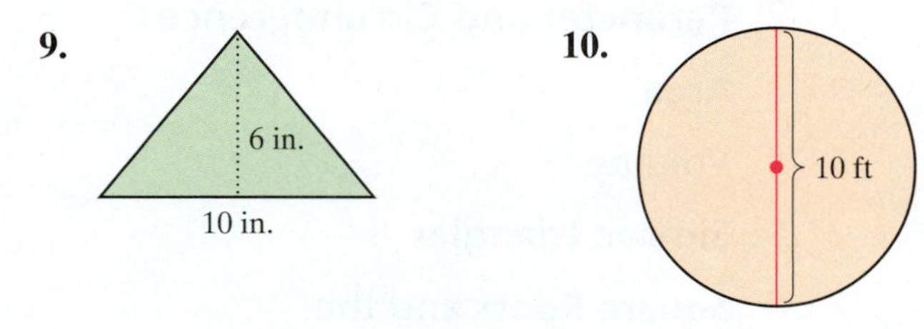

11. **12.**

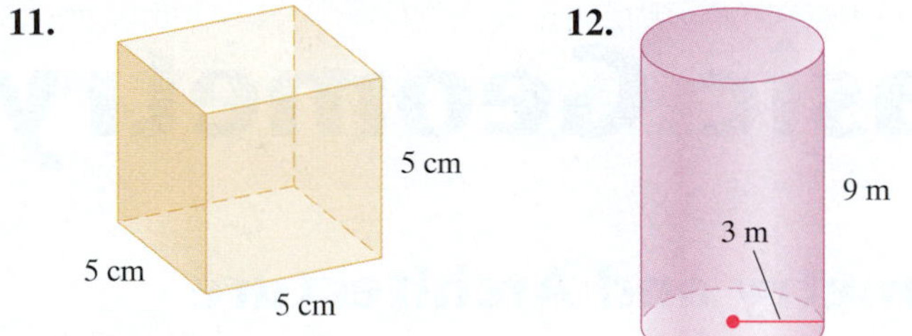

Find the value of each unknown.

13.

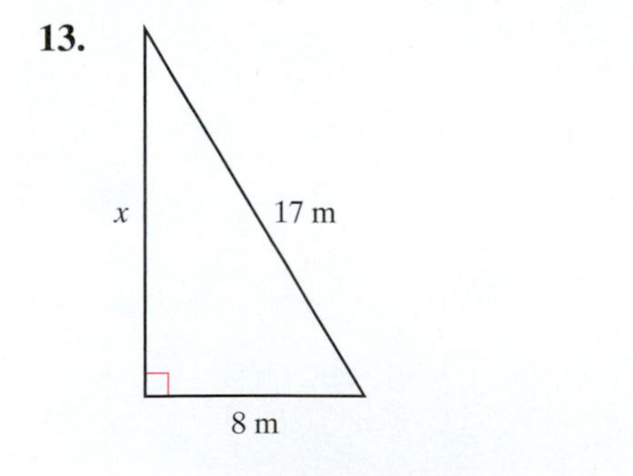

14.

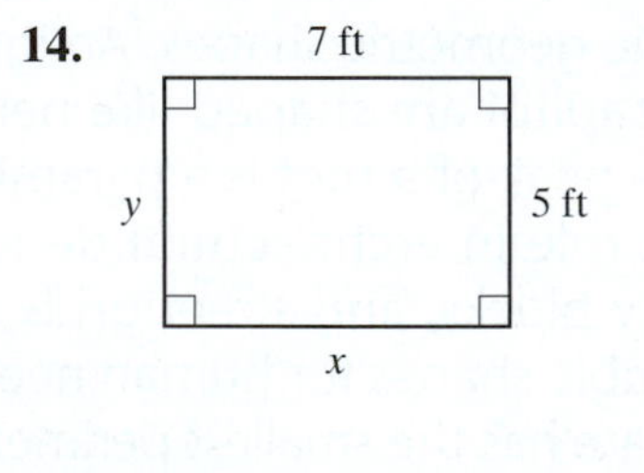

15.

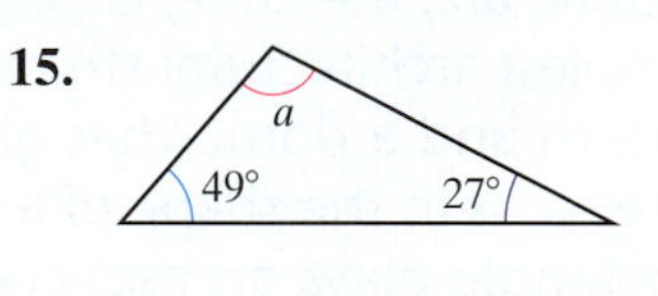

16. For the diagrams shown, ΔABC is similar to ΔDEF. Find y.

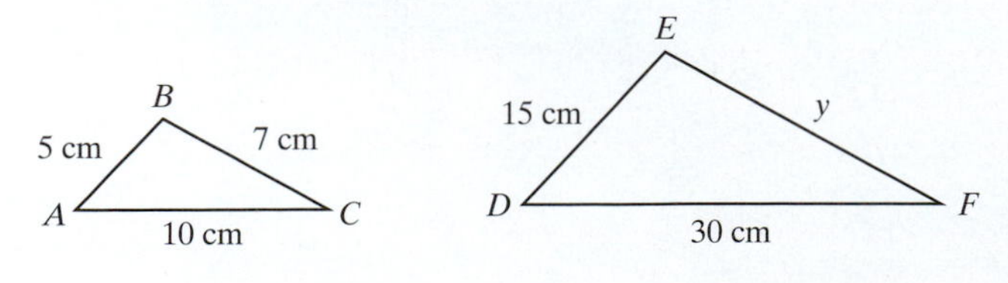

17. On the front of the brick shown, does each corner form an acute angle?

Solve.

18. Rescuers are searching for survivors of a shipwreck that took place within a mile of a rock. What is the area of the region that would be most appropriate to search for survivors? Round to the nearest square mile.

19. An entertainment center has just enough room to hold a 52-inch television, measured diagonally. Will an LCD HDTV with sides 44.4 inches and 30.2 inches fit in the entertainment center?

20. In constructing the foundation for a house, a contractor digs a hole 6 feet deep, 54 feet long, and 25 feet wide. How many cubic feet of earth are removed?

• Check your answers on page A-14.

Cultural Note

A *tessellation* is any repeating pattern of interlocking shapes. Some shapes, depending on their geometric properties, will tessellate, that is, go on indefinitely, covering the plane without overlapping and without gaps. These shapes include squares and equilateral triangles. Many other shapes will not tessellate. Tessellations are commonly found in the home—for instance, in the design of wall, ceiling, and floor coverings. More elaborate tessellations are found in mosaics that survive from ancient times. The tile mosaic shown to the left is found at the Alhambra, the summer residence of Moorish kings built in the fourteenth century in Granada, Spain.

(*Source:* The Metropolitan Museum of Art, *Islamic Art and Geometric Design: Activities for Learners*, New York: The Metropolitan Museum of Art, 2004)

11.1 Introduction to Basic Geometry

OBJECTIVES

- **A** To identify basic geometric concepts or figures
- **B** To find missing sides or angles
- **C** To solve applied problems involving basic geometric concepts or figures

What Geometry Is and Why It Is Important

The word *geometry*, which dates back thousands of years, means "measurement of the Earth." Today, we use the term to mean the branch of mathematics that deals with concepts such as point, line, angle, perimeter, area, and volume.

Ancient peoples, including the Egyptians, used the principles of geometry in their construction projects. They understood these principles because of observations they made in their daily lives and their studies of the physical forms in nature.

Geometry also has many practical applications in such diverse fields as art and design, multimedia, architecture, physics, and engineering. In city planning, geometric concepts, relationships, and notation are often used when designing the layout of a city. Note below how the use of geometric thinking helps to transform the street plan on the left to the geometric diagram on the right, making it easier to focus on the key features of the street plan.

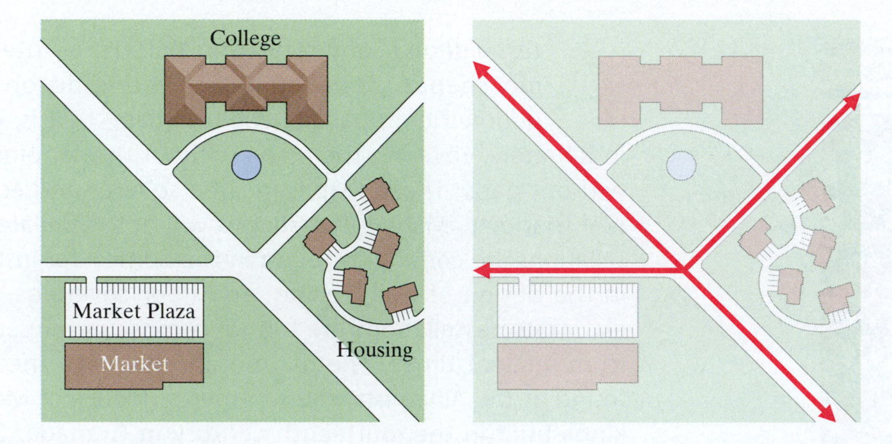

Basic Geometric Concepts

Let's first consider some of the basic concepts that underlie the study and application of geometry. The following table lists and explains some basic geometric terms illustrated in the preceding street plan. We use these terms throughout this chapter.

Definition	Example
A **point** is an exact location in space. A point has no dimension.	A • (read "point A")
A **line** is a collection of points along a straight path that extends endlessly in both directions. A line has only one dimension.	$\overleftrightarrow{CB}$ (read "line CB")
A **line segment** is a part of a line having two endpoints. Every line segment has a length.	$\overline{AB}$ (read "line segment AB") The length of $\overline{AB}$ is denoted AB.
A **ray** is a part of a line having only one endpoint.	$\overrightarrow{CD}$ (read "ray CD") (The endpoint is always the first letter.)

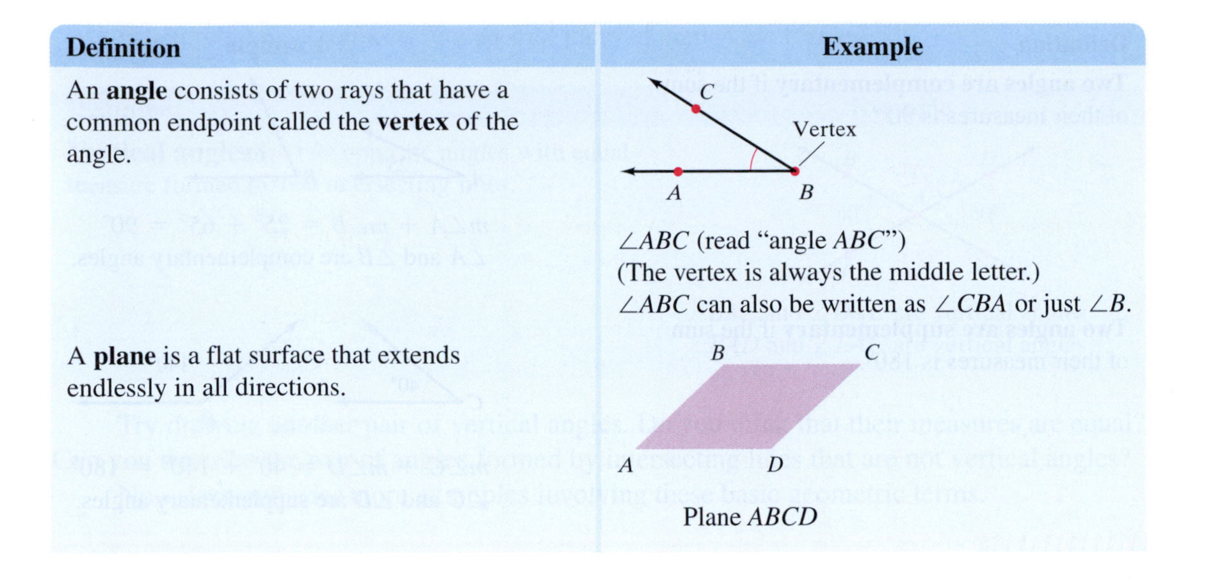

Definition	Example
An **angle** consists of two rays that have a common endpoint called the **vertex** of the angle.	$\angle ABC$ (read "angle ABC") (The vertex is always the middle letter.) $\angle ABC$ can also be written as $\angle CBA$ or just $\angle B$.
A **plane** is a flat surface that extends endlessly in all directions.	Plane $ABCD$

The unit in which angles are commonly measured is the degree (°). Angles are classified according to their measures. To indicate the measure of $\angle ABC$, we write $m\angle ABC$.

Definition	Example
A **straight angle** is an angle whose measure is 180°.	$\angle ABC$ is a straight angle, $m\angle ABC = 180°$.
A **right angle** is an angle whose measure is 90°.	Symbol for right angle $\angle DEF$ is a right angle; $m\angle DEF = 90°$.
An **acute angle** is an angle whose measure is less than 90°.	$\angle XYZ$ is an acute angle.
An **obtuse angle** is an angle whose measure is more than 90° and less 180°.	$\angle CDE$ is an obtuse angle.

continued

EXAMPLE 4

In the diagram to the right, $\angle ABC$ is a straight angle. Find y.

Solution Because $\angle ABC$ is a straight angle, $y + 39° = 180°$. We solve this equation for y.

$$y + 39° = 180°$$
$$y + 39° - 39° = 180° - 39°$$
$$y = 141°$$

PRACTICE 4

In the diagram shown, find x.

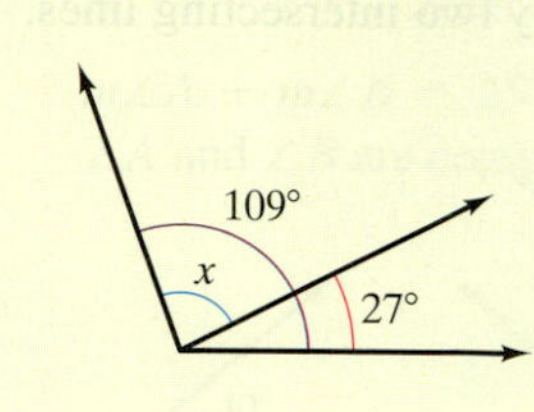

EXAMPLE 5

Find the values of x and y in the diagram to the right.

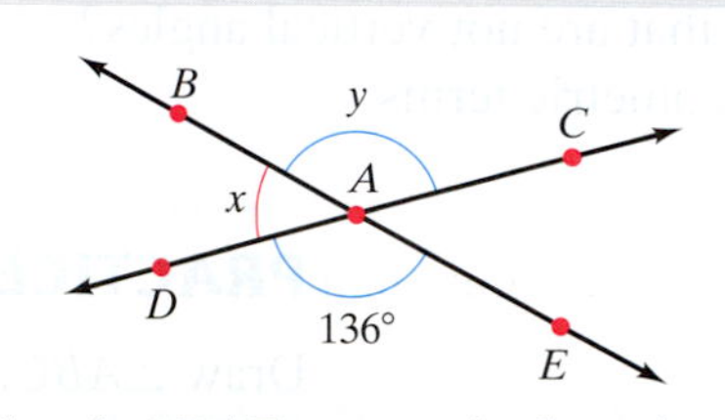

Solution Because $\angle BAC$ and $\angle DAE$ are vertical angles and $\angle DAE = 136°$, $\angle BAC = 136°$, or $y = 136°$. Because $\angle DAC$ is a straight angle, the sum of x and y is 180°.

$$x + y = 180°$$
$$x + 136° = 180°$$
$$x + 136° - 136° = 180° - 136°$$
$$x = 44°$$

So $x = 44°$ and $y = 136°$.

PRACTICE 5

In the following diagram, what are the values of a and b?

Basic Geometric Figures

Here we use the concepts just discussed to define some basic geometric figures: triangles, trapezoids, parallelograms, rectangles, squares, and circles. Except for circles, these figures are *polygons*.

DEFINITION

A **polygon** is a closed plane figure made up of line segments.

Closed: A polygon **Not closed: Not a polygon**

Polygons are classified according to the number of their sides. Here we examine two types of polygons—triangles and quadrilaterals.

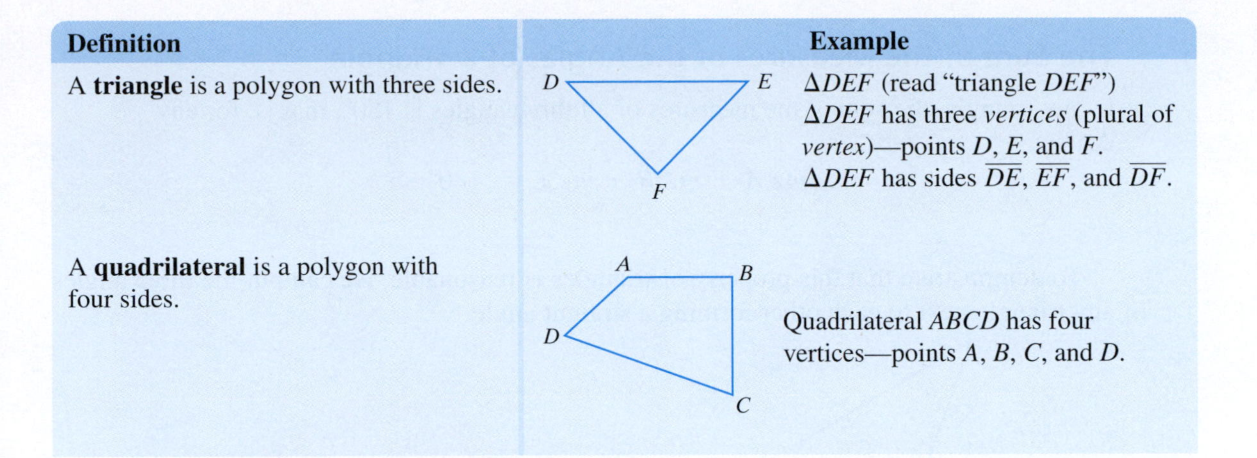

Definition	Example
A **triangle** is a polygon with three sides.	ΔDEF (read "triangle DEF") ΔDEF has three *vertices* (plural of *vertex*)—points D, E, and F. ΔDEF has sides $\overline{DE}$, $\overline{EF}$, and $\overline{DF}$.
A **quadrilateral** is a polygon with four sides.	Quadrilateral $ABCD$ has four vertices—points A, B, C, and D.

Triangles are classified by the lengths of their sides or the measures of their angles.

Definition	Example
An **equilateral triangle** is a triangle with three sides equal in length.	$\overline{PQ}$, $\overline{QR}$, and $\overline{PR}$ have equal lengths.
An **isosceles triangle** is a triangle with two or more sides equal in length.	$\overline{AB}$ and $\overline{BC}$ have equal lengths.
A **scalene triangle** is a triangle with no sides equal in length.	The three sides have unequal lengths.
An **acute triangle** is a triangle with three *acute* angles.	$\angle R$, $\angle S$, and $\angle T$ are acute angles.
A **right triangle** is a triangle with one right angle.	$\angle P$ is a right angle.
An **obtuse triangle** is a triangle with one obtuse angle.	$\angle Y$ is an obtuse angle.

The Sum of the Measures of the Angles of a Triangle

In any triangle, the sum of the measures of all three angles is 180°, that is, for any ΔABC,

$$m\angle A + m\angle B + m\angle C = 180°$$

To demonstrate that this property of triangles is reasonable, we can put the three angles of any triangle next to each other forming a straight angle.

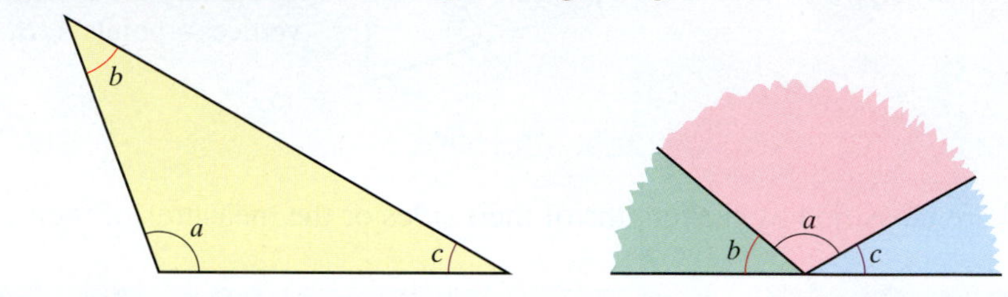

We have already seen that a polygon with four sides is called a *quadrilateral*. Let's consider special types of quadrilaterals.

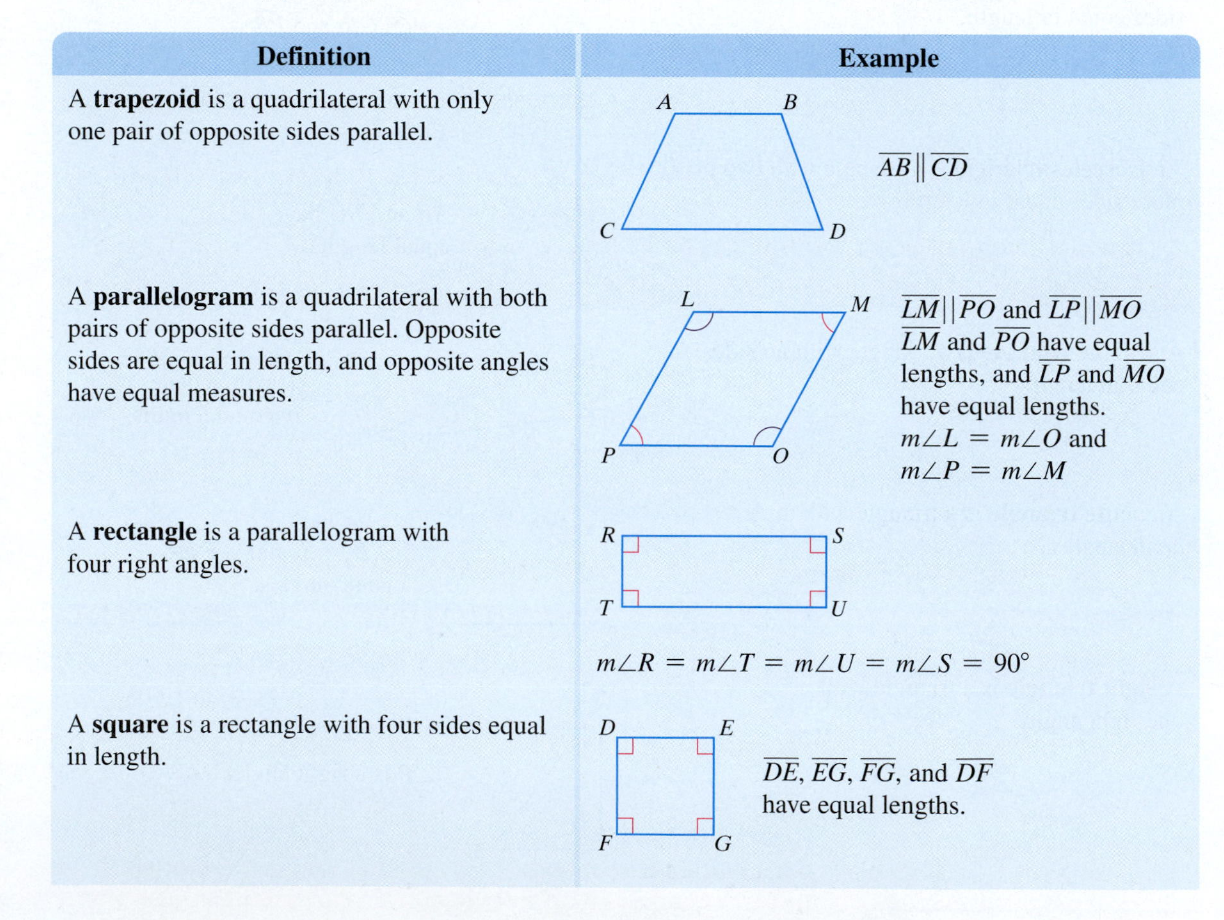

Definition	Example
A **trapezoid** is a quadrilateral with only one pair of opposite sides parallel.	$\overline{AB} \parallel \overline{CD}$
A **parallelogram** is a quadrilateral with both pairs of opposite sides parallel. Opposite sides are equal in length, and opposite angles have equal measures.	$\overline{LM} \parallel \overline{PO}$ and $\overline{LP} \parallel \overline{MO}$ $\overline{LM}$ and $\overline{PO}$ have equal lengths, and $\overline{LP}$ and $\overline{MO}$ have equal lengths. $m\angle L = m\angle O$ and $m\angle P = m\angle M$
A **rectangle** is a parallelogram with four right angles.	$m\angle R = m\angle T = m\angle U = m\angle S = 90°$
A **square** is a rectangle with four sides equal in length.	$\overline{DE}$, $\overline{EG}$, $\overline{FG}$, and $\overline{DF}$ have equal lengths.

The Sum of the Measures of the Angles of a Quadrilateral

In any quadrilateral, the sum of the measures of the angles is 360°. That is, for any quadrilateral $ABCD$,

$$m\angle A + m\angle B + m\angle C + m\angle D = 360°$$

We can see why this property of quadrilaterals is true by cutting a quadrilateral into two triangles. In each triangle, the sum of the measures of the three angles is 180°.

Triangle 1

Triangle 2

The last basic geometric figure we consider here is the circle.

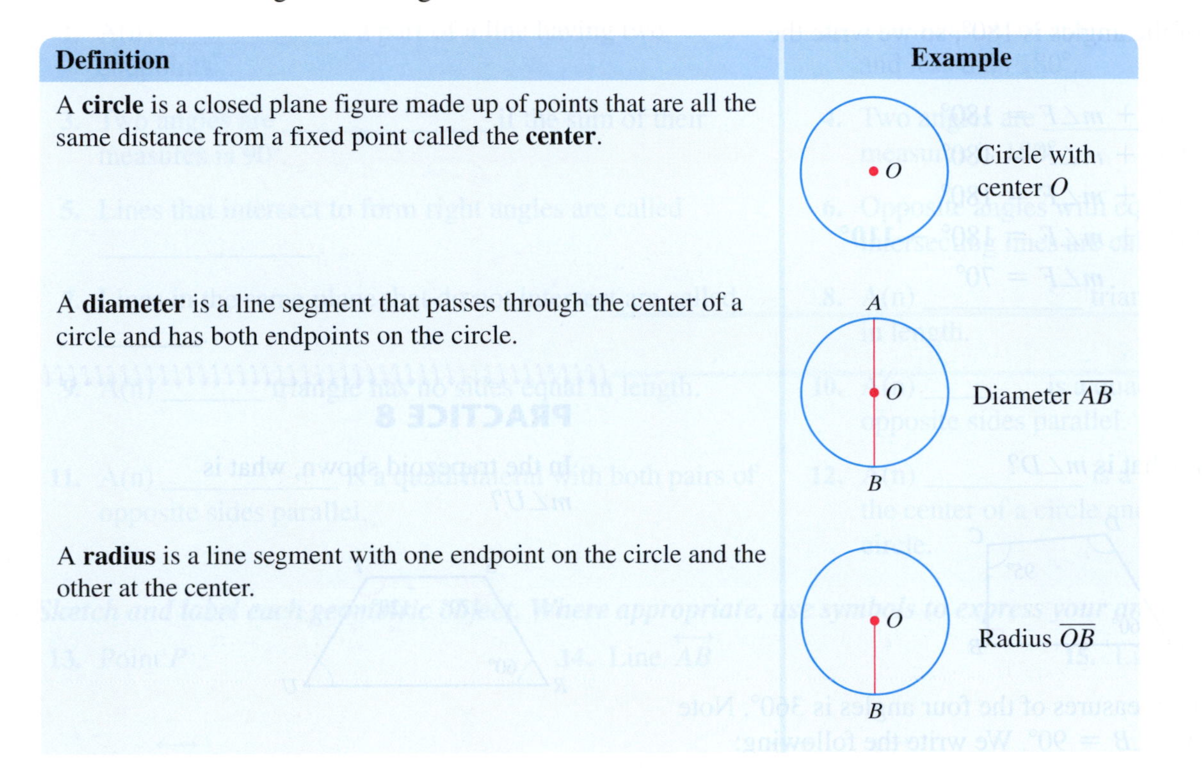

Definition	Example
A **circle** is a closed plane figure made up of points that are all the same distance from a fixed point called the **center**.	O Circle with center O
A **diameter** is a line segment that passes through the center of a circle and has both endpoints on the circle.	A, O, B Diameter $\overline{AB}$
A **radius** is a line segment with one endpoint on the circle and the other at the center.	O, B Radius $\overline{OB}$

Note that the diameter (d) of a circle is twice the radius (r), or $d = 2r$.

EXAMPLE 6

Sketch and label isosceles triangle ABC. Name the equal sides.

Solution

B

A C

$\overline{AB}$ and $\overline{BC}$ have equal lengths.

PRACTICE 6

Draw and label quadrilateral $ABCD$ that has at least one right angle with opposite sides equal and parallel. Name both pairs of parallel sides.

25. Acute $\angle FGH$

26. Vertical angles

B *Solve.*

27. Find x.

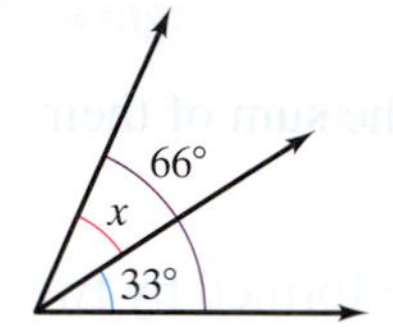

28. $\angle PQR$ is a straight angle. Find the measure of $\angle PQS$.

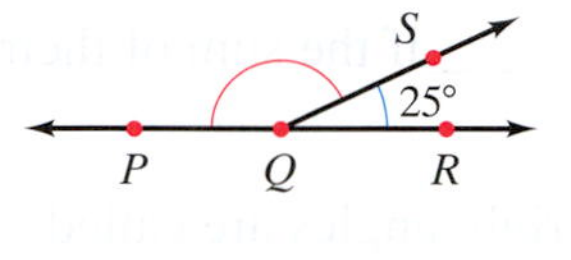

29. $\angle DEF$ is a right angle. Find the measure of $\angle DEG$.

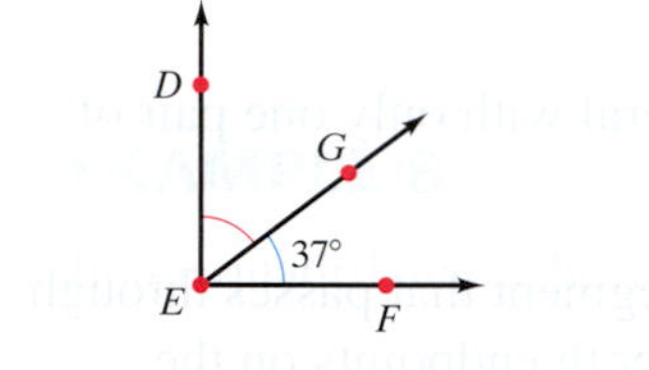

30. Solve for x and y.

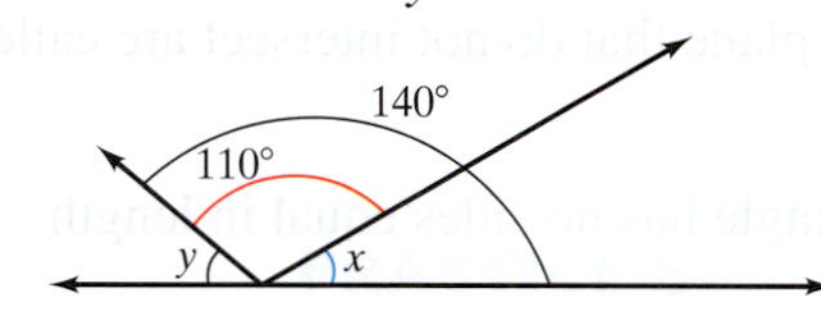

In the diagram shown, $\overleftrightarrow{AB} \perp \overleftrightarrow{CD}$ ***and*** $m\angle CPE = 35°$***. Find the measure of each angle.***

31. $\angle CPD$

32. $\angle APD$

33. $\angle BPD$

34. $\angle CPB$

35. $\angle APB$

36. $\angle BPE$

37. $\angle FPD$

38. $\angle APF$

In each figure, find the measure of the unknown angle(s).

39.

40.

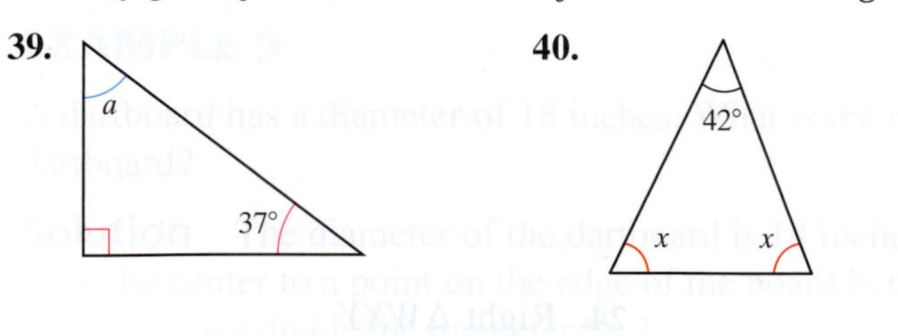

41.

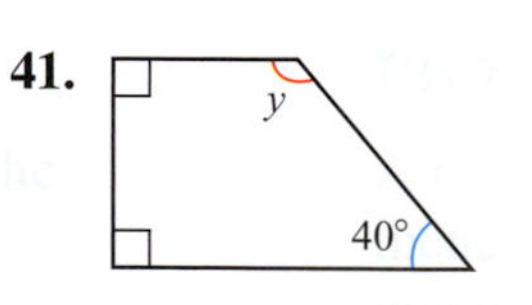

42.

Solve.

43. Find the complement of 35°.

44. What is the measure of an angle that is complementary to itself?

45. Find the supplement of 105°.

46. What is the measure of an angle that is supplementary to 88°?

47. In ΔABC, $m\angle A = 35°$ and $m\angle B = 75°$. Find $m\angle C$.

48. In ΔDEF, $m\angle E = 90°$ and $m\angle F = 19°$. Find $m\angle D$.

49. In a parallelogram, the sum of three of the angles is 275°. What is the measure of the fourth angle?

50. In a triangle in which all angles are equal, what is the measure of each angle?

Mixed Practice

Solve.

51. Sketch and label the diameter of a circle

52. In the following diagram, find the measure of $\angle DBE$.

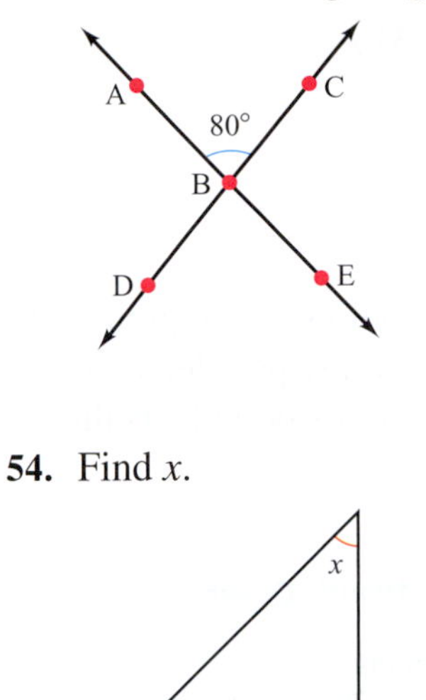

53. Find x.

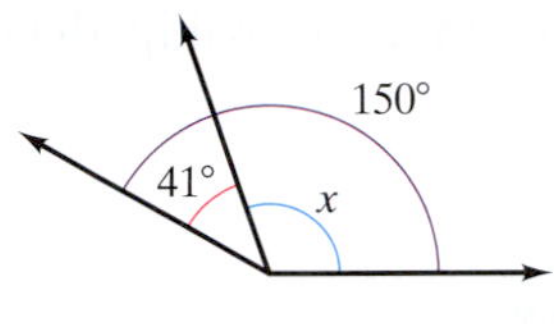

54. Find x.

55. Sketch and label obtuse ΔRST.

56. What is the measure of an angle that is complementary to 20°?

Applications

Solve.

57. An ancient circular medicine wheel made with rocks, as shown in the diagram, was built by Native Americans in Wyoming. What is the wheel's radius? (*Source:* Works Projects Administration, *Wyoming: A Guide to Its History, Highways, and People*)

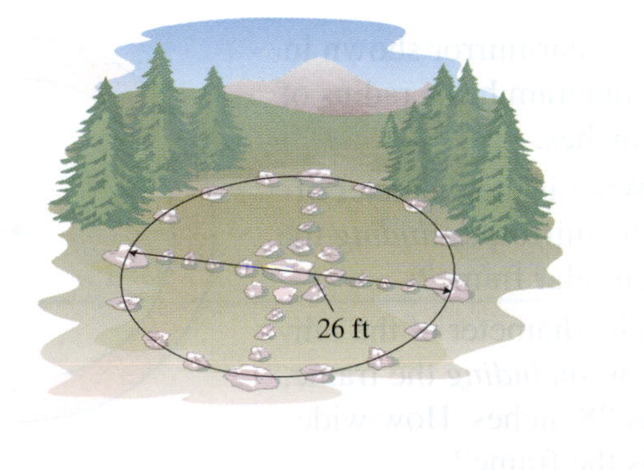

58. The Human Centrifuge, located outside of Philadelphia, is a sphere at the end of a 50-foot-long metal pole spinning in a circular room. The centrifuge was used to sling around early astronauts in order to understand the effect on their bodies of high pressure. What is the diameter of the circle formed by the spinning centrifuge? (*Sources:* nadcmuseum.org and hq.nasa.gov)

11.2 Perimeter and Circumference

OBJECTIVES

- **A** To find the perimeter of a polygon or the circumference of a circle
- **B** To find the perimeter or circumference of a composite geometric figure
- **C** To solve applied problems involving perimeter or circumference

The Perimeter of a Polygon

One of the most basic features of a plane geometric figure is its *perimeter*. The length of a fence around a plot of land, the length of a state's border, and the length of a picture frame are examples of perimeters.

DEFINITION

The **perimeter** of a polygon is the distance around it.

To find the perimeter of any polygon, we add the lengths of its sides. Note that perimeters are measured in linear units such as feet or meters.

Suppose that we want to build a fence around the garden shown below:

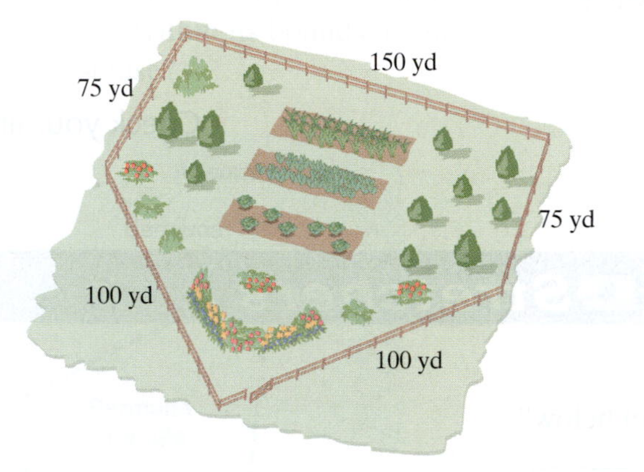

How much fencing do we need? Using the definition of perimeter, we obtain the distance around this garden.

$$75 + 150 + 75 + 100 + 100 = 500$$

So we need 500 yards of fencing.

For some polygons, we can also use a *formula* to find the perimeter. Let's consider the formulas for the perimeter of a triangle, a rectangle, and a square.

Figure	Formula	Example
Triangle	$P = a + b + c$ Perimeter equals the sum of the lengths of the three sides.	$a = 12$ cm, $b = 20$ cm, $c = 24$ cm $P = a + b + c$ $= 12 + 20 + 24$ $= 56$, or 56 cm

Figure	Formula	Example
Rectangle	$P = 2l + 2w$ Perimeter equals twice the length plus twice the width.	$l = 10$ m, $w = 5$ m $P = 2l + 2w$ $= 2 \cdot 10 + 2 \cdot 5$ $= 20 + 10$ $= 30$, or 30 m
Square	$P = 4s$ Perimeter equals 4 times the length of a side.	$s = 6$ ft $P = 4s$ $= 4 \cdot 6$ $= 24$, or 24 ft

EXAMPLE 1

Find the perimeter of the polygon shown.

2 m, 1 m, 1 m, 3 m, 3 m, 4 m

Solution To find the perimeter, we add the lengths of the sides.

$$2 + 1 + 1 + 3 + 3 + 4 = 14$$

So the perimeter is 14 meters.

PRACTICE 1

What is the perimeter of this polygon?

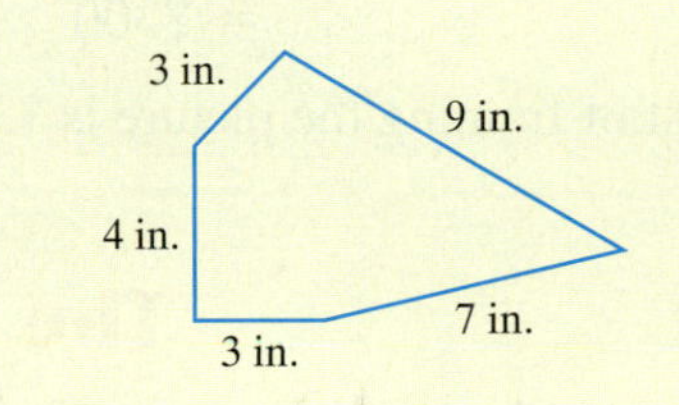

EXAMPLE 2

Find the perimeter of an equilateral triangle with side 1.4 meters long.

Solution Recall that all three sides of an equilateral triangle are equal.

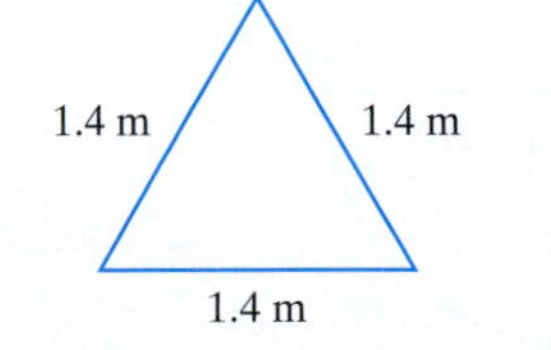

We use the formula for the perimeter of a triangle.

$$\begin{aligned} P &= a + b + c \\ &= 1.4 + 1.4 + 1.4 \\ &= 4.2 \end{aligned}$$

Therefore, the perimeter of the triangle is 4.2 meters. Because all three sides are equal in length, we could have used the formula

$$P = 3s = 3(1.4) = 4.2, \text{ or } 4.2 \text{ meters}$$

PRACTICE 2

Find the perimeter of a square with side $\frac{3}{4}$ miles long.

EXAMPLE 3

A rectangular picture is 35 inches long and 25 inches wide. To frame the picture costs \$2.50 per inch. What is the cost of framing the picture?

Solution Let's draw a diagram.

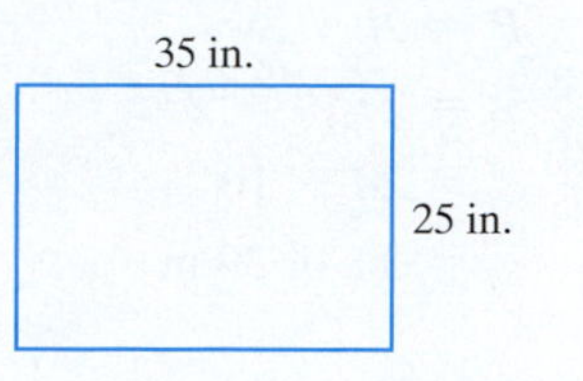

The picture is rectangular, so we use the formula $P = 2l + 2w$ to find its perimeter.

$$\begin{aligned} P &= 2l + 2w \\ &= 2(35) + 2(25) \\ &= 70 + 50 \\ &= 120 \end{aligned}$$

The distance around the picture is 120 inches. To find the cost of framing the picture, we multiply this perimeter by the cost per inch.

$$\begin{aligned} \text{Cost} &= 120 \text{ in.} \times \frac{\$2.50}{\text{in.}} \\ &= \$300 \end{aligned}$$

So the cost of framing the picture is \$300.

PRACTICE 3

A square garden has sides 10 feet long. How much will it cost to install a fence around the garden if the fence costs \$1.75 per foot?

The Circumference of a Circle

We refer to the perimeter of a polygon but the *circumference* of a circle.

DEFINITION

The distance around a circle is called its **circumference.**

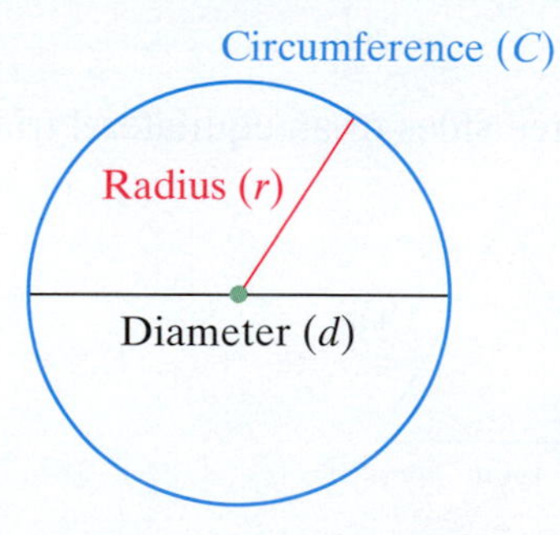

For every circle, the ratio of the circumference C to the diameter d is the same number, which is written as π (read "pi"). This relationship $\frac{C}{d} = \pi$ can also be written as $C = \pi d$ or $C = 2\pi r$. Do you see why πd and $2\pi r$ are equal? Explain.

The value of π is 3.1415926. . . . It is an *irrational number*, so that when π is expressed as a decimal, the digits go on indefinitely without any pattern being repeated. For convenience, we often use an approximate value of π, such as 3.14 or $\frac{22}{7}$, when calculating a circumference by hand.

Figure	Formula	Example
Circumference	$C = \pi d$, or $C = 2\pi r$ Circumference equals π times the diameter, or 2 times π times the radius r.	10 cm $C = \pi d$ $\approx 3.14(10)$ ≈ 31.4, or 31.4 cm

EXAMPLE 4

Find the circumference of the circle shown. Use $\pi \approx 3.14$.

4 m

Solution The radius of the circle is 4 meters. We use the formula for the circumference of a circle in terms of the radius.

$$
\begin{aligned}
C &= 2\pi r \\
&\approx 2(3.14)(4) \quad \text{Substitute 4 for } r. \\
&\approx 25.12
\end{aligned}
$$

So the circumference is approximately 25.12 meters.

PRACTICE 4

What is the circumference of the circle shown? Use $\pi \approx \frac{22}{7}$.

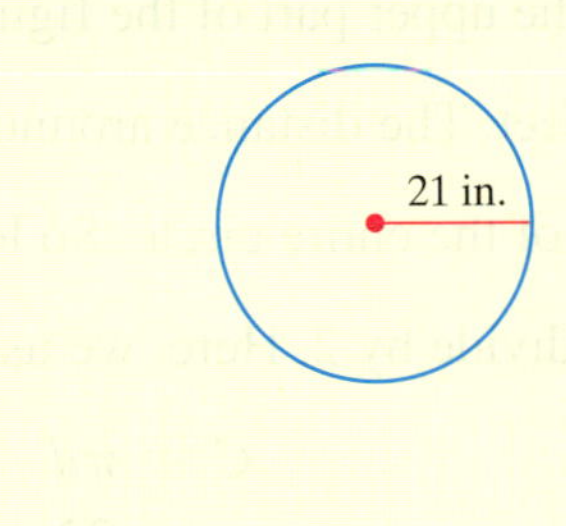

EXAMPLE 5

The diameter of a rolling wheel is 20 inches. How far does it travel in one complete turn?

Solution First, let's draw a diagram.

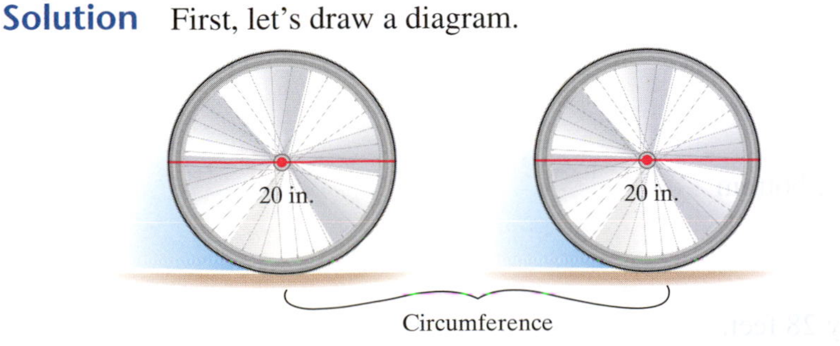

PRACTICE 5

A circular swimming pool has a radius of 18 feet. If a metal rail is to be placed around the edge of the pool, how many feet of railing are needed?

11.2 Exercises

FOR EXTRA HELP MyMathLab Math XL PRACTICE WATCH READ REVIEW

Mathematically Speaking

Fill in each blank with the most appropriate term or phrase from the given list.

simple	rectangle	composite
circle	circumference	square
perimeter	length	

1. The __________ of a polygon is the distance around it.

2. The perimeter of a __________ is equal to the sum of twice the length and twice the width.

3. The perimeter of a _______ is equal to 4 times the length of a side.

4. The distance around a circle is called its __________.

5. A formula for the circumference of a _______ is $C = 2\pi r$.

6. Two or more basic geometric figures are combined in a __________ figure.

A ***Find the perimeter or circumference of each figure. Use*** $\pi \approx 3.14$***, when needed.***

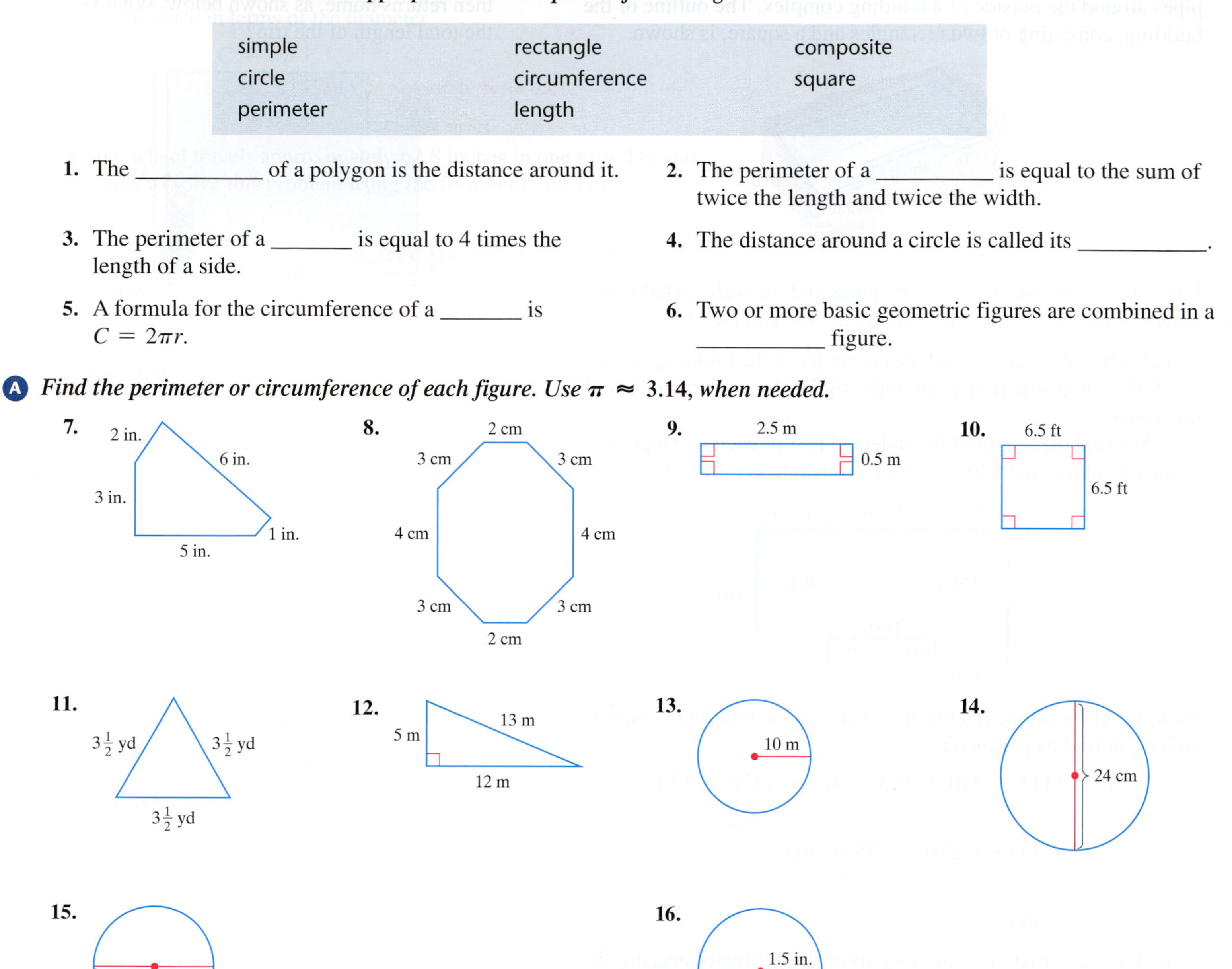

Find the perimeter or circumference. Use $\pi \approx 3.14$***, when needed.***

17. A square with side $5\frac{1}{4}$ yards long

18. A circle whose radius is 20 inches long

19. A rectangle of length $5\frac{3}{4}$ feet and width $3\frac{1}{4}$ feet

20. A triangle whose side lengths are 2 inches, $1\frac{1}{2}$ inches, and $\frac{7}{8}$ inches.

21. An isosceles triangle whose equal sides are $7\frac{1}{2}$ centimeters long and whose third side is 4 centimeters long

22. A rectangle with length 8 meters and width $4\frac{1}{2}$ meters

23.

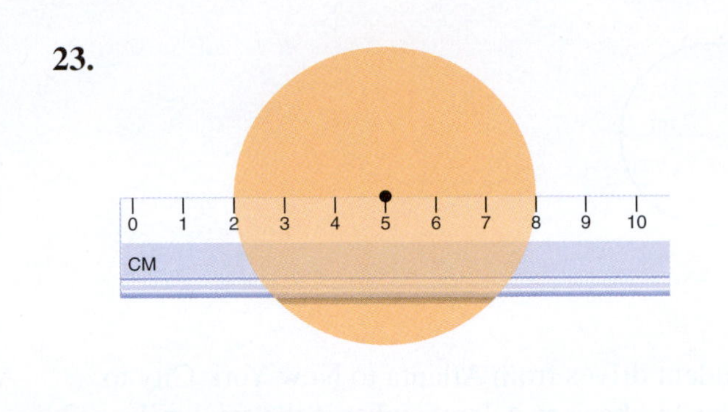

24.

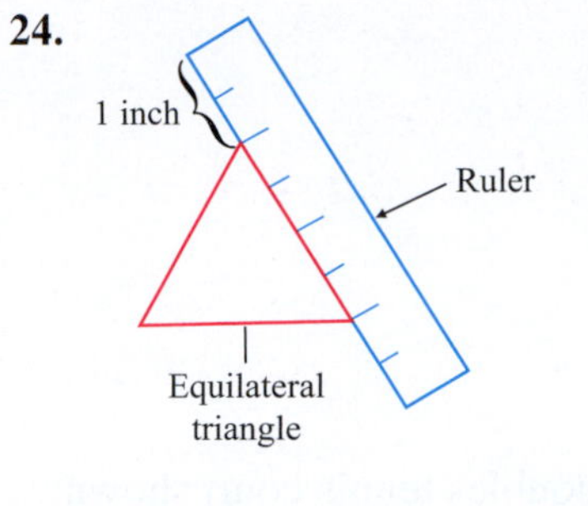

25. A circle whose diameter is 3.54 meters long

26. A polygon whose side lengths are 22.75 feet, 25.73 feet, 15.94 feet, 18.23 feet, 21.65 feet, and 34.98 feet

B *Find the perimeter of each composite geometric figure. Use* $\pi \approx 3.14$*, when needed.*

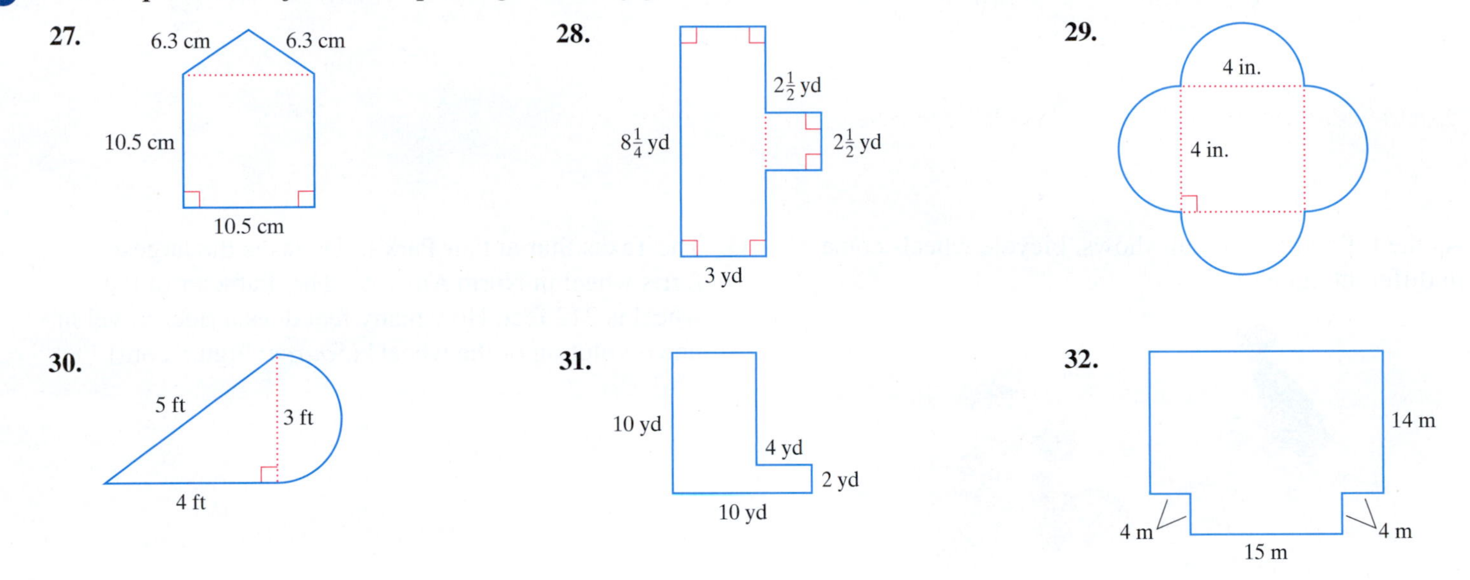

Mixed Practice

Find the perimeter or circumference of each figure. Use $\pi \approx 3.14$*, when needed.*

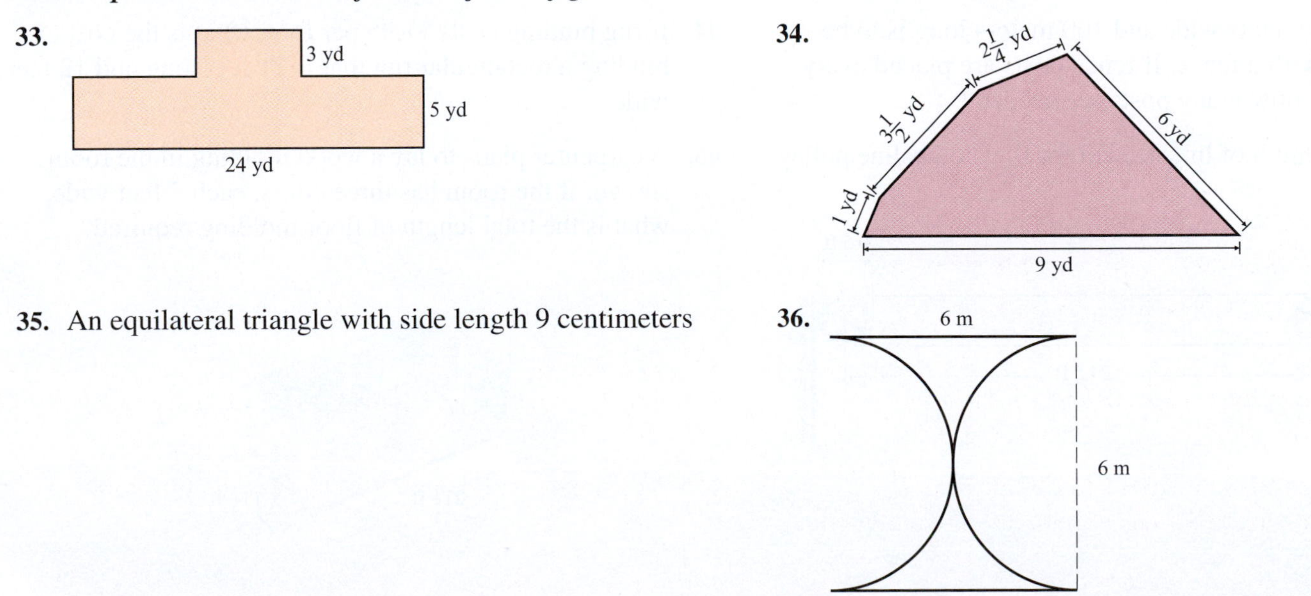

37.

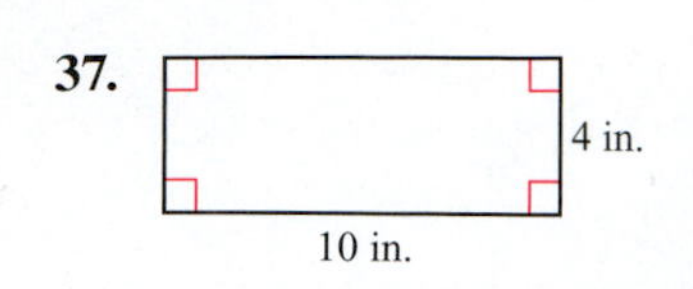

38.

7 mi

Applications

C *Solve.*

39. Find the perimeter of the doubles tennis court shown below.

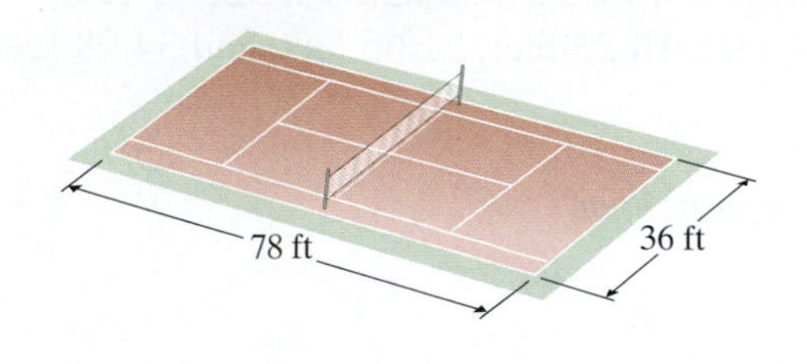

40. If a student drives from Atlanta to New York City to Chicago and back to Atlanta, what is the total mileage?

41. As the following diagram shows, bicycle wheels come in different diameters.

In one wheel rotation, how much farther does the 27-inch bicycle wheel go than the 25-inch bicycle wheel?

42. The Texas Star at Fair Park in Dallas is the largest ferris wheel in North America. The diameter of the wheel is 212 feet. How many feet does a rider travel in one revolution of the wheel? (*Source:* bigtex.com)

43. A field 50 meters wide and 100 meters long is to be enclosed with a fence. If fence posts are placed every 10 meters, how many posts are needed?

44. If rug binding costs $1.95 per foot, what is the cost of binding a rectangular rug that is 21 feet long and 12 feet wide?

45. Find the length of line needed for the clothesline pulley.

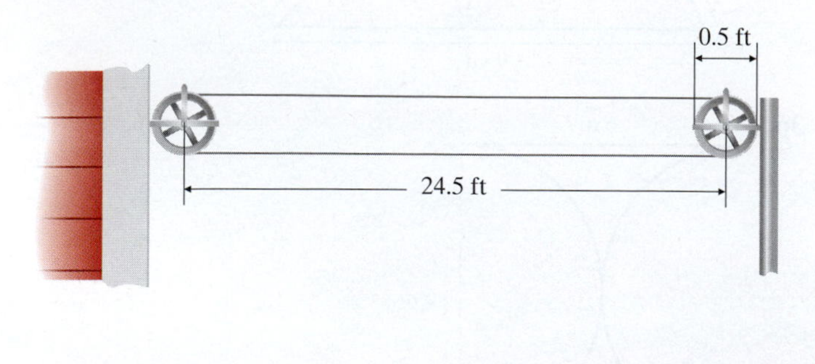

46. A carpenter plans to lay a wood molding in the room shown. If the room has three doors, each 3 feet wide, what is the total length of floor molding required?

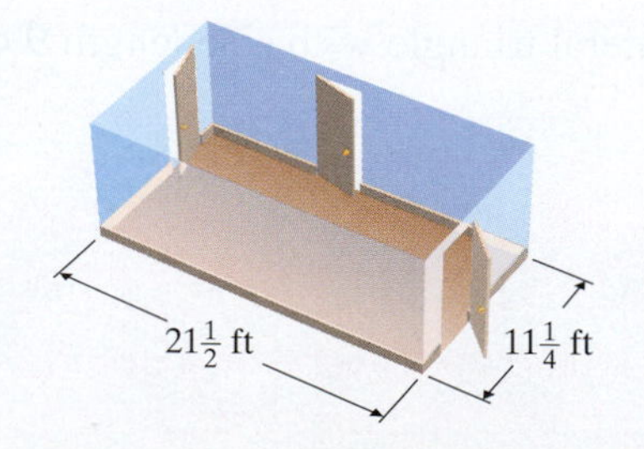

47. The radius of Earth is about 6,400 kilometers. If a satellite is orbiting 400 kilometers above Earth, find the distance that the satellite travels in one orbit.

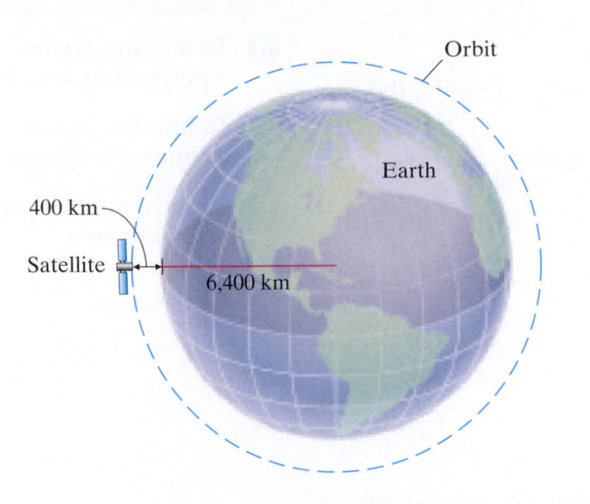

48. A circular crater on the Moon has a circumference of about 214.66 miles. What is the radius of the crater?

• Check your answers on page A-14.

MIND*Stretchers*

Investigation

1. Draw three triangles. Label the sides of each triangle *a*, *b*, and *c*. Measure each side, writing the measurements in the following table. Compare the sum of any two sides of a triangle with its third side.

	a	b	c	$a + b$	$a + c$	$b + c$
Triangle 1						
Triangle 2						
Triangle 3						

How does the length of the side of a triangle compare to the sum of the lengths of the other two sides?

Mathematical Reasoning

2. Consider the cart pictured. Which wheel do you think is likely to wear out more quickly? Justify your answer.

Groupwork

3. Explain how you can approximate the circumference of a circular room with a ruler. Compare your method with those of other members of the group.

11.3 Area

OBJECTIVES

- **A** To find the area of a polygon or a circle
- **B** To find the area of a composite figure
- **C** To solve applied problems involving area

The Area of a Polygon and a Circle

Area is a measure of the size of a plane geometric figure. The size of a piece of paper, the size of a volleyball court, and the size of a lawn are all examples of areas.

To find the area of the rectangle shown, we split it into little 1-inch by 1-inch squares, each representing 1 square inch.

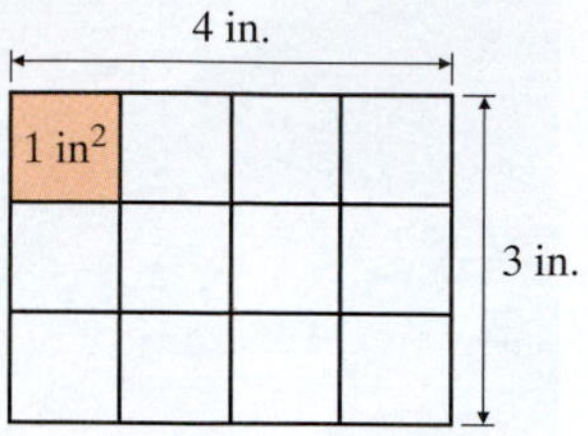

Then, we count the number of square inches within the rectangle, which is 12 square inches.

Each row of the rectangle contains 4 square inches, and the rectangle has 3 rows. So a shortcut to counting the total number of square inches is to multiply 3 by 4, getting 12 square inches in all. Note that areas are measured in square units, such as square inches (sq in. or in^2), square miles (sq mi or mi^2), or square meters (m^2).

DEFINITION

Area is the number of square units that a figure contains.

In this section, we focus on finding the area of common polygons and circles. First, we consider the areas of polygons.

Figure	Formula	Example
Rectangle	$A = lw$ Area equals the length times the width. 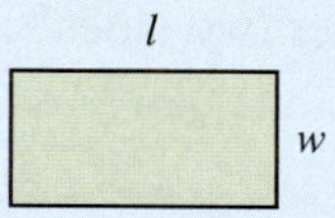	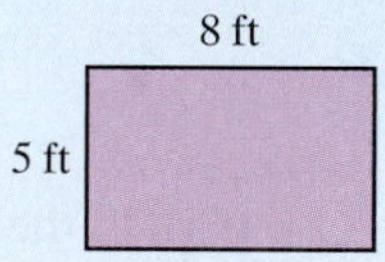 $A = lw$ $= 8 \cdot 5$ $= 40$, or 40 ft^2
Square	$A = s \cdot s = s^2$ Area equals the square of a side. 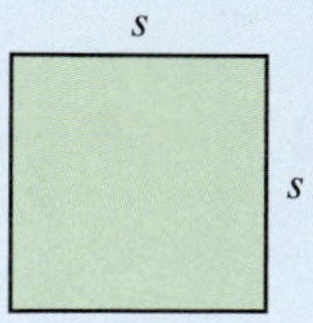	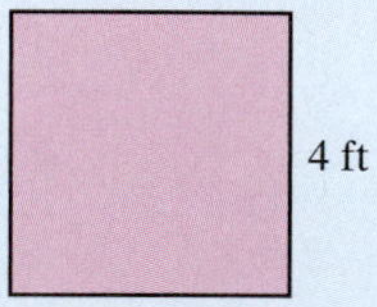 $A = s^2$ $= (4)^2$ $= 4 \cdot 4$ $= 16$, or 16 ft^2

continued

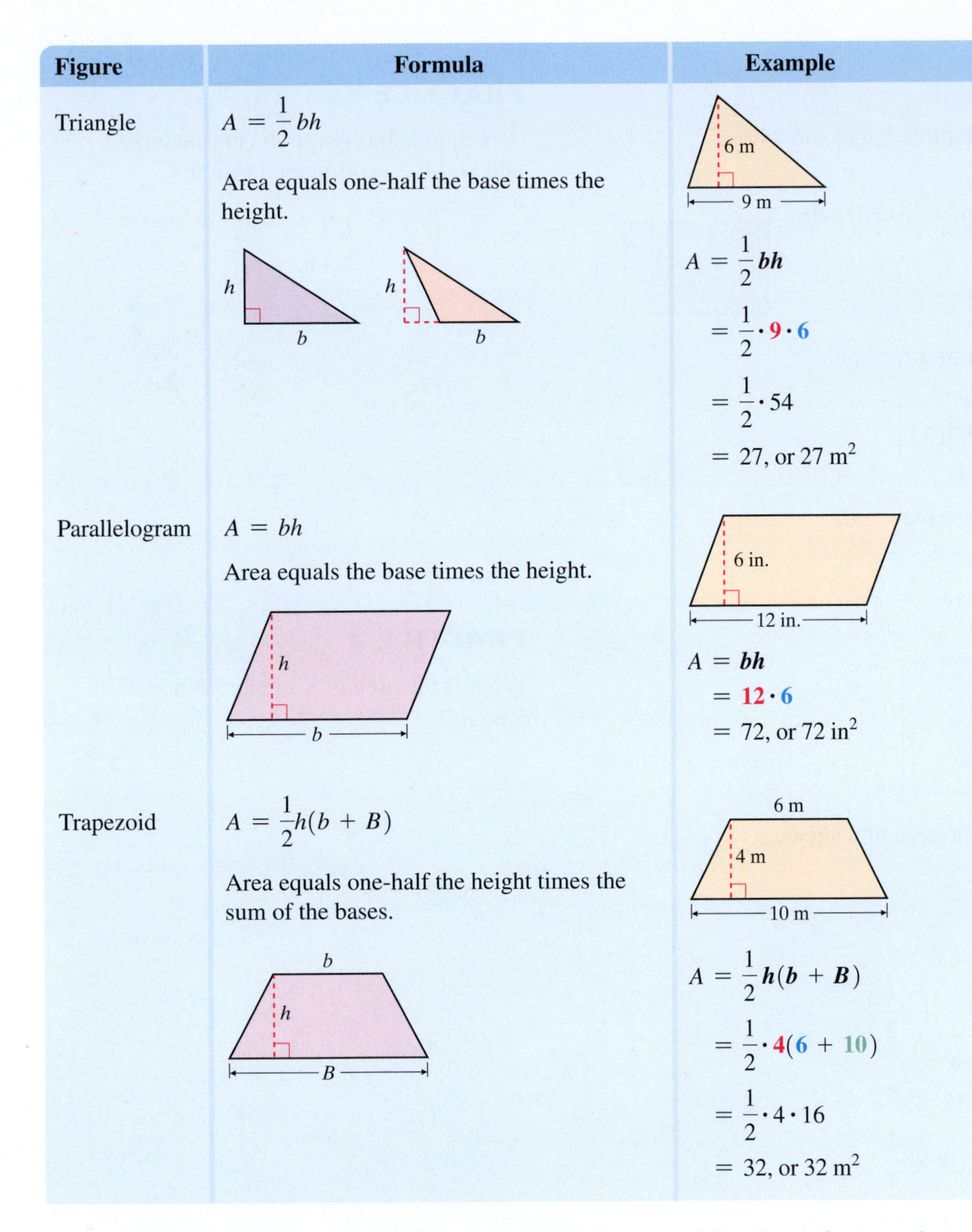

Figure	Formula	Example
Triangle	$A = \frac{1}{2}bh$ Area equals one-half the base times the height.	$A = \frac{1}{2}bh$ $= \frac{1}{2} \cdot 9 \cdot 6$ $= \frac{1}{2} \cdot 54$ $= 27$, or 27 m^2
Parallelogram	$A = bh$ Area equals the base times the height.	$A = bh$ $= 12 \cdot 6$ $= 72$, or 72 in^2
Trapezoid	$A = \frac{1}{2}h(b + B)$ Area equals one-half the height times the sum of the bases.	$A = \frac{1}{2}h(b + B)$ $= \frac{1}{2} \cdot 4(6 + 10)$ $= \frac{1}{2} \cdot 4 \cdot 16$ $= 32$, or 32 m^2

Now, let's consider the area of a circle. As in the case of the circumference, the area of a circle is expressed in terms of π. Recall that π is approximately 3.14 or $\frac{22}{7}$.

Figure	Formula	Example
Circle	$A = \pi r^2$ Area equals π times the square of the radius.	$A = \pi r^2$ $\approx 3.14\,(3)^2$ $\approx 3.14\,(9)$ ≈ 28.26, or 28.26 cm^2

EXAMPLE 1

Find the area of a rectangle whose length is 5 feet and whose width is 3 feet.

Solution First, we draw a diagram to visualize the problem.

Then, we use the formula for the area of a rectangle.

$$\begin{aligned} A &= lw \\ &= (5)(3) \\ &= 15 \end{aligned}$$

So the area of the rectangle is 15 square feet.

PRACTICE 1

A rectangle has length 6 centimeters and width 2 centimeters. Find its area.

EXAMPLE 2

Find the area of the square.

Solution We use the formula for the area of a square.

$$\begin{aligned} A &= s^2 \\ &= \left(4\frac{1}{2}\right)^2 \\ &= \left(4\frac{1}{2}\right)\left(4\frac{1}{2}\right) \\ &= \frac{9}{2} \cdot \frac{9}{2} \\ &= \frac{81}{4}, \text{ or } 20\frac{1}{4} \end{aligned}$$

So the area of the square is $20\frac{1}{4}$ square inches.

PRACTICE 2

What is the area of a square with side 3.6 cm?

EXAMPLE 3

Find the area of a triangle with base 8 centimeters and height 5.9 centimeters.

Solution First, we draw a diagram.

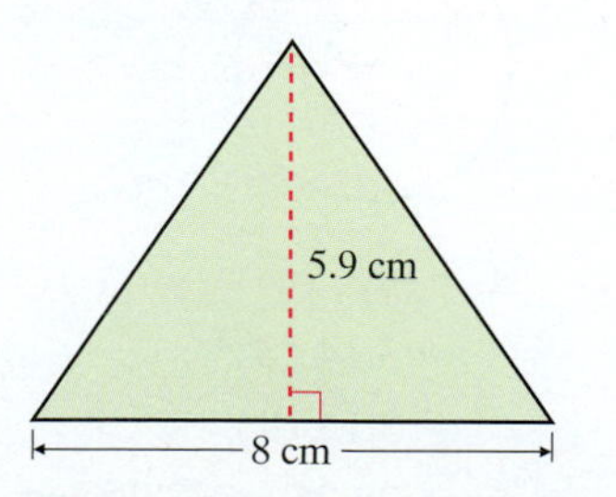

PRACTICE 3

A triangle has a height of 3 inches and a base of 5 inches. What is its area?

Next, we use the formula for finding the area of a triangle.

$$A = \frac{1}{2}bh$$
$$= \frac{1}{2}(8)(5.9)$$
$$= 23.6$$

So the area of the triangle is 23.6 square centimeters.

EXAMPLE 4

What is the area of a parallelogram with base $6\frac{1}{2}$ meters and height 3 meters?

Solution We draw a diagram and then use the formula for the area of a parallelogram.

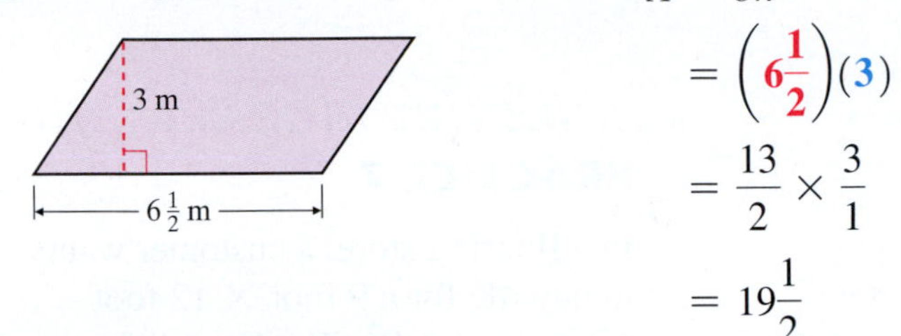

$$A = bh$$
$$= \left(6\frac{1}{2}\right)(3)$$
$$= \frac{13}{2} \times \frac{3}{1}$$
$$= 19\frac{1}{2}$$

So the area of the parallelogram is $19\frac{1}{2}$ square meters.

PRACTICE 4

Find the area of a parallelogram whose base is 5 feet and height is $2\frac{1}{2}$ feet.

EXAMPLE 5

What is the area of the trapezoid shown?

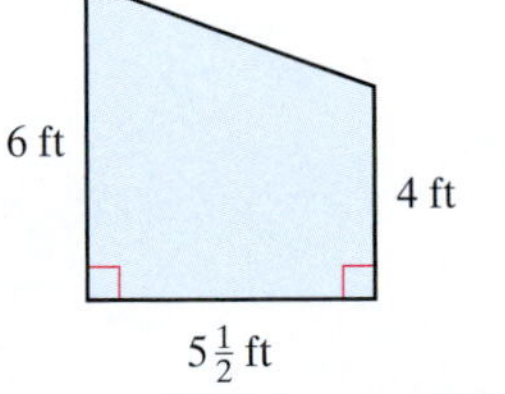

Solution This polygon is a trapezoid, so we use the following formula to find its area.

$$A = \frac{1}{2}h(b + B)$$
$$= \frac{1}{2} \cdot 5\frac{1}{2}(6 + 4)$$
$$= \frac{1}{2} \cdot \frac{11}{2} \cdot 10$$
$$= \frac{55}{2}, \text{ or } 27\frac{1}{2}$$

So the area of the trapezoid is $27\frac{1}{2}$ square feet.

PRACTICE 5

Find the area of the following trapezoid.

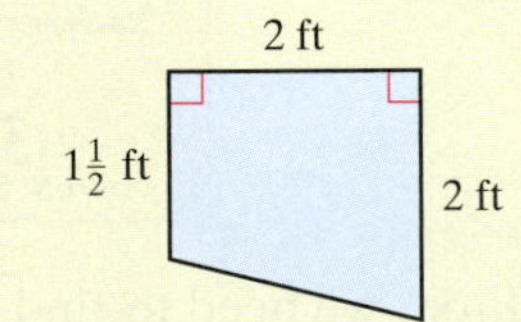

EXAMPLE 6

What is the area of a circle whose diameter is 8 meters?

Solution First, we draw a diagram.

We know that the radius is one-half of 8 meters, or 4 meters, which we substitute in the formula for the area of a circle.

$$\begin{aligned} A &= \pi r^2 \\ &\approx 3.14(4)^2 \\ &\approx 3.14(16) \\ &\approx 50.24 \end{aligned}$$

So the area of the circle is approximately 50.24 square meters.

PRACTICE 6

Find the area of a circle whose radius is 5 yards.

EXAMPLE 7

An artist wants to buy an ad in a magazine that charges $1,000 per square inch for advertising space. Will the cost of the ad be greater than $6,000?

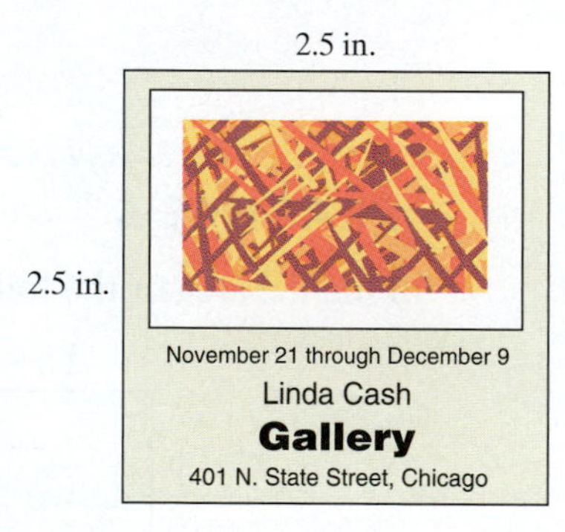

Solution First, we need to find the area of the ad, which is square.

$$\begin{aligned} A &= s^2 \\ &= (2.5)^2 \\ &= (2.5)(2.5) \\ &= 6.25, \text{ or } 6.25 \text{ square inches} \end{aligned}$$

To find the cost of the ad, we multiply 6.25 by 1,000, getting 6,250. The cost of the ad is $6,250. So the cost is greater than $6,000.

PRACTICE 7

In a flooring store, a customer wants to buy tile for a 9 foot × 12 foot room using 1 ft^2 tiles that sell for $4.99 apiece. Will $500 be enough to pay for the tiles?

EXAMPLE 8

In a certain town, only students living outside a 2-mile radius of their school must pay a fee for bus transportation. To the nearest square mile, what is the area of the region in which students do not pay a fee for bus transportation?

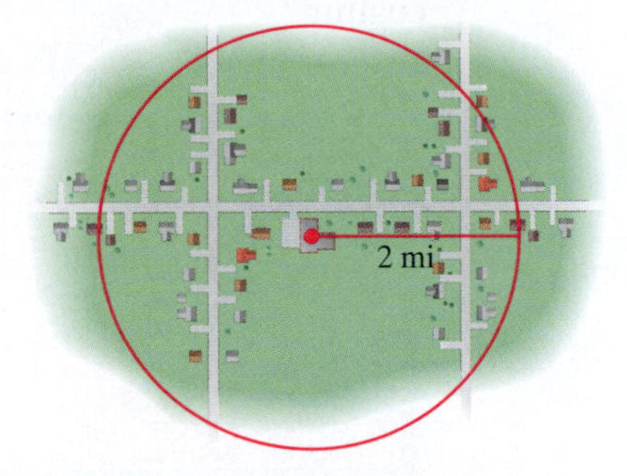

 We need to find the area of the region, which is a circle.

$$\begin{aligned} A &= \pi r^2 \\ &\approx \pi(2)^2 \\ &\approx 3.14\,(4) \\ &\approx 12.56 \end{aligned}$$

So the area of the region is approximately 13 square miles.

PRACTICE 8

The following diagram shows the region in which the beam of light from a lighthouse can be seen in any direction in the fog. What is the area of this region?

Composite Figures

Recall that a composite figure comprises two or more simple figures. Let's consider finding areas of such figures.

EXAMPLE 9

Find the area of the shaded portion of the figure.

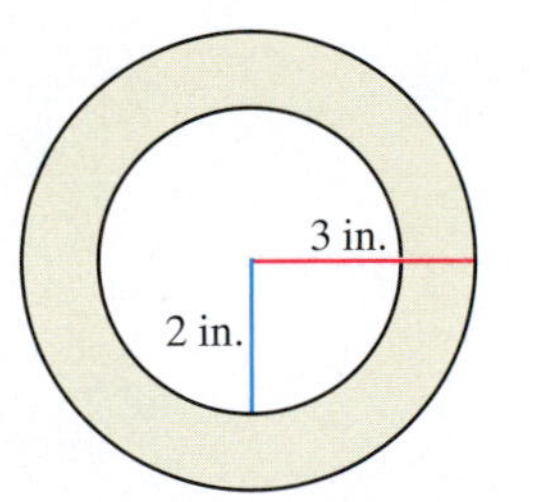

Solution To find the area of the shaded portion, we subtract the area of the small (inner) circle from the area of the large (outer) circle.

$$\begin{aligned} \text{Shaded Area} &= \textbf{Area of large circle} - \textbf{Area of small circle} \\ &\approx 3.14\,(3)^2 - 3.14\,(2)^2 \\ &\approx 3.14\,(9) - 3.14\,(4) \\ &\approx 28.26 - 12.56 \\ &\approx 15.70 \end{aligned}$$

So the area of the shaded figure is approximately 15.7 square inches. How could the distributive property be used to solve this problem?

PRACTICE 9

Find the shaded area.

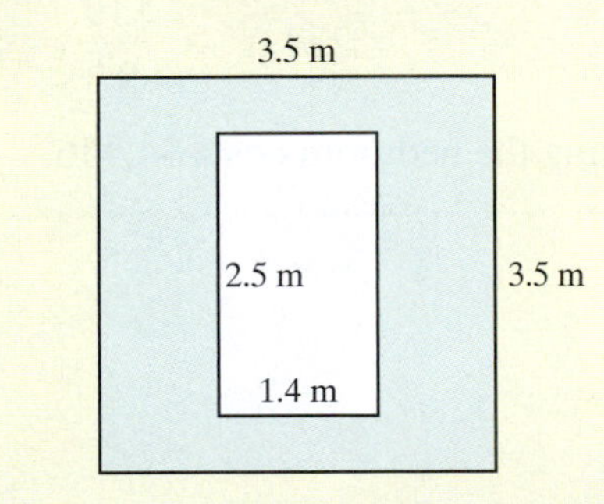

EXAMPLE 10

At \$19 per square foot, how much will it cost to carpet the bedroom pictured?

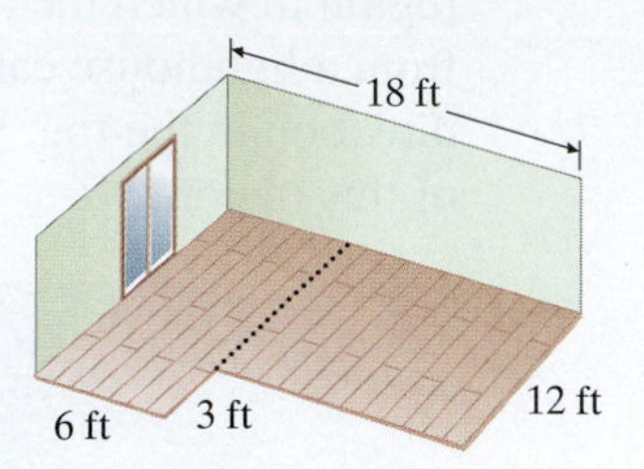

Solution First, we must find the area of the room. Note that the room consists of a 15 foot × 6 foot rectangle, and a square 12 feet on a side.

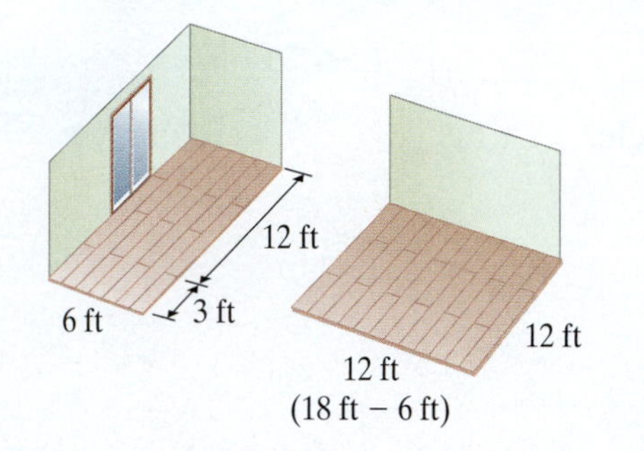

$$\begin{aligned}\text{Total area} &= \text{Area of rectangle} + \text{Area of square}\\ &= l \cdot w + s^2\\ &= 15 \cdot 6 + (12)^2\\ &= 90 + 144\\ &= 234, \text{ or } 234 \text{ square feet}\end{aligned}$$

The total area of the room is 234 square feet.

Since the carpet costs \$19 per square foot, we calculate the total cost as follows:

$$234 \text{ ft}^2 \times \frac{\$19}{\text{ft}^2} = \$4{,}446$$

So carpeting the bedroom costs \$4,446.

PRACTICE 10

A coating of polyurethane is applied to the central circle on the gymnasium floor shown below. What is the area of the part of the floor that still needs coating?

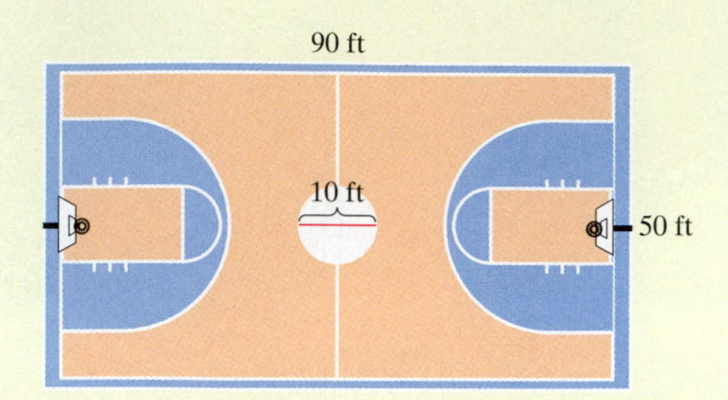

11.3 Exercises

FOR EXTRA HELP MyMathLab Math XL PRACTICE WATCH READ REVIEW

Mathematically Speaking

Fill in each blank with the most appropriate term or phrase from the given list.

trapezoid	volume	square meters
circle	meters	triangle
square	area	

1. The number of square units that a figure contains is called its __________.

2. Areas are measured in square units, such as __________.

3. The area of a(n) __________ is equal to one-half the product of the base and the height.

4. The area of a(n) __________ is equal to the square of a side.

5. A formula for the area of a(n) __________ is $A = \pi r^2$.

6. The formula $A = \frac{1}{2}h(b + B)$ is used to find the area of a(n) __________.

A *Find the area of each figure. Use* $\pi \approx 3.14$*, when needed.*

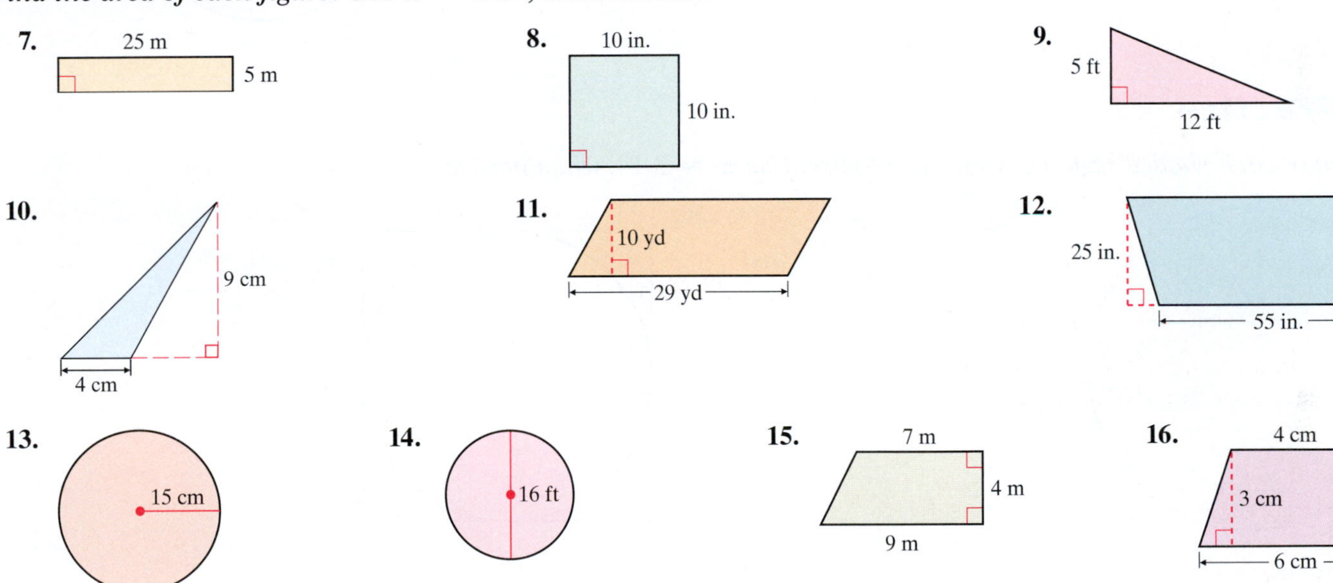

17. A parallelogram with base 4 meters and height 3.9 meters

18. A parallelogram with base 6.5 inches and height 4 inches

19. A circle with diameter 20 inches

20. A circle with radius 100 feet

21. A triangle with height 2.5 feet and base 5 feet

22. A triangle with base 8 inches and height $6\frac{1}{2}$ inches

23. A trapezoid with height 4.2 yards and bases 7 yards and 14 yards

24. A trapezoid with height 3.5 meters and bases 4 meters and 6.5 meters

25. A rectangle with length 2.6 meters and width 1.4 meters

26. A rectangle with length $\frac{1}{2}$ foot and width $\frac{2}{3}$ foot

27. A square with side $\frac{1}{4}$ yard long

28. A square with side 15.5 centimeters long

B *Find the shaded area.*

29. **30.** **31.** **32.** **33.** **34.**

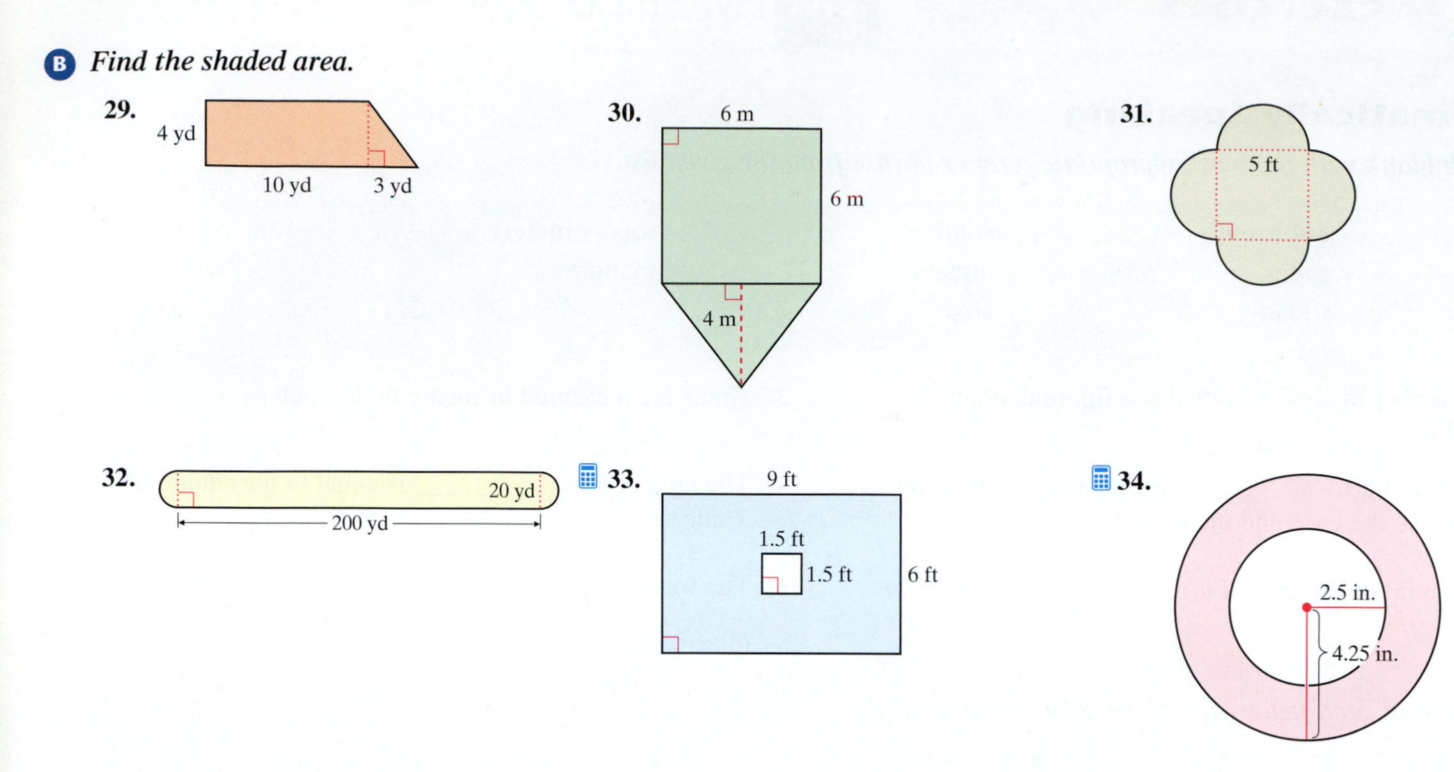

Mixed Practice

Find the area of each shaded region or described figure. Use $\pi \approx 3.14$, *when needed.*

35. **36.** **37.** **38.**

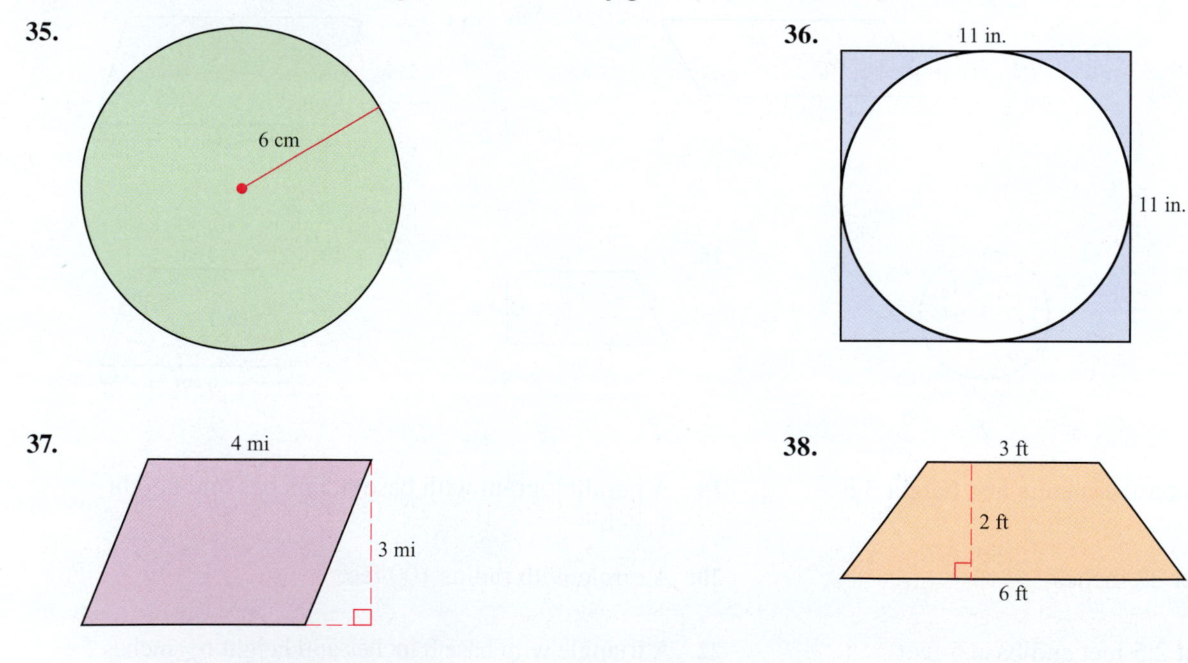

39. A right triangle with legs 6.5 meters and 10 meters

40. A parallelogram with base 8.5 yards and height 7 yards

Applications

C *Solve. Use $\pi \approx 3.14$, when needed.*

41. The boxing ring below is a square that measures 18 feet on a side inside the ropes. The area outside the ropes, called the apron, extends 2 feet beyond the ropes on each side. What is the area of the apron?

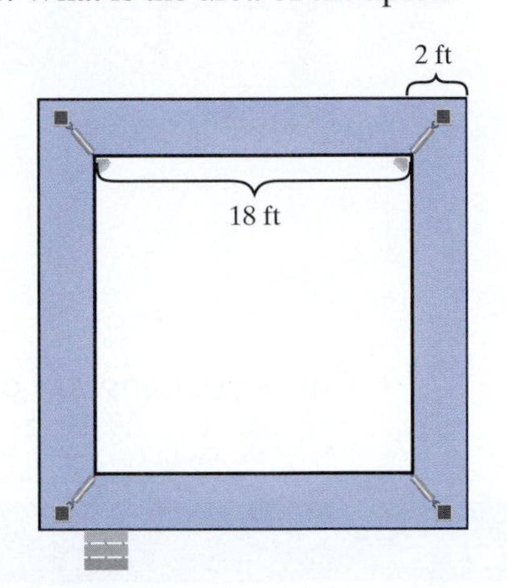

42. The base of the United Nations Secretariat building is a rectangle with length 88 meters and width 22 meters. The Empire State Building has a rectangular base measuring 129 meters by 57 meters. What is the difference in the area between the two bases? (*Sources:* docomomo-us.org and newyorktransportation.com)

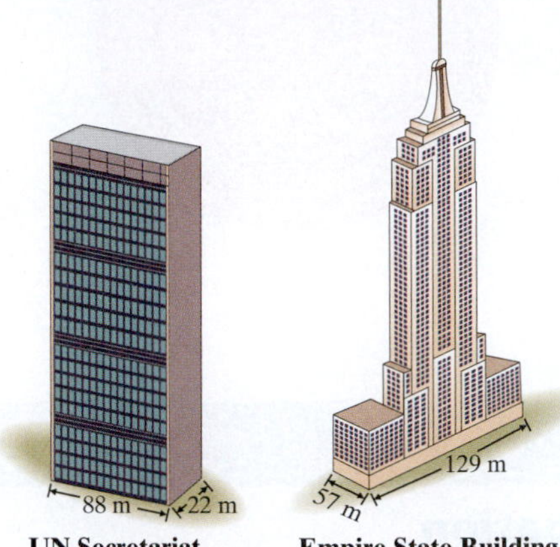

43. A microscope allows a scientist to see a circular region that is 0.25 millimeters in diameter. What is the area of this region?

44. An air-traffic control tower can identify an airplane within 10 miles of the tower in any direction. What area does the tower cover?

45. Suppose that an L-shaped house is located on the rectangular lot shown. How much yard space is there?

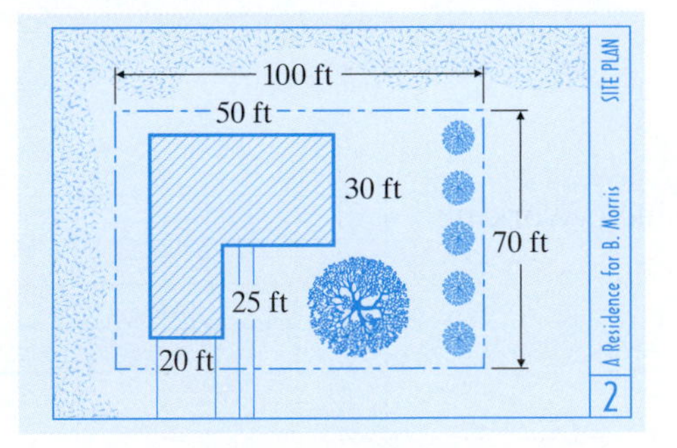

46. A walkway 2 yards wide, shown below, is built around the entire building below. Find the area of the walkway.

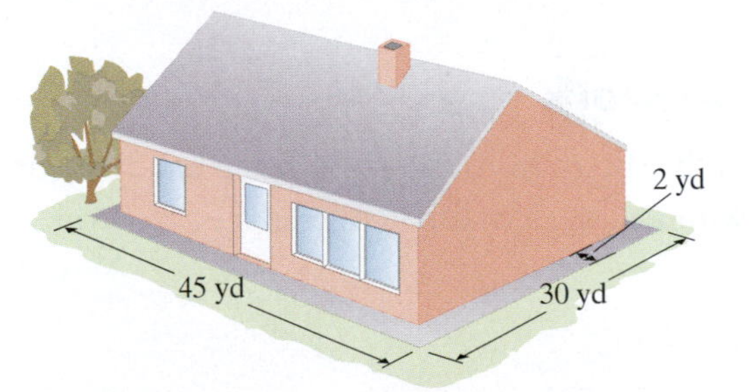

47. An online furniture company is advertising two sizes of a trapezoid-shaped computer workstation. For the smaller workstation, the two parallel sides are 24 inches and 48 inches, and the depth is 24 inches. The larger has parallel sides of length 30 inches and 60 inches and a depth of 30 inches. What is the difference between the area of the two workstations? (*Source:* csnlibraryfurniture.com)

48. The John Hancock Tower, the tallest skyscraper in New England, has a floor plan shaped like a parallelogram. If the parallelogram's base is 293 feet and its height is 107 feet, find the area of the floor plan. (*Source:* pcfandp.com)

49. A circular rug comes in two sizes. In one, the diameter is 41 inches, and in the other only 32 inches. If the smaller rug sells for \$79.95 and the prices are proportional to the areas, find the selling price of the larger rug, rounded to the nearest cent.

50. Consider the two tables pictured. How much larger in area to the nearest square meter is the semicircular table than the rectangular table?

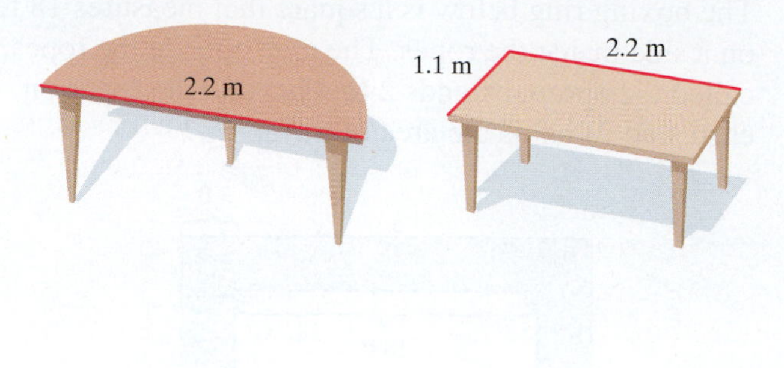

• Check your answers on page A-14.

MINDStretchers

Investigation

1. • Draw a square.
   • Measure its side lengths.
   • Find its area.
   • Double the side length of the square.
   • Find the area of the new square.
   • Start with another square and repeat this process several times.
   • How does doubling the side length of a square affect its area?

Groupwork

2. In the diagram to the right, each small square represents 1 square inch. Working with a partner, estimate the area of the oval.

Mathematical Reasoning

3. In the diagram below, $\overline{AC}$ is parallel to $\overline{ED}$.

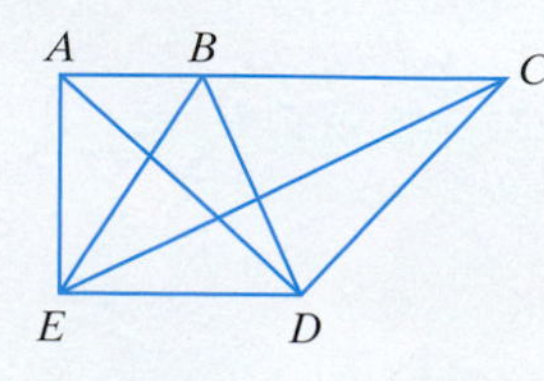

What relationship do you see between the areas of ΔEAD, ΔEBD, and ΔECD? Justify your answer.

11.4 Volume

OBJECTIVES

- **A** To find the volume of a geometric solid
- **B** To find the volume of a composite geometric solid
- **C** To solve applied problems involving volume

The Volume of a Geometric Solid

Volume is a measure of the amount of space inside a three-dimensional figure. The amount of water in an aquarium, the amount of juice in a can, or the amount of grain in a bin are all examples of volumes.

To find the volume of the box shown, we can split it into little 1-inch by 1-inch by 1-inch cubes, each representing 1 cubic inch. Then, we count the number of cubic inches within the box, which is 24 cubic inches.

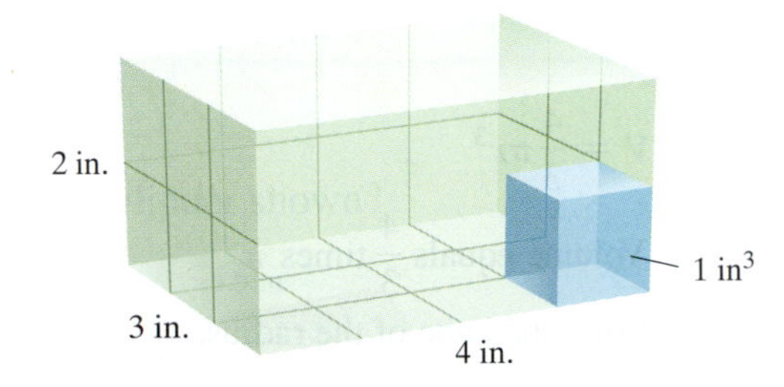

A shortcut to counting the total number of cubic inches is to multiply the length, the width, and the height: $4 \times 3 \times 2$, getting 24, or 24 cubic inches. Note that volumes are measured in cubic units, such as cubic inches (cu in. or in^3), cubic miles (cu mi or mi^3), or cubic meters (cu m or m^3).

DEFINITION
Volume is the number of cubic units required to fill a three-dimensional figure.

In this section, we consider basic three-dimensional objects and find their volume by using the following formulas:

Definition	Formula	Example
A **rectangular solid** is a solid in which all six faces are rectangles.	$V = lwh$ Volume equals length times width times height.	$V = lwh$ $= 6 \cdot 4 \cdot 3$ $= 72$, or 72 ft^3
A **cube** is a solid in which all six faces are squares.	$V = e^3$ Volume equals the cube of the edge.	$V = e^3$ $= (3)^3$ $= 27$, or 27 cm^3

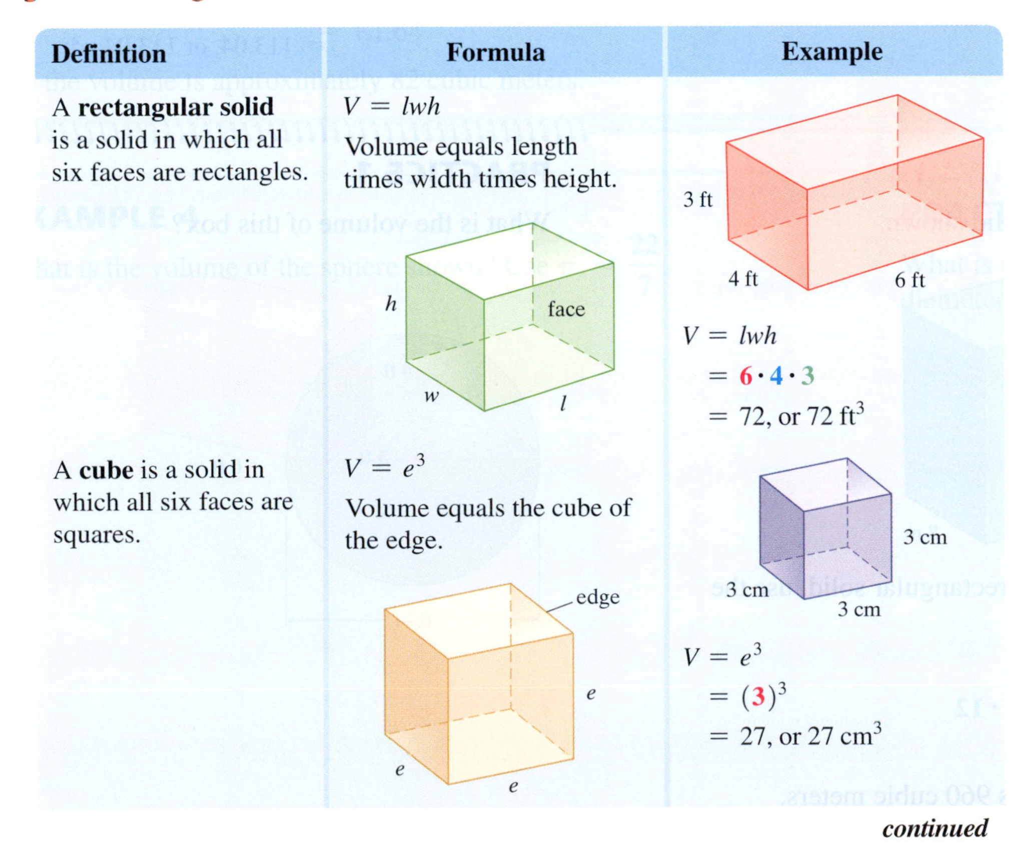

continued

11.5 Similar Triangles

OBJECTIVES

- **A** To find the missing sides of similar triangles
- **B** To solve applied problems involving similar triangles

Identifying Corresponding Sides of Similar Triangles

Some figures have the same shape but different size. For example, when a photograph is enlarged, everything in the enlargement is the same shape as in the original—only larger. In this section, we focus on triangles that have this relationship, which are called *similar triangles.*

DEFINITION

Similar triangles are triangles that have the same shape but not necessarily the same size.

When two triangles are similar, for each angle of the first triangle there corresponds an angle of the second triangle with the same measure. The sides opposite these *corresponding angles* are called *corresponding sides.*

In similar triangles, the measures of corresponding angles are equal and corresponding sides are in proportion. For example, the following triangles *ABC* and *DEF* are similar:

Since these triangles are similar, the measures of their corresponding angles are equal. So we write:

$$m\angle A = m\angle D$$
$$m\angle B = m\angle E$$
$$m\angle C = m\angle F$$

Also the lengths of the corresponding sides are in proportion, that is:

$$\frac{AB}{DE} = \frac{BC}{EF} = \frac{AC}{DF}$$
$$\frac{6}{3} = \frac{8}{4} = \frac{4}{2} = \frac{2}{1}$$

The ratio of the corresponding sides is $\frac{2}{1}$.

When we write that two triangles are similar, we name them so that the order of corresponding angles in both triangles is the same. In this case,

$$\Delta ABC \sim \Delta DEF$$

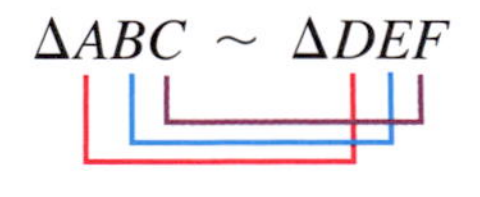

where the symbol "$\sim$" means "is similar to."

EXAMPLE 1

$\Delta RST \sim \Delta XYZ$. Name the corresponding sides of these triangles.

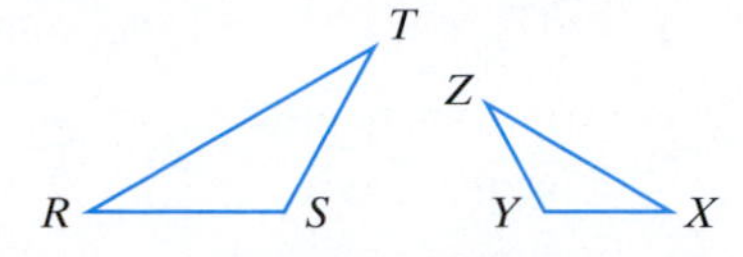

Solution Because $\Delta RST \sim \Delta XYZ$, $m\angle R = m\angle X$, $m\angle S = m\angle Y$, and $m\angle T = m\angle Z$. We know that the corresponding sides are opposite the angles with equal measure. So we write the following:

Because $m\angle R = m\angle X$, $\overline{ST}$ corresponds to $\overline{YZ}$. — $\overline{ST}$ is opposite $\angle R$, and $\overline{YZ}$ is opposite $\angle X$.

Because $m\angle S = m\angle Y$, $\overline{RT}$ corresponds to $\overline{XZ}$. — $\overline{RT}$ is opposite $\angle S$, and $\overline{XZ}$ is opposite $\angle Y$.

Because $m\angle T = m\angle Z$, $\overline{RS}$ corresponds to $\overline{XY}$. — $\overline{RS}$ is opposite $\angle T$, and $\overline{XY}$ is opposite $\angle Z$.

PRACTICE 1

$\Delta ABC \sim \Delta GHI$. List the corresponding sides of these triangles.

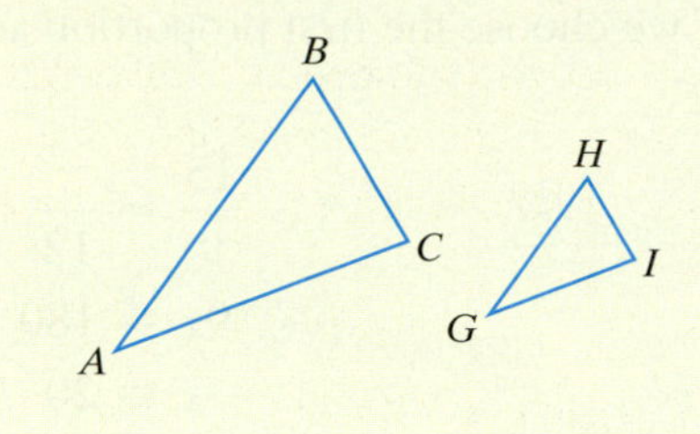

Finding the Missing Sides of Similar Triangles

Since corresponding sides in similar triangles are in proportion, we can use these proportions to find the length of a missing side.

To Find a Missing Side of a Similar Triangle

- Write the ratios of the lengths of the corresponding sides.
- Write a proportion using a ratio with known terms and a ratio with an unknown term.
- Solve the proportion for the unknown term.

EXAMPLE 2

In the following diagram, $\Delta TAP \sim \Delta RUN$. Find x.

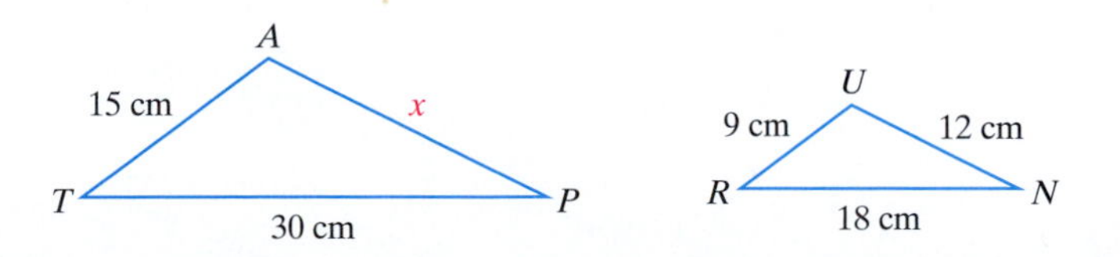

Solution Because $\Delta TAP \sim \Delta RUN$, we write the ratios of the lengths of the corresponding sides as follows:

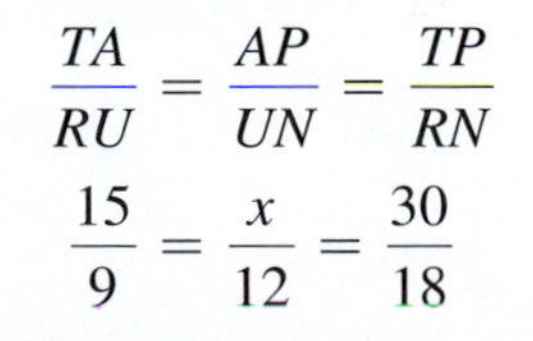

$$\frac{TA}{RU} = \frac{AP}{UN} = \frac{TP}{RN}$$

$$\frac{15}{9} = \frac{x}{12} = \frac{30}{18}$$

PRACTICE 2

In the following diagram, $\Delta DOT \sim \Delta PAN$. Find y.

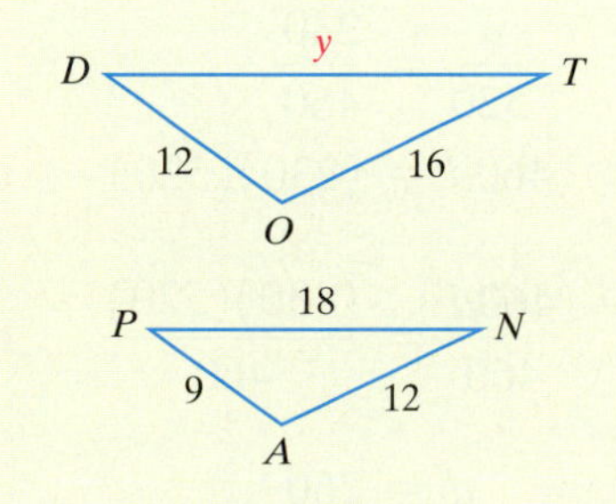

EXAMPLE 2 (continued)

To solve for x, we can consider either the proportion $\frac{15}{9} = \frac{x}{12}$ or the proportion $\frac{x}{12} = \frac{30}{18}$. Note that each proportion contains the unknown term x. If we choose the first proportion and solve for x, we get the following:

$$\frac{15}{9} = \frac{x}{12}$$
$$9x = 180$$
$$x = 20$$

So x is 20 centimeters.

Similar triangles are useful in finding lengths that cannot be measured directly, as Example 3 illustrates.

EXAMPLE 3

A surveyor took the measurements shown. If $\Delta ABC \sim \Delta EFC$, find d, the distance across the river.

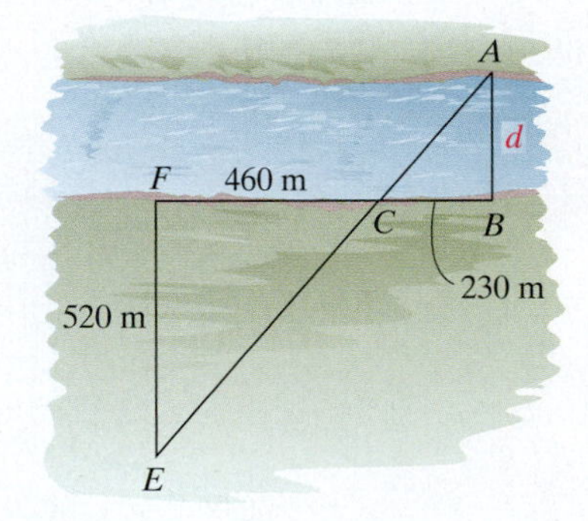

Solution Here, $\Delta ABC \sim \Delta EFC$, so we write the proportion $\frac{AB}{EF} = \frac{BC}{FC}$.

Then, we substitute the values given in the diagram.

$$\frac{d}{520} = \frac{230}{460}$$
$$460d = (230)(520) \quad \text{Cross multiply.}$$
$$\frac{\overset{1}{\cancel{460}}d}{\underset{1}{\cancel{460}}} = \frac{(\overset{1}{\cancel{230}})(520)}{\underset{2}{\cancel{460}}} \quad \text{Divide both sides by 460.}$$
$$d = 260$$

So the distance across the river is 260 meters.

PRACTICE 3

The height of a man and his shadow form a triangle similar to that formed by a nearby tree and its shadow. What is the height of the tree?

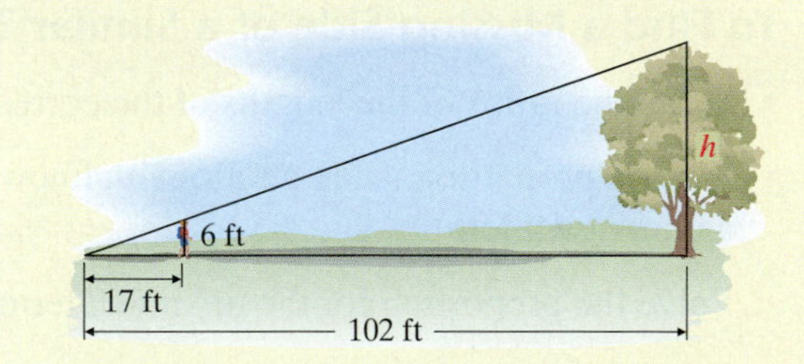

11.5 Exercises

FOR EXTRA HELP MyMathLab MathXL PRACTICE WATCH READ REVIEW

Mathematically Speaking

Fill in each blank with the most appropriate term or phrase from the given list.

equal	shape	similar
area	in proportion	corresponding

1. The symbol ~ is used to indicate that triangles are ____________.

2. Similar triangles have the same ____________ but not necessarily the same size.

3. In similar triangles, ____________ sides are opposite angles with equal measure.

4. Corresponding sides of similar triangles are ____________.

A *Find the value of each unknown.*

5. $\triangle DEF \sim \triangle ABC$

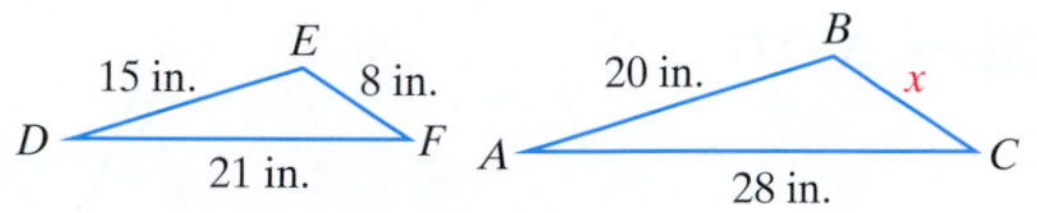

6. $\triangle LOM \sim \triangle RST$

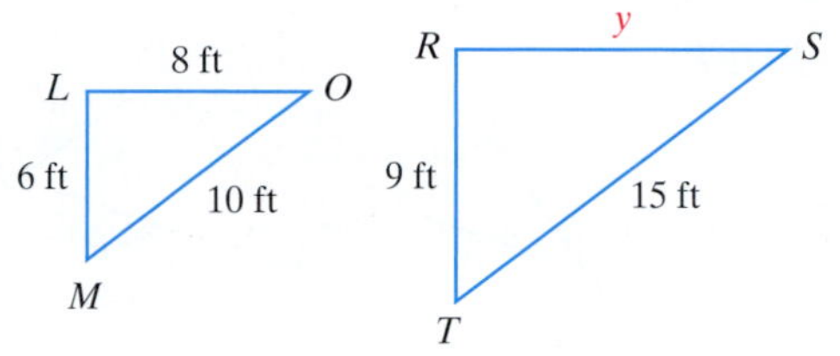

7. $\triangle DOT \sim \triangle PAN$

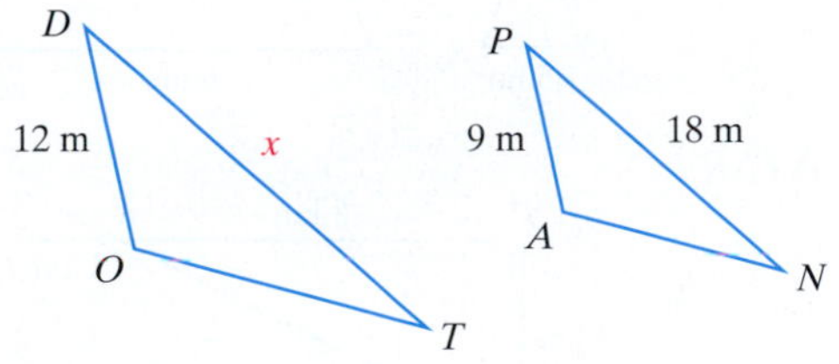

8. $\triangle ACT \sim \triangle MLK$

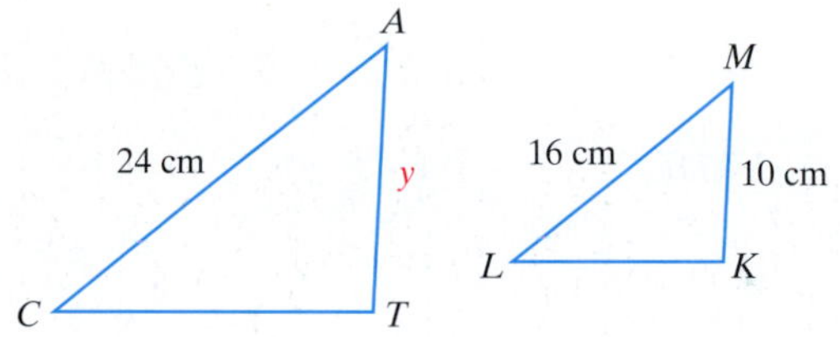

9. $\triangle DEF \sim \triangle ABC$

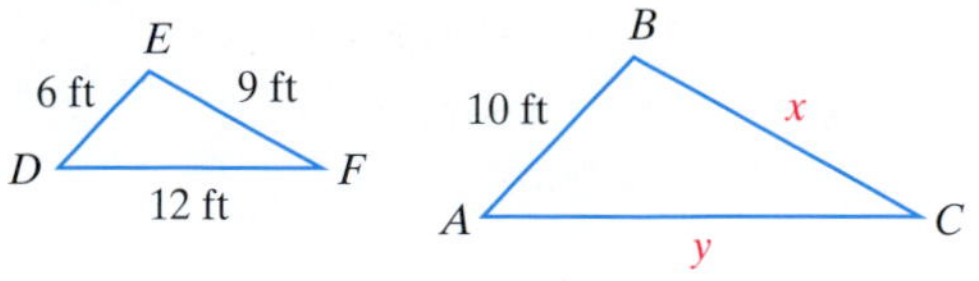

10. $\triangle DOT \sim \triangle PIN$

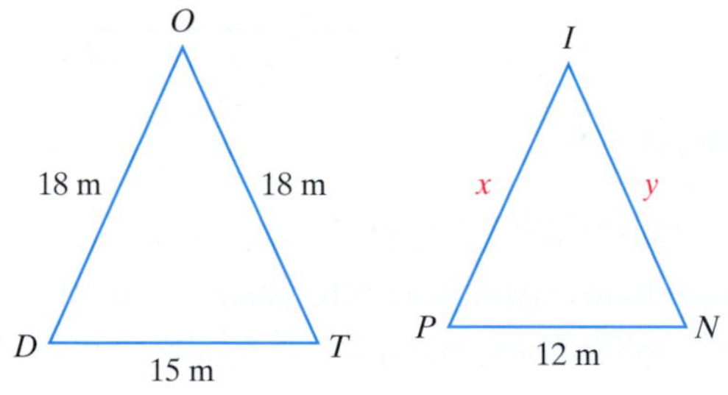

11. $\triangle TAP \sim \triangle RON$

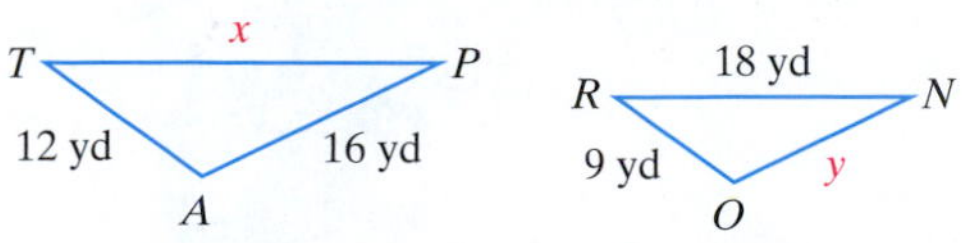

12. $\triangle FEG \sim \triangle CBD$

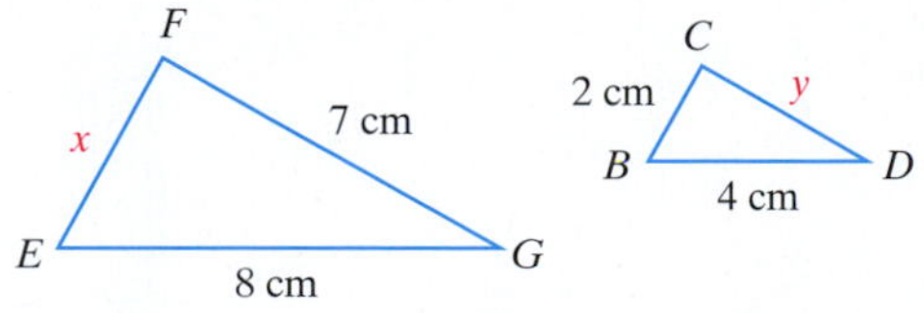

13. $\triangle ABC \sim \triangle DEC$

14. $\triangle DEF \sim \triangle DGH$

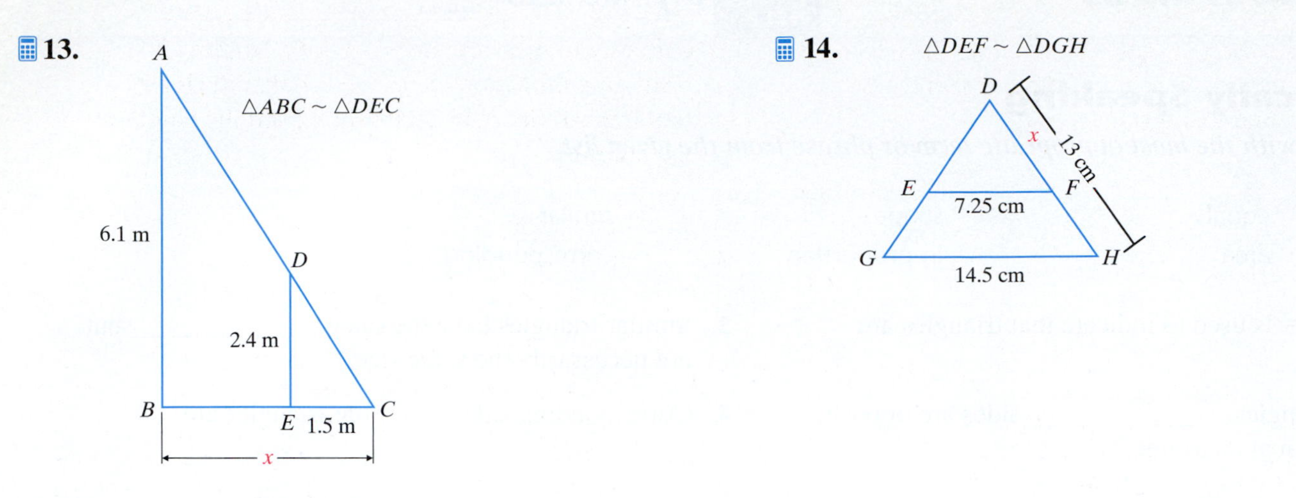

Mixed Practice

Find the value of each unknown.

15. $\Delta ABC \sim \Delta DEF$

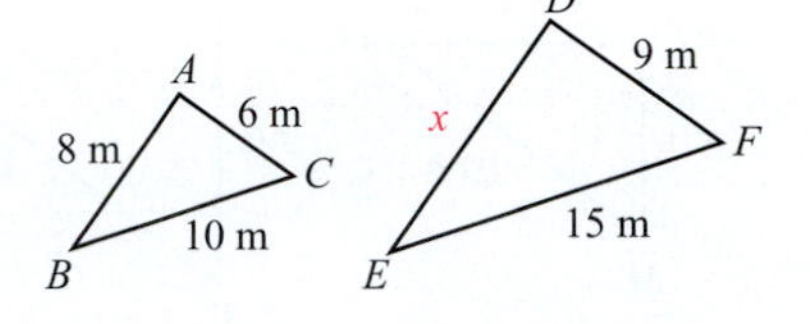

16. $\Delta PQR \sim \Delta STU$

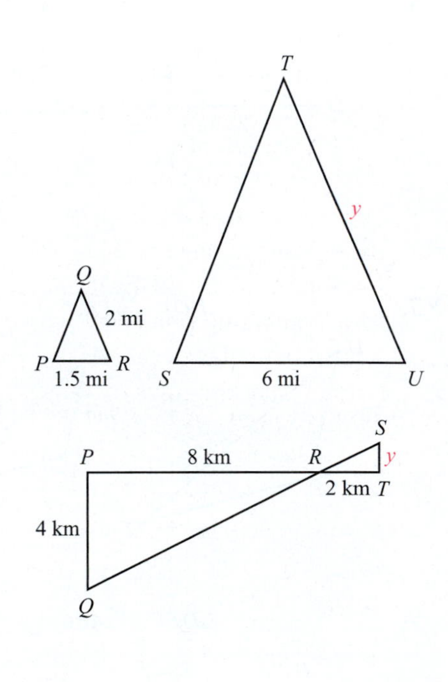

17. $\Delta ABC \sim \Delta ADE$

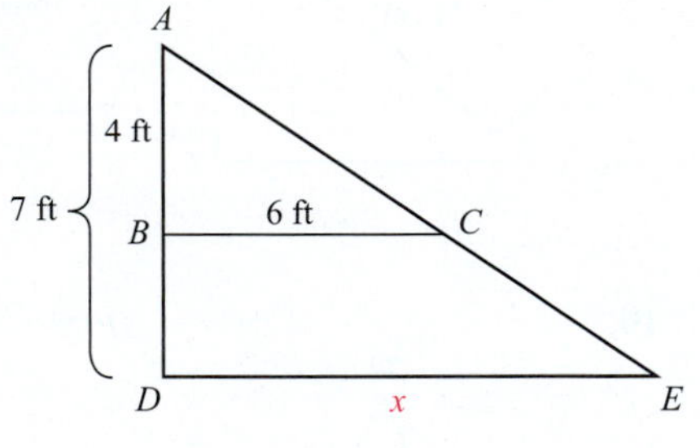

18. $\Delta PQR \sim \Delta TSR$

P
8 km
R
S
y
2 km T
4 km
Q

Applications

A *Solve. Assume the triangles are similar.*

19. The diagram below (not drawn to scale) shows the shadows cast by a column and a ruler. Find the height of the column.

1 ft
3 ft
27 ft

20. Light from a flashlight shines through a transparent dragon puppet onto a screen behind it, as shown below. Find the height of the puppet's image.

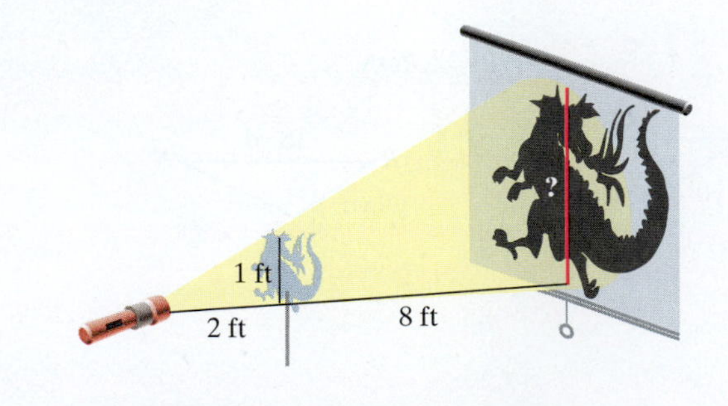

21. A Coast Guard observer sees a boat out on the ocean and wants to know how far it is from the shore. Use the diagram to find that distance.

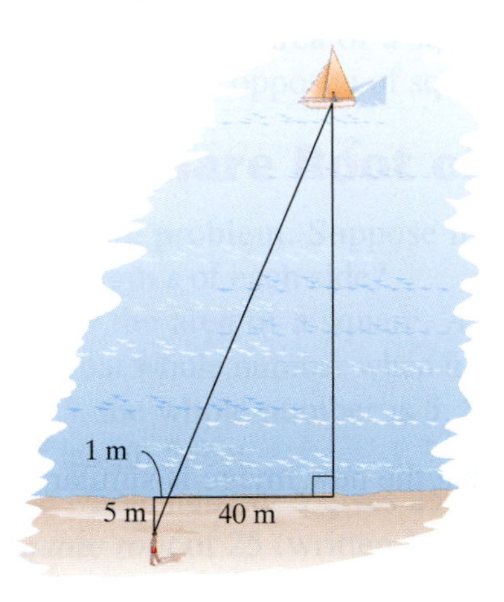

22. One way to measure the height of a building is to position a mirror on the ground so that the top of the building's reflection can be seen. Find the height of the building.

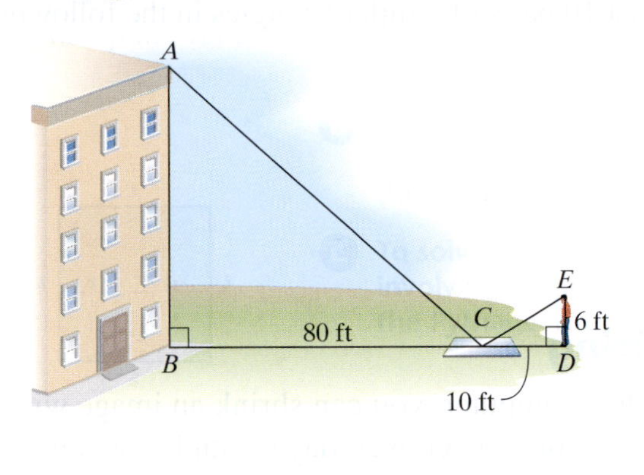

23. To make paintings appear to be three-dimensional, artists draw parallel lines such as railroad tracks getting closer to one another the further away they get. The "vanishing point" refers to the point in the distance where the lines meet, forming the vertex of a triangle. Find the missing dimension in the image at right. (*Source:* math.vcu.edu)

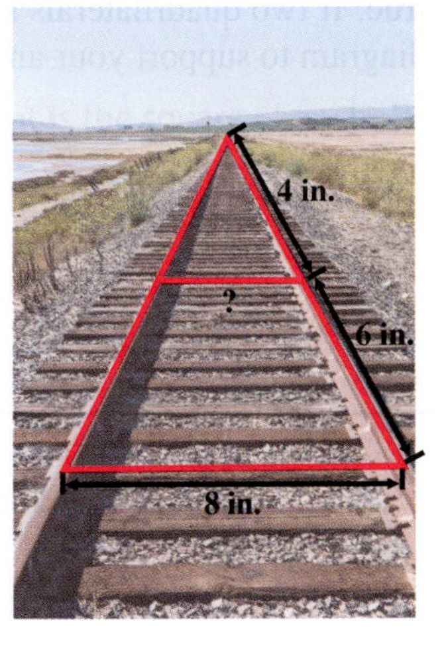

24. The diagram below shows the side view of an escalator, where ΔABC is similar to ΔCBD. Find the length of $\overline{CD}$.

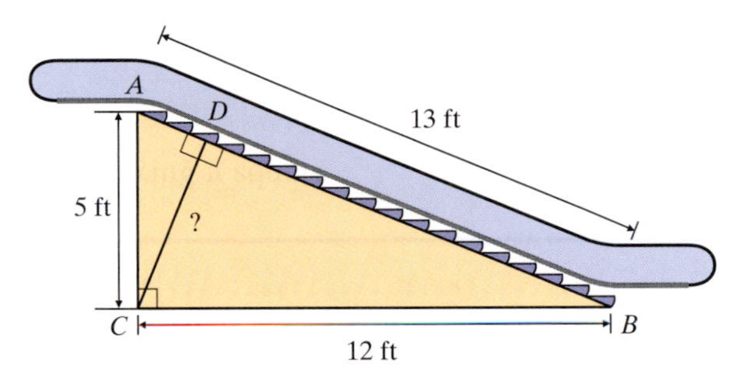

25. Two support wires are attached to a utility pole as shown in the diagram. If $\Delta ABC \sim \Delta ADE$, find AD.

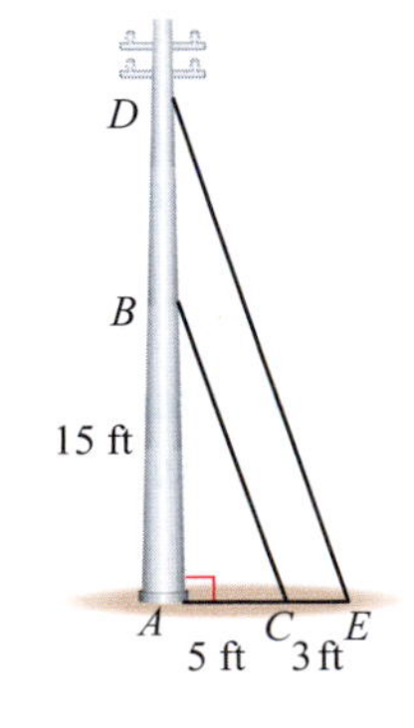

26. To measure $\overline{AB}$, the distance across a certain lake, the length of line segments $\overline{BC}$, $\overline{DC}$, and $\overline{ED}$ were "staked out," as shown in the diagram below. If $\Delta ABC \sim \Delta EDC$, how wide is the lake?

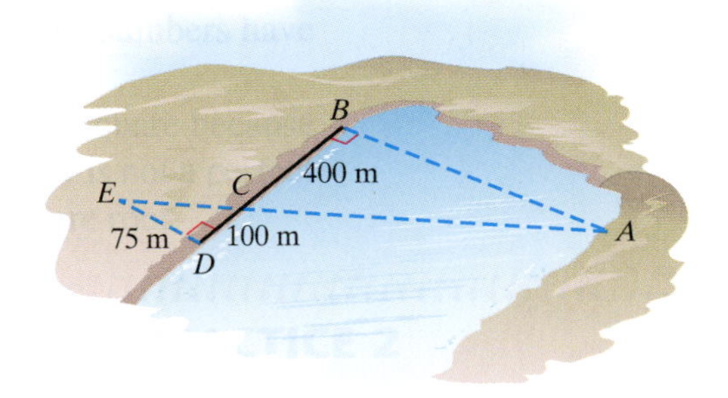

• Check your answers on page A-14.

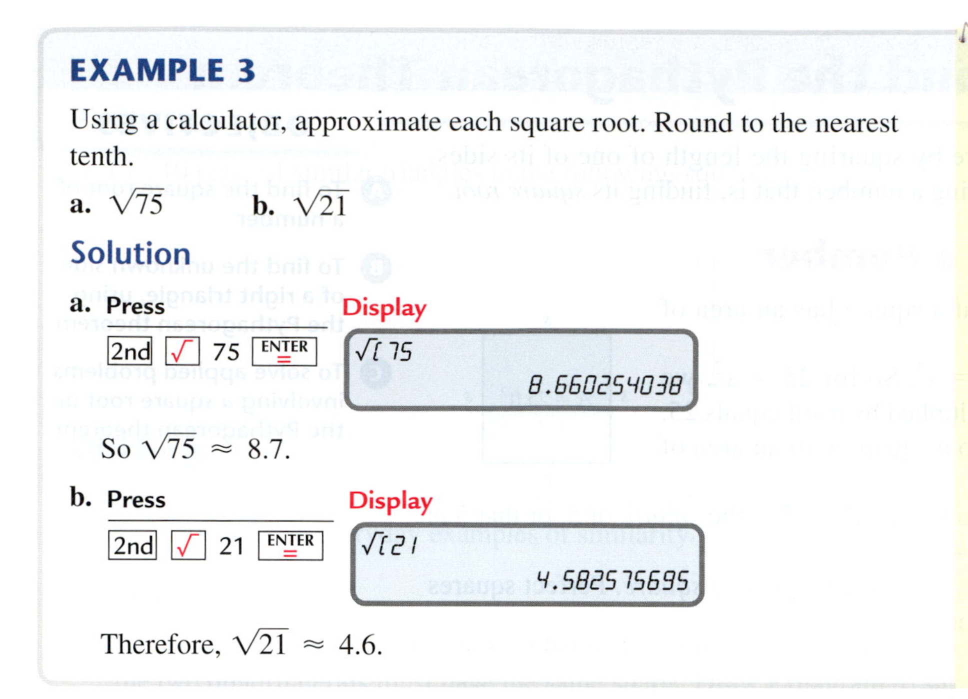

EXAMPLE 3

Using a calculator, approximate each square root. Round to the nearest tenth.

a. $\sqrt{75}$ **b.** $\sqrt{21}$

Solution

a. Press

2nd √ 75 ENTER

Display

So $\sqrt{75} \approx 8.7$.

b. Press

2nd √ 21 ENTER

Display

Therefore, $\sqrt{21} \approx 4.6$.

PRACTICE 3

Using a calculator, approximate each square root. Round to the nearest hundredth.

a. $\sqrt{56}$

b. $\sqrt{12}$

The Pythagorean Theorem

Recall that a right triangle is a triangle that has one 90° angle. In a right triangle, the side opposite the right angle is called the **hypotenuse**. The other two sides are called **legs**.

Hypotenuse

Leg

Leg

The lengths of the three sides of a right triangle are related in a special way. To understand this relationship, consider the areas of the squares on the legs and on the hypotenuse, as in the following example:

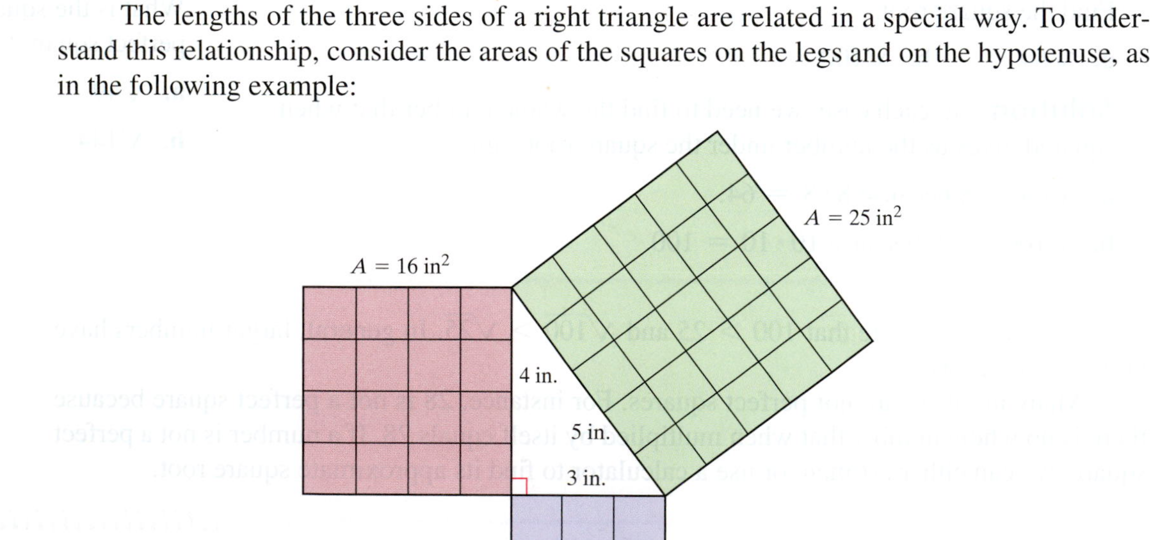

$$9 \text{ in}^2 + 16 \text{ in}^2 = 25 \text{ in}^2$$

Area of the square on one leg + Area of the square on the other leg = Area of the square on the hypotenuse

In general, if we let a and b represent the lengths of the legs and c represent the length of the hypotenuse, then $a^2 + b^2 = c^2$.

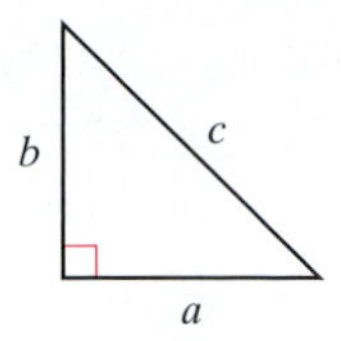

This relationship is called the *Pythagorean theorem.*

The Pythagorean Theorem

For every right triangle, the sum of the squares of the lengths of the two legs equals the square of the length of the hypotenuse, that is,

$$a^2 + b^2 = c^2$$

where a and b are the lengths of the legs, and c is the length of the hypotenuse.

We can use the Pythagorean theorem to find the third side of a right triangle if we know the other two sides.

EXAMPLE 4

Find the length of the hypotenuse.

Solution To find the length of the hypotenuse, we use the Pythagorean theorem.

$$a^2 + b^2 = c^2$$
$$5^2 + 12^2 = c^2$$
$$25 + 144 = c^2$$
$$169 = c^2$$
$$\sqrt{169} = c$$

Taking a square root is the opposite of squaring.

$$13 = c, \text{ or } c = 13$$

So the hypotenuse is 13 centimeters long.

PRACTICE 4

Find the length of the unknown side in ΔABC.

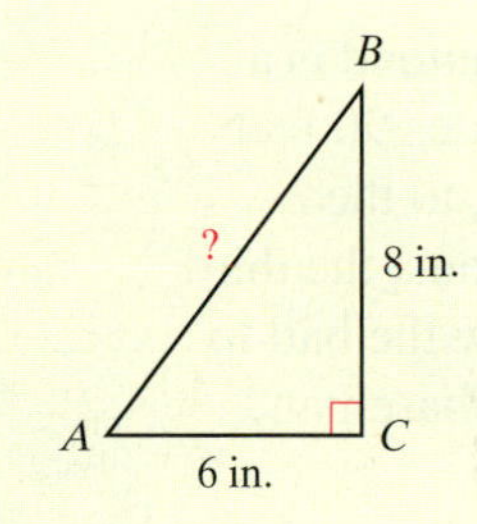

EXAMPLE 5

If b equals 1 centimeter and c equals 2 centimeters, what is a in a right triangle, where a and b are the lengths of the legs and c is the length of the hypotenuse? Round the answer to the nearest tenth of a centimeter.

PRACTICE 5

In a right triangle, one leg equals 2 feet and the hypotenuse equals 4 feet. Approximate the length of the missing leg. Round the answer to the nearest tenth of a foot.

EXAMPLE 5 (continued)

Solution First, we draw a diagram.

Then, we use the Pythagorean theorem and substitute the given values to obtain the following:

$$a^2 + b^2 = c^2$$
$$a^2 + 1^2 = 2^2$$
$$a^2 + 1 = 4$$
$$a^2 + 1 - 1 = 4 - 1$$
$$a^2 = 3$$
$$a = \sqrt{3}$$

Taking a square root is the opposite of squaring.

To express the answer as a decimal, we use a calculator. Rounding to the nearest tenth, we find that $\sqrt{3}$ is 1.7. So $a \approx 1.7$ centimeters.

EXAMPLE 6

A baseball diamond is a square with sides 90 feet long. How far, to the nearest foot, must the third baseman throw the ball to reach the first baseman?

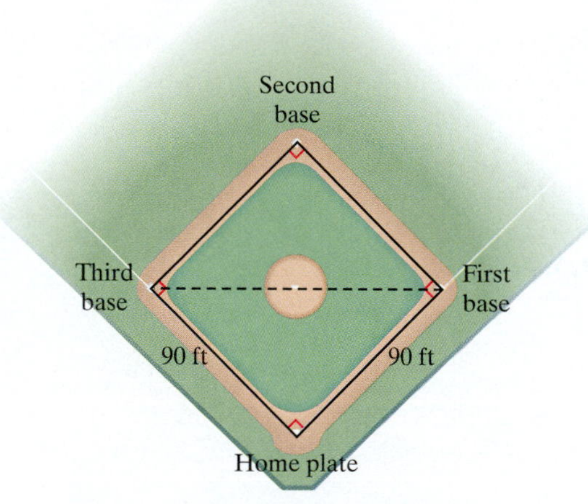

Solution The sides of the diamond together with the diagonal from third base to first base form a right triangle. To find the distance from third base to first base, we use the Pythagorean theorem.

$$a^2 + b^2 = c^2$$
$$90^2 + 90^2 = c^2$$
$$8{,}100 + 8{,}100 = c^2$$
$$16{,}200 = c^2$$
$$\sqrt{16{,}200} = c$$
$$c = \sqrt{16{,}200}$$

We use a calculator and round to find that $\sqrt{16{,}200}$ is approximately 127 feet. So the third baseman must throw the ball approximately 127 feet to reach the first baseman.

PRACTICE 6

Stair stringers, the structural supporting parts of staircases, are used by carpenters in building stairs. What is the length, to the nearest tenth of a foot, of the stair stringer shown in the diagram below? (*Source:* wikipedia.org)

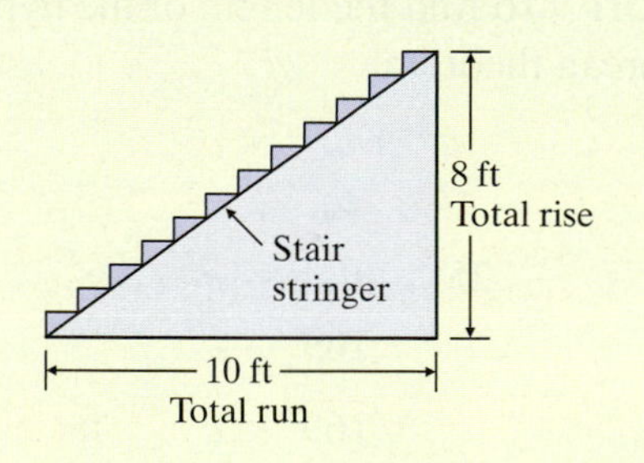

11.6 Exercises

FOR EXTRA HELP MyMathLab MathXL PRACTICE WATCH READ REVIEW

Mathematically Speaking

Fill in each blank with the most appropriate term or phrase from the given list.

squaring	leg	hypotenuse
multiple	Area of three squares	prime
consecutive		perfect square
doubling	Pythagorean theorem	square root

1. The number 5 is the __________ of 25.

2. Finding a square root is the opposite of __________ the number.

3. The square of a whole number is said to be a(n) ______________.

4. The whole numbers 5 and 6 are ______________.

5. In a right triangle, the longest side is called the __________.

6. If a and b are the lengths of the legs of a right triangle and c is the length of the hypotenuse, then the ________________ states that $a^2 + b^2 = c^2$.

A *Find each square root.*

7. $\sqrt{9}$ **8.** $\sqrt{4}$ **9.** $\sqrt{16}$ **10.** $\sqrt{36}$

11. $\sqrt{81}$ **12.** $\sqrt{64}$ **13.** $\sqrt{169}$ **14.** $\sqrt{121}$

15. $\sqrt{400}$ **16.** $\sqrt{225}$ **17.** $\sqrt{256}$ **18.** $\sqrt{900}$

Determine between which two consecutive whole numbers each square root lies.

19. $\sqrt{50}$ **20.** $\sqrt{7}$ **21.** $\sqrt{80}$ **22.** $\sqrt{31}$

23. $\sqrt{39}$ **24.** $\sqrt{2}$ **25.** $\sqrt{14}$ **26.** $\sqrt{105}$

Approximate each square root. Round to the nearest tenth, if needed.

27. $\sqrt{5}$ **28.** $\sqrt{11}$ **29.** $\sqrt{37}$ **30.** $\sqrt{74}$

31. $\sqrt{139}$ **32.** $\sqrt{165}$ **33.** $\sqrt{9,801}$ **34.** $\sqrt{8,649}$

B *Find each missing length. Round to the nearest tenth, if needed.*

35.

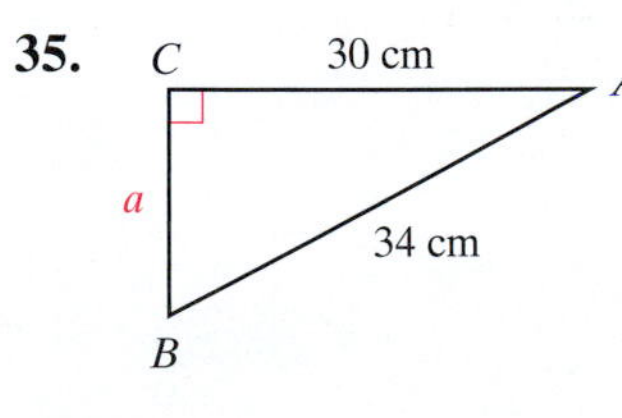

36.

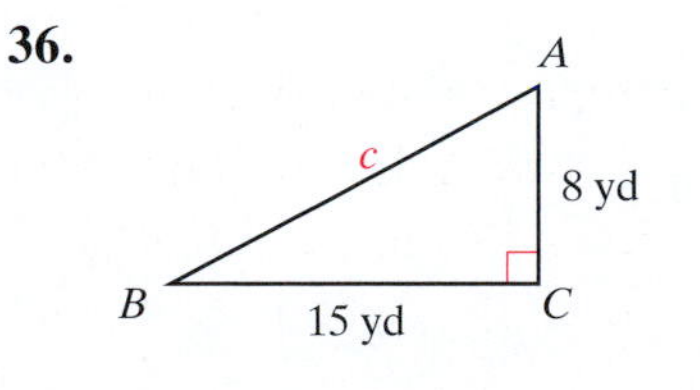

37.

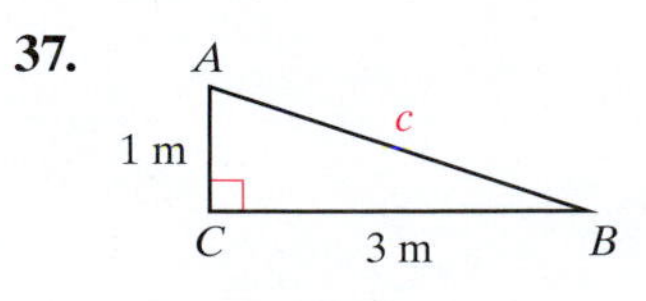

38.

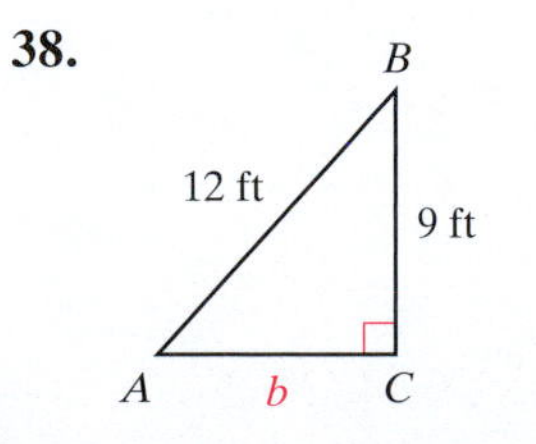

Given a right triangle with legs a and b, and hypotenuse c, find the missing side. Round to the nearest tenth, if needed.

	a	b	c
39.	24 m		25 m
40.	5 in.	12 in.	
41.	6 ft		10 ft
42.		4 cm	5 cm
43.	12 m	16 m	
44.		9 in.	15 in.
45.	7 cm	9 cm	
46.	2 yd	5 yd	
47.		18 ft	20 ft
48.	2 in.	2 in.	

Mixed Practice

Solve.

49. Find $\sqrt{196}$.

50. Determine between which two consecutive whole numbers $\sqrt{95}$ lies.

51. Find the missing length.

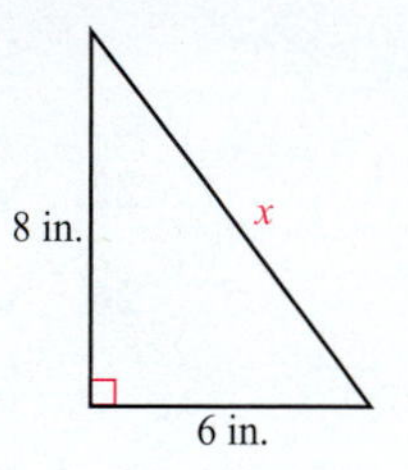

52. Find the missing length.

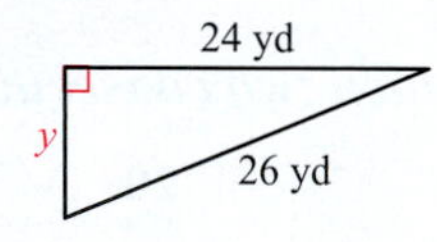

53. Find the missing length. Round to the nearest tenth.

n
10 cm
2 cm

54. Find $\sqrt{41}$ to the nearest tenth.

Applications

C *Solve. Use a calculator, if needed.*

55. A contractor leans a ladder against the side of a building. How high up the building does the ladder reach?

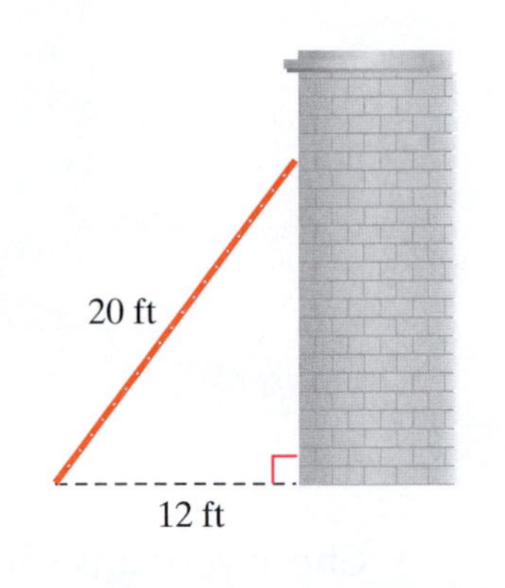

56. A scuba diver swims away from the boat and then dives, as shown. How far from the boat, to the nearest foot, will he be?

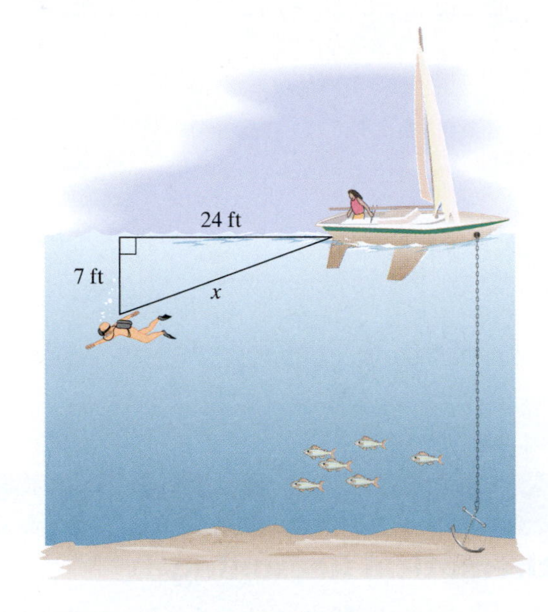

57. What is the length of the rectangular plot of land shown?

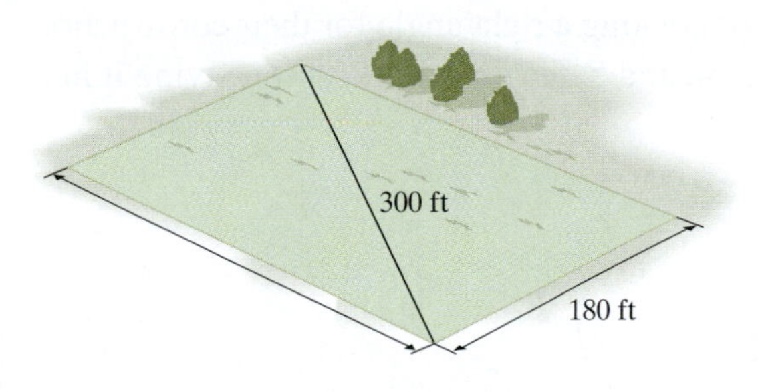

58. *ABCE* is a rectangular picnic area, with a picnic table at point *B* and the entrance at point *D*. The lengths *BC*, *CD*, and *BE* are as shown below:

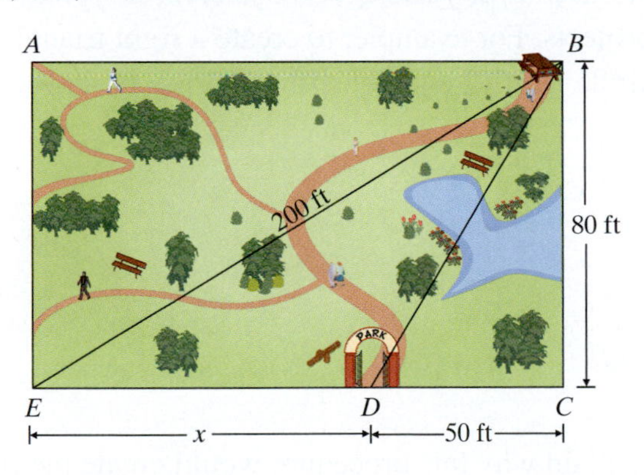

Find the distance between the entrance and point *E*.

59. A builder constructed a roof of wooden beams. According to the diagram, what is the length of the sloping beam?

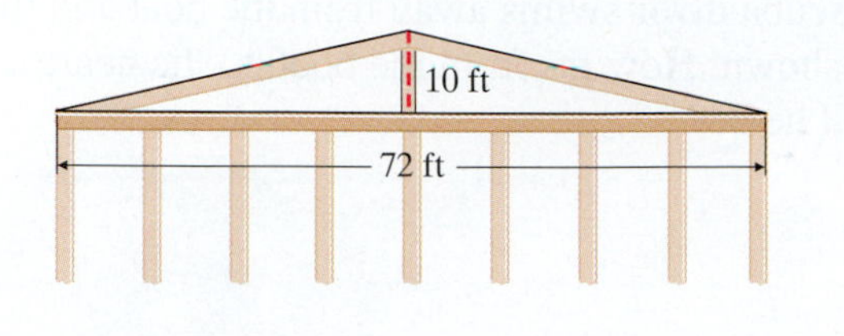

60. A college is constructing an access ramp to a door in one of its buildings, as shown. Find the length of the ramp.

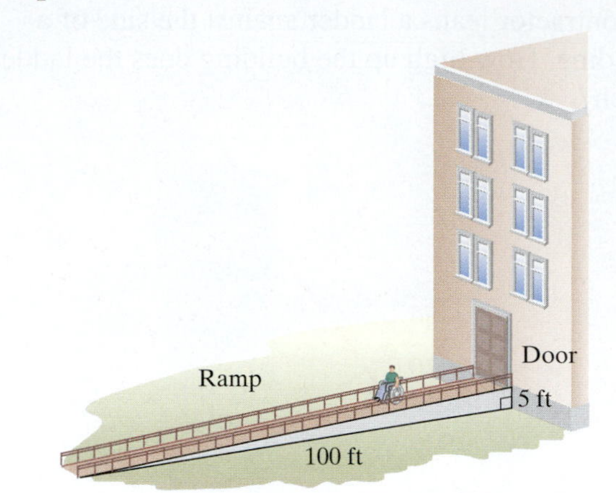

• Check your answers on page A-14.

MINDStretchers

Mathematical Reasoning

1. Give an example of a number that is smaller than its square root.

Writing

2. Thousands of years ago, the ancient Egyptians used a clever way of creating a right angle for their construction projects. For example, to create a right triangle with side lengths 3, 4, and 5, they would use a rope tying it in a circle with 12 equally spaced knots, as shown:

Explain why this procedure would create the right triangle. (*Source:* Peter Tompkins, *Secrets of the Great Pyramids*)

Investigation

3. Choose a whole number. Use a calculator to determine whether it is a perfect square.

Key Concepts and Skills

CONCEPT SKILL

Concept/Skill	Description	Example
[11.1] Point	An exact location in space, with no dimension.	A
[11.1] Line	A collection of points along a straight path, that extends endlessly in both directions.	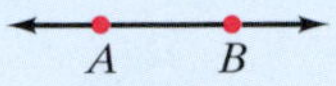 $\overleftrightarrow{AB}$
[11.1] Line segment	A part of a line having two endpoints.	 $\overline{BC}$
[11.1] Ray	A part of a line having only one endpoint.	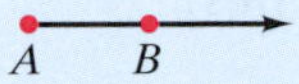 $\overrightarrow{AB}$
[11.1] Angle	Two rays that have a common endpoint called the *vertex* of the angle.	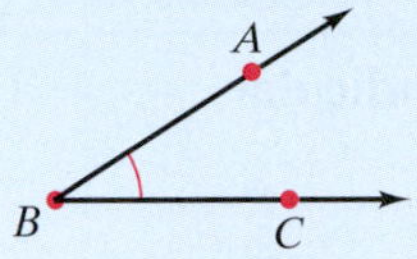 $\angle ABC$
[11.1] Plane	A flat surface that extends endlessly in all directions.	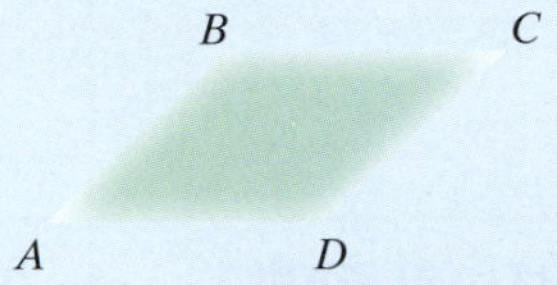 B C A D
[11.1] Straight angle	An angle whose measure is 180°.	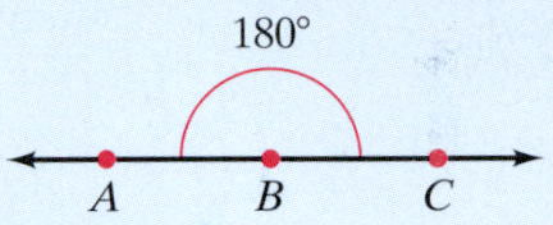 180° A B C
[11.1] Right angle	An angle whose measure is 90°.	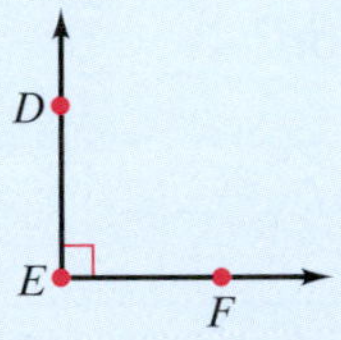 D E F
[11.1] Acute angle	An angle whose measure is less than 90°.	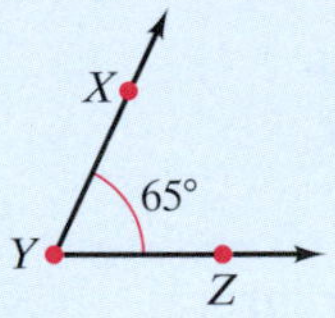 X 65° Y Z
[11.1] Obtuse angle	An angle whose measure is more than 90° and less than 180°.	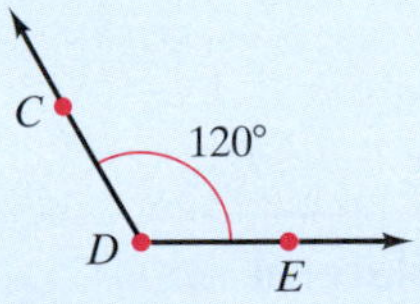 C 120° D E
[11.1] Complementary angles	Two angles the sum of whose measures is 90°.	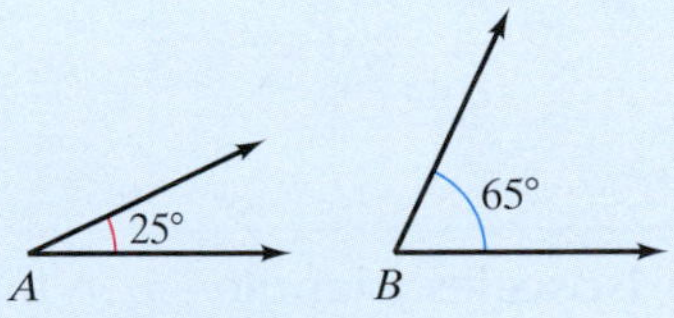 25° A 65° B
[11.1] Supplementary angles	Two angles the sum of whose measures is 180°.	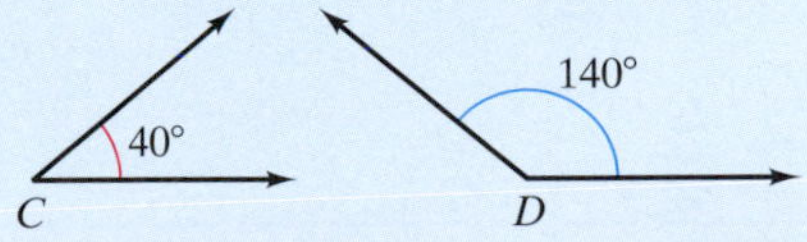 40° C 140° D

continued

Concept/Skill	Description	Example
[11.1] **Intersecting lines**	Two lines that cross.	D, C, B, A, E
[11.1] **Parallel lines**	Two lines on the same plane that do not intersect.	E, F, G, H; $\overleftrightarrow{EF} \parallel \overleftrightarrow{GH}$
[11.1] **Perpendicular lines**	Two lines that intersect to form right angles.	R, S, P, Q, T; $\overleftrightarrow{RT} \perp \overleftrightarrow{PQ}$
[11.1] **Vertical angles**	Two opposite angles with equal measure formed by two intersecting lines.	B, D, A, C, E; 120°, 60°, 60°, 120°
[11.1] **Polygon**	A closed plane figure made up of line segments.	
[11.1] **Triangle**	A polygon with three sides.	D, E, F
[11.1] **Quadrilateral**	A polygon with four sides.	A, B, C, D
[11.1] **Equilateral triangle**	A triangle with three sides equal in length.	Q, P, R; $\overline{PQ}$, $\overline{QR}$, and $\overline{PR}$ have equal lengths.
[11.1] **Isosceles triangle**	A triangle with two or more sides equal in length.	B, A, C; $\overline{AB}$ and $\overline{BC}$ have equal lengths.

Concept/Skill	Description	Example
[11.1] **Scalene triangle**	A triangle with no sides equal in length.	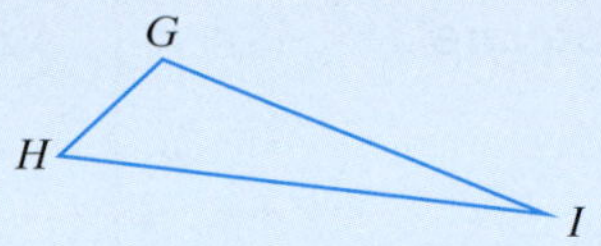 G, H, I; $\overline{HG}$, $\overline{GI}$, and $\overline{HI}$ have unequal lengths.
[11.1] **Acute triangle**	A triangle with three acute angles.	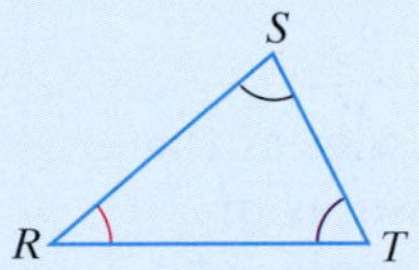 S, R, T
[11.1] **Right triangle**	A triangle with one right angle.	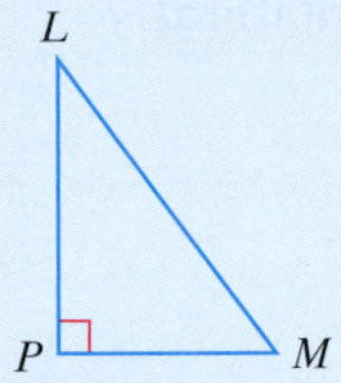 L, P, M
[11.1] **Obtuse triangle**	A triangle with one obtuse angle.	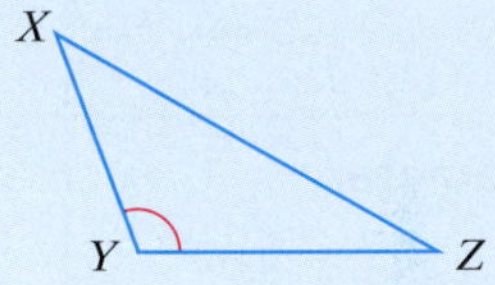 X, Y, Z
[11.1] **The sum of the measures of the angles of a triangle**	In any triangle, the sum of the measures of all three angles is 180°.	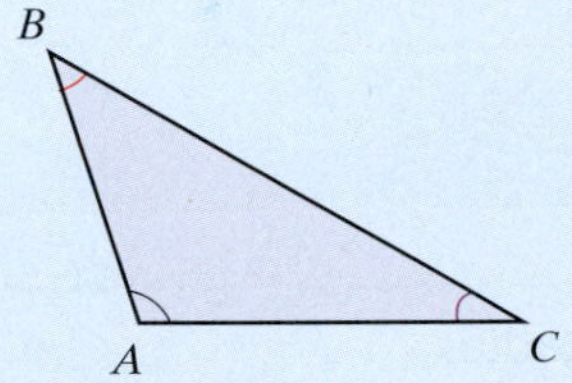 B, A, C; $m\angle A + m\angle B + m\angle C = 180°$
[11.1] **Trapezoid**	A quadrilateral with only one pair of opposite sides parallel.	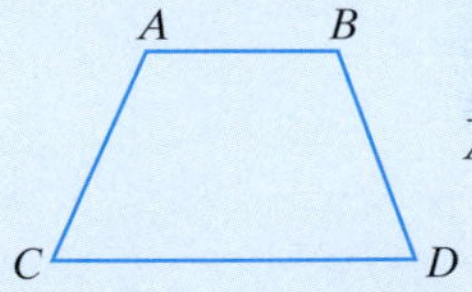 A, B, C, D; $\overline{AB} \parallel \overline{CD}$
[11.1] **Parallelogram**	A quadrilateral with both pairs of opposite sides parallel. Opposite sides are equal in length, and opposite angles have equal measures.	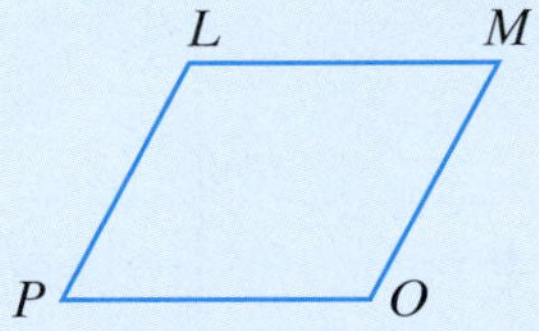 L, M, P, O; $\overline{LM} \parallel \overline{PO}$; $\overline{LP} \parallel \overline{MO}$; $\overline{LM}$ and $\overline{PO}$ have equal lengths, and $\overline{LP}$ and $\overline{MO}$ have equal lengths.
[11.1] **Rectangle**	A parallelogram with four right angles.	R, S, T, U

continued

CONCEPT SKILL

Concept/Skill	Description	Example
[11.1] **Square**	A rectangle with four sides equal in length.	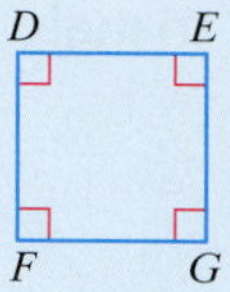 D E F G $\overline{DE}$, $\overline{EG}$, $\overline{FG}$, and $\overline{DF}$ have equal lengths.
[11.1] **The sum of the measures of the angles of a quadrilateral**	In any quadrilateral, the sum of the measures of the angles is 360°.	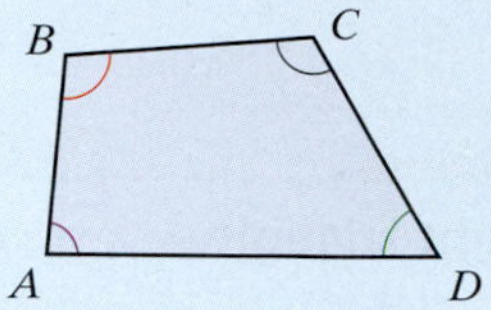 B C A D $m\angle A + m\angle B + m\angle C + m\angle D = 360°$
[11.1] **Circle**	A closed plane figure made up of points that are all the same distance from a fixed point called the center.	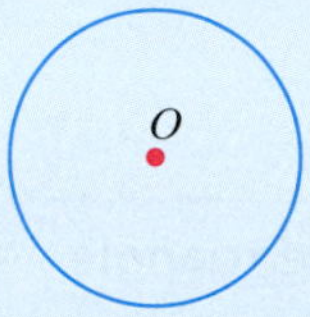 O
[11.1] **Diameter**	A line segment that passes through the center of a circle and has both endpoints on the circle.	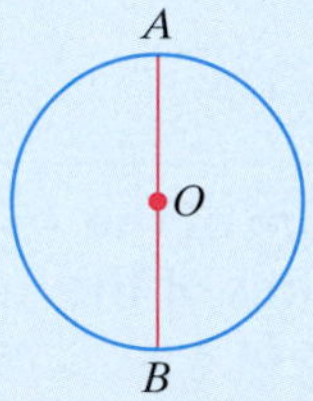 A O B Diameter $\overline{AB}$
[11.1] **Radius**	A line segment with one endpoint on the circle and the other at the center.	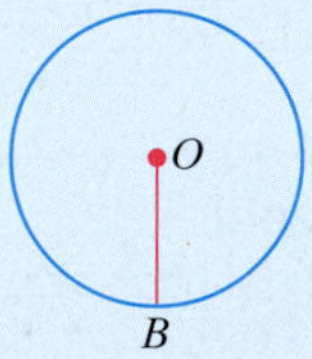 O B Radius $\overline{OB}$
[11.2] **Perimeter**	The distance around a polygon.	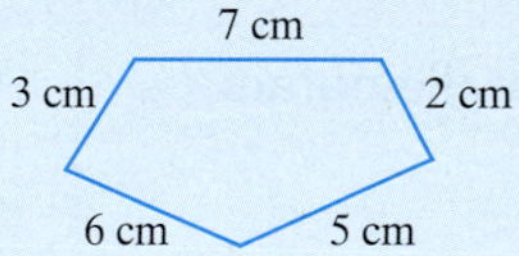 7 cm, 3 cm, 2 cm, 6 cm, 5 cm $P = 3 + 7 + 2 + 5 + 6 = 23$ $P = 23$ cm
[11.2] **Circumference**	The distance around a circle.	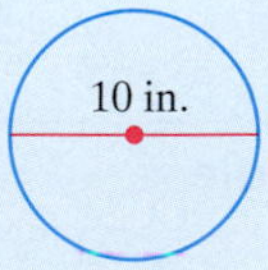 10 in. $C = 10\pi \approx 31.4$ $C \approx 31.4$ in.

Concept/Skill	Description	Example
[11.3] **Area**	The number of square units that a figure contains.	4 in.; 3 in.; 1 in^2 $A = 4 \cdot 3 = 12$ $A = 12 \text{ in}^2$
[11.4] **Volume**	The number of cubic units required to fill a three-dimensional figure.	2 in.; 3 in.; 4 in.; 1 in^3; $\Delta ABC \sim \Delta DEF$ $V = 2 \cdot 3 \cdot 4 = 24$ $V = 24 \text{ in}^3$
[11.5] **Similar triangles**	Triangles that have the same shape but not necessarily the same size.	B, A, C, E, D, F
[11.5] **Corresponding sides**	In similar triangles, the sides opposite the equal angles.	In the similar triangles pictured, $\overline{AB}$ corresponds to $\overline{DE}$, $\overline{BC}$ corresponds to $\overline{EF}$, and $\overline{AC}$ corresponds to $\overline{DF}$.
[11.5] **To find a missing side of a similar triangle**	• Write the ratios of the lengths of the corresponding sides. • Write a proportion using a ratio with known terms and a ratio with an unknown term. • Solve the proportion for the unknown term.	$\Delta TRS \sim \Delta XYW$ Find a. S, R, T: 6 in., 4 in., 8 in.; W, Y, X: 9 in., 6 in., a $\frac{ST}{WX} = \frac{TR}{XY}$ $\frac{4}{6} = \frac{8}{a}$ $4a = 48$ $a = 12$, or 12 in.
[11.6] **Perfect square**	A number that is the square of a whole number.	49 and 144
[11.6] **(Principal) square root of *n***	The positive number, written $\sqrt{n}$, whose square is n.	$\sqrt{36}$ and $\sqrt{8}$

continued

Concept/Skill	Description	Example
[11.6] **Pythagorean theorem**	For every right triangle, the sum of the squares of the lengths of the two legs equals the square of the length of the hypotenuse, that is, $a^2 + b^2 = c^2$ where a and b are the lengths of the legs, and c is the length of the hypotenuse.	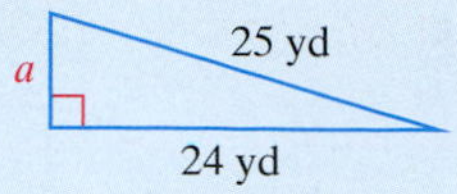 Find a. $a^2 + b^2 = c^2$ $a^2 + (24)^2 = (25)^2$ $a^2 + 576 = 625$ $a^2 + 576 - 576 = 625 - 576$ $a^2 = 49$ $a = \sqrt{49}$ $= 7$, or 7 yd

Key Formulas

Figure	Formula	Example
[11.2]–[11.3] **Triangle**	*Perimeter* $P = a + b + c$ Perimeter equals the sum of the lengths of the three sides. *Area* $A = \frac{1}{2}bh$ Area equals one-half the base times the height.	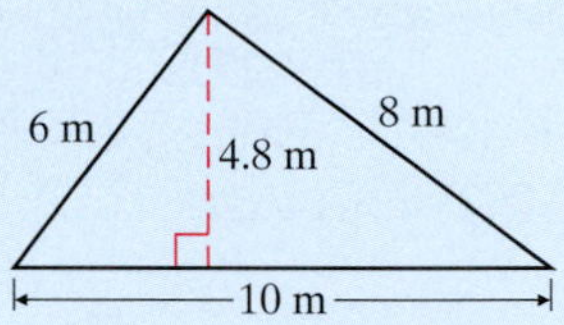 $P = a + b + c$ $= 6 + 10 + 8$ $= 24$, or 24 m $A = \frac{1}{2}bh$ $= \frac{1}{\underset{1}{\cancel{2}}} \cdot \overset{5}{\cancel{10}} \cdot 4.8$ $= 24$, or 24 m^2
[11.2]–[11.3] **Rectangle**	*Perimeter* $P = 2l + 2w$ Perimeter equals twice the length plus twice the width. *Area* $A = lw$ Area equals the length times the width.	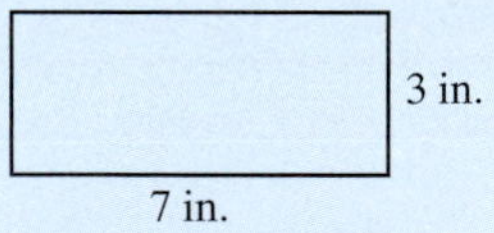 $P = 2l + 2w$ $= 2(7) + 2(3)$ $= 14 + 6$ $= 20$, or 20 in. $A = lw$ $= 7 \cdot 3$ $= 21$, or 21 in^2

Figure	Formula	Example
[11.2]–[11.3] **Square**	*Perimeter* $$P = 4s$$ Perimeter equals four times the length of a side. *Area* $$A = s^2$$ Area equals the square of a side.	$\frac{1}{2}$ in. $P = 4s$ $= 4 \cdot \frac{1}{2}$ $= 2$, or 2 in. $A = s^2$ $= \left(\frac{1}{2}\right)^2$ $= \frac{1}{4}$, or $\frac{1}{4}$ in^2
[11.3] **Parallelogram**	*Area* $$A = bh$$ Area equals the base times the height.	3 ft 6 ft $A = bh$ $= 6 \cdot 3$ $= 18$, or 18 ft^2
[11.3] **Trapezoid**	*Area* $$A = \frac{1}{2}h(b + B)$$ Area equals one-half the height times the sum of the bases.	3 in. 4 in. 5 in. $A = \frac{1}{2}h(b + B)$ $= \frac{1}{2} \cdot 4(3 + 5)$ $= \frac{1}{\not{2}_1} \cdot \not{4}^{2} \cdot 8$ $= 16$, or 16 in^2
[11.2]–[11.3] **Circle**	*Circumference* $$C = \pi d, \text{ or } C = 2\pi r$$ Circumference equals π times the diameter, or 2 times π times the radius. *Area* $$A = \pi r^2$$ Area equals π times the square of the radius.	8 cm $C = \pi d$ $\approx 3.14(8)$ ≈ 25.12, or 25.12 cm $A = \pi r^2$ $\approx 3.14\,(4)^2$ $\approx 3.14\,(16)$ ≈ 50.24, or 50.24 cm^2 Note: $d = 8$ cm, so $r = 4$ cm.

continued

Figure	Formula	Example
[11.4] **Rectangular solid**	*Volume* $$V = lwh$$ Volume equals length times width times height.	5 cm, 7 cm, 15 cm $V = lwh$ $= 15 \cdot 7 \cdot 5$ $= 525$, or $525\ \text{cm}^3$
[11.4] **Cube**	*Volume* $$V = e^3$$ Volume equals the cube of the edge.	2 in. $V = e^3$ $= (2)^3$ $= 2 \cdot 2 \cdot 2$ $= 8$, or $8\ \text{in}^3$
[11.4] **Cylinder**	*Volume* $$V = \pi r^2 h$$ Volume equals π times the square of the radius times the height.	12 m, 4 m $V = \pi r^2 h$ $\approx 3.14\,(4)^2(12)$ $\approx 3.14\,(16)(12)$ ≈ 603, or $603\ \text{m}^3$
[11.4] **Sphere**	*Volume* $$V = \frac{4}{3}\pi r^3$$ Volume equals $\frac{4}{3}$ times π times the cube of the radius.	2 ft $V = \frac{4}{3}\pi r^3$ $\approx \frac{4}{3}(3.14)(2)^3$ $\approx \frac{4}{3}(3.14)(8)$ $\approx \frac{100.48}{3}$ ≈ 33, or $33\ \text{ft}^3$

CHAPTER 11 Review Exercises

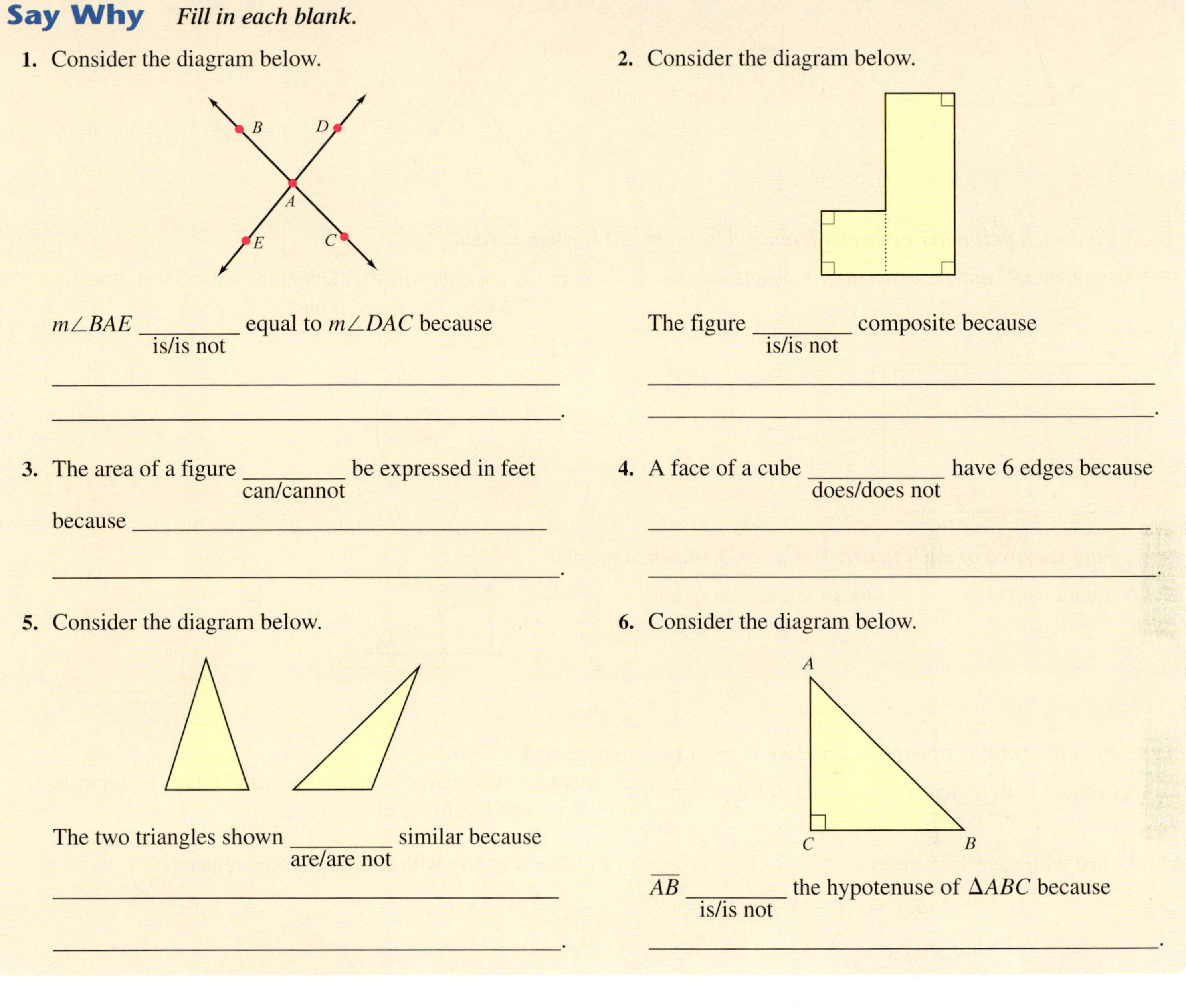

Say Why *Fill in each blank.*

1. Consider the diagram below.

$m\angle BAE$ ________ equal to $m\angle DAC$ because
is/is not

____

____.

2. Consider the diagram below.

The figure ________ composite because
is/is not

____

____.

3. The area of a figure ________ be expressed in feet
can/cannot

because ________________________________

____.

4. A face of a cube ________ have 6 edges because
does/does not

____

____.

5. Consider the diagram below.

The two triangles shown ________ similar because
are/are not

____

____.

6. Consider the diagram below.

$\overline{AB}$ ________ the hypotenuse of ΔABC because
is/is not

____.

[11.1] *Sketch and label an example of each of the following.*

7. $\overline{AB}$

8. $\angle PQR$

9. Parallel lines $\overleftrightarrow{ST}$ and $\overleftrightarrow{UV}$

10. Obtuse ΔABC

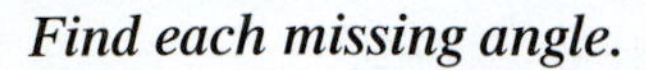

Find each missing angle.

11.

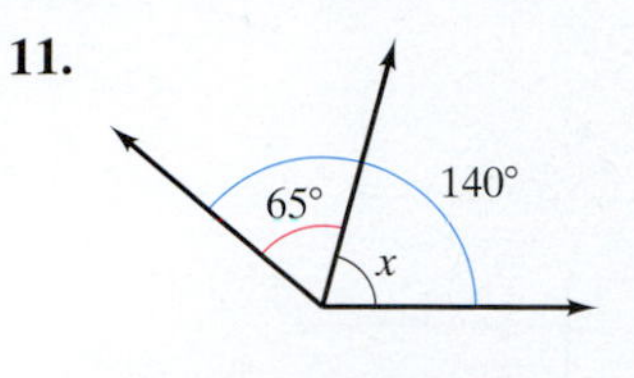

12.

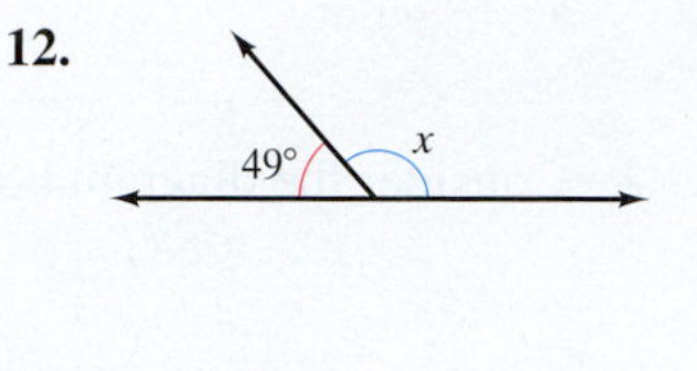

13.

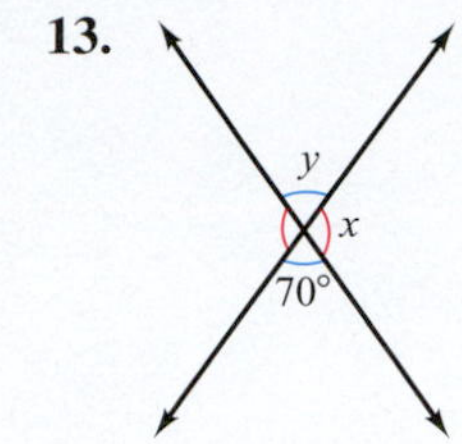

14.

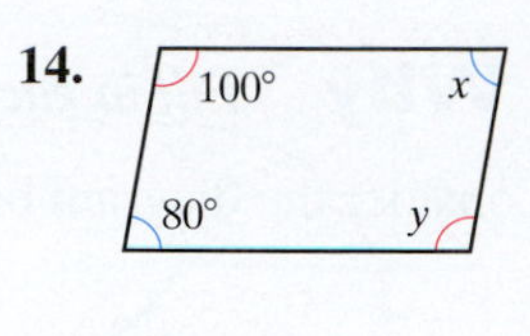

[11.2] *Find each perimeter or circumference. Use* $\pi \approx$ **3.14,** *when needed.*

15. An equilateral triangle with side 1.8 meters

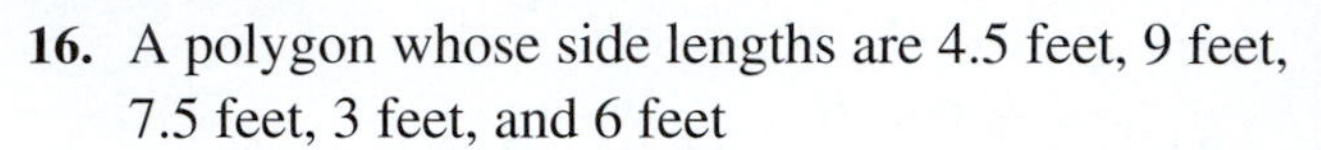

16. A polygon whose side lengths are 4.5 feet, 9 feet, 7.5 feet, 3 feet, and 6 feet

17.

18.

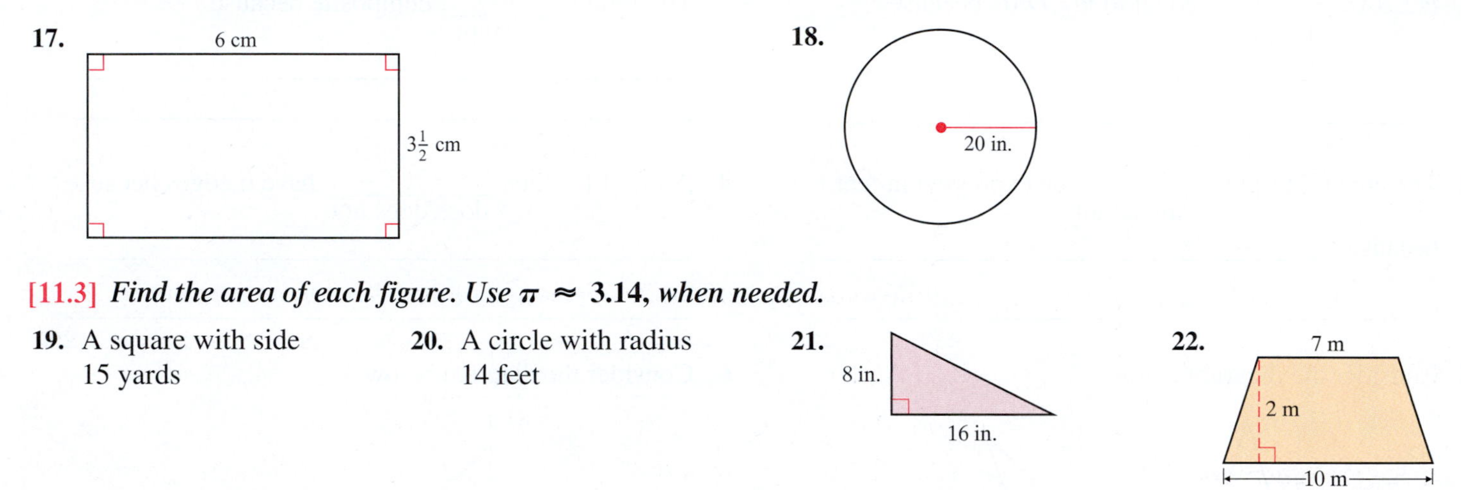

[11.3] *Find the area of each figure. Use* $\pi \approx$ **3.14,** *when needed.*

19. A square with side 15 yards

20. A circle with radius 14 feet

21.

8 in.
16 in.

22.

7 m
2 m
10 m

[11.4] *Find the volume of each figure. Use* $\pi \approx$ **3.14,** *when needed.*

23. A cylinder with radius 10 inches and height 4.2 inches

24. A rectangular solid with length 16 feet, width $4\frac{1}{2}$ feet, and height 3 feet

25. A cube with edge 1.25 meters

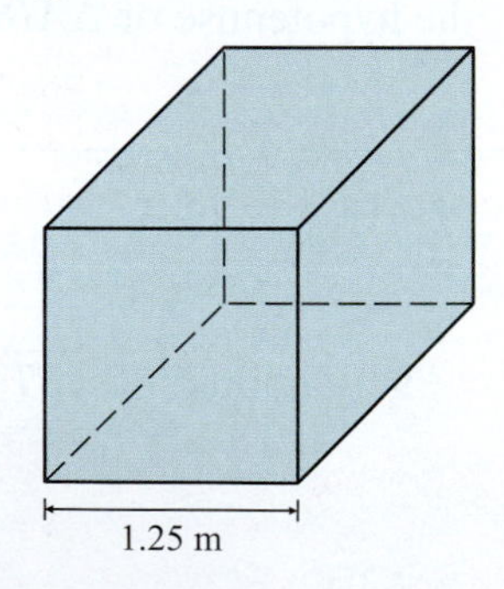

26. A sphere with diameter 2.5 centimeters

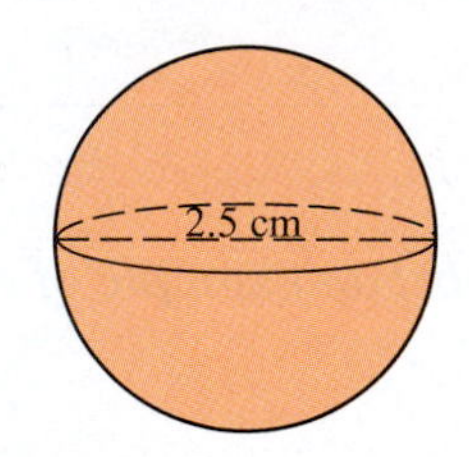

[11.2]–[11.4] *Solve.*

27. Find the perimeter of the figure shown, which is made up of a semicircle and a trapezoid. Use $\pi \approx 3.14$.

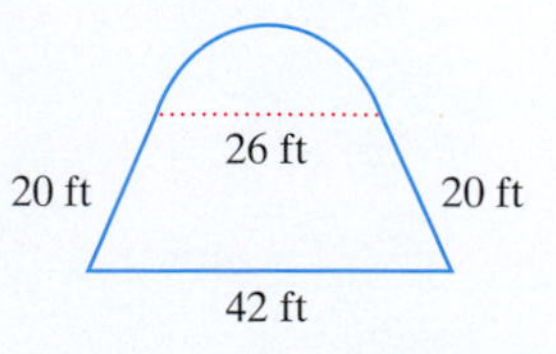

28. Find the area of the shaded portion of the figure. Use $\pi \approx 3.14$.

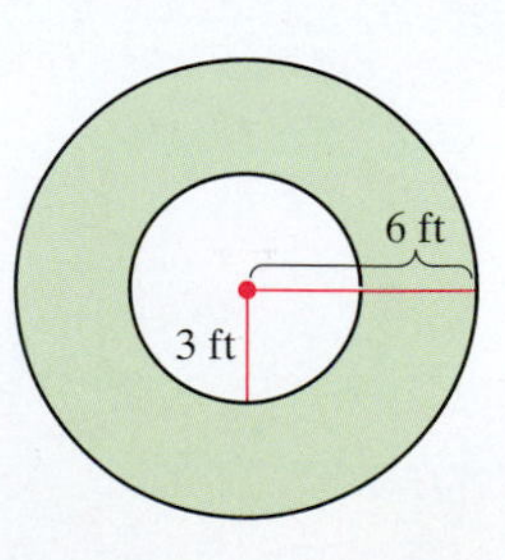

29. What is the area of the figure that consists of a square and two semicircles?

30. Find the volume of the shaded region between the sphere and the cube.

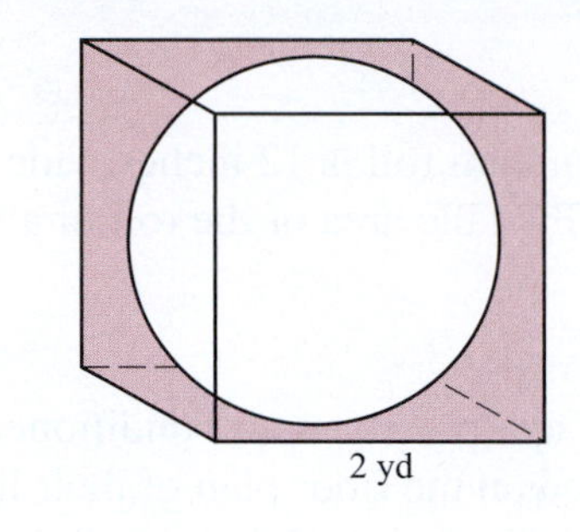

[11.5] *Find the value of each unknown.*

31. $\Delta DEF \sim \Delta HGF$

32. $\Delta ABC \sim \Delta DEC$

[11.6] *Find the square root.*

33. $\sqrt{9}$ **34.** $\sqrt{64}$ **35.** $\sqrt{121}$ **36.** $\sqrt{900}$

Determine between which two consecutive whole numbers each square root lies.

37. $\sqrt{3}$ **38.** $\sqrt{84}$ **39.** $\sqrt{40}$ **40.** $\sqrt{10}$

Find the square root. Round to the nearest hundredth.

41. $\sqrt{8}$ **42.** $\sqrt{1,235}$ **43.** $\sqrt{195}$ **44.** $\sqrt{29}$

For a given triangle, the lengths of the legs are a and b, and the length of the hypotenuse is c. Find the length of the missing side. Round to the nearest tenth, if needed.

	a	*b*	*c*
45.	9 ft		15 ft
46.		24 in.	26 in.
47.	8 yd	5 yd	
48.	2 ft	2 ft	

Mixed Applications

Solve.

49. A roll of aluminum foil is 12 inches wide and 2,400 inches long. Find the area of the roll of aluminum foil.

50. Six weeks after an underwater oil well exploded, oil pouring into the Gulf of Mexico spread throughout a circular region with radius 200 miles. How big was the affected area? (*Source:* myfoxdc.com)

51. In a couple's apartment, an air conditioner can cool a room up to 3,000 cubic feet in volume. Based on the floor plan of their living room and a ceiling height of 10 feet, can the air conditioner cool the room?

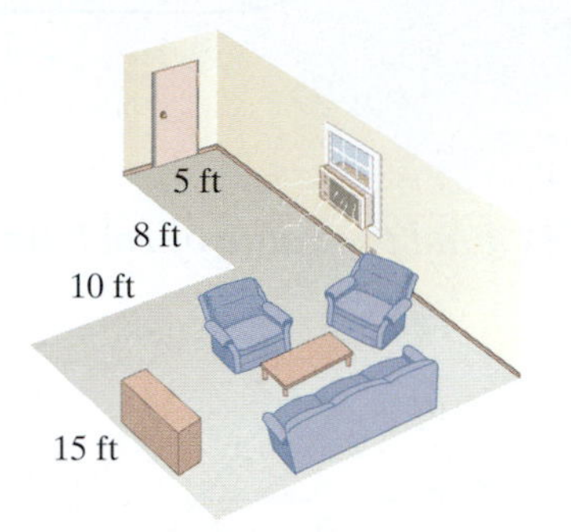

52. Of the two high-definition TV screens shown below, how much greater is the area of the larger screen?

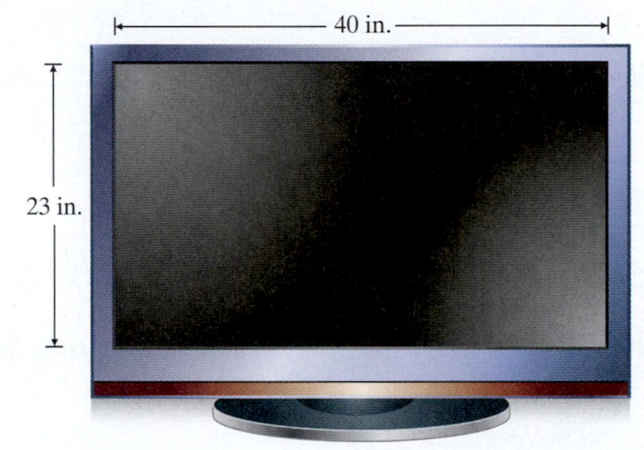

53. A pilot flies 12 miles west from city A to city B. Then, he flies 5 miles south from city B to city C. What is the straight-line distance from city A to city C?

54. How high up on a wall will a 12-foot ladder reach if the bottom of the ladder is placed 6 feet from the wall? Round to the nearest foot.

55. On the pool table shown, a player hits the ball at point E. It ricochets off point C and winds up in the pocket at point A. If $\Delta ABC \sim \Delta EDC$, find CD.

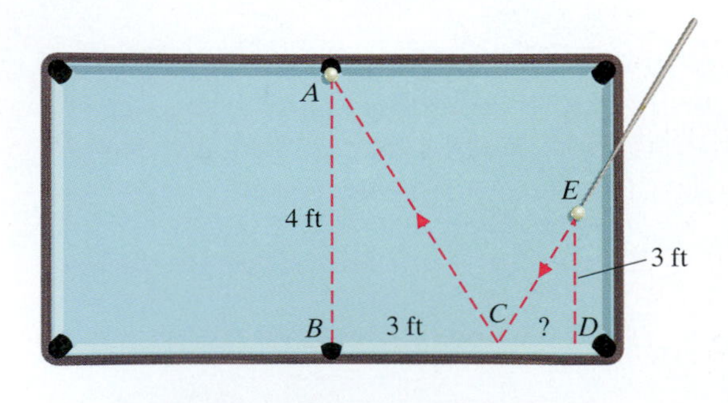

56. On the campus map below, the two triangles are similar. Find the distance between the Athletic Center and the Student Center.

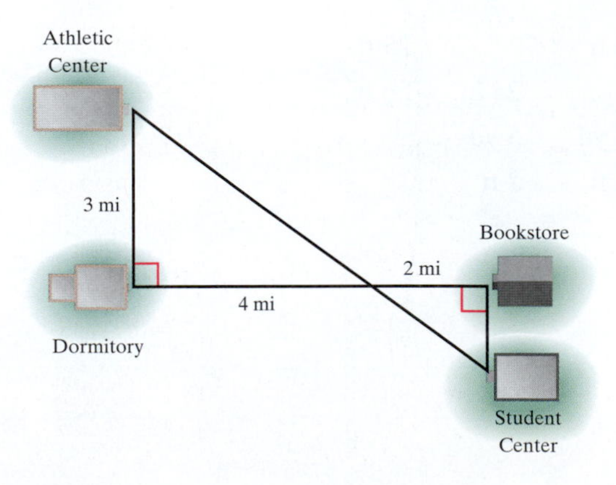

57. From the following drawing, find the total length of the building's walls.

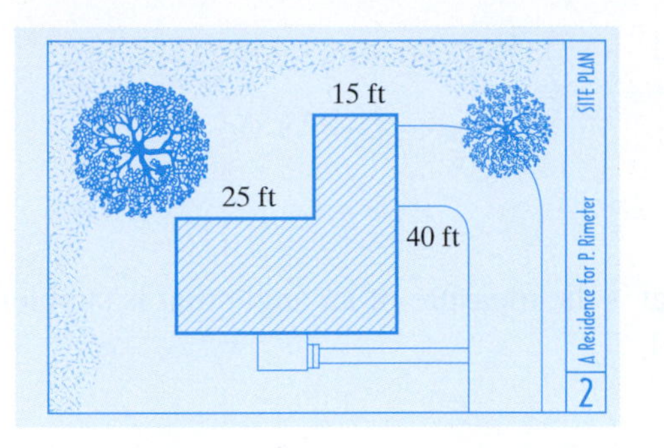

58. According to interior designers, the distances between the refrigerator, stove, and sink usually form a *work triangle*. To be efficient, the perimeter of a work triangle must be no more than 22 feet. Determine whether the model kitchen shown is efficient.

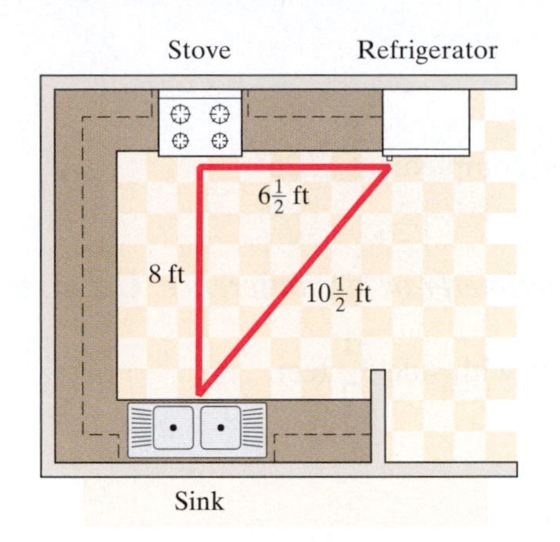

59. The coffee in this cylindrical can weighs 13 ounces.

What is the weight of a cubic inch of coffee, to the nearest tenth of an ounce?

60. How much soil is needed to fill the flower box shown to 1 centimeter from the top? Round to the nearest cubic centimeter.

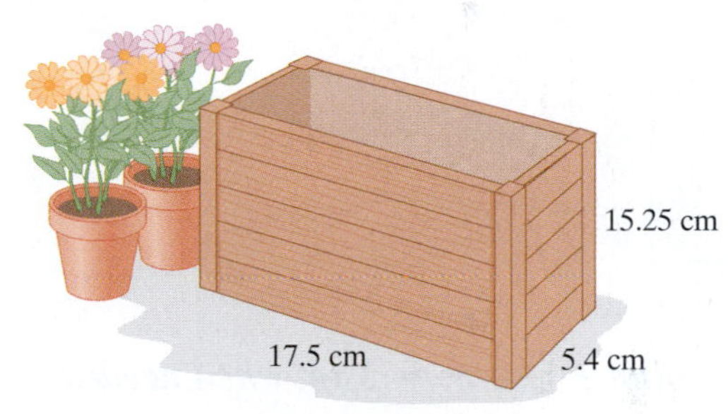

61. Find the area of the picture matting shown.

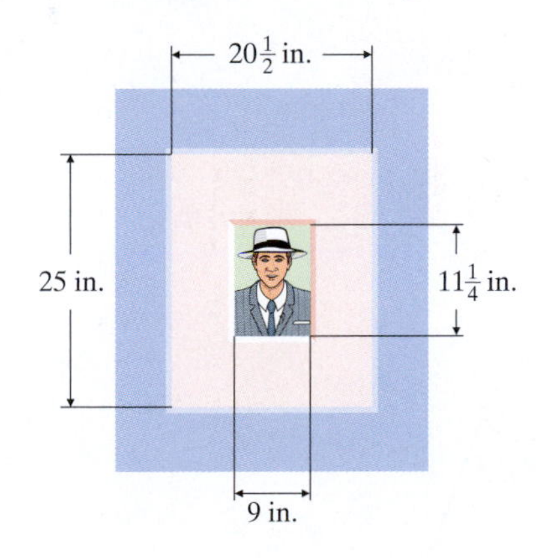

62. The game of racquetball is played with a small, hollow rubber ball, as shown.

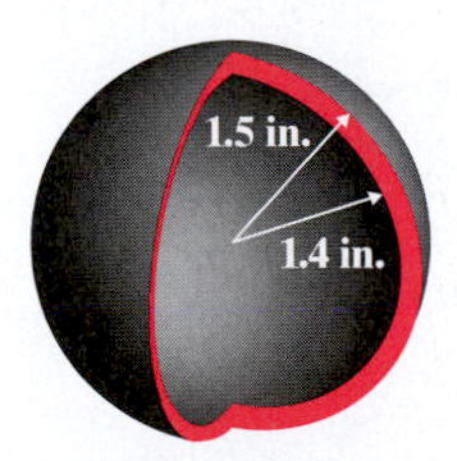

How much rubber, to the nearest tenth of a cubic inch, does a racquetball contain?

• Check your answers on page A-14.

CHAPTER 11 Posttest

FOR EXTRA HELP

The Chapter Test Prep Videos with test solutions are available on DVD, in MyMathLab, and on YouTube (search "AkstBasicMath" and click on "Channels").

To see if you have mastered the topics in this chapter, take this test.

1. Sketch and label an example of acute $\angle PQR$.

2. Find: $\sqrt{225}$

3. Find the complement of 25°.

4. What is the measure of an angle that is supplementary to 91°?

Find each perimeter or circumference. Use $\pi \approx$ 3.14, when needed.

5. A square with side $3\frac{1}{2}$ feet

6. An equilateral triangle with side 1.5 meters

7.

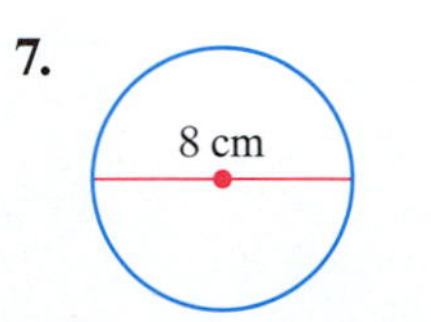

8.

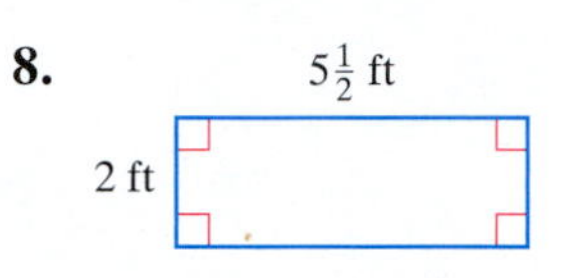

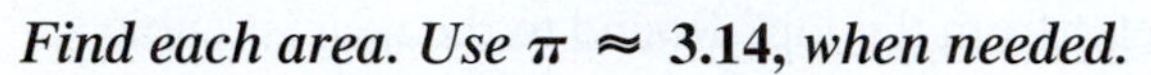

Find each area. Use $\pi \approx$ 3.14, when needed.

9.

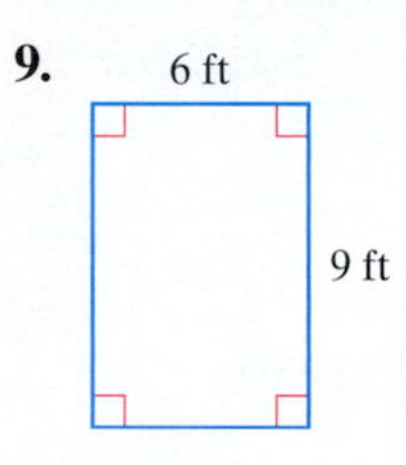

10.

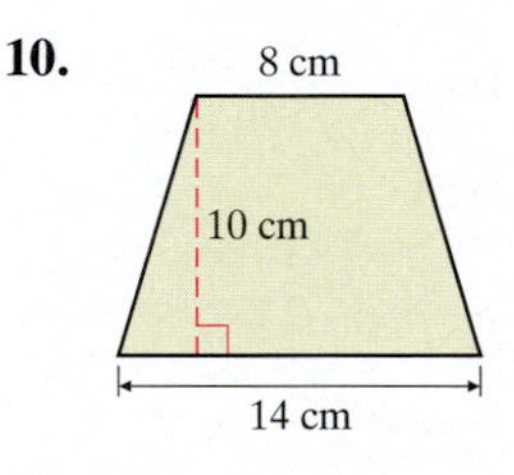

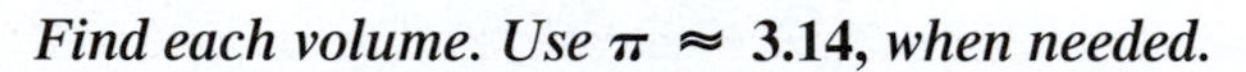

Find each volume. Use $\pi \approx$ 3.14, when needed.

11.

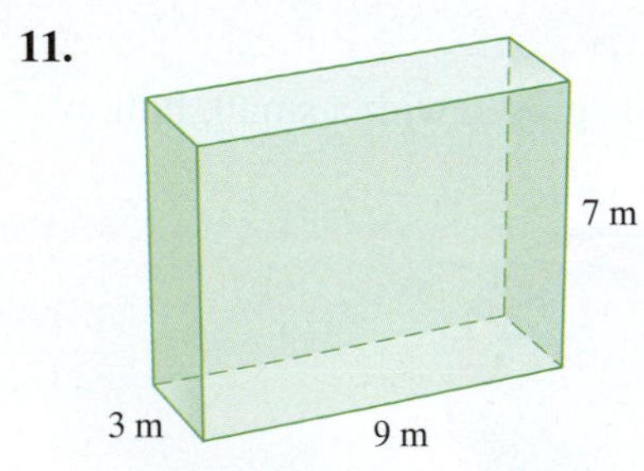

12.

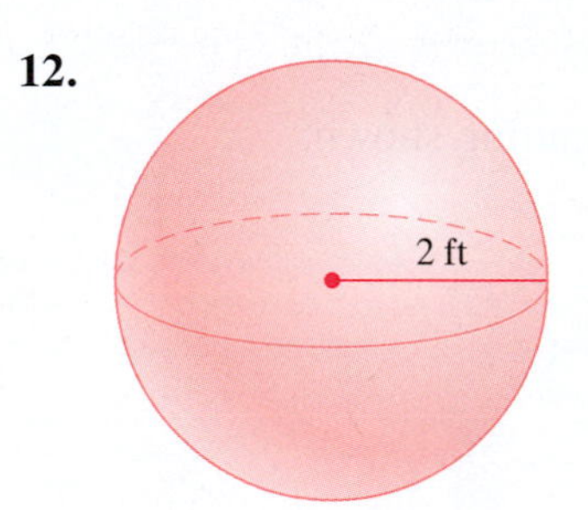

Find the value of each unknown.

13.

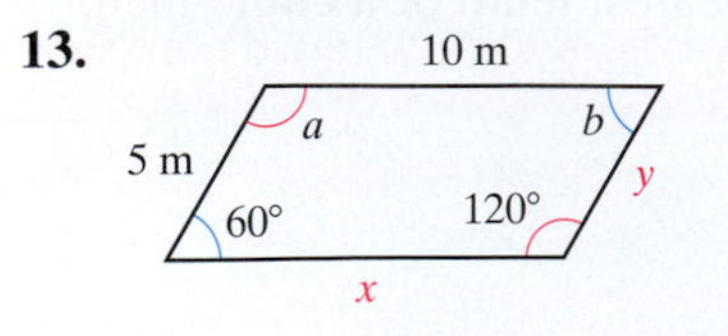

14.

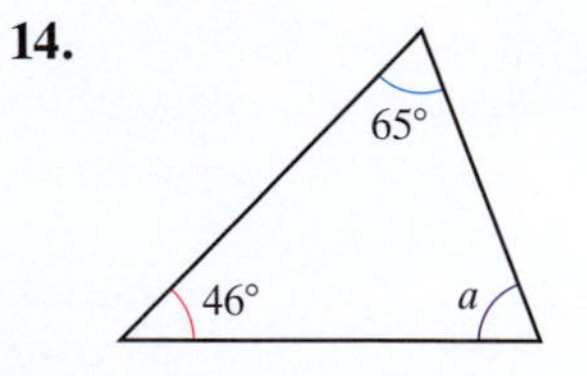

15.

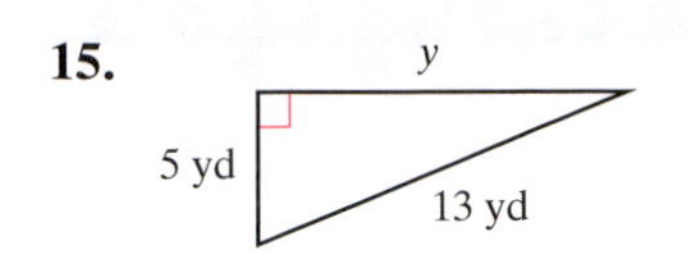

16. In the following diagram, $\Delta ABC \sim \Delta DEF$. Find x and y.

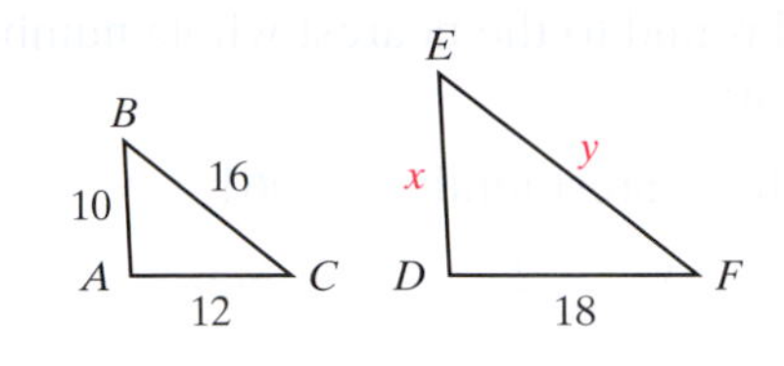

17. Dupont Circle, a major traffic circle and park located in northwest Washington, DC, is formed by the intersection of Massachusetts Avenue NW, Connecticut Avenue NW, New Hampshire Avenue NW, P Street NW, and 19th Street NW.

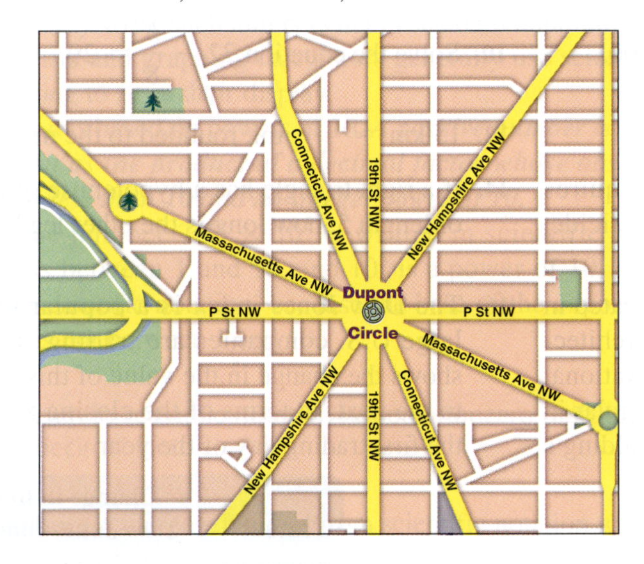

Is the angle formed by Massachusetts Avenue NW and New Hampshire Avenue NW below P Street NW acute? (*Source:* maps.google.com)

Solve.

18. A board foot is a special measure of volume used in the lumber industry. If a board foot contains 144 cubic inches of wood, how many board feet rounded to the nearest tenth are there in the board shown?

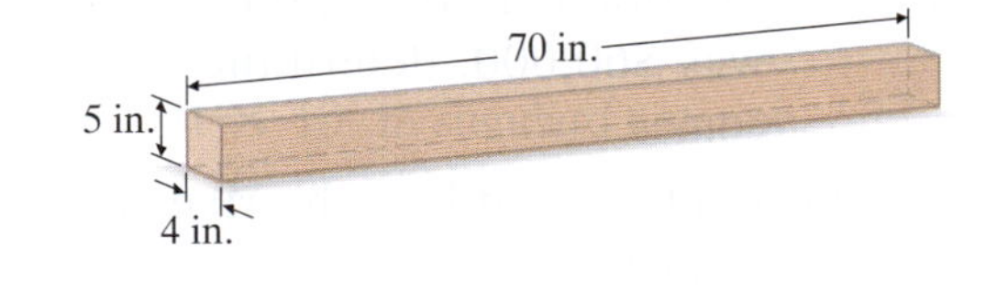

19. Suppose that the leash on a dog is 5 meters long. To the nearest square meter, what is the area of the dog's "run"? Use $\pi \approx 3.14$.

20. An airplane flying due north is 200 miles from the airport. At the same time, another airplane flying due east is 150 miles from the airport. How far apart are the two airplanes at that time?

• Check your answers on page A-15.

EXAMPLE 2

Rewrite 3,700,000,000 in scientific notation.

Solution We know that for a number to be written in scientific notation, it must be the product of a decimal factor whose absolute value is greater than or equal to 1 but less than 10 and an integer power of 10. Recall that 3,700,000,000 and 3,700,000,000. are the same. We move the decimal point *to the left* so that there is one nonzero digit to its left. The number of places moved is the power of 10 by which we need to multiply.

$$3{,}700{,}000{,}000. = 3\underbrace{.700000000}_{9 \text{ places}} \times 10^9 = 3.7 \times 10^9$$

Note that we dropped the extra zeros in 3.700000000. So 3,700,000,000 expressed in scientific notation is 3.7×10^9.

PRACTICE 2

Write 8,000,000,000,000 in scientific notation.

Now, let's turn our attention to writing *small numbers* in scientific notation. The key is an understanding of *negative exponents*. Until now, we have only considered exponents that were positive integers. What meaning should we attach to a negative exponent? To the exponent 0? The following pattern, in which each number is $\frac{1}{10}$ of the previous number, suggests an answer.

$$10^3 = 1{,}000$$
$$10^2 = 100$$
$$10^1 = 10$$
$$10^0 = 1$$
$$10^{-1} = \frac{1}{10}$$
$$10^{-2} = \frac{1}{100}$$
$$10^{-3} = \frac{1}{1{,}000}$$

Notice that 10^{-1}, or $\frac{1}{10}$, is the reciprocal of 10^1 or 10. So in general, ***a number raised to a negative exponent is defined to be the reciprocal of that number raised to the corresponding positive exponent. Also, a number raised to the power 0 is 1.***

When written in scientific notation, large numbers have positive powers of 10, whereas small numbers have negative powers of 10. For instance, 3×10^5 is large, whereas 3×10^{-5} is small.

Next, let's look at how we change *small* numbers from scientific notation to standard notation and vice versa.

EXAMPLE 3

Convert 3×10^{-5} to standard notation.

Solution Using the meaning of negative exponents, we get:

$$3 \times 10^{-5} = 3 \times \frac{1}{10^5}, \quad \text{or} \quad \frac{3}{10^5}$$

Since $10^5 = 100{,}000$, dividing 3 by 10^5 gives us:

$$\frac{3}{10^5} = \frac{3}{100{,}000} = 0.00003$$

So 3×10^{-5} written in standard notation is 0.00003.

PRACTICE 3

Change 4.3×10^{-9} to standard notation.

In Example 3, note that the power of 10 is *negative*, and the decimal point, which is understood to be at the right end of a whole number, was moved five places *to the left*. So just as with 7.34×10^5, there is a shortcut for expressing 3×10^{-5} in standard notation. To do this, we move the decimal point five places *to the left*:

$$3 \times 10^{-5} = 3. \times 10^{-5} = .00003 = .00003, \text{ or } 0.00003.$$

5 places

TIP When converting a number from scientific notation to standard notation, move the decimal point to the *left* if the power of 10 is *negative* and to the *right* if the power of 10 is *positive.* The number of places the decimal point is moved is the absolute value of the power of 10.

EXAMPLE 4

Write 0.00000000000000002 in scientific notation.

Solution To write 0.00000000000000002 in scientific notation, we move the decimal point *to the right* until there is one nonzero digit to the left of the decimal point. The number of places moved, preceded by a *negative* sign, is the power of 10 that we need.

$$0.00000000000000002 = 00000000000000002. \times 10^{-17}$$

17 places

$$= 2 \times 10^{-17}$$

PRACTICE 4

Express 0.000000000071 in scientific notation.

Computation Involving Scientific Notation

Now, let's consider how to perform calculations on numbers written in scientific notation. We focus on the operations of multiplication and division.

Multiplying and dividing numbers written in scientific notation can best be understood in terms of two *laws of exponents*—the *product rule* and the *quotient rule*.

- The *product rule of exponents* states that when we multiply a base raised to a power by the same base raised to another power, we add the exponents and leave the base the same. For example,

$$10^3 \cdot 10^2 = 10^{3+2} = 10^5$$

Add the exponents.

Keep the base.

This result is reasonable, since $10^3 \times 10^2 = 1{,}000 \times 100 = 100{,}000 = 10^5$.

- The *quotient rule of exponents* states that when we divide a base raised to a power by the same base to another power, we subtract the second power from the first power, and leave the base the same. For instance,

Subtract the exponents.

$$10^5 \div 10^2 = 10^{5-2} = 10^3$$

Keep the base.

We would have expected this result even if we did not know the quotient rule, since $\frac{10^5}{10^2} = \frac{100{,}000}{100} = \frac{1{,}000}{1} = 1{,}000 = 10^3$.

EXAMPLE 5

Calculate, writing the result in scientific notation.

a. $(4 \times 10^{-1})(2.1 \times 10^6)$

b. $(1.2 \times 10^5) \div (2 \times 10^{-4})$

Solution

a. $(4 \times 10^{-1})(2.1 \times 10^6)$

$= (4 \times 2.1)(10^{-1} \times 10^6)$ Change the order of the factors and regroup.

$= (8.4)(10^{-1} \times 10^6)$ Multiply the decimal factors.

$= 8.4 \times 10^{-1+6}$ Use the product rule of exponents.

$= 8.4 \times 10^5$ Simplify.

b. $(1.2 \times 10^5) \div (2 \times 10^{-4})$

$= \frac{1.2 \times 10^5}{2 \times 10^{-4}}$

$= \frac{1.2}{2} \times \frac{10^5}{10^{-4}}$ Write as the product of fractions.

$= 0.6 \times \frac{10^5}{10^{-4}}$ Divide the decimal factors.

$= 0.6 \times 10^{5-(-4)}$ Use the quotient rule of exponents.

$= 0.6 \times 10^9$ Simplify.

Note that 0.6×10^9 is not written in scientific notation, because 0.6 is not between 1 and 10, that is, it does not have one nonzero digit to the left of the decimal point. To write 0.6×10^9 in scientific notation, we convert 0.6 to scientific notation and simplify the product.

$0.6 \times 10^9 = 6 \times 10^{-1} \times 10^9$

$= 6 \times 10^{-1+9}$ Use the product rule of exponents.

$= 6 \times 10^8$

So the quotient, written in scientific notation, is 6×10^8.

PRACTICE 5

Calculate, expressing the answer in scientific notation.

a. $(7 \times 10^{-2})(3.52 \times 10^3)$

b. $(5.01 \times 10^3) \div (6 \times 10^{-9})$

Exercises

FOR EXTRA HELP MyMathLab MathXL PRACTICE WATCH READ REVIEW

Express in standard notation.

1. 3.17×10^8
2. 9.1×10^5
3. 1×10^{-6}
4. 8.013×10^{-4}
5. 4.013×10^{-5}
6. 2.1×10^{-3}

Express in scientific notation.

7. 400,000,000
8. 10,000,000
9. 0.0000035
10. 0.00017
11. 0.00000000031
12. 218,000,000,000

Multiply, and write the result in scientific notation.

13. $(3 \times 10^2)(3 \times 10^5)$
14. $(5 \times 10^6)(1 \times 10^3)$
15. $(2.5 \times 10^{-2})(8.3 \times 10^{-3})$
16. $(2.1 \times 10^4)(8 \times 10^{-4})$

Divide, and write the result in scientific notation.

17. $(2.5 \times 10^8) \div (2 \times 10^{-2})$
18. $(3.0 \times 10^4) \div (1 \times 10^3)$
19. $(1.2 \times 10^5) \div (3 \times 10^3)$
20. $(4.88 \times 10^{-3}) \div (8 \times 10^2)$

Answers

CHAPTER 1

Chapter 1 Pretest, p. 2

1. Two hundred five thousand, seven **2.** 1,235,000 **3.** Hundred thousands **4.** 8,100 **5.** 8,226 **6.** 4,714 **7.** 185 **8.** 29,124 **9.** 260 **10.** 308 R6 **11.** 2^3 **12.** 36 **13.** 5 **14.** 43 **15.** 75 years old **16.** $55 **17.** 68 **18.** 324 sec **19.** $36 **20.** Room C, which measures 126 sq ft

Section 1.1 Practices, pp. 4–8

1, ***p. 4:*** **a.** Thousands **b.** Hundred thousands **c.** Ten millions **2,** ***p. 4:*** Eight billion, three hundred seventy-six thousand, fifty-two **3,** ***p. 4:*** $7,372,050 Seven million, three hundred seventy-two thousand, fifty dollars **4,** ***p. 5:*** $95,000,003 **5,** ***p. 5:*** $375,000 **6,** ***p. 6:*** **a.** 2 ten thousands + 7 thousands + 0 hundreds + 1 ten + 3 ones = 20,000 + 7,000 + 0 + 10 + 3 or 20,000 + 7,000 + 10 + 3 **b.** 1 million + 2 hundred thousands + 7 ten thousands + 9 tens + 3 ones = 1,000,000 + 200,000 + 70,000 + 90 + 3 **7,** ***p. 7:*** **a.** 52,000 **b.** 50,000 **8,** ***p. 8:*** 420,000,000 **9,** ***p. 8:*** **a.** One million, six hundred ninety-nine thousand, two hundred **b.** 1,960,000

Exercises 1.1, pp. 9–13

1. whole numbers **3.** odd **5.** standard form **7.** placeholder **9.** expanded form **11.** 4,867 **13.** 316 **15.** 28,461,013 **17.** Hundred thousands **19.** Hundreds **21.** Billions **23.** Four hundred eighty-seven thousand, five hundred **25.** Two million, three hundred fifty thousand **27.** Nine hundred seventy-five million, one hundred thirty-five thousand **29.** Two billion, three hundred fifty-two **31.** One billion **33.** 10,120 **35.** 150,856 **37.** 6,000,055 **39.** 50,600,195 **41.** 400,072 **43.** 3 ones = 3 **45.** 8 hundreds + 5 tens + 8 ones = 800 + 50 + 8 **47.** 2 millions + 5 hundred thousands + 4 ones = 2,000,000 + 500,000 + 4 **49.** 670 **51.** 7,100 **53.** 30,000 **55.** 700,000 **57.** 30,000

59.

To the nearest	135,842	2,816,533
Hundred	135,800	2,816,500
Thousand	136,000	2,817,000
Ten thousand	140,000	2,820,000
Hundred thousand	100,000	2,800,000

61. 1 ten thousand + 2 thousands + 5 tens + 1 one = 10,000 + 2,000 + 50 + 1 **63.** 40,059 **65.** 1,056,100; one million, fifty-six thousand, one hundred **67.** Nine hundred thousand **69.** forty-eight thousand, three hundred eighty-one **71.** Three hundred million **73.** 100,000,000,000 **75.** 3,288 **77.** 3,233,300,000,000 **79.** 150 ft **81.** 20,000 mi **83.** 1,900 **85.** **a.** Three million, six hundred thousand, nine hundred thirty sq mi **b.** 301,000 sq mi

Section 1.2 Practices, pp. 15–23

1, ***p. 15:*** 385 **2,** ***p. 16:*** 10,436 **3,** ***p. 17:*** 16 mi **4,** ***p. 18:*** 651 **5,** ***p. 19:*** 4,747 **6,** ***p. 20:*** 750 plant species **7,** ***p. 20:*** **a.** 286,000 **b.** 193,000 **c.** Less **8,** ***p. 22:*** 9,477 **9,** ***p. 22:*** 2,791 **10,** ***p. 23:*** 20,000 ft

Calculator Practices, pp. 23–24

11, ***p. 23:*** 49,532 **12,** ***p. 24:*** 31,899 **13,** ***p. 24:*** 2,499 ft

Exercises 1.2, pp. 25–31

1. right **3.** sum **5.** Associative Property of Addition **7.** subtrahend **9.** 177,778 **11.** 14,710 **13.** 14,002 **15.** 56,188 **17.** 6,978 **19.** 4,820 **21.** 413 **23.** 14,865 **25.** 15,509 m **27.** 82 hr **29.** $104,831 **31.** $12,724 **33.** 31,200 tons **35.** 13,296,657 **37.** 1,662,757

39.

+	400	200	1,200	300	Total
300	700	500	1,500	600	3,300
800	1,200	1,000	2,000	1,100	5,300
Total	1,900	1,500	3,500	1,700	8,600

41.

+	389	172	1,155	324	Total
255	644	427	1,410	579	3,060
799	1,188	971	1,954	1,123	5,236
Total	1,832	1,398	3,364	1,702	8,296

43. a; possible estimate: 12,800 **45.** a; possible estimate: $900,000 **47.** 217 **49.** 90 **51.** 362 **53.** 68,241 **55.** 2,285 **57.** 52,999 **59.** 2,943 **61.** 203,465 **63.** 368 **65.** 4,996 **67.** 982 **69.** 1,995 mi **71.** $669 **73.** $3,609 **75.** 273 books **77.** 209 m **79.** 2,001,000 **81.** 813,429 **83.** c; possible estimate: 40,000,000 **85.** a; possible estimate: $200,000 **87.** 7,065 **89.** 1,676 **91.** 5,186 **93.** 281,000,000 **95.** 3,400,000 sq mi **97.** **a.** Austria, 16; Canada, 26; Germany 30; Norway, 23; United States, 37 **b.** United States **99.** About 43 years old **101.** No, the elevator is not overloaded. The total weight of passengers is 963 lb. **103.** 180°F **105.** 151 mi **107.** 19,403,000 **109.** 1,454 **111.** **a.** Less (2,804) **b.** 2,932 seats **113.** **a.** 280,000 species **b.** 1,060,000 species **c.** 260,000 species **115.** $28,576

Section 1.3 Practices, pp. 34–37

1, ***p. 34:*** 608 **2,** ***p. 34:*** 4,230 **3,** ***p. 35:*** 480,000 **4,** ***p. 35:*** 205,296 **5,** ***p. 36:*** 107 sq ft **6,** ***p. 36:*** 112,840 **7,** ***p. 37:*** No; possible estimate = 20,000

Calculator Practices, p. 38

8, ***p. 38:*** 1,026,015 **9,** ***p. 38:*** 345,546

Exercises 1.3, pp. 39–43

1. product **3.** Identity Property of Multiplication **5.** addition **7.** 400 **9.** 142,000 **11.** 170,000 **13.** 7,000,000 **15.** 12,700 **17.** 418 **19.** 3,248,000 **21.** 65,268 **23.** 817 **25.** 34,032 **27.** 3,003 **29.** 3,612 **31.** 57,019 **33.** 243,456 **35.** 200,120 **37.** 149,916 **39.** 144,500 **41.** 123,830 **43.** 3,312 **45.** 2,106 **47.** 40,000 **49.** 23,085 **51.** 3,274,780 **53.** 54,998,850 **55.** c; possible estimate: 480,000 **57.** b; possible estimate: 80,000 **59.** 2,880 **61.** 230,520 **63.** 1,071,000 **65.** 300,000 **67.** 3,300 yr **69.** **a.** 3,000,000 **b.** 1,000,000 **71.** Yes **73.** 5,775 sq in. **75.** 1,750 mi **77.** $442 **79.** **a.** 294 mi **b.** 1,470 mi **81.** Colorado; area ≈ 106,700 sq mi

Section 1.4 Practices, pp. 46–50

1, ***p. 46:*** 807 **2,** ***p. 46:*** 7,002 **3,** ***p. 47:*** 5,291 R1 **4,** ***p. 48:*** 79 R1 **5,** ***p. 48:*** 94 R10 **6,** ***p. 49:*** 607 R3 **7,** ***p. 49:*** 200 **8,** ***p. 50:*** 967 **9,** ***p. 50:*** 5 times

Calculator Practice, p. 51

10, ***p. 51:*** 603

Exercises 1.4, pp. 52–54

1. divisor **3.** multiplication **5.** 400 **7.** 2,560 **9.** 301 **11.** 3,003 **13.** 8,044 **15.** 500 **17.** 30 **19.** 14 **21.** 42 **23.** 400 **25.** 159 **27.** 5,353 **29.** 1,002 **31.** 6,944 **33.** 1,001 **35.** 3,050 **37.** 907 **39.** 1,201 **41.** 651 R2 **43.** 11 R7 **45.** 116 R83 **47.** 700 R2 **49.** 723 R19 **51.** 428 R8 **53.** 1,010 R10 **55.** 928 R24 **57.** 721 **59.** 155 **61.** c; possible estimate: 800 **63.** a; possible estimate: 7,000 **65.** 907 R1 **67.** 2,000 **69.** 2,400 **71.** 370 **73.** $135 **75.** 2 times **77.** 300 people per square mile **79.** 6 calories **81.** **a.** 304 tiles **b.** 26 boxes **c.** $468

Section 1.5 Practices, pp. 55–59

1, ***p. 55:*** $5^5 \cdot 2^2$ **2,** ***p. 56:*** **a.** 1 **b.** 1,331 **3,** ***p. 56:*** 784 **4,** ***p. 56:*** 10^9 **5,** ***p. 57:*** 28 **6,** ***p. 58:*** 146 **7,** ***p. 58:*** 4 **8,** ***p. 58:*** 130 **9,** ***p. 59:*** 60 ft **10,** ***p. 59:*** $40 **11,** ***p. 59:*** **a.** 61 fatalities **b.** 2006 and 2009

Calculator Practices, p. 60

12, ***p. 60:*** 140,625; **13,** ***p. 60:*** 131

Exercises 1.5, pp. 61–65

1. base **3.** adding

5.

n	0	2	4	6	8	10	12
n^2	0	4	16	36	64	100	144

7.

n	0	2	4	6	8
n^3	0	8	64	216	512

9. 10^2 **11.** 10^4 **13.** 10^6 **15.** $2^2 \cdot 3^2$ **17.** $4^3 \cdot 5^1$ **19.** 900 **21.** 1,568 **23.** 18 **25.** 4 **27.** 13 **29.** 14 **31.** 35 **33.** 225 **35.** 250 **37.** 36 **39.** 5 **41.** 28 **43.** 6 **45.** 99 **47.** 99 **49.** 4 **51.** 39 **53.** 16 **55.** 93 **57.** 67 **59.** 18 **61.** 529 **63.** 419 **65.** 137,088 **67.** $\boxed{4} \cdot 3 + \boxed{6} \cdot 5 + \boxed{6} \cdot 7 = 98$ **69.** $(\boxed{8})(3 + \boxed{4}) - 2 \cdot \boxed{6} = 44$ **71.** $\boxed{8} + 10 \times \boxed{4} - \boxed{6} \div 2 = 45$ **73.** $(5 + 2) \cdot 4^2 = 112$ **75.** $(5 + 2 \cdot 4)^2 = 169$ **77.** $(8 - 4) \div 2^2 = 1$ **79.** 242 sq cm **81.** 3,120 sq in.

83.

Input	Output
0	$21 + 3 \times \mathbf{0} = 21$
1	$21 + 3 \times \mathbf{1} = 24$
2	$21 + 3 \times \mathbf{2} = 27$

85. 25 **87.** 40 **89.** 4 **91.** 2,412 mi **93.** 8 **95.** 10^8 **97.** 289 **99.** 48 **101.** 8 **103.** 625 sq ft **105.** $5^2 + 12^2 = 13^2$; $25 + 144 = 169$ **107.** 10^6 **109.** **a.** $21,500 **b.** $1,050 **111.** **a.** 69 **b.** At home; the average score for home games was higher than the average score for away games. **113.** **a.** 108,000 workers **b.** 1,969,000 workers; below average **115.** Yes, because the average number of customers is 502.

Section 1.6 Practices, pp. 67–69

1, ***p. 67:*** 10,670 employees **2,** ***p. 68:*** 2 yr **3,** ***p. 68:*** 1,551 students **4,** ***p. 69:*** 180 lb

Exercises 1.6, pp. 70–71

1. $2,150 **3.** 27 mi **5.** 75 times **7.** 5,882 mi **9.** 528,179 immigrants **11.** 300¢, or $3 **13.** $17,000 **15.** $6,036 **17.** $1,458 **19.** 8 extra pens **21.** 1952 was closer by 31 votes. **23.** $983

Chapter 1 Review Exercises, pp. 75–79

1. are; possible answer: both digits are in the hundreds place **2.** is not; possible answer: the critical digit 6 is greater than 5 **3.** is not; possible answer: the perimeter is the sum of the lengths of the figure's sides **4.** is not; possible answer: 8 times 7 is 56 **5.** is; possible answer: the area of a rectangle is the product of its length and its width **6.** is not; possible answer: 9 is the base which is raised to the power 2 **7.** is; possible answer: of the distributive property **8.** is; possible answer: 10 divided by 5 is 2 **9.** is not; possible answer: the sum of the numbers should be divided by 3 and not by 2 since there are three numbers **10.** before; possible answer: of the order of operations rule **11.** Ones **12.** Ten thousands **13.** Hundred millions **14.** Ten billions **15.** Four hundred ninety-seven **16.** Two thousand, fifty **17.** Three million, seven **18.** Eighty-five billion **19.** 251 **20.** 9,002 **21.** 14,000,025 **22.** 3,000,003,000 **23.** 2 millions + 5 hundred thousands = 2,000,000 + 500,000 **24.** 4 ten thousands + 2 thousands + 7 hundreds + 7 ones = 40,000 + 2,000 + 700 + 7 **25.** 600 **26.** 1,000 **27.** 380,000 **28.** 70,000 **29.** 9,486 **30.** 65,692 **31.** 173,543 **32.** 150,895 **33.** 1,957,825 **34.** $223,067 **35.** 445 **36.** 10,016 **37.** 11,109 **38.** 5,510 **39.** 11,042,223 **40.** $2,062,852 **41.** 11,006 **42.** 2,989 **43.** 432 **44.** 1,200 **45.** 149,073 **46.** 12,000,000 **47.** 477,472 **48.** 1,019,000 **49.** 1,397,508 **50.** 188,221,590 **51.** 39 **52.** 307 R3 **53.** 37 R10 **54.** 680 R8 **55.** 25,625 **56.** 957 **57.** 343 **58.** 1 **59.** 72 **60.** 300,000 **61.** 5 **62.** 169 **63.** 5 **64.** 19 **65.** 12 **66.** 18 **67.** 10,833,312 **68.** 2,694 **69.** $7^2 \cdot 5^2$ **70.** $2^2 \cdot 5^3$ **71.** 39 **72.** 7 **73.** 6 **74.** 5 **75.** Two million, four hundred thousand **76.** 150,000,000 **77.** $3,009,000,000,000 **78.** 1985 **79.** 300,000 sq mi **80.** 32,000,000 iPods **81.** 9 **82.** possible answer; 24 times **83.** 2,717 **84.** 23 flats **85.** $307 per week

86.

Net sales	$430,000
− Cost of merchandise sold	− 175,000
Gross margin	$255,000
− Operating expenses	− 135,000
Net profit	$120,000

87. 6,675 sq m **88.** 272 legs **89.** 1968 to 1972 (15,385,031 votes) **90.** 4,341 points **91.** **a.** 1,949 km **b.** 1,683 km **92.** **a.** 14,994,000 **b.** The average would increase by 62,000. **93.** 29 sq mi **94.** 162 cm

Chapter 1 Posttest, p. 80

1. 225,067 **2.** 1,7̲68,405 **3.** One million, two hundred five thousand, seven **4.** 200,000 **5.** 1,894 **6.** 607 **7.** 147 **8.** 297,496 **9.** 509 **10.** 622 R19 **11.** 625 **12.** $4^3 \cdot 5^2$ **13.** 84 **14.** 2 **15.** 5,600,000 sq mi **16.** 46,177,500 acres **17.** $469 **18.** $123 **19.** $1,380 **20.** 26 g of fat

CHAPTER 2

Chapter 2 Pretest, p. 82

1. 1, 2, 4, 5, 10, 20 **2.** $2 \times 2 \times 2 \times 3 \times 3$, or $2^3 \times 3^2$ **3.** $\frac{2}{5}$ **4.** $\frac{61}{3}$ **5.** $1\frac{1}{30}$ **6.** $\frac{3}{4}$ **7.** 20 **8.** $\frac{1}{8}$ **9.** $1\frac{1}{5}$ **10.** $12\frac{5}{6}$ **11.** $2\frac{1}{4}$ **12.** $4\frac{5}{8}$ **13.** $3\frac{1}{2}$ **14.** 60 **15.** $\frac{2}{3}$ **16.** $3\frac{2}{3}$ **17.** $\frac{2}{21}$ **18.** 6 students **19.** 6 mi **20.** 69 g

Section 2.1 Practices, pp. 83–89

1, ***p. 83:*** 1, 7 **2,** ***p. 84:*** 1, 3, 5, 15, 25, 75 **3,** ***p. 85:*** 1, 2, 3, 5, 6, 9, 10, 15, 18, 30, 45, 90 **4,** ***p. 85:*** Yes; 24 is a multiple of 3. **5,** ***p. 86:*** **a.** Prime **b.** Composite **c.** Prime **d.** Composite **e.** Prime **6,** ***p. 87:*** $2^3 \times 7$ **7,** ***p. 87:*** 3×5^2 **8,** ***p. 88:*** 18 **9,** ***p. 89:*** 66 **10,** ***p. 89:*** 12 **11,** ***p. 89:*** 6 yr

Exercises 2.1, pp. 90–91

1. factors **3.** prime **5.** prime factorization **7.** 1, 3, 7, 21 **9.** 1, 17 **11.** 1, 2, 3, 4, 6, 12 **13.** 1, 31 **15.** 1, 2, 3, 4, 6, 9, 12, 18, 36 **17.** 1, 29 **19.** 1, 2, 4, 5, 10, 20, 25, 50, 100 **21.** 1, 2, 4, 7, 14, 28 **23.** Prime **25.** Composite (2, 4, 8) **27.** Composite (7) **29.** Prime **31.** Composite (3, 9, 27) **33.** 2^3 **35.** 7^2 **37.** $2^3 \times 3$ **39.** 2×5^2 **41.** 7×11 **43.** 3×17 **45.** 5^2 **47.** 2^5 **49.** 3×7 **51.** $2^3 \times 13$ **53.** 11^2 **55.** 2×71 **57.** $2^2 \times 5^2$ **59.** 5^3 **61.** $3^3 \times 5$ **63.** 15 **65.** 40 **67.** 90 **69.** 110 **71.** 72 **73.** 360 **75.** 300 **77.** 84 **79.** 105 **81.** 60 **83.** 3×5^2 **85.** 1, 2, 3, 4, 6, 8, 9, 12, 18, 24, 36, and 72 **87. a.** No, because 2015 is not a multiple of 10 **b.** Yes, because 2020 is a multiple of 10 **89.** No **91.** 30 students **93.** 30 days

Section 2.2 Practices, pp. 93–102

1, *p. 94:* $\frac{5}{8}$ **2, *p. 94:*** $\frac{7}{30}$ **3, *p. 94:*** $\frac{3}{4}$
4, *p. 95:*

5, *p. 95:* **a.** $\frac{16}{3}$ **b.** $\frac{102}{5}$ **6, *p. 96:*** **a.** 2 **b.** $5\frac{5}{9}$ **c.** $2\frac{2}{3}$ **7, *p. 98:*** Possible answer: $\frac{4}{10}, \frac{6}{15}, \frac{8}{20}$ **8, *p. 98:*** $\frac{45}{72}$ **9, *p. 99:*** $\frac{2}{3}$ **10, *p. 99:*** $\frac{7}{3}$ **11, *p. 99:*** $\frac{5}{16}$ **12, *p. 101:*** $\frac{11}{16}$ **13, *p. 102:*** $\frac{8}{15}, \frac{23}{30}, \frac{9}{10}$ **14, *p. 102:*** Country stations

Exercises 2.2, pp. 103–108

1. proper fraction **3.** equivalent **5.** like fractions **7.** $\frac{1}{3}$ **9.** $\frac{3}{6}$ **11.** $1\frac{1}{4}$ **13.** $3\frac{2}{4}$ **15.** **17.** **19.** **21.** **23.** **25.**

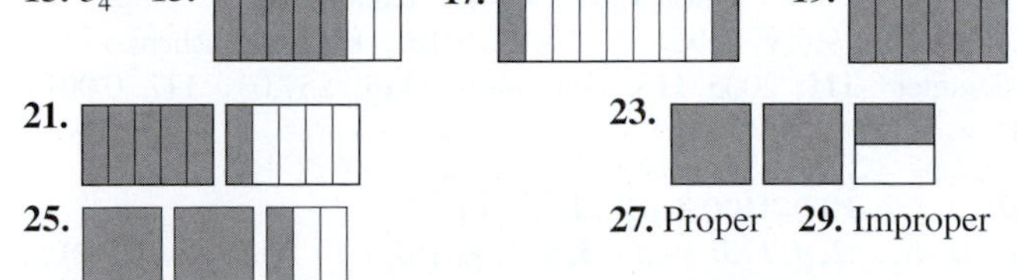

27. Proper **29.** Improper **31.** Mixed **33.** Improper **35.** Proper **37.** Mixed **39.** $\frac{13}{5}$ **41.** $\frac{55}{9}$ **43.** $\frac{57}{5}$ **45.** $\frac{5}{1}$ **47.** $\frac{59}{8}$ **49.** $\frac{88}{9}$ **51.** $\frac{27}{2}$ **53.** $\frac{98}{5}$ **55.** $\frac{14}{1}$ **57.** $\frac{54}{11}$ **59.** $\frac{115}{14}$ **61.** $\frac{202}{25}$ **63.** $1\frac{1}{3}$ **65.** $1\frac{1}{9}$ **67.** 3 **69.** 1 **71.** $19\frac{4}{5}$ **73.** $9\frac{1}{9}$ **75.** 1 **77.** $8\frac{2}{9}$ **79.** $13\frac{1}{2}$ **81.** $11\frac{1}{9}$ **83.** 27 **85.** 8 **87.** Possible answers: $\frac{2}{16}, \frac{3}{24}$ **89.** Possible answers: $\frac{4}{22}, \frac{6}{33}$ **91.** Possible answers: $\frac{6}{8}, \frac{9}{12}$ **93.** Possible answers: $\frac{2}{18}, \frac{3}{27}$ **95.** 9 **97.** 15 **99.** 40 **101.** 36 **103.** 40 **105.** 54 **107.** 36 **109.** 42 **111.** 6 **113.** 49 **115.** 32 **117.** 30 **119.** $\frac{2}{3}$ **121.** 1 **123.** $\frac{1}{3}$ **125.** $\frac{9}{20}$ **127.** $\frac{1}{4}$ **129.** $\frac{1}{8}$ **131.** $\frac{5}{4}$, or $1\frac{1}{4}$ **133.** $\frac{33}{16}$, or $2\frac{1}{16}$ **135.** $\frac{9}{16}$ **137.** $\frac{7}{24}$ **139.** 3 **141.** $\frac{1}{7}$ **143.** $3\frac{2}{3}$ **145.** 3 **147.** $<$ **149.** $>$ **151.** $=$ **153.** $<$ **155.** $\frac{1}{4}, \frac{1}{3}, \frac{1}{2}$ **157.** $\frac{7}{12}, \frac{2}{3}, \frac{5}{6}$ **159.** $\frac{3}{5}, \frac{2}{3}, \frac{8}{9}$ **161.** $\frac{5}{6}$ **163.** Possible answers: $\frac{4}{18}, \frac{6}{27}$ **165.** $\frac{12}{15}$ **167.** $2\frac{1}{5}$ hr per day **169. a.** $\frac{20}{401}$ **b.** $\frac{381}{401}$ **171.** $\frac{50}{103}$ **173.** The plain yogurt, because $\frac{2}{5}$ is greater than $\frac{1}{10}$. **175. a.** $\frac{1}{16}$ **b.** $\frac{1}{2}$ **177.** $250\frac{1}{2}$ lb

Section 2.3 Practices, pp. 109–124

1, *p. 109:* $\frac{2}{3}$ **2, *p. 110:*** $1\frac{7}{40}$ **3, *p. 110:*** $\frac{2}{5}$ **4, *p. 110:*** **a.** $\frac{3}{5}$ g **b.** $\frac{2}{5}$ g **5, *p. 112:*** $1\frac{2}{3}$ **6, *p. 112:*** $\frac{3}{10}$ **7, *p. 112:*** $\frac{71}{72}$ **8, *p. 113:*** $2\frac{1}{30}$ mi **9, *p. 114:*** $34\frac{2}{5}$ **10, *p. 114:*** $7\frac{1}{2}$ **11, *p. 115:*** 4 lengths **12, *p. 115:*** $7\frac{5}{8}$ **13, *p. 116:*** $11\frac{5}{24}$ **14, *p. 117:*** $4\frac{2}{5}$ **15, *p. 117:*** $1\frac{2}{5}$ in. **16, *p. 118:*** $4\frac{7}{12}$ **17, *p. 118:*** $1{,}439\frac{7}{10}$ mi **18, *p. 119:*** $1\frac{2}{7}$ **19, *p. 120:*** $5\frac{1}{6}$ **20, *p. 121:*** $12\frac{1}{2}$ **21, *p. 121:*** No, there will be only $1\frac{5}{8}$ yd left. **22, *p. 122:*** $10\frac{11}{20}$ **23, *p. 123:*** $1\frac{3}{8}$ **24, *p. 124:*** $6\frac{3}{4}$

Exercises 2.3, pp. 125–128

1. numerators **3.** regroup **5.** $1\frac{1}{4}$ **7.** $1\frac{1}{2}$ **9.** $\frac{4}{5}$ **11.** $\frac{3}{5}$ **13.** $1\frac{1}{6}$ **15.** $\frac{7}{8}$ **17.** $\frac{77}{100}$ **19.** $\frac{37}{40}$ **21.** $1\frac{5}{18}$ **23.** $1\frac{17}{100}$ **25.** $\frac{3}{4}$ **27.** $\frac{53}{80}$ **29.** $1\frac{7}{72}$ **31.** $1\frac{13}{40}$ **33.** $3\frac{1}{3}$ **35.** $15\frac{2}{5}$ **37.** $14\frac{1}{5}$ **39.** 15 **41.** $10\frac{5}{12}$ **43.** $3\frac{11}{15}$ **45.** $13\frac{13}{15}$ **47.** $6\frac{19}{24}$ **49.** $20\frac{1}{4}$ **51.** $10\frac{3}{100}$ **53.** $11\frac{3}{8}$ **55.** $36\frac{3}{50}$ **57.** $91\frac{7}{12}$ **59.** $6\frac{1}{2}$ **61.** $10\frac{33}{40}$ **63.** $11\frac{3}{8}$ **65.** $\frac{1}{5}$ **67.** $\frac{2}{5}$ **69.** $\frac{4}{25}$ **71.** $\frac{1}{2}$ **73.** 2 **75.** $\frac{1}{12}$ **77.** $\frac{5}{18}$ **79.** $\frac{1}{20}$ **81.** $\frac{1}{14}$ **83.** $\frac{5}{72}$ **85.** $\frac{1}{4}$ **87.** $4\frac{2}{7}$ **89.** $1\frac{3}{4}$ **91.** 20 **93.** $4\frac{1}{10}$ **95.** $3\frac{1}{3}$ **97.** $3\frac{3}{10}$ **99.** $6\frac{1}{3}$ **101.** $5\frac{1}{2}$ **103.** $4\frac{1}{2}$ **105.** $3\frac{1}{4}$ **107.** $11\frac{4}{5}$ **109.** $6\frac{2}{3}$ **111.** $7\frac{5}{6}$ **113.** $3\frac{13}{24}$ **115.** $15\frac{7}{18}$ **117.** $2\frac{29}{30}$ **119.** $\frac{1}{4}$ **121.** $5\frac{5}{12}$ **123.** $13\frac{39}{40}$ **125.** $\frac{3}{8}$ **127.** $1\frac{11}{40}$ **129.** $16\frac{23}{30}$ **131.** $5\frac{1}{5}$ **133.** $18\frac{11}{20}$ **135.** $4\frac{1}{8}$ **137.** $8\frac{1}{3}$ **139.** $2\frac{14}{15}$ **141.** $\frac{1}{8}$ in. **143. a.** $1\frac{1}{2}$ mi **b.** $\frac{1}{4}$ mi **145.** 5 hr **147.** $790\frac{1}{4}$ ft **149.** $\frac{1}{10}$ **151.** 1 lb

Section 2.4 Practices, pp. 131–140

1, *p. 131:* $\frac{15}{28}$ **2, *p. 131:*** $\frac{81}{100}$ **3, *p. 131:*** 20 **4, *p. 131:*** $\frac{7}{22}$ **5, *p. 132:*** $\frac{2}{9}$ **6, *p. 132:*** $5\frac{1}{4}$ hr **7, *p. 132:*** $20,769 **8, *p. 133:*** $7\frac{7}{8}$ **9, *p. 133:*** 28 **10, *p. 134:*** $25\frac{1}{2}$ sq in. **11, *p. 135:*** $18\frac{1}{4}$ **12, *p. 136:*** 6 **13, *p. 137:*** 8 **14, *p. 137:*** $2\frac{2}{3}$ yr **15, *p. 138:*** $1\frac{3}{5}$ **16, *p. 138:*** $\frac{7}{16}$ **17, *p. 138:*** 6 lb **18, *p. 139:*** 6 **19, *p. 140:*** $4\frac{1}{2}$

Exercises 2.4, pp. 141–144

1. multiply **3.** reciprocal **5.** invert **7.** $\frac{2}{15}$ **9.** $\frac{5}{12}$ **11.** $\frac{9}{16}$ **13.** $\frac{8}{25}$ **15.** $\frac{35}{32} = 1\frac{3}{32}$ **17.** $\frac{45}{16} = 2\frac{13}{16}$ **19.** $\frac{2}{9}$ **21.** $\frac{7}{12}$ **23.** $\frac{3}{40}$ **25.** $\frac{31}{30} = 1\frac{1}{30}$ **27.** $\frac{40}{3} = 13\frac{1}{3}$ **29.** $\frac{40}{3} = 13\frac{1}{3}$ **31.** 16 **33.** 4 **35.** 4 **37.** $\frac{35}{4} = 8\frac{3}{4}$ **39.** $1\frac{5}{16}$ **41.** $2\frac{1}{8}$ **43.** $\frac{25}{27}$ **45.** $2\frac{2}{3}$ **47.** 1 **49.** $\frac{7}{8}$ **51.** $1\frac{13}{35}$ **53.** $4\frac{41}{100}$ **55.** $7\frac{4}{5}$ **57.** 375 **59.** 8 **61.** 3 **63.** $41\frac{2}{3}$ **65.** $113\frac{1}{3}$ **67.** $1\frac{1}{6}$ **69.** $\frac{7}{12}$ **71.** $\frac{77}{100}$ **73.** $3\frac{3}{8}$ **75.** $\frac{9}{10}$ **77.** $\frac{32}{35}$ **79.** $3\frac{1}{2}$ **81.** $4\frac{4}{9}$ **83.** $1\frac{1}{2}$ **85.** $2\frac{1}{3}$ **87.** $1\frac{1}{5}$ **89.** $\frac{1}{4}$ **91.** $\frac{2}{21}$ **93.** $\frac{1}{9}$ **95.** 40 **97.** $16\frac{1}{3}$ **99.** $13\frac{1}{3}$ **101.** 7 **103.** $6\frac{11}{18}$ **105.** $1\frac{2}{3}$ **107.** $9\frac{22}{27}$ **109.** $100\frac{1}{2}$ **111.** $\frac{7}{90}$ **113.** $\frac{5}{26}$ **115.** $3\frac{1}{5}$ **117.** $\frac{21}{200}$ **119.** $\frac{35}{44}$ **121.** $1\frac{47}{115}$ **123.** $\frac{14}{27}$ **125.** $2\frac{1}{4}$ **127.** $1\frac{7}{18}$ **129.** $4\frac{13}{15}$ **131.** $\frac{87}{160}$ **133.** $3\frac{19}{27}$ **135.** $4\frac{1}{5}$ **137.** $3\frac{1}{8}$ **139.** $11\frac{1}{6}$ **141.** $\frac{1}{20}$ **143.** $\frac{1}{9}$ **145.** $20\frac{13}{16}$ **147.** $2\frac{5}{22}$ **149.** $\frac{21}{40}$ **151.** 8 **153.** $\frac{7}{12}$ **155.** $1,340 **157.** $1,116 **159.** $6\frac{1}{4}$ **161.** $\frac{27}{64}$ **163.** 7 times **165. a.** The scented candle **b.** The unscented candle

Chapter 2 Review Exercises, pp. 148–154

1. is; possible answer: it has more than two factors: 1, 3, 9, and 27 **2.** is not; possible answer: 4 is not a prime number **3.** is not; possible answer: the denominator of a fraction must be nonzero **4.** is; possible answer: the numerator, 12, is greater than the denominator, 11 **5.** is; possible answer: if you divide both the numerator and the denominator by the same number, 16, you get $\frac{1}{3}$ **6.** are; possible answer: they have different denominators **7.** is; possible answer: 24 is the least common multiple of 8 and 12 **8.** is not; possible answer: the reciprocal, $\frac{8}{6}$, is formed by switching the numerator and the denominator **9.** 1, 2, 3, 5, 6, 10, 15, 25, 30, 50, 75, 150 **10.** 1, 2, 3, 4, 5, 6, 9, 10, 12, 15, 18, 20, 30, 36, 45, 60, 90, 180 **11.** 1, 3, 19, 57 **12.** 1, 2, 5, 7, 10, 14, 35, 70 **13.** Prime **14.** Composite **15.** Composite **16.** Prime **17.** $2^2 \times 3^2$ **18.** 3×5^2 **19.** $3^2 \times 11$ **20.** 2×3^3 **21.** 42 **22.** 10 **23.** 72 **24.** 60 **25.** $\frac{2}{4}$ **26.** $\frac{6}{12}$ **27.** $1\frac{1}{6}$ **28.** $2\frac{3}{5}$ **29.** Mixed **30.** Proper **31.** Improper **32.** Improper **33.** $\frac{23}{3}$ **34.** $\frac{9}{5}$ **35.** $\frac{91}{10}$ **36.** $\frac{59}{7}$ **37.** $6\frac{1}{2}$ **38.** $4\frac{2}{3}$ **39.** $2\frac{3}{4}$ **40.** 1 **41.** 84 **42.** 4 **43.** 5 **44.** 27 **45.** $\frac{1}{2}$ **46.** $\frac{5}{7}$ **47.** $\frac{2}{3}$ **48.** $\frac{3}{4}$ **49.** $5\frac{1}{2}$ **50.** $8\frac{2}{3}$ **51.** $6\frac{2}{7}$ **52.** $8\frac{5}{7}$ **53.** $>$ **54.** $>$ **55.** $<$ **56.** $>$ **57.** $>$ **58.** $>$ **59.** $>$ **60.** $>$ **61.** $\frac{2}{7}, \frac{3}{8}, \frac{1}{2}$ **62.** $\frac{2}{15}, \frac{1}{5}, \frac{1}{3}$ **63.** $\frac{3}{4}, \frac{4}{5}, \frac{9}{10}$ **64.** $\frac{13}{18}, \frac{7}{9}, \frac{7}{8}$ **65.** $\frac{6}{5} = 1\frac{1}{5}$ **66.** $\frac{3}{4}$ **67.** $\frac{15}{8} = 1\frac{7}{8}$ **68.** $\frac{3}{5}$ **69.** $\frac{11}{15}$ **70.** $1\frac{17}{24}$ **71.** $1\frac{4}{5}$ **72.** $1\frac{37}{40}$ **73.** $5\frac{7}{8}$ **74.** $9\frac{1}{2}$ **75.** $10\frac{3}{5}$ **76.** 8 **77.** $12\frac{1}{3}$ **78.** $4\frac{3}{10}$ **79.** $5\frac{7}{10}$ **80.** $17\frac{13}{24}$ **81.** $23\frac{5}{12}$ **82.** $46\frac{3}{8}$ **83.** $20\frac{3}{4}$ **84.** $56\frac{1}{24}$ **85.** $\frac{1}{4}$ **86.** $\frac{2}{3}$ **87.** 1 **88.** 0 **89.** $\frac{1}{4}$ **90.** $\frac{3}{8}$ **91.** $\frac{7}{20}$ **92.** $\frac{7}{30}$ **93.** $7\frac{1}{2}$ **94.** $2\frac{3}{10}$ **95.** $3\frac{3}{4}$ **96.** $18\frac{1}{2}$ **97.** $6\frac{1}{2}$ **98.** $1\frac{1}{10}$ **99.** $2\frac{2}{3}$ **100.** $\frac{1}{5}$ **101.** $1\frac{4}{5}$ **102.** $\frac{3}{4}$ **103.** $2\frac{1}{2}$ **104.** $3\frac{1}{3}$ **105.** $\frac{3}{10}$ **106.** $2\frac{7}{8}$ **107.** $\frac{7}{12}$ **108.** $3\frac{8}{9}$ **109.** $\frac{2}{3}$ **110.** $9\frac{9}{20}$ **111.** $\frac{3}{16}$ **112.** $\frac{7}{16}$ **113.** $\frac{5}{8}$ **114.** $\frac{1}{6}$ **115.** $5\frac{1}{3}$

116. $\frac{7}{10}$ **117.** $\frac{1}{125}$ **118.** $\frac{8}{27}$ **119.** $\frac{1}{4}$ **120.** $\frac{7}{120}$ **121.** $\frac{24}{25}$ **122.** $1\frac{5}{9}$ **123.** $2\frac{2}{3}$
124. $\frac{2}{3}$ **125.** 6 **126.** $18\frac{5}{12}$ **127.** $8\frac{7}{16}$ **128.** $21\frac{1}{4}$ **129.** $\frac{9}{20}$ **130.** $1\frac{9}{16}$
131. $37\frac{1}{27}$ **132.** $3\frac{3}{8}$ **133.** $3\frac{1}{8}$ **134.** $1\frac{41}{90}$ **135.** $2\frac{1}{10}$ **136.** $7\frac{1}{5}$ **137.** $\frac{3}{2}$
138. $\frac{2}{3}$ **139.** $\frac{1}{8}$ **140.** 4 **141.** $\frac{7}{40}$ **142.** $\frac{5}{81}$ **143.** $\frac{2}{15}$ **144.** $\frac{1}{200}$ **145.** $\frac{3}{4}$
146. $1\frac{1}{3}$ **147.** 30 **148.** $8\frac{3}{4}$ **149.** $1\frac{1}{6}$ **150.** $1\frac{4}{5}$ **151.** 2 **152.** 4 **153.** $1\frac{3}{4}$
154. $\frac{4}{7}$ **155.** $1\frac{7}{12}$ **156.** $\frac{12}{19}$ **157.** $5\frac{1}{2}$ **158.** $2\frac{11}{20}$ **159.** 2 **160.** 3 **161.** $9\frac{3}{4}$
162. $1\frac{3}{10}$ **163.** $5\frac{1}{3}$ **164.** $7\frac{5}{9}$ **165.** No **166.** 50¢ **167.** $\frac{1}{4}$ **168.** $\frac{2}{9}$
169. The Filmworks camera **170.** $\frac{7}{12}$ **171.** The patient got back more than $\frac{1}{3}$, because $\frac{275}{700} = \frac{11}{28} = \frac{33}{84}$, which is greater than $\frac{1}{3} = \frac{28}{84}$. **172.** Yes it should, because $\frac{23}{32}$ is greater than $\frac{2}{3}$. $\frac{23}{32} = \frac{69}{96}$, whereas $\frac{2}{3} = \frac{64}{96}$ **173. a.** $\frac{12}{23}$ **b.** $\frac{3}{4}$ **174. a.** Lisa Gregory **b.** Monica Yates **175.** $\frac{3}{4}$ **176.** $\frac{11}{12}$ oz
177. $\frac{1}{4}$ carat **178.** $\frac{3}{5}$ **179.** 12 women **180.** 2,685 undergraduate students
181. 2 awardees **182.** 7 lb **183.** $1,050 **184.** 2 times **185.** 19 fish
186. $11\frac{3}{4}$ mi **187.** $11\frac{1}{2}$ ft **188.** $7\frac{5}{12}$ hr **189.** 1,500 fps **190.** 500 lb/sq in.
191. $281\frac{1}{4}$ lb **192.** 3,100,000 **193.** 8 orbits **194.** $25\frac{5}{8}$ sq mi

195.

Employee	Saturday	Sunday	Total
L. Chavis	$7\frac{1}{2}$	$4\frac{1}{4}$	$11\frac{3}{4}$
R. Young	$5\frac{3}{4}$	$6\frac{1}{2}$	$12\frac{1}{4}$
Total	$13\frac{1}{4}$	$10\frac{3}{4}$	24

196.

Worker	Hours per Day	Days Worked	Total Hours	Wage per Hour	Gross Pay
Maya	5	3	15	$7	$105
Noel	$7\frac{1}{4}$	4	29	$10	$290
Alisa	$4\frac{1}{2}$	$5\frac{1}{2}$	$24\frac{3}{4}$	$9	$\$222\frac{3}{4}$

197. $10\frac{10}{11}$ lb **198.** $22\frac{1}{2}$ cups **199.** 3,200 mi **200.** 6 times

Chapter 2 Posttest, p. 155

1. 1, 3, 7, 9, 21, 63 **2.** 2×3^3 **3.** $\frac{4}{9}$ **4.** $\frac{12}{1}$ **5.** $10\frac{1}{4}$ **6.** $\frac{7}{8}$ **7.** $\frac{5}{10}$ **8.** 24
9. $1\frac{13}{24}$ **10.** $8\frac{7}{40}$ **11.** $4\frac{2}{7}$ **12.** $5\frac{23}{30}$ **13.** $\frac{1}{81}$ **14.** 12 **15.** $\frac{7}{9}$ **16.** $7\frac{5}{6}$ **17.** $\frac{10}{11}$
18. $19\frac{1}{5}$ mi **19.** $\frac{5}{6}$ hr **20.** $94

Chapter 2 Cumulative Review, p. 156

1. Five million, three hundred fifteen **2.** 1,900,000 **3.** 581,400 **4.** 908
5. 1 **6.** $\frac{1}{25}$ **7.** $2^2 \times 3 \times 7$ **8.** 120 **9.** $\frac{3}{4}$ **10.** $\frac{3}{8}$ **11.** $1\frac{11}{24}$ **12.** $6\frac{2}{5}$ **13.** 7
14. $1\frac{5}{11}$ **15.** $37 billion **16.** 1 million times **17.** 12 above
18. Yes. The room has 370 square feet of wall area. **19.** $\frac{1}{3}$ **20.** 4 pieces

CHAPTER 3

Chapter 3 Pretest, p. 158

1. Hundredths **2.** Four and twelve thousandths **3.** 3.1 **4.** 0.0029
5. 21.52 **6.** 7.3738 **7.** 11.69 **8.** 9.81 **9.** 8,300 **10.** 18.423
11. 0.0144 **12.** 7.1 **13.** 0.00605 **14.** 32.7 **15.** 0.875 **16.** 2.83
17. One with a pH value of 2.95 **18.** $58.44 billion **19.** 3 times
20. $3.74

Section 3.1 Practices: pp. 160–167

1, *p. 160:* **a.** The tenths place **b.** The ten-thousandths place **c.** The thousandths place **2, *p. 161:*** $\frac{7}{8}$ **3, *p. 161:*** $2\frac{3}{100}$ **4, *p. 162:*** **a.** $5\frac{3}{5}$ **b.** $5\frac{3}{5}$ **5, *p. 162:*** **a.** $7\frac{3}{1{,}000}$ **b.** $4\frac{1}{10}$ **6, *p. 162:*** **a.** Sixty-one hundredths **b.** Four and nine hundred twenty-three thousandths **c.** Seven and five hundredths **7, *p. 163:*** **a.** 0.043 **b.** 10.26 **8, *p. 163:*** 3.14 **9, *p. 164:*** 0.8297 **10, *p. 164:*** 3.51, 3.5, 3.496 **11, *p. 165:*** The one with the rating of 8.1, because 9 > 8.2 > 8.1 **12, *p. 166–167:*** **a.** 748.1 **b.** 748.08 **c.** 748.077 **d.** 748 **e.** 700 **13, *p. 167:*** 7.30 **14, *p. 167:*** 11.7 m

Exercises 3.1, pp. 168–172

1. right **3.** hundredths **5.** greater **7.** $2.\underline{7}8$ **9.** $9.0\underline{1}$ **11.** $2.001\underline{7}5$
13. $823.00\underline{1}$ **15.** Tenths **17.** Hundredths **19.** Thousandths **21.** Ones
23. Fifty-three hundredths **25.** Three hundred five thousandths
27. Six tenths **29.** Five and seventy-two hundredths **31.** Twenty-four and two thousandths **33.** 0.8 **35.** 1.041 **37.** 60.01 **39.** 4.107
41. 3.2 m **43.** $\frac{3}{5}$ **45.** $\frac{39}{100}$ **47.** $1\frac{1}{2}$ **49.** 8 **51.** $5\frac{3}{250}$ **53.** > **55.** <
57. > **59.** = **61.** < **63.** 7, 7.07, 7.1 **65.** 4.9, 5.001, 5.2
67. 9.1 mi, 9.38 mi, 9.6 mi **69.** 17.4 **71.** 3.591 **73.** 37.1 **75.** 0.40
77. 7.06 **79.** 9 mi

81.

To the Nearest	8.0714	0.9916
Tenth	8.1	1.0
Hundredth	8.07	0.99
Ten	10	0

83. $0.\underline{0}24$ **85.** 870.06 **87.** 2.04 m, 2.14 m, 2.4 m **89.** Twenty-three and nine hundred thirty-four thousandths **91.** Eighteen and seven tenths; eighteen and eight tenths **93.** Two hundred eleven and seven tenths, sixty-nine and four tenths, one hundred eighty-nine and eight tenths, one hundred ninety-three and five tenths, forty-seven and five tenths **95.** One hundred-thousandth; eight hundred-thousandths **97.** 1.2 acres **99.** 74.59 mph
101. 14.7 lb **103.** 9.6 V **105.** 352.1 kWh **107.** Evgeni Plushenko
109. Last winter **111.** 2005 **113.** Husband's **115.** $57.03 **117.** 0.001
119. 1.4

Section 3.2 Practices: pp. 173–177

1, *p. 173:* 10.387 **2, *p. 173:*** 39.3 **3, *p. 174:*** 102.1°F **4, *p. 174:*** 46.2125
5, *p. 175:* $485.43 **6, *p. 175:*** 16.9 mi **7, *p. 175:*** 22.13 mi **8, *p. 176:*** 0.863
9, *p. 176:* 0.079 **10, *p. 176:*** 0.5744 **11, *p. 177:*** Possible estimate: $480
12, *p. 177:* Possible estimate: $2 million

Calculator Practices, p. 178

13, *p. 178:* 79.23; **14, *p. 178:*** 0.00002

Exercises 3.2, pp. 179–182

1. decimal points **3.** sum **5.** 9.33 **7.** 0.9 **9.** 8.13 **11.** 21.45
13. 7.67 **15.** $77.21 **17.** 1.08993 **19.** 24.16 **21.** 44.422
23. 20.32 mm **25.** 16.682 kg **27.** 23.30595 **29.** 0.7 **31.** 16.8
33. 18.41 **35.** 75.63 **37.** 22.324 **39.** 0.17 **41.** 0.1142 **43.** 6.2
45. 15.37 **47.** 5.9 **49.** 6.21 **51.** 1.85 lb **53.** 4.9°F **55.** 39.752
57. 27.9 mg **59.** 3.205 **61.** 21.19896 **63.** c; possible estimate: 0.084
65. b; possible estimate: 0.06 **67.** 7.771 **69.** 7.75 lb **71.** 11.6013
73. $1.03 **75.** 56.8 centuries **77.** $1.7 million **79.** 6.84 in.
81. Yes; 2.8 + 2.9 + 2.6 + 1.6 = 9.9

83. a.

Gymnast	VT	UB	BB	FX	AA
Nastia Liukin (U.S.)	15.1	15.95	15.975	15.35	62.375
Yang Yilin (China)	15.2	16.65	15.5	15	62.35
Shawn Johnson (U.S)	16	15.325	15.975	15.425	62.725

b. Shawn Johnson **85.** A total of 16.2 mg of iron; no, she needs 1.8 mg more

Section 3.3 Practices: pp. 183–186

1, *p. 183:* 9.835 **2, *p. 184:*** 1.4 **3, *p. 184:*** 0.01 **4, *p. 184:*** 0.024
5, *p. 184:* 9.91 **6, *p. 185:*** 325 **7, *p. 185:*** 327,000 **8, *p. 185:*** **a.** 18.015 **b.** 18 **9, *p. 186:*** 0.0003404; possible estimate: 0.004 × 0.09 = 0.00036
10, *p. 186:* 3.6463 **11, *p. 186:*** Possible answer: 1,200 mi

ANSWERS

Calculator Practices, p. 187

12, ***p. 187:*** 815.6 **13,** ***p. 187:*** 9.261

Exercises 3.3, pp. 188–191

1. multiplication **3.** two **5.** square **7.** 2.99212 **9.** 204.360 **11.** 2,492.0 **13.** 0.0000969 **15.** 2,870.00 **17.** $0.73525 **19.** 0.54 **21.** 0.4 **23.** 0.02 **25.** 0.0028 **27.** 0.765 **29.** 2.016 **31.** 7.602 **33.** 0.5 **35.** 5.852 **37.** 151.14 **39.** 3.7377 **41.** 1.7955 **43.** 8,312.7 **45.** 23 **47.** 0.09 **49.** 1.05 **51.** 0.000000001 **53.** 42.5 ft **55.** 1.4 mi **57.** 42.77325 **59.** 272,593.75 **61.** 70 **63.** 25.75 **65.** 1.09 **67.** 2.86 **69.** 3.952 **71.** 0.14

73.

Input	Output
1	$3.8 \times \mathbf{1} - 0.2 = 3.6$
2	$3.8 \times \mathbf{2} - 0.2 = 7.4$
3	$3.8 \times \mathbf{3} - 0.2 = 11.2$
4	$3.8 \times \mathbf{4} - 0.2 = 15$

75. a; possible estimate: 50 **77.** b; possible estimate: 0.014 **79.** 8.75 **81.** 0.068 **83.** 4.48 **85.** 2,900 fps **87.** 57,900,000 km **89.** 254.3 sq ft **91.** 1.25 mg **93.** 1,308 calories

95. a.

Purchase	Quantity	Unit Price	Price
Belt	1	$11.99	$11.99
Shirt	3	$16.95	$50.85
Total Price			$62.84

b. $17.16 **97.** 88.81 in.

Section 3.4 Practices: pp. 193–199

1, ***p. 193:*** 0.375 **2,** ***p. 193:*** 7.625 **3,** ***p. 194:*** 83.3 **4.** ***p. 194:*** 0.1 **5,** ***p. 196:*** 18.04 **6,** ***p. 196:*** 2,050 **7,** ***p. 197:*** 73.4 **8,** ***p. 197:*** 0.0341 **9,** ***p. 198:*** 0.00086 **10,** ***p. 198:*** 1.5 **11,** ***p. 199:*** 21.1; possible estimate: 20 **12,** ***p. 199:*** 295.31 **13,** ***p. 199:*** 8 times as great

Calculator Practices, p. 200

14, ***p. 200:*** 0.2 **15,** ***p. 200:*** 4.29

Exercises 3.4, pp. 201–204

1. decimal **3.** right **5.** quotient **7.** 0.5 **9.** 0.375 **11.** 3.7 **13.** 1.625 **15.** 6.2 **17.** 21.03 **19.** 0.67 **21.** 0.78 **23.** 3.11 **25.** 5.06 **27.** 4.25 **29.** 4.2 **31.** 1.375 **33.** 8.5 **35.** 3.286 **37.** 0.273 **39.** 6.571 **41.** 70.077 **43.** 58.82 **45.** 0.0663 **47.** 2.8875 **49.** 0.286 **51.** 4.3 **53.** 0.0015 **55.** 1.73 **57.** 2.875 **59.** 4 **61.** 70.4 **63.** 94 **65.** 12.5 **67.** 0.3 **69.** 0.2 **71.** 0.952 **73.** 0.00082 **75.** 383.88 **77.** 0.01 **79.** 9.23 **81.** 9,666.67 **83.** 1,952.38 **85.** 325.18 **87.** 67.41 **89.** 41.61 **91.** 2.765 **93.** 32.9 **95.** 52.2 **97.** 4.05 **99.** 396.5 **101.** 49.9

103.

Input	Output
1	$\mathbf{1} \div 5 - 0.2 = 0$
2	$\mathbf{2} \div 5 - 0.2 = 0.2$
3	$\mathbf{3} \div 5 - 0.2 = 0.4$
4	$\mathbf{4} \div 5 - 0.2 = 0.6$

105. c; possible estimate: 50 **107.** b; possible estimate: 0.2 **109.** 0.8 **111.** 1.17 **113.** 0.45 **115.** 0.0037 in. per yr **117. a.** 0.6 **b.** 0.55 **c.** The women's team has a better record. The team won $\frac{3}{5}$, or 0.6, of the games played, and the men's team won $\frac{11}{20}$, or 0.55, of the games played.

119. a.

SUV	Distance Driven (in miles)	Gasoline Used (in gallons)	Gasoline Mileage (miles per gallon)
Honda CR-V	40.5	1.9	21
Ford Escape Hybrid	62.4	2.4	26
GMC Terrain	42.6	2.4	18

b. Ford Escape Hybrid **121.** 2,000 shares **123.** 13 times **125.** 0.4 lb **127.** .366

Chapter 3 Review Exercises, pp. 207–210

1. is not; possible answer: it does not have a decimal point (or it is written in fractional form). **2.** is not; possible answer: decimal places are to the right of the decimal point, and this digit is to the left. **3.** is not; possible answer: the number to the right of the critical digit 2 is 6, so we round up, getting 48.73 **4.** can; possible answer: then each column will contain digits with the same place value. **5.** is not; possible answer: no digits repeat indefinitely. **6.** is; possible answer: the dividend is smaller than the divisor. **7.** Hundredths **8.** Tenths **9.** Tenths **10.** Ten-thousandths **11.** $\frac{7}{20}$ **12.** $8\frac{1}{5}$ **13.** $4\frac{7}{1,000}$ **14.** 10 **15.** Seventy-two hundredths **16.** Five and six tenths **17.** Three and nine ten-thousandths **18.** Five hundred ten and thirty-six thousandths **19.** 0.007 **20.** 2.1 **21.** 0.09 **22.** 7.041 **23.** $<$ **24.** $>$ **25.** $>$ **26.** $>$ **27.** 1.002, 0.8, 0.72 **28.** 0.004, 0.003, 0.00057 **29.** 7.3 **30.** 0.039 **31.** 4.39 **32.** $899 **33.** 12.11 **34.** 52.75 **35.** $24.13 **36.** 12 m **37.** 28.78 **38.** 87.752 **39.** 1.834 **40.** 48.901 **41.** 98.2033 **42.** $90,948.80 **43.** 2.912 **44.** 1,008 **45.** 0.00001 **46.** 13.69 **47.** 2,710 **48.** 0.034 **49.** 5.75 **50.** 13.5 **51.** 1,569.36846 **52.** 441.760662 **53.** 0.625 **54.** 90.2 **55.** 4.0625 **56.** 0.045 **57.** 0.17 **58.** 0.29 **59.** 8.33 **60.** 11.22 **61.** 0.65 **62.** 1.6 **63.** 0.175 **64.** 0.277 **65.** 5.2 **66.** 3.2 **67.** 23.7 **68.** 16,358.3 **69.** 1.9 **70.** 360.7 **71.** 3.0 **72.** 0.3 **73.** 1.18 **74.** 117 **75.** 34.375 **76.** 1.4 **77.** 54.49 sec **78.** $14.14 **79.** Four ten-millionths **80.** 14.22 in. **81.** 1.5 AU. **82.** 4.35 times **83.** $0.06 **84.** $250 **85.** 7.19 g **86.** 3.5°C **87.** 36,162.45

88.

Period Ending	Google	Yahoo!
June 30	5.523	1.573
September 30	5.945	1.575
December 31	6.674	1.732
March 31	6.775	1.597

$18.440 billion

Chapter 3 Posttest, p. 211

1. 6 **2.** Five and one hundred two thousandths **3.** 320.15 **4.** 0.00028 **5.** $3\frac{1}{25}$ **6.** 0.004 **7.** 4.354 **8.** $5.66 **9.** 20.9 **10.** 5.72 **11.** 0.001 **12.** 3.36 **13.** 0.0029 **14.** 32.7 **15.** 0.375 **16.** 4.17 **17.** 0.01 lb **18.** 2.6 ft **19.** Belmont Stakes **20.** $32.40

Chapter 3 Cumulative Review, p. 212

1. 1,000,000 **2.** 2,076 **3.** 27,403 **4.** 900 sq m **5.** 42 **6.** 1, 2, 3, 4, 5, 6, 10, 12, 15, 20, 30, 60 **7.** $1\frac{1}{2}$ **8.** $2\frac{2}{3}$ **9.** 32 **10.** $\frac{17}{30}$ **11.** $4\frac{18}{25}$ **12.** 38.4 **13.** 60.213 **14.** 610 **15.** $0.17 **16.** $\frac{2}{15}$ acre **17.** 325 **18.** 26,000 mi **19.** $193.86 **20.** 2.6

CHAPTER 4

Chapter 4 Pretest, p. 214

1. Possible answer: four less than t **2.** Possible answer: quotient of y and three **3.** $m + 8$ **4.** $2n$ **5.** 4 **6.** $1\frac{1}{2}$ **7.** $x + 3 = 5$ **8.** $4y = 12$ **9.** $x = 6$ **10.** $t = 10$ **11.** $n = 13$ **12.** $a = 12$ **13.** $m = 6.1$ **14.** $n = 30$ **15.** $m = 13\frac{1}{2}$, or 13.5 **16.** $n = 15$ **17.** $63 = x + 36$; 27 moons **18.** $6.75 = x - 2.75$; \$9.50 **19.** $\frac{2}{5}x = 39{,}900$; 99,750 sq mi **20.** $40 = 10x$; 4 mg

Section 4.1 Practices: pp. 216–218

1, *p. 216:* Answers may vary **a.** One-half of p **b.** x less than 5 **c.** y divided by 4 **d.** 3 more than n **e.** $\frac{3}{5}$ of b **2, *p. 216:*** **a.** $x + 9$ **b.** $10y$ **c.** $n - 7$ **d.** $p \div 5$ **e.** $\frac{2}{5}v$ **3, *p. 217:*** **a.** $q + 12$, where q represents the quantity **b.** $\frac{9}{a}$, where a represents the account balance **c.** $\frac{2}{7}c$, where c represents the cost **4, *p. 217:*** $\frac{h}{4}$ hr **5, *p. 217:*** $s - 3$ **6, *p. 218:*** **a.** 25 **b.** 0.38 **c.** 4.8 **d.** 26.6 **7, *p, 218:*** $\frac{1}{5}n$ dollars; \$750 **8, *p. 218:*** The total amount was $(18.45 + t)$ dollars; \$21.45 for $t =$ \$3

Exercises 4.1, pp. 219–221

1. variable **3.** algebraic **5.** 9 more than t; t plus 9 **7.** c minus 12; 12 subtracted from c **9.** c divided by 3; the quotient of c and 3; **11.** 10 times s; the product of 10 and s **13.** y minus 10; 10 less than y **15.** 7 times a; the product of 7 and a **17.** x divided by 6; the quotient of x and 6 **19.** x minus $\frac{1}{2}$; $\frac{1}{2}$ less than x **21.** $\frac{1}{4}$ times w; $\frac{1}{4}$ of w **23.** 2 minus x; the difference between 2 and x **25.** 1 increased by x; x added to 1 **27.** 3 times p; the product of 3 and p **29.** n decreased by 1.1; n minus 1.1 **31.** y divided by 0.9; the quotient of y and 0.9 **33.** $x + 10$ **35.** $n - 1$ **37.** $y + 5$ **39.** $t \div 6$ **41.** $10y$ **43.** $w - 5$ **45.** $n + \frac{4}{5}$ **47.** $z \div 3$ **49.** $\frac{2}{7}x$ **51.** $k - 6$ **53.** $n + 12$ **55.** $n - 5.1$ **57.** 26 **59.** 2.5 **61.** 15 **63.** $1\frac{1}{6}$ **65.** 1.1 **67.** $\frac{1}{5}$

69.

x	$x + 8$
1	9
2	10
3	11
4	12

71.

n	$n - 0.2$
1	0.8
2	1.8
3	2.8
4	3.8

73.

x	$\frac{3}{4}x$
4	3
8	6
12	9
16	12

75.

z	$\frac{z}{2}$
2	1
4	2
6	3
8	4

77. $x - 7$ **79.** Possible answers: n over 2; n divided by 2 **81.** $3.5t$ **83.** Possible answers: 6 more than x; the sum of x and 6 **85.** $(m - 25)$ mg **87.** $25° + 90° + d°$, or $115° + d°$ **89.** 220 mi **91.** **a.** $2.5w$ dollars **b.** \$22.50

Section 4.2 Practices: pp. 223–227

1, *p. 223:* **a.** $n - 5.1 = 9$ **b.** $y + 2 = 12$ **c.** $n - 4 = 11$ **d.** $n + 5 = 7\frac{3}{4}$ **2, *p. 224:*** $p - 6 = 49.95$, where p is the regular price. **3, *p. 225:*** $x = 9$; **4, *p. 225:*** $t = 2.7$; **5, *p. 226:*** $m = 5\frac{1}{4}$; **6, *p. 226:*** **a.** $11 = m - 4$; $m = 15$ **b.** $12 + n = 21$; $n = 9$ **7, *p. 227:*** $x + 3.99 = 27.18$; \$23.19 **8, *p. 227:*** $269{,}000 = x - 394{,}000$; 663,000 sq mi

Exercises 4.2, pp. 228–231

1. equation **3.** subtract **5.** $z - 9 = 25$ **7.** $7 + x = 25$ **9.** $t - 3.1 = 4$ **11.** $\frac{3}{2} + y = \frac{9}{2}$ **13.** $n - 3\frac{1}{2} = 7$ **15.** **a.** Yes **b.** No **c.** Yes **d.** No **17.** Subtract 4. **19.** Add 6. **21.** Add 7. **23.** Subtract 21. **25.** $a = 31$ **27.** $y = 2$ **29.** $x = 12$ **31.** $n = 4$ **33.** $m = 2$ **35.** $y = 90$ **37.** $z = 2.9$ **39.** $n = 8.9$ **41.** $y = 0.9$ **43.** $x = 8\frac{2}{3}$ **45.** $m = 5\frac{1}{3}$ **47.** $x = 3\frac{3}{4}$ **49.** $c = 47\frac{1}{5}$ **51.** $x = 13$ **53.** $y = 6\frac{1}{4}$ **55.** $a = 3\frac{5}{12}$ **57.** $x = 8.2$ **59.** $y = 19.91$ **61.** $x = 4.557$ **63.** $y = 10.251$ **65.** $n + 3 = 11$; $n = 8$ **67.** $y - 6 = 7$; $y = 13$ **69.** $n + 10 = 19$; $n = 9$ **71.** $x + 3.6 = 9$; $x = 5.4$ **73.** $n - 4\frac{1}{3} = 2\frac{2}{3}$; $n = 7$ **75.** Equation c **77.** Equation a **79.** $a = 14.5$ **81.** Equation b **83.** Yes **85.** $4.2 + n = 8$ **87.** Add 1.9. **89.** $x + 12 = 106$; \$94 **91.** $40° + x = 90°$; 50° **93.** $621{,}000 = x - 13{,}000$; \$634,000 **95.** $45 = x - 20$; 65 mph **97.** $m + 1{,}876{,}674{,}000 = 4{,}023{,}362{,}895$; \$2,146,688,895

Section 4.3 Practices, pp. 232–237

1, *p. 232:* **a.** $2x = 14$ **b.** $\frac{a}{6} = 1.5$ **c.** $\frac{n}{0.3} = 1$ **d.** $10 = \frac{1}{2}n$ **2, *p. 232:*** $15 = 3w$ **3, *p. 233:*** $x = 5$ **4, *p. 233:*** $a = 6$ **5, *p. 234:*** $x = 4$ **6, *p. 234:*** $a = 2.88$ **7, *p. 235:*** $x = 16$ **8, *p. 235:*** **a.** $12 = \frac{z}{6}, z = 72$; $12 \stackrel{?}{=} \frac{72}{6}$, $12 \stackrel{\checkmark}{=} 12$ **b.** $16 = 2x$, $8 = x$, *or* $x = 8$; $16 \stackrel{?}{=} 2(8)$, $16 \stackrel{\checkmark}{=} 16$ **9, *p. 236:*** $1.6 = 5x$; 0.32 km **10, *p. 236:*** $\frac{x}{25.5} = 87$; \$2,218.50 **11, *p. 237:*** $\frac{3}{8}p = 150{,}000$; \$400.000

Exercises 4.3, pp. 238–241

1. divide **3.** substituting **5.** equation **7.** $\frac{3}{4}y = 12$ **9.** $\frac{x}{7} = \frac{7}{2}$ **11.** $\frac{1}{3}x = 2$ **13.** $\frac{n}{3} = \frac{1}{3}$ **15.** $9a = 27$ **17.** **a.** Yes **b.** No **c.** No **d.** No **19.** Divide by 3. **21.** Multiply by 2. **23.** Divide by $\frac{3}{4}$, or multiply by $\frac{4}{3}$. **25.** Divide by 1.5. **27.** $x = 6$ **29.** $x = 18$ **31.** $n = 4$ **33.** $x = 91$ **35.** $y = 4$ **37.** $b = 20$ **39.** $m = 157.5$ **41.** $t = 0.4$ **43.** $x = \frac{3}{2}$, or $1\frac{1}{2}$ **45.** $x = 36$ **47.** $t = 3$ **49.** $y = \frac{2}{5}$ **51.** $n = 700$ **53.** $x = 12.5$ **55.** $x = \frac{1}{2}$ **57.** $m = 6$ **59.** $x \approx 6.8$ **61.** $x \approx 4.9$ **63.** $8n = 56$; $n = 7$ **65.** $\frac{3}{4}y = 18$; $y = 24$ **67.** $\frac{x}{5} = 11$; $x = 55$ **69.** $2x = 36$; $x = 18$ **71.** $\frac{1}{2}a = 4$; $a = 8$ **73.** $\frac{n}{5} = 1\frac{3}{5}$; $n = 8$ **75.** $\frac{n}{2.5} = 10$; $n = 25$ **77.** Equation d **79.** Equation a **81.** $x = 5.5$ **83.** Equation d **85.** Yes **87.** $2x = 5$ **89.** Multiply by 2. **91.** $4s = 60$; 15 units **93.** $56 = \frac{1}{2}x$; 112 mi **95.** $\frac{c}{3} = 8.99$; \$26.97 **97.** **a.** $\frac{2}{5}x = 60$; 150 ml **b.** 90 ml **99.** $\frac{1}{3}x = 128$; \$384 million **101.** $\frac{p}{3{,}537{,}438} = 86.8$; 307,000,000 people

Chapter 4 Review Exercises, pp. 243–245

1. is; possible answer: x represents an unknown number **2.** is not; possible answer: constants are known numbers such as 6 or -5 **3.** can; possible answer: algebraic expressions combine constants, variables, and algebraic operations **4.** cannot; possible answer: an equation is a mathematical statement that two expressions are equal **5.** is; possible answer: substituting 28 for x makes the equation $72 - 28 = 44$ a true statement **6.** is; possible answer: x is alone on one side of the equation **7.** x plus 1 **8.** Four more than y **9.** w minus 1 **10.** Three less than s **11.** c divided by 7 **12.** The quotient of a and 10 **13.** Two times x **14.** The product of 6 and y **15.** y divided by 0.1 **16.** The quotient of n and 1.6 **17.** One-third of x **18.** One-tenth of w **19.** $m + 9$ **20.** $b + \frac{1}{2}$ **21.** $y - 1.4$ **22.** $z - 3$ **23.** $\frac{3}{x}$ **24.** $n \div 2.5$ **25.** $3n$ **26.** $12n$ **27.** 12 **28.** 19 **29.** 0 **30.** 6 **31.** 0.3 **32.** 6.5 **33.** $1\frac{1}{2}$ **34.** $\frac{5}{12}$ **35.** 0.4 **36.** $4\frac{1}{2}$ **37.** 1.6 **38.** 9 **39.** $x = 9$ **40.** $y = 9$ **41.** $n = 26$ **42.** $b = 20$ **43.** $a = 3.5$ **44.** $c = 7.5$ **45.** $x = 11$ **46.** $y = 2$ **47.** $w = 1\frac{1}{2}$ **48.** $s = \frac{1}{3}$ **49.** $c = 6\frac{3}{4}$ **50.** $p = 11\frac{2}{3}$ **51.** $m = 5$ **52.** $n = 0$ **53.** $c = 78$ **54.** $y = 90$ **55.** $n = 11$ **56.** $x = 25$ **57.** $x = 31.0485$ **58.** $m = 26.6225$ **59.** $n - 19 = 35$ **60.** $a - 37 = 234$ **61.** $9 + n = 15\frac{1}{2}$ **62.** $n + 26 = 30\frac{1}{3}$ **63.** $2y = 16$ **64.** $25t = 175$ **65.** $34 = \frac{n}{19}$ **66.** $17 = \frac{z}{13}$ **67.** $\frac{1}{3}n = 27$ **68.** $\frac{2}{5}n = 4$ **69.** **a.** No **b.** Yes **c.** Yes **d.** No **70.** **a.** Yes **b.** No **c.** No **d.** Yes **71.** $x = 5$ **72.** $t = 2$ **73.** $a = 105$ **74.** $n = 54$ **75.** $y = 9$ **76.** $r = 10$ **77.** $w = 90$ **78.** $x = 100$ **79.** $y = 20$ **80.** $a = 120$ **81.** $n = 32$ **82.** $b = 32$ **83.** $m = 3.15$ **84.** $z = 0.57$ **85.** $x = \frac{2}{5}$, or 0.4 **86.** $t = \frac{1}{2}$, or 0.5 **87.** $m = 1.2$ **88.** $b = 9.8$ **89.** $x = 12.5$ **90.** $x = 1.4847$ **91.** $2h$ degrees; 6 degrees **92.** $\frac{d}{20}$ dollars per hr; \$9.55 per hr **93.** \$0.89p; \$2.67 **94.** $(3{,}000 + d)$ dollars; \$3,225 **95.** $x + 238 = 517$; \$279 **96.** $225 = x + 50$; 175 **97.** $2.9x = 100$; 34 L

98. $\frac{1}{4}x = 500{,}000$; 2,000,000 people **99.** $\frac{x}{6} = 30$; 180 lb **100.** $2.5x = 3{,}000{,}000{,}000{,}000$; \$1,200,000,000,000 or \$1.2 trillion **101.** $98.6 + x = 101$; 2.4°F **102.** $x - 256 = 8{,}957$; 9,213 applications

Chapter 4 Posttest, p. 246

1. Possible answer: x plus $\frac{1}{2}$ **2.** Possible answer: the quotient of a and 3 **3.** $n - 10$ **4.** $\frac{8}{p}$ **5.** 0 **6.** $\frac{1}{4}$ **7.** $x - 6 = 4\frac{1}{2}$ **8.** $\frac{y}{8} = 3.2$ **9.** $x = 0$ **10.** $y = 12$ **11.** $n = 27$ **12.** $a = 738$ **13.** $m = 7.8$ **14.** $n = 50$ **15.** $x = \frac{11}{20}$ **16.** $n = 760$ **17.** $1\frac{3}{4} + x = 2\frac{1}{4}$; $\frac{1}{2}$ lb **18.** $\frac{1}{4}x = 500$; 2,000 wolves **19.** $1.5x = 9$; 6 billion **20.** $x - 19.8 = 7.6$; 27.4°C

Chapter 4 Cumulative Review, pp. 247–248

1. 314,200 **2.** c **3.** 23,316 **4.** 1,030 **5.** $\frac{84}{96}$ **6.** $1\frac{2}{5}$ **7.** $5\frac{3}{8}$ **8.** Five and two hundred thirty-nine thousandths **9.** $<$ **10.** 3.89 **11.** 0.0075 **12.** Yes **13.** $n = 7.8$ **14.** $x = 32$ **15.** 7,200 images **16.** He got back $\frac{2}{7}$ of his money, which is less than $\frac{1}{3}$. **17.** $\frac{3}{5}x = 300{,}000{,}000$; 500 million tons **18.** 92.4 lb **19.** 55,000 beehives **20. a.** $4.5x = 2{,}900{,}000$ **b.** 600,000 personnel

CHAPTER 5

Chapter 5 Pretest, p. 250

1. $\frac{3}{4}$ **2.** $\frac{2}{5}$ **3.** $\frac{5}{3}$ **4.** $\frac{19}{51}$ **5.** $\frac{16 \text{ gal}}{5 \text{ min}}$ **6.** $\frac{5 \text{ mg}}{3 \text{ hr}}$ **7.** $\frac{2 \text{ dental assistants}}{1 \text{ dentist}}$ **8.** $\frac{1 \text{ calculator}}{1 \text{ student}}$ **9.** $\frac{\$230}{\text{box}}$ **10.** $\frac{\$0.50}{\text{bottle}}$ **11.** True **12.** False **13.** $x = 9$ **14.** $x = 31\frac{1}{2}$ **15.** $x = 16$ **16.** $x = 160$ **17.** $\frac{4}{5}$ **18.** 200 lb/min **19.** 4.5 in. **20.** 76 mi

Section 5.1 Practices, pp. 251–255

1, ***p. 251:*** $\frac{2}{3}$ **2,** ***p. 252:*** $\frac{9}{5}$ **3,** ***p. 252:*** $\frac{20}{19} \approx 1.05 > 1$; yes **4,** ***p. 253:*** **a.** $\frac{5 \text{ mL}}{2 \text{ min}}$ **b.** $\frac{3 \text{ lb}}{2 \text{ wk}}$ **5,** ***p. 253:*** **a.** 48 ft/sec **b.** 0.375 hit per time at bat **6,** ***p. 254:*** 1.5 min/city block **7,** ***p. 254:*** **a.** \$174/flight **b.** \$2.75/hr **c.** \$0.99/download **8,** ***p. 255:*** The 150-caplet bottle

Exercises 5.1, pp. 256–260

1. quotient **3.** simplest form **5.** denominator **7.** $\frac{2}{3}$ **9.** $\frac{2}{3}$ **11.** $\frac{11}{7}$ **13.** $\frac{3}{2}$ **15.** $\frac{1}{4}$ **17.** $\frac{4}{3}$ **19.** $\frac{1}{1}$ **21.** $\frac{5}{3}$ **23.** $\frac{7}{24}$ **25.** $\frac{20}{1}$ **27.** $\frac{8}{7}$ **29.** $\frac{4}{5}$ **31.** $\frac{5 \text{ calls}}{2 \text{ days}}$ **33.** $\frac{36 \text{ cal}}{5 \text{ min}}$ **35.** $\frac{1 \text{ million hits}}{3 \text{ mo}}$ **37.** $\frac{17 \text{ baskets}}{30 \text{ attempts}}$ **39.** $\frac{37 \text{ points}}{2 \text{ games}}$ **41.** $\frac{100 \text{ sq ft}}{\$329}$ **43.** $\frac{16 \text{ males}}{3 \text{ females}}$ **45.** $\frac{8 \text{ Democrats}}{7 \text{ Republicans}}$ **47.** $\frac{1 \text{ lb}}{8 \text{ servings}}$ **49.** $\frac{307 \text{ flights}}{3 \text{ days}}$ **51.** $\frac{1 \text{ lb}}{200 \text{ sq ft}}$ **53.** 225 revolutions/min **55.** 8 gal/day **57.** 0.3 tank/acre **59.** 1.6 yd/dress **61.** 2 hr/day **63.** 0.25 km/min **65.** 70 fat calories/tbsp **67.** \$0.45/bar **69.** \$2.95/roll **71.** \$66.67/plant **73.** \$99/night

75.

Number of Units	Total Price	Unit Price
30	\$1.69	\$0.06
100	\$5.49	\$0.05

100 cough drops

77.

Number of Units (Sheets)	Total Price	Unit Price
500	9.69	\$0.019
2,500	\$42.99	\$0.017

2,500 sheets of paper

79.

Number of Units	Total Price	Unit Price
14	\$8.49	\$0.61
25	\$11.49	\$0.46
28	\$7.49	\$0.27

28 trash bags

81. \$0.16/oz **83.** 2 tutors/15 students **85.** $\frac{5}{1}$ **87.** $\frac{2}{3}$ **89.** 170 cal/oz **91.** \$0.03 per page **93.** $\frac{1}{2}$ **95.** $\frac{1}{6}$ **97.** In the Senate **99. a.** $\frac{63}{68}$ **b.** $\frac{8}{9}$ **101.** 0.51 to 1

Section 5.2 Practices, pp. 261–265

1, ***p. 261:*** Yes **2,** ***p. 261:*** Not a true proportion **3,** ***p. 262:*** No **4,** ***p. 263:*** $x = 8$ **5,** ***p. 263:*** $x = 12$ **6,** ***p. 264:*** 64,000 flowers **7,** ***p. 264:*** Mach 1.06 **8,** ***p. 265:*** 300 faculty members

Exercises 5.2, pp. 266–269

1. proportion **3.** as **5.** True **7.** False **9.** True **11.** False **13.** True **15.** True **17.** $x = 20$ **19.** $x = 38$ **21.** $x = 4$ **23.** $x = 13$ **25.** $x = 8$ **27.** $x = 4$ **29.** $x = 20$ **31.** $x = 15$ **33.** $x = 21$ **35.** $x = 13\frac{1}{3}$ **37.** $x = 100$ **39.** $x = 1.8$ **41.** $x = 21$ **43.** $x = 280$ **45.** $x = 300$ **47.** $x = 20$ **49.** $x = 10$ **51.** $x = 5.4$ **53.** $x = \frac{1}{5}$ **55.** $x = 0.005$ **57.** $x = \frac{6}{5}$ **59.** $x = 1\frac{3}{5}$ **61.** False **63.** Not the same **65.** $1\frac{7}{8}$ gal **67.** 54.5 g **69.** 100 oxygen atoms **71.** $41\frac{2}{3}$ in. **73.** 0.25 ft **75.** \$600 **77.** 12,000 fish **79.** 280 times **81.** 90 mg and 50 mg **83. a.** 92 g **b.** 4 g **85.** 835,000 gal

Chapter 5 Review Exercises, pp. 271–272

1. is not; possible answer: when simplifying a ratio, drop the units if the ratio has like quantities **2.** is; possible answer: it is a comparison of two unlike quantities **3.** is not; possible answer: the denominator is not 1 **4.** is; possible answer: it is the price of one foot **5.** is not; possible answer: a proportion must have an equal sign **6.** are: possible answer: cross products are found by multiplying diagonally **7.** $\frac{2}{3}$ **8.** $\frac{1}{2}$ **9.** $\frac{3}{4}$ **10.** $\frac{25}{8}$ **11.** $\frac{8}{5}$ **12.** $\frac{3}{4}$ **13.** $\frac{44 \text{ ft}}{5 \text{ sec}}$ **14.** $\frac{9 \text{ applicants}}{2 \text{ positions}}$ **15.** 0.0025 lb/sq ft **16.** 500,000,000 calls/day **17.** 8 yd/down **18.** 400 sq ft/gal **19.** 10,500,000 vehicles/yr **20.** 76,000 commuters/day **21.** \$118.75/night **22.** \$3.89/rental **23.** \$1,250/station **24.** \$93.64/share

25.

Number of Units	Total Price	Unit Price
47	\$39.95	\$0.85
94	\$69.95	\$0.74

94 issues

26.

Number of Units	Total Price	Unit Price
100	\$13.95	\$0.14
250	\$37.95	\$0.15

100 checks

27.

Number of Units	Total Price	Unit Price
30	\$16.95	\$0.57
90	\$44.85	\$0.50
180	\$77.70	\$0.43

180 capsules

28.

Number of Units (fluid ounces)	Total Price	Unit Price
4	\$2.78	\$0.70
14	\$3.92	\$0.28
20	\$3.74	\$0.19

20 fl oz

29. True **30.** False **31.** False **32.** True **33.** $x = 6$ **34.** $x = 3$ **35.** $x = 32$ **36.** $x = 30$ **37.** $x = 2$ **38.** $x = 8$ **39.** $x = \frac{7}{10}$ **40.** $x = \frac{2}{5}$ **41.** $x = 67\frac{1}{2}$ **42.** $x = 45$ **43.** $x = \frac{3}{4}$, or 0.75 **44.** $x = \frac{3}{20}$, or 0.15 **45.** $x = 28$ **46.** $x = 0.14$ **47.** $\frac{1}{15}$ **48.** $\frac{23}{45}$ **49.** \$90/day **50.** 0.125 in./mo **51.** $\frac{2}{3}$ **52.** 50,000 books **53.** No **54.** $2\frac{1}{7}$ hr **55.** 55 cc **56.** 1.25 in. **57.** 0.68 g/cc **58.** 0.07 admitted students per applicant

Chapter 5 Posttest, p. 273

1. $\frac{2}{3}$ **2.** $\frac{5}{14}$ **3.** $\frac{55}{31}$ **4.** $\frac{12}{1}$ **5.** $\frac{13 \text{ revolutions}}{12 \text{ sec}}$ **6.** $\frac{1 \text{ cm}}{25 \text{ km}}$ **7.** 68 mph
8. 8 m/sec **9.** $136/day **10.** $0.80/greeting card **11.** False
12. True **13.** $x = 25$ **14.** $x = 6$ **15.** $x = 28$ **16.** $x = 1$
17. 5 million e-mail addresses **18.** $\frac{9}{8}$ **19.** 25 ft **20.** 48 beats/min

Chapter 5 Cumulative Review, p. 274

1. $108,411 **2.** 189 **3.** 52 **4.** $2^3 \cdot 3 \cdot 7$ **5.** $\frac{2}{5}$ **6.** $\frac{4}{39}$ **7.** $\frac{3}{4}$
8. 8,200 **9.** Possible answer: 3 **10.** $x = 2.5$ **11.** $n = 70$ **12.** $\frac{1}{4}$
13. $4 per yd **14.** $x = \frac{3}{4}$ **15.** 2,106 sq ft **16.** $445.88 **17.** 0.19 in.
18. $r \cdot t$ miles; 260 mi **19.** 7.5 in. **20.** $\frac{1}{29}$

CHAPTER 6

Chapter 6 Pretest, p. 276

1. $\frac{1}{20}$ **2.** $\frac{3}{8}$ **3.** 2.5 **4.** 0.03 **5.** 0.7% **6.** 800% **7.** 67%
8. 110% **9.** $37\frac{1}{2}$ ft **10.** 55 **11.** 32 **12.** 250 **13.** 40% **14.** 250%
15. $10.50 **16.** 8% **17.** $\frac{6}{25}$ **18.** 25% **19.** $61.11 **20.** $10,000

Section 6.1 Practices, pp. 278–284

1, *p. 278:* $\frac{21}{100}$ **2, *p. 278:*** $\frac{9}{4}$, or $2\frac{1}{4}$ **3, *p. 279:*** $\frac{2}{300}$ **4, *p. 279:*** $\frac{1}{8}$
5, *p. 279:* $\frac{14}{25}$ **6, *p. 280:*** 0.31 **7, *p. 280:*** 0.05 **8, *p. 281:*** 0.482
9, *p. 281:* 0.6225 **10, *p. 281:*** 1.637 **11, *p. 282:*** 2.5% **12, *p. 282:*** 9%
13, *p. 282:* 70% **14, *p. 282:*** 300% **15, *p. 283:*** 71% **16, *p. 283:***
Nitrogen; 78% > 0.93%, or 0.78 > 0.0093. **17, *p. 284:*** 16%
18, *p. 284:* True. $\frac{2}{3} \approx 67\% > 60\%$ **19, *p. 284:*** 27%

Exercises 6.1, pp. 285–288

1. percent **3.** left **5.** $\frac{2}{25}$ **7.** $2\frac{1}{2}$ **9.** $\frac{33}{100}$ **11.** $\frac{9}{50}$ **13.** $\frac{7}{50}$ **15.** $\frac{13}{20}$
17. $\frac{3}{400}$ **19.** $\frac{3}{1,000}$ **21.** $\frac{3}{40}$ **23.** $\frac{1}{7}$ **25.** 0.06 **27.** 0.72 **29.** 0.001
31. 1.02 **33.** 0.425 **35.** 5 **37.** 1.069 **39.** 0.035 **41.** 0.009
43. 0.0075 **45.** 31% **47.** 17% **49.** 30% **51.** 4% **53.** 12.5%
55. 129% **57.** 290% **59.** 287% **61.** 101.6% **63.** 900%
65. 30% **67.** 10% **69.** 16% **71.** 90% **73.** 6% **75.** $55\frac{5}{9}\%$
77. $11\frac{1}{9}\%$ **79.** 600% **81.** 150% **83.** $216\frac{2}{3}\%$ **85.** < **87.** <
89. 44% **91.** 225%

93.

Fraction	Decimal	Percent
$\frac{1}{3}$	0.333 . . .	$33\frac{1}{3}\%$
$\frac{2}{3}$	0.666 . . .	$66\frac{2}{3}\%$
$\frac{1}{4}$	0.25	25%
$\frac{3}{4}$	0.75	75%
$\frac{1}{5}$	0.2	20%
$\frac{2}{5}$	0.4	40%
$\frac{3}{5}$	0.6	60%

95. $1\frac{1}{25}$ **97.** $316\frac{2}{3}\%$ **99.** 0.275 **101.** 310% **103.** 254% **105.** 0.96
107. $\frac{17}{50}$ **109.** $\frac{1}{10}$ **111.** 1,005% **113.** $1\frac{7}{20}$ **115.** 0.515 **117.** Among men; $\frac{1}{4} = 25\% < 42\%$ **119. a.** 0.4% **b.** 99.6% **121.** 69%

Section 6.2 Practices, pp. 291–297

1, *p. 291:* **a.** $x = 0.7 \cdot 80$ **b.** $0.5 \cdot x = 10$ **c.** $x \cdot 40 = 20$ **2, *p. 291:*** 8
3, *p. 291:* 12 **4, *p. 292:*** 200 **5, *p. 292:*** 51 workers **6, *p. 293:*** 50
7, *p. 293:* 7.2 **8, *p. 293:*** 2,500,000 sq ft **9, *p. 294:*** $83\frac{1}{3}\%$
10, *p. 294:* $112\frac{1}{2}\%$ **11, *p. 295:*** 30% **12, *p. 295:*** 270 **13, *p. 296:*** 1,080
14, *p. 296:* $33\frac{1}{3}\%$ **15, *p. 296:*** $98 **16, *p. 297:*** $340,000
17, *p. 297:* 105% **18, *p. 297:*** $5.97

Exercises 6.2, pp. 298–301

1. base **3.** percent **5.** 6 **7.** 23 **9.** 2.87 **11.** $140 **13.** 0.62
15. 0.045 **17.** 0.1 **19.** 4 **21.** $18.32 **23.** 32 **25.** $120 **27.** 2.5
29. $250 **31.** 45 **33.** 1.75 **35.** 4,600 **37.** 3,000 **39.** $49,230.77
41. 50% **43.** 75% **45.** $83\frac{1}{3}\%$ **47.** 25% **49.** 150% **51.** $112\frac{1}{2}\%$
53. 62.5% **55.** 50% **57.** 31% **59.** 60 **61.** $66\frac{2}{3}\%$ **63.** 175 mi
65. 5% **67.** $500 **69.** 10 **71.** $66\frac{2}{3}\%$ **73.** 40 questions **75.** $600
77. 18,750,000 people **79.** 72 free throws **81.** 10.2 gal
83. $30,000,000 **85.** $9,000 **87.** 30% **89.** 6.8 million **91.** 5,100 employees **93. a.** 10 million **b.** 80%

Section 6.3 Practices, pp. 302–307

1, *p. 302:* 300% **2, *p. 303:*** 1929 **3, *p. 304:*** $462.50
4, *p. 304:* **a.** $1,125 **b.** $2,625 **5, *p. 305:*** $69.60 **6, *p. 305:*** $1,450
7, *p. 306:* $1,792 **8, *p. 307:*** $2,524.95

Exercises 6.3, pp. 308–312

1. original **3.** discount

5.

Original Value	New Value	Percent Increase or Decrease
$10	$12	20% increase
$10	$8	20% decrease
$6	$18	200% increase
$35	$70	100% increase
$14	$21	50% increase
$10	$1	90% decrease
$8	$6.50	$18\frac{3}{4}\%$ decrease
$6	$5.25	$12\frac{1}{2}\%$ decrease

7.

Selling Price	Rate of Sales Tax	Sales Tax
$30.00	5%	$1.50
$24.88	3%	$0.75
$51.00	$7\frac{1}{2}\%$	$3.83
$196.23	4.5%	$8.83

9.

Sales	Rate of Commission	Commission
$700	10%	$70.00
$450	2%	$9.00
$870	$4\frac{1}{2}\%$	$39.15
$922	7.5%	$69.15

11.

Original Price	Rate of Discount	Discount	Sale Price
$700.00	25%	$175.00	$525.00
$18.00	10%	$1.80	$16.20
$43.50	20%	$8.70	$34.80
$16.99	5%	$0.85	$16.14

13.

Principal	Interest Rate	Time (in years)	Interest	Final Balance
$300	4%	2	$24.00	$324.00
$600	7%	2	$84.00	$684.00
$500	8%	2	$80.00	$580.00
$375	10%	4	$150.00	$525.00
$1,000	3.5%	3	$105.00	$1,105.00
$70,000	6.25%	30	$131,250.00	$201,250.00

15.

Principal	Interest Rate	Time (in years)	Final Balance
$500	4%	2	$540.80
$6,200	3%	5	$7,187.50
$300	1%	8	$324.86
$20,000	4%	2	$21,632.00
$145	3.8%	3	$162.17
$810	2.9%	10	$1,078.05

17.

Original Value	New Value	Percent Decrease
$5	$4.50	10%

19.

Original Price	Rate of Discount	Discount	Sale Price
$87.33	40%	$34.93	$52.40

21.

Selling Price	Rate of Sales Tax	Sales Tax
$200	7.25%	$14.50

23.

Principal	Interest Rate	Kind of Interest	Time (in years)	Interest	Final Balance
$3,000	5%	simple	5	$750.00	$3,750.00

25. 28% **27.** 550% **29.** 300% **31.** $84.95 **33.** 6.5% **35.** $2,700 **37.** $9 **39.** 15% **41.** $259.35 **43.** $150 **45.** $250 **47. a.** $144 **b.** $152.64 **49.** $3,244.80 **51.** 5,856

Chapter 6 Review Exercises, pp. 315–318

1. are; possible answer: a percent can be written as a fraction with denominator 100 **2.** are not; possible answer: $\frac{1}{2}$% is one-hundredth of $\frac{1}{2}$ **3.** is; possible answer: 8 is the number that we are taking the percent of **4.** is not; possible answer: the amount (4) is the percent (50%) of the base (8) **5.** is not; possible answer: when the percent is greater than 100%, the base is smaller than the amount **6.** is; possible answer: if the number were smaller than or equal to 20, then 5% of it would be still smaller **7.** is; possible answer: taking 100% of a number is the same as multiplying it by 1 **8.** is not; possible answer: the difference between the new and the original values is 100% of the original value

9.

Fraction	Decimal	Percent
$\frac{1}{4}$	0.25	25%
$\frac{7}{10}$	0.7	70%
$\frac{3}{400}$	0.0075	$\frac{3}{4}$%
$\frac{5}{8}$	0.625	62.5%
$\frac{41}{100}$	0.41	41%
$1\frac{1}{100}$	1.01	101%
$2\frac{3}{5}$	2.6	260%
$3\frac{3}{10}$	3.3	330%
$\frac{3}{25}$	0.12	12%
$\frac{2}{3}$	0.66 . . .	$66\frac{2}{3}$%
$\frac{1}{6}$	0.166 . . .	$16\frac{2}{3}$%

10.

Fraction	Decimal	Percent
$\frac{3}{8}$	0.375	37.5%
$\frac{49}{100}$	0.49	49%
$\frac{1}{1,000}$	0.001	0.1 %
$1\frac{1}{2}$	1.5	150%
$\frac{7}{8}$	0.875	87.5%
$\frac{5}{6}$	0.833 . . .	$83\frac{1}{3}$%
$2\frac{3}{4}$	2.75	275%
$1\frac{1}{5}$	1.2	120%
$\frac{3}{4}$	0.75	75%
$\frac{1}{10}$	0.1	10%
$\frac{1}{3}$	0.33 . . .	$33\frac{1}{3}$%

11. 12 **12.** 120% **13.** 50% **14.** 20 **15.** 43.75% **16.** 5.5 **17.** $6 **18.** 20% **19.** 0.3 **20.** 460 **21.** $70 **22.** 1,000 **23.** 2000% **24.** 25% **25.** $200 **26.** $44\frac{4}{9}$% **27.** $12 **28.** $1,600 **29.** 17% **30.** 42.42

31.

Original Value	New Value	Percent Decrease
24	16	$33\frac{1}{3}$%

32.

Original Value	New Value	Percent Decrease
360 mi	300 mi	$16\frac{2}{3}$%

33.

Selling Price	Rate of Sales Tax	Sales Tax
$50	6%	$3.00

Chapter 8 Review Exercises, pp. 392–395

1. cannot; possible answer: the mean is the sum of the numbers divided by the number of numbers **2.** can; possible answer: there may be two or more numbers that occur with the highest frequency **3.** is; possible answer: the difference between the largest and smallest numbers in each set is the same **4.** does; possible answer: reading a graph generally involves estimating **5.** is not; possible answer: they are difficult to distinguish on a pictograph **6.** is not; possible answer: it is used to represent parts of a whole

7.

List of Numbers	Mean	Median	Mode	Range
6, 7, 4, 10, 4, 5, 6, 8, 7, 4, 5	6	6	4	6
1, 3, 4, 4, 2, 3, 1, 4, 5, 1	2.8	3	1 and 4	4

8. a. The husbands (83 yr) lived longer than the wives (70 yr). **b.** 13 yr **c.** 23 yr **9. a.** Half of the people who listen to NPR are younger than 55, and half are older. **b.** 30 years below the median **10.** This machine is reliable because the range is 1.7 fl oz. **11.** Since the average was 13,109, the zoo was profitable. **12. a.** 70,720,000 tons **b.** Beaumont, TX **c.** 41,440,000 tons more **d.** All except the Port of South Louisiana **13. a.** 830,000 monthly domestic flights **b.** In 2005 **c.** 80,000 more monthly domestic flights **14. a.** 10 million volumes **b.** 5 million volumes more **c.** Harvard University and the Boston Public Library **15. a.** 61 million licensed drivers **b.** 10% **c.** Possible answer: It increases for younger age groups, peaks for 40–44, and decreases for older age groups. **16. a.** Run number 3 **b.** Approximately 3 min **c.** With practice, the rat ran through the maze more quickly. **17. a.** 2003 **b.** $\frac{1}{2}$ **c.** Both out-of-pocket expenditures and insurance expenditures increased each year from 2000 through 2010 (with the exception of the period 2005–2006). **18. a.** 27,578 transplants **b.** 4,429 more liver transplants **c.** 58% **19.** 76 **20.** Internet users

Chapter 8 Posttest, pp. 396–397

1. 3.2 **2.** 162 **3.** 77 **4.** Mode: 200 lb; range: 90 lb **5.** Toyota Prius, Ford Fusion Hybrid, Mercury Milan Hybrid, and Nissan Altima Hybrid **6.** 30% greater **7.** 3 million more victims **8. a.** 2007 **b.** Overall music purchases are the sum of digital track and album sales. \$600 million + \$600 million = \$1,200 million, or \$1.2 billion, in sales for overall music purchases in the year 2006 **9.** 60% **10.** 1961–1975; U.S. scientists received approximately 40 and U.K. scientists received approximately 20 Nobel prizes.

Chapter 8 Cumulative Review, pp. 398–399

1. 23 **2.** $\frac{4}{5}$ **3.** $1\frac{87}{100}$ **4.** 40 **5.** 2.81 **6.** 3,010 **7.** $x = 17\frac{1}{4}$ **8.** No **9.** $n = 21$ **10.** $4\frac{5}{8}$ **11.** 2 **12.** Negative; $\frac{7}{4}$ **13.** −18 **14.** Mean: 3.7; median: 3.8; mode: 3.9; range: 0.6 **15.** 10.6 times **16.** 8 jobless people per opening **17.** 2% **18.** 5,000 yr **19.** Higher in the American League; the medians were 48 and 50. **20.** Males, age 45–54

CHAPTER 9

Chapter 9 Pretest, p. 401

1. $y = -6$ **2.** $x = -2$ **3.** $x = -4$ **4.** $a = -5$ **5.** $x = 3$ **6.** $y = 1$ **7.** $c = -2$ **8.** $a = -10$ **9.** $x = 2$ **10.** $y = 2$ **11.** $x = -1$ **12.** $x = -4$ **13.** $3(n + 7) = 30$ **14.** $j = \frac{2}{3}f$ **15.** $1{,}050x + 350 = 3{,}500$; 3 elephants **16.** $475 = 250 + 75(x - 3)$; 6 hr **17.** $15x - 125 = 400$; \$35 **18.** $\frac{1}{2}x + 7.8 = 72.8$; \$130 **19.** $m = \frac{1}{6}E$ **20.** 171 beats per minute

Section 9.1 Practices: pp. 402–406

1, *p. 402:* $x = -3$ **2, *p. 403:*** $m = -12$ **3, *p. 403:*** $y = -2$ **4, *p. 403:*** $y = 6$ **5, *p. 404:*** 7 wk **6, *p. 404:*** $x = -7$ **7, *p. 405:*** $k = 15$ **8, *p. 405:*** $d = 4$ **9, *p. 406:*** $x = 80$ **10, *p. 406:*** **a.** $2 + 0.2x = 10$ **b.** $x = 40$ oranges

Exercises 9.1, pp. 407–408

1. $a = -14$ **3.** $b = -11$ **5.** $z = -7$ **7.** $x = -2$ **9.** $c = -19$ **11.** $z = -12$ **13.** $z = -7.7$ **15.** $x = 8.2$ **17.** $y = -\frac{2}{3}$ **19.** $n = \frac{1}{6}$ **21.** $t = -6\frac{1}{4}$ **23.** $x = -6$ **25.** $n = 4$ **27.** $w = -240$ **29.** $x = 12$ **31.** $t = -30$ **33.** $y = -\frac{2}{5}$ **35.** $m = -1.5$ **37.** $y = -15$ **39.** $n = 14$ **41.** $x = 2$ **43.** $k = -3$ **45.** $x = 0$ **47.** $h = -7$ **49.** $p = -5\frac{1}{4}$ **51.** $b = 2$ **53.** $a = -33$ **55.** $y = -36$ **57.** $x = 48$ **59.** $c = -21$ **61.** $x = 18$ **63.** $x = -19$ **65.** $x = -12$ **67.** $n \approx -1{,}434.3$ **69.** $x \approx -15.2$ **71.** $r \approx -0.4$ **73.** $x \approx -0.3$ **75.** $t = 9.6$ **77.** $n = -1$ **79.** $c = -2\frac{1}{2}$ **81.** $y = -1\frac{4}{5}$ **83.** $-1.5x = -750$; 500 shares **85.** $-9{,}046 + x = -15{,}671$; a loss of \$6,625 **87.** $18x + 5 = 365$; 20 days **89.** $8x + 3 = 43$; 5 people **91.** $0.1374x + 6.29 = 94.98$; 645 kwh

Section 9.2 Practices, pp. 411–414

1, *p. 411:* **a.** $12x$ **b.** $\frac{1}{8}n$ **c.** $-z + 6$ **2, *p. 411:*** $y = -8$ **3, *p. 412:*** $x = -21$ **4, *p. 412:*** $x = 1$ **5, *p. 413:*** **a.** $x + 3x = x + 9$ **b.** $x = 3$ **6, *p. 413:*** $x = -1$ **7, *p. 414:*** $x = 3$ **8, *p. 414:*** 4 toppings

Exercises 9.2, pp. 415–417

1. $7x$ **3.** $3a$ **5.** $-3y$ **7.** $-n$ **9.** $\frac{5}{6}x$ **11.** $3c + 12$ **13.** $8 - 6x$ **15.** $2y + 5$ **17.** $m = 4$ **19.** $y = 9$ **21.** $a = 0$ **23.** $b = -1\frac{3}{4}$ **25.** $b = 20$ **27.** $n = 13$ **29.** $x = 3$ **31.** $n = 9$ **33.** $a = -\frac{2}{3}$ **35.** $x = 4$ **37.** $p = -2$ **39.** $x = -1$ **41.** $p = 1\frac{1}{2}$ **43.** $n = 12$ **45.** $t = 2$ **47.** $y = 1$ **49.** $r = 30$ **51.** $x = \frac{5}{6}$ **53.** $n = 3$ **55.** $x = -2$ **57.** $x = 2$ **59.** $n = 35$ **61.** $y = 15$ **63.** $y = -\frac{3}{8}$ **65.** $n = 21$ **67.** $r = 2$ **69.** $y = 5$ **71.** $6x - 9 = -3x + 18$; $x = 3$ **73.** $3(x + 5) = 2x$; $x = -15$ **75.** $r = -0.5$ **77.** $n = 2.7$ **79.** $y = 19.1$ **81.** $4x + 6$ **83.** $y = -1\frac{1}{2}$ **85.** $x = -\frac{1}{2}$ **87.** $m = -\frac{1}{5}$ **89.** $a - 0.83a = 255$; 1,500 interviewees **91.** $140 + 2w = 200$; 30 ft **93.** $\frac{1}{6}x + \frac{3}{4}x = 440$; 480 passenger seats **95.** $3{,}000 + 120x = 2{,}400 + 145x$; 24 mo **97.** Let x = B1; $2x + 3(20 - x) = 40$; 20 **99.** $49.99(x - 2) + 119.99 = 269.96$; 5 lines

Section 9.3 Practices, pp. 418–419

1, *p. 418:* $t = g - \frac{1}{200}a$ **2, *pp. 418–419:*** $I = \$360$ **3, *p. 419:*** $F = 39.5°$ **4, *p. 419:*** 21 kg per sq m

Exercises 9.3, pp. 420–421

1. $d = \frac{t}{3}$ **3.** $a = \frac{h}{n}$ **5.** $A = 6e^2$ **7.** 14° **9.** −3° **11.** 150 cents **13.** 120 **15.** 17.9 in. **17.** −1 **19.** $\frac{7}{9}$ **21.** 13.5 sq in. **23.** $v = \$13{,}500$ **25.** $C = 60$ mg **27.** $RE = 66$

Chapter 9 Review Exercises, pp. 424–426

1. is not; possible answer: substituting 18 for b does not make the equation a true statement. **2.** are not; possible answer: the variable, y, has different exponents. **3.** are; possible answer: they have different variables. **4.** is not; possible answer: it is not an equation. **5.** does; possible answer: we can get the second equation by dividing both sides of the first equation by −2. **6.** does; possible answer: of the distributive property. **7.** $x = -6$ **8.** $y = -9$ **9.** $d = -9$ **10.** $w = -11$ **11.** $x = -9$ **12.** $y = -16$ **13.** $p = 0$ **14.** $x = 0$ **15.** $a = -9.7$ **16.** $b = -9$ **17.** $y = -5\frac{1}{2}$ **18.** $d = -6\frac{1}{3}$ **19.** $x = -7$ **20.** $c = -1$ **21.** $p = -\frac{1}{4}$ **22.** $a = -\frac{2}{3}$ **23.** $y = 20$ **24.** $w = 30$ **25.** $b = -180$ **26.** $x = -450$ **27.** $x = -7.5$ **28.** $a = -5.4$ **29.** $x = 4$ **30.** $a = 2$ **31.** $y = 5$

32. $w = 2$ **33.** $y = 3\frac{1}{2}$ **34.** $y = 2\frac{2}{3}$ **35.** $a = 24$ **36.** $w = 16$ **37.** $x = -25$ **38.** $b = -10$ **39.** $c = 0$ **40.** $d = 0$ **41.** $c = -6$ **42.** $a = -18$ **43.** $y = 9$ **44.** $b = 6$ **45.** $c = -2$ **46.** $x = 2\frac{1}{5}$ **47.** $y = 8$ **48.** $t = 3$ **49.** $x = 8$ **50.** $m = 3$ **51.** $n = \frac{1}{3}$ **52.** $a = -9$ **53.** $r = -8$ **54.** $y = -1$ **55.** $x = 9$ **56.** $y = -1$ **57.** $s = 2$ **58.** $x = -2$ **59.** $m = 2$ **60.** $x = 6$ **61.** $s = 10$ **62.** $z = 3.8$ **63.** $w = 3$ **64.** $z = 0$ **65.** $x = 6$ **66.** $y = -17$ **67.** $y = -8$ **68.** $m = 2\frac{2}{3}$ **69.** $b = -15$ **70.** $d = -40$ **71.** $6x + 3 = x - 17; -4$ **72.** $3x + 2(x - 1) = 33; 7$ **73.** $R = F + 460$ **74.** $s = fw$ **75.** $A = 9\left(\frac{E}{7}\right)$ **76.** $m = \frac{1}{2}(s + l)$ **77.** \$130.90 **78.** 144 ft **79.** 8 wk **80.** −20°C **81.** p = \$5.38/unit **82.** 7 wk **83.** 9 ft **84.** 4 hr **85.** D = 0.4 g/cc **86.** P = \$55

Chapter 9 Posttest, p. 427

1. $x = -5$ **2.** $y = 0$ **3.** $a = -3$ **4.** $b = -10$ **5.** $x = \frac{1}{2}$ **6.** $y = -2\frac{1}{2}$ **7.** $y = -9$ **8.** $a = -60$ **9.** $x = -6$ **10.** $c = -2$ **11.** $x = 2$ **12.** $y = -6$ **13.** $F = ma$ **14.** $4(n + 3) = 5n$ **15.** $0.25 + 0.1(x - 1) = 2.75$; 26 min **16.** $100 + \frac{8}{9}P = 12{,}906$; \$14,406.75. **17.** $15 + 25x = 165$; 6 mo **18.** $x - \frac{1}{3}x - 10 = 12$; \$33 **19.** $C \approx \frac{P}{750{,}000} + 2$ **20.** 66 fps

Chapter 9 Cumulative Review, pp. 428–429

1. 1,512 **2.** $(8 + 4) \cdot 2^2$ **3.** $2\frac{5}{6}$ **4.** $5\frac{1}{4}$ **5.** 0.001 **6.** $x + 15.7 = 84.6; x = 68.9$ **7.** $t = 64$ **8.** $83\frac{1}{3}\%$ **9.** −19 **10.** −2 **11.** Mean: $2\frac{3}{5}$; median: $2\frac{1}{2}$; no mode; range: $1\frac{1}{4}$ **12.** $x = 6$ **13.** $a = -1\frac{1}{5}$ **14.** 405 **15.** This change represented an increase because 0.1 is larger than 0.075 (by 0.025). **16.** $\frac{1}{17}$ **17.** 1,200 stocks **18.** −\$38.75 million **19.** 300 million more users **20.** 40 mph

CHAPTER 10

Chapter 10 Pretest, p. 431

1. $\frac{1}{2}$ **2.** 72 **3.** 48 **4.** 3 hr 52 min **5.** b **6.** d **7.** 3,500 **8.** 2.1 **9.** 8 **10.** 0.0000015 **11.** 0.002 **12.** 4.5 **13.** 195 **14.** About 2.2 qt **15.** 334 billion lb of trash **16.** 5 times **17.** 440 oz **18.** 180 cm **19.** 300 mi **20.** 7 cm

Section 10.1 Practices, pp. 434–437

1, ***p. 434:*** 2 **2,** ***p. 434:*** $1\frac{2}{3}$ yd **3,** ***p. 434:*** 86,400 sec **4,** ***p. 435:*** No **5,** ***p. 435:*** 19 in. **6,** ***p. 435:*** 1 yr 8 mo **7,** ***p. 436:*** 5 lb 8 oz **8,** ***p. 436:*** 13 lb 4 oz **9,** ***p. 437:*** 3 qt **10,** ***p. 437:*** 6 sec

Exercises 10.1, pp. 438–440

1. length **3.** larger **5.** unit **7.** 4 **9.** 3 **11.** 720 **13.** 420 **15.** 30 **17.** 2 **19.** 3,520 **21.** 512 **23.** 2 **25.** 5 **27.** $3\frac{1}{2}$ **29.** 3,000 **31.** $\frac{3}{4}$ **33.** 12 **35.** 310 **37.** 7 ft 6 in.

39.	192	$5\frac{1}{3}$
41.	180	15
43.	48	$\frac{3}{2000}$
45.	128,000	8,000
47.	32	2
49.	1	$\frac{1}{2}$
51.	$\frac{5}{6}$	$\frac{1}{72}$
53.	1,320	$\frac{11}{30}$

55. 1 lb 14 oz **57.** 29 lb 15 oz **59.** 9 yr 6 mo **61.** 2 gal 3 qt **63.** 4 qt **65.** 8 min 15 sec **67.** 14 lb 1 oz **69.** 247 sec **71.** 5 yd **73.** Equal to **75.** 6 in. **77.** 13 min 43 sec **79.** 1 lb 5 oz **81.** More; he has 1 yr 10 mo left on the lease. **83.** 13 lb 14 oz **85. a.** Width: 3 ft, depth: $2\frac{1}{2}$ ft **b.** $7\frac{1}{2}$ sq ft **87.** 5.5 mi

Section 10.2 Practices, pp. 444–448

1, ***p. 444:*** 3.1 **2,** ***p. 444:*** 25 **3,** ***p. 445:*** 25 m **4,** ***p. 445:*** 5 kL **5,** ***p. 446:*** Large intestine **6,** ***p. 447:*** 5,300,000 **7,** ***p. 447:*** 0.005 mg **8,** ***p. 448:*** 38 L **9,** ***p. 448:*** 4 lb

Exercises 10.2, pp. 449–452

1. length **3.** kilo- **5.** centi- **7.** c **9.** b **11.** b **13.** b **15.** a **17.** 1 **19.** 0.75 **21.** 80 **23.** 3,500 **25.** 0.005 **27.** 4 **29.** 7 **31.** 4.13 **33.** 2,000 **35.** 0.0075 **37.** 30,000 **39.** 4,000,000 **41.** 7 **43.** 1.28

45.	2,380	2.38
47.	1	0.01
49.	0.3	0.0003
51.	450,000	0.45
53.	0.709	0.000709
55.	17,000,000	17,000

57. 3,250 m, or 3.25 km **59.** 98,025.6 g, or 98.0256 kg **61.** 840 **63.** 4 **65.** 1.2 **67.** 2.4 **69.** 3.5 m or 350 cm **71.** 2,500 **73.** a **75.** Approximately 3.3 qt **77.** 6g **79.** Yes **81.** 3 in. **83.** 2,400 mm **85.** 750 cm **87.** 200,000 mL **89.** 5.6 mg **91.** 10.1 kg **93.** 1,350 g **95.** 3.2 L

Chapter 10 Review Exercises, pp. 455–457

1. do not; Possible answer: the unit factor must have the desired unit in the numerator and the original unit in the denominator: $\frac{24 \text{ hr}}{1 \text{ day}}$ **2.** cannot; Possible answer: a quart is a unit of capacity and not of weight **3.** is; Possible answer: the prefix "centi-" means $\frac{1}{100}$, the prefix "milli-" means $\frac{1}{1{,}000}$, and the first fraction is larger than the second **4.** is; Possible answer: on the metric conversion line, we move three units to the right to go from kilometers to meters **5.** is; Possible answer: the prefix "mega-" means one million **6.** is not; Possible answer: the prefix "micro-" means one millionth and not one billionth **7.** 15 **8.** $1\frac{2}{3}$ **9.** 2 **10.** $3\frac{1}{3}$ **11.** 3,000 **12.** 136 **13.** 48 **14.** $2\frac{1}{2}$ **15.** 435 **16.** 4 ft 2 in. **17.** 2 **18.** 125 **19.** 8 hr 10 min **20.** 18 ft 9 in. **21.** 1 gal 3 qt **22.** 7 lb 2 oz **23.** c **24.** a **25.** b **26.** c **27.** b **28.** a **29.** a **30.** c **31.** 0.037 **32.** 4,000 **33.** 800 **34.** 2,100 **35.** 0.6 **36.** 5.1 **37.** 0.005 **38.** 4,000 **39.** 112 **40.** 2 **41.** 20 **42.** 15 **43.** 1.2 hr **44.** 11,440 pt **45.** *Frankenstein* **46.** 0.6 L **47.** 2 g **48.** 0.072 g **49.** 0.2 g **50.** 750,000 g **51.** 6 hr **52.** 25,000 mg **53.** 33 ft **54.** 9.104 g **55.** 199,900 mcg **56.** 50,000 MB **57.** 4 min 30 sec **58.** 10 in. **59.** 26 mg **60.** 5 MB **61.** 24 in. **62.** 20,000 L **63.** 16 wk **64.** 178–208 km/hr

Chapter 10 Posttest, p. 458

1. 2 **2.** 21 **3.** 24 **4.** 3 ft 3 in. **5.** 1 hr 45 min **6.** c **7.** c **8.** 4 **9.** 6,000 **10.** 4,600,000 **11.** 2,000 **12.** 5,000 **13.** 84 **14.** 3.3 **15.** 14 cm **16.** 6 mi **17.** Finback whale **18.** 0.000003 sec **19.** 2.1 times as much memory **20.** 22,000 mi

Chapter 10 Cumulative Review, pp. 459–460

1. 729 **2.** $5\frac{1}{2}$ **3.** $\frac{1}{4}$ **4.** 14 **5.** $k = \frac{4}{3}$ **6.** 24 mg of sodium/cracker **7.** $66\frac{2}{3}\%$ **8.** \$18 **9.** $-4\frac{3}{5}$ **10.** $\frac{2}{3}$ **11.** 8 **12.** $\frac{5}{3}$ **13.** 84 **14.** 40 g **15.** 30 bushels per acre **16.** −\$1.47 **17.** \$500,000 **18.** 8,000 more golden retrievers **19.** 8 lb **20.** 400 mi

CHAPTER 11

Chapter 11 Pretest, pp. 462–463

1.

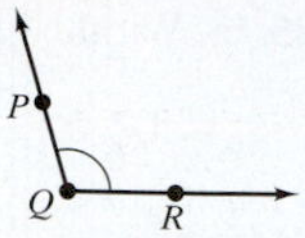

2. 11 **3.** 80° **4.** 54° **5.** 14 in. **6.** 20 ft **7.** Approximately 12.56 in. **8.** 10.4 m **9.** 30 in^2 **10.** 78.5 ft^2 **11.** 125 cm^3 **12.** Approximately 254.34 m^3 **13.** 15 m **14.** $x = 7$ ft; $y = 5$ ft **15.** $a = 104°$ **16.** 21 cm **17.** No, a right angle. **18.** 3 mi^2 **19.** No, because its diagonal is more than 53 in. **20.** 8,100 ft^3

Section 11.1 Practices, pp. 467–472

1, *p. 467:*

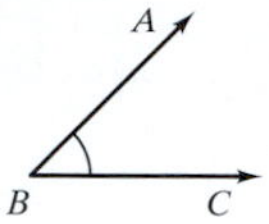

2, *p. 467:* 53° **3, *p. 467:*** 165°

4, *p. 468:* $x = 82°$ **5, *p. 468:*** $a = b = 153°$

6, *p. 471:*

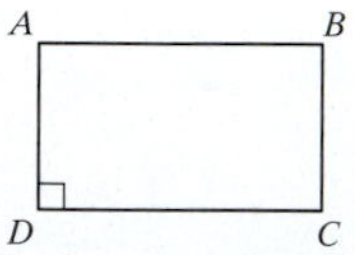

$\overline{AB} \| \overline{DC}$; $\overline{AD} \| \overline{BC}$

7, *p. 472:* 60° **8, *p. 472:*** $m\angle U = 60°$ **9, *p. 472:*** 24 in.

Exercises 11.1, pp. 473–477

1. line segment **3.** complementary **5.** perpendicular **7.** parallel **9.** scalene **11.** parallelogram **13.**

15.

17.

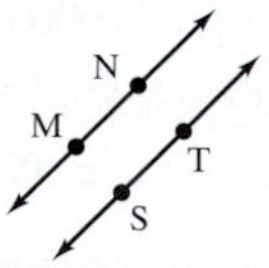

19.

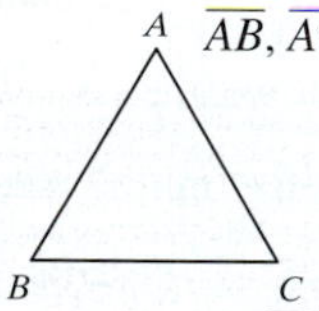

$\overline{AB}$, $\overline{AC}$, and $\overline{BC}$ are the same length.

21.

23.

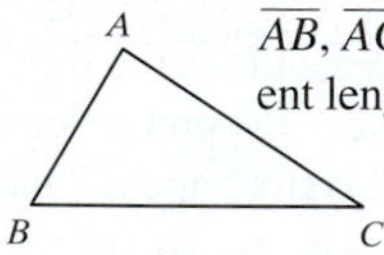

$\overline{AB}$, $\overline{AC}$, and $\overline{BC}$ have different lengths.

25.

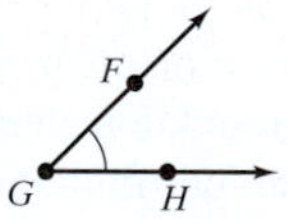

27. $x = 33°$ **29.** 53° **31.** 180° **33.** 90° **35.** 180° **37.** 35° **39.** $a = 53°$ **41.** $y = 140°$ **43.** 55° **45.** 75° **47.** 70° **49.** 85° **51.**

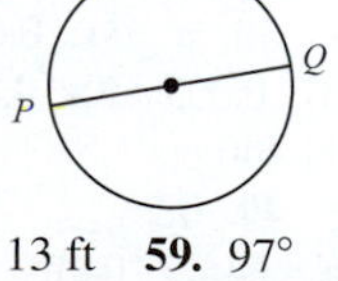

53. $x = 109°$ **55.**

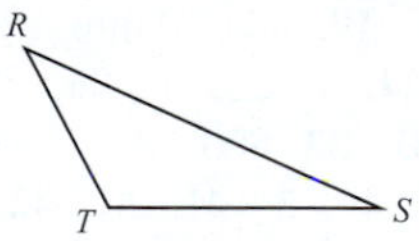

57. 13 ft **59.** 97° **61.** Acute **63.** 45 ft by 32 ft **65. a.** 6 in. **b.** 60° **67. a.** A trapezoid **b.** 80°

Section 11.2 Practices, pp. 479–483

1, *p. 479:* 26 in. **2, *p. 479:*** 3 mi **3, *p. 480:*** \$70 **4, *p. 481:*** Approximately 132 in. **5, *p. 481:*** Approximately 113 ft **6, *p. 482:*** Approximately 164.5 yd **7, *p. 483:*** 18 mi

Exercises 11.2, pp. 484–487

1. perimeter **3.** square **5.** circle **7.** 17 in **9.** 6 m **11.** $10\frac{1}{2}$ yd **13.** Approximately 62.8 m **15.** Approximately 21.98 ft **17.** 21 yd **19.** 18 ft **21.** 19 cm **23.** Approximately 18.84 cm **25.** Approximately 11.12 m **27.** 44.1 cm **29.** Approximately 25.12 in. **31.** 40 yd **33.** 64 yd **35.** 27 cm **37.** 28 in. **39.** 228 ft **41.** Approximately 6 in. **43.** 30 posts **45.** Approximately 50.57 ft **47.** Approximately 42,700 km

Section 11.3 Practices, pp. 490–494

1, *p. 490:* 12 cm^2 **2, *p. 490:*** 12.96 cm^2 **3, *p. 490:*** 7.5 in^2 **4, *p. 491:*** $12\frac{1}{2}$ ft^2 **5, *p. 491:*** $3\frac{1}{2}$ ft^2 **6, *p. 492:*** Approximately 78.5 yd^2 **7, *p. 492:*** No. A customer needs 108 tiles at a total cost of \$538.92. **8, *p. 493:*** Approximately 79 mi^2 **9, *p. 493:*** 8.75 m^2 **10, *p. 494:*** 4,421.5 ft^2

Exercises 11.3, pp. 495–498

1. area **3.** triangle **5.** circle **7.** 125 m^2 **9.** 30 ft^2 **11.** 290 yd^2 **13.** Approximately 706.5 cm^2 **15.** 32 m^2 **17.** 15.6 m^2 **19.** Approximately 314 in^2 **21.** 6.25 ft^2 **23.** 44.1 yd^2 **25.** 3.64 m^2 **27.** $\frac{1}{16}$ yd^2 **29.** 46 yd^2 **31.** Approximately 64.25 ft^2 **33.** 51.75 ft^2 **35.** Approximately 113.04 cm^2 **37.** 12 mi^2 **39.** 32.5 m^2 **41.** 160 ft^2 **43.** Approximately 0.05 mm^2 **45.** 5,000 ft^2 **47.** 486 in^2 **49.** \$131.25

Section 11.4 Practices, pp. 500–503

1, *p. 500:* 72 ft^3 **2, *p. 501:*** 3,375 cm^3 **3, *p. 501:*** Approximately 11 in^3 **4, *p. 501:*** Approximately 113 in^3 **5, *p. 502:*** Approximately 4.19 m^3 **6, *p. 502:*** The ball's volume is about 523 in^3, whereas the box's volume is 1,000 in^3. Therefore, the peanut packing occupies approximately 477 in^3. **7, *p. 503:*** 784 in^3

Exercises 11.4, pp. 504–507

1. volume **3.** rectangular solid **5.** cylinder **7.** 216 in^3 **9.** 2,560 m^3 **11.** Approximately 63 ft^3 **13.** Approximately 125.1 ft^3 **15.** Approximately 2,144 in^3 **17.** Approximately 1.95 m^3 **19.** Approximately 92.1 ft^3 **21.** 30,000 ft^3 **23.** Approximately 20.3 m^3 **25.** Approximately 33.5 in^3 **27.** 0.01 kg/cm^3 **29.** 4,187,000,000 mi^3 **31.** Approximately 14 in^3 **33.** 15.19 m^3 **35.** 1,400 times **37.** Approximately 2.9 g/cm^3

Section 11.5 Practices, pp. 509–510

1, *p. 509:* $\overline{AB}$ corresponds to $\overline{GH}$; $\overline{AC}$ corresponds to $\overline{GI}$; $\overline{BC}$ corresponds to $\overline{HI}$. **2, *p. 509:*** $y = 24$ **3, *p. 510:*** $h = 36$ ft

Exercises 11.5, pp. 511–513

1. similar **3.** corresponding **5.** $x = 10\frac{2}{3}$ in. **7.** $x = 24$ m **9.** $x = 15$ ft, $y = 20$ ft **11.** $x = 24$ yd, $y = 12$ yd **13.** $x = 3.8125$ m **15.** $x = 12$ m **17.** $x = 10.5$ ft **19.** 9 ft tall **21.** 200 m **23.** 3.2 in. **25.** 24 ft

Section 11.6 Practices, pp. 515–518

1, *p. 515:* **a.** 7 **b.** 12 **2, *p. 515:*** Between 6 and 7 **3, *p. 516:*** **a.** 7.48 **b.** 3.46 **4, *p. 517:*** 10 in. **5, *p. 517:*** 3.5 ft **6, *p. 518:*** Approximately 12.8 ft

Exercises 11.6, pp. 519–522

1. square root **3.** perfect square **5.** hypotenuse **7.** 3 **9.** 4 **11.** 9 **13.** 13 **15.** 20 **17.** 16 **19.** 7 and 8 **21.** 8 and 9 **23.** 6 and 7 **25.** 3 and 4 **27.** 2.2 **29.** 6.1 **31.** 11.8 **33.** 99 **35.** $a = 16$ cm **37.** $c \approx 3.2$ m **39. b.** 7 m **41. b.** 8 ft **43. c.** 20 m **45. c.** 11.4 cm **47. a.** 8.7 ft **49.** 14 **51.** $x = 10$ in. **53.** $n \approx 9.8$ cm **55.** 16 ft **57.** 240 ft **59.** Approximately 37.4 ft

Chapter 11 Review Exercises, pp. 531–535

1. is; Possible answer: they are vertical angles **2.** is; Possible answer: it is formed by combining a square and a rectangle **3.** cannot; Possible answer: a foot is a unit of length and not of area **4.** does not; Possible answer: it is a square and therefore has 4 sides **5.** are not; Possible answer: they have different shapes **6.** is; Possible answer: it is opposite right angle C **7.**

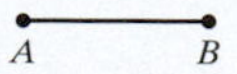

8.

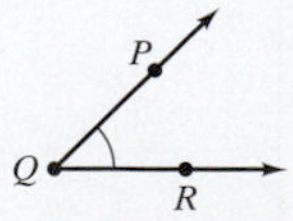

9.

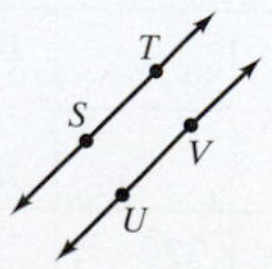

10. 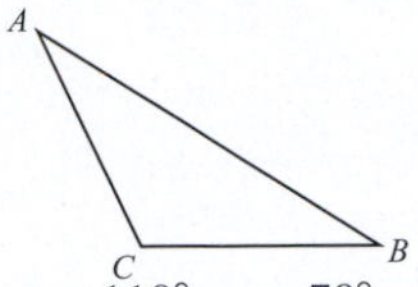

11. $x = 75°$ **12.** $x = 131°$ **13.** $x = 110°, y = 70°$ **14.** $x = 80°; y = 100°$ **15.** 5.4 m **16.** 30 ft **17.** 19 cm **18.** Approximately 125.6 in. **19.** 225 yd^2 **20.** Approximately 615 ft^2 **21.** 64 in^2 **22.** 17 m^2 **23.** Approximately 1,318.8 in^3 **24.** 216 ft^3 **25.** Approximately 1.95 m^3 **26.** Approximately 8.18 cm^3 **27.** Approximately 122.82 ft **28.** Approximately 84.78 ft^2 **29.** Approximately 17,850 ft^2 **30.** Approximately 3.81 yd^3 **31.** $x = 8$ ft, $y = 7.875$ ft **32.** $x = 4.5$ m **33.** 3 **34.** 8 **35.** 11 **36.** 30 **37.** 1 and 2 **38.** 9 and 10 **39.** 6 and 7 **40.** 3 and 4 **41.** 2.83 **42.** 35.14 **43.** 13.96 **44.** 5.39 **45. b.** 12 ft **46. a.** 10 in. **47. c.** 9.4 yd **48. c.** 2.8 ft **49.** 28,800 in^2 **50.** 125,600 mi^2 **51.** Yes, it can; the volume of the room is 2,650 ft^3. **52.** 220 in^2 **53.** 13 mi **54.** 10 ft **55.** $CD = 2\frac{1}{4}$ ft **56.** 7.5 mi **57.** 160 ft **58.** It is not efficient because the perimeter of the work triangle is 25 ft. **59.** 0.2 oz **60.** 1,347 cm^3 of soil **61.** 411.25 in^2 **62.** 2.6 in^3

Chapter 11 Posttest, pp. 536–537

1. 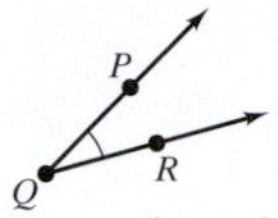

2. 15 **3.** 65° **4.** 89° **5.** 14 ft **6.** 4.5 m **7.** Approximately 25 cm **8.** 15 ft **9.** 54 ft^2 **10.** 110 cm^2 **11.** 189 m^3 **12.** Approximately 33 ft^3 **13.** $a = 120°; b = 60°; y = 5$ m; $x = 10$ m **14.** $a = 69°$ **15.** $y = 12$ yd **16.** $x = 15; y = 24$ **17.** No, it is obtuse. **18.** 9.7 board ft **19.** 79 m^2 **20.** 250 mi

Chapter 11 Cumulative Review, p. 538

1. 60 **2.** $\frac{4}{5}$ **3.** 3.1 **4.** 15.5036 **5.** $y = 8$ **6.** 50 **7.** $583.20 **8.** 15 **9.** $61,400 **10.** $b = 24$ **11.** 2 qt **12.** 9.4 **13.** $\frac{1}{51}$ **14.** 12 **15.** 448 ft **16.** −637, that is, 637 points down **17.** Approximately 707 mi^2 **18.** $2,000 and $3,000 **19.** Approximately 180,000 more complaints **20.** 1 mi

APPENDIX

Appendix Practices, pp. 539–542

1, ***p. 539:*** 25,390,000 **2,** ***p. 540:*** 8.0×10^{12}, or 8×10^{12} **3.** ***p. 541:*** 0.0000000043 **4.** ***p. 541:*** 7.1×10^{-11} **5.** ***p. 542:*** **a.** 2.464×10^2 **b.** 8.35×10^{11}

Exercises p. 543

1. 317,000,000 **3.** 0.000001 **5.** 0.00004013 **7.** 4×10^8 **9.** 3.5×10^{-6} **11.** 3.1×10^{-10} **13.** 9×10^7 **15.** 2.075×10^{-4} **17.** 1.25×10^{10} **19.** 4×10^1

Glossary

The numbers in brackets following each glossary term represent the section in which that term is discussed.

absolute value [7.1] The absolute value of a number is its distance from zero on the number line. The absolute value of a number is represented by the symbol $| \ |$.

acute angle [11.1] An acute angle is an angle whose measure is less than 90°.

acute triangle [11.1] An acute triangle is a triangle with three *acute* angles.

addends [1.2] In an addition problem, the numbers being added are called addends.

algebraic expression [4.1] An algebraic expression is an expression that combines variables, constants, and arithmetic operations.

amount (percent) [6.2] The amount is the result of taking the percent of the base.

angle [11.1] An angle consists of two rays that have a common endpoint.

area [11.3] Area is the number of square units that a figure contains.

associative property of addition [1.2] The associative property of addition states that when adding three numbers, regrouping the addends gives the same sum.

associative property of multiplication [1.3] The associative property of multiplication states that when multiplying three numbers, regrouping the factors gives the same product.

average (or mean) [1.5, 8.1] The average of a set of numbers is the sum of those numbers divided by however many numbers are in the set.

bar graph [8.2] A bar graph is a graph in which quantities are represented by thin, parallel rectangles called bars. The length of each bar is proportional to the quantity that it represents.

base (exponent) [1.5] The base is the number that is a repeated factor when written with an exponent.

base (percent) [6.2] The base is the number that we are taking the percent of. It always follows the word *of* in the statement of the problem.

circle [11.1] A circle is a closed plane figure made up of points that are all the same distance from a fixed point called the center.

circle graph [8.2] A circle graph is a graph that resembles a pie (the whole amount) that has been cut into slices (the parts).

circumference [5.1, 11.2] The distance around a circle is called its circumference.

commission [6.3] Salespeople may work on commission instead of receiving a fixed salary. This means that the amount of money that they earn is a specified percent of the total sales for which they are responsible.

commutative property of addition [1.2] The commutative property of addition states that changing the order in which two numbers are added does not affect the sum.

commutative property of multiplication [1.3] The commutative property of multiplication states that changing the order in which two numbers are multiplied does not affect the product.

complementary angles [11.1] Two angles are complementary if the sum of their measures is 90°.

composite figure [11.2] A composite figure is the combination of two or more basic geometric figures.

composite number [2.1] A composite number is a whole number that has more than two factors.

constant [4.1] A constant is a known number.

corresponding sides [11.5] Corresponding sides are the sides opposite the equal angles in similar triangles.

cube [11.4] A cube is a solid in which all six faces are squares.

cylinder [11.4] A cylinder is a solid in which the bases are circles and are perpendicular to the height.

decimal [3.1] A decimal is a number written with three parts: a whole number, the decimal point, and a fraction whose denominator is a power of 10.

decimal place [3.1] The decimal places are the places to the right of the decimal point.

denominator [2.2] The number below the fraction line in a fraction is called the denominator. It stands for the number of parts into which the whole is divided.

diameter [5.1, 11.1] A diameter is a line segment that passes through the center of a circle and has both endpoints on the circle.

difference [1.2] The result of a subtraction problem is called the difference.

digits [1.1] Digits are the numbers 0, 1, 2, 3, 4, 5, 6, 7, 8, and 9.

discount [6.3] In buying or selling merchandise, the term *discount* refers to a reduction on the merchandise's original price.

distributive property [1.3] The distributive property states that multiplying a factor by the sum of two numbers gives the same result as multiplying the factor by each of the two numbers and then adding.

dividend [1.4] In a division problem, the number being divided is called the dividend.

divisor [1.4] In a division problem, the number that is being used to divide another number is called the divisor.

equation [4.2] An equation is a mathematical statement that two expressions are equal.

equilateral triangle [11.1] An equilateral triangle is a triangle with three sides equal in length.

equivalent fractions [2.2] Equivalent fractions are fractions that represent the same quantity.

evaluate [4.1] To evaluate an algebraic expression, substitute the given value for each variable and carry out the computation.

exponent (or power) [1.5] An exponent (or power) is a number that indicates how many times another number is used as a factor.

exponential form [1.5] Exponential form is a shorthand method for representing a repeated multiplication of the same factor.

factors [1.3, 2.1] In a multiplication problem, the numbers being multiplied are called factors.

formula [9.3] A formula is an equation that indicates how the variables are related to one another.

fraction [2.2] A fraction is any number that can be written in the form $\frac{a}{b}$, where a and b are whole numbers and b is nonzero.

fraction line [2.2] The fraction line separates the numerator from the denominator and stands for "out of" or "divided by."

graph [8.2] A graph is a picture or diagram of the data in a table.

gram [10.2] A unit of weight or, more precisely, of mass.

hexagon [11.1] A hexagon is a polygon with six sides and six angles.

histogram [8.2] A histogram is a graph of a frequency table.

hypotenuse [11.6] In a right triangle, the hypotenuse is the side opposite the right angle.

identity property of addition [1.2] The identity property of addition states that the sum of a number and zero is the original number.

identity property of multiplication [1.3] The identity property of multiplication states that the product of any number and 1 is that number.

improper fraction [2.2] An improper fraction is a fraction greater than or equal to 1, that is, a fraction whose numerator is greater than or equal to its denominator.

integers [7.1] The integers are the numbers $\ldots, -4, -3, -2, -1, 0, +1, +2, +3, +4, \ldots,$ continuing indefinitely in both directions.

intersecting lines [11.1] Intersecting lines are two lines that cross.

isosceles triangle [11.1] An isosceles triangle is a triangle with two or more sides equal in length.

least common denominator (LCD) [2.2] The least common denominator (LCD) for two or more fractions is the least common multiple of their denominators.

least common multiple (LCM) [2.1] The least common multiple (LCM) of two or more whole numbers is the smallest nonzero whole number that is a multiple of each number.

legs [11.6] In a right triangle, the legs are the two sides that form the right angle.

like fractions [2.2] Like fractions are fractions with the same denominator.

like quantities [5.1] Like quantities are quantities that have the same unit.

like terms [9.2] Like terms are terms that have the same variables with the same exponents.

line [11.1] A line is a collection of points along a straight path that extends endlessly in both directions. A line has only one dimension.

line graph (broken-line graph) [8.2] A line graph (broken-line graph) is a graph in which quantities are represented as points connected by straight line segments. The position of any point on a line graph is read against the vertical axis and the horizontal axis.

line segment [11.1] A line segment is part of a line having two endpoints. Every line segment has a length.

liter [10.2] A unit of capacity, that is, of liquid volume.

magic square [1.2] A magic square is a square array of numbers in which the sum of every row, column, and diagonal is the same number.

mean (average) [1.5, 8.1] The mean of a set of numbers is the sum of those numbers divided by however many numbers are in the set.

median [8.1] In a set of numbers arranged in numerical order, the median of the numbers is the number in the middle. If there are two numbers in the middle, the median is the mean of the two middle numbers.

meter [10.2] A unit of length, which gives the metric system its name.

minuend [1.2] In a subtraction problem, the number from which we subtract from is called the minuend.

mixed number [2.2] A mixed number is a number greater than 1 with a whole number part and a proper fraction part.

mode [8.1] The mode of a set of numbers is the number (or numbers) occurring most frequently in the set.

multiplication property of 0 [1.3] The multiplication property of 0 states that the product of any number and 0 is 0.

negative number [7.1] A negative number is a number less than 0.

numerator [2.2] The number above the fraction line in a fraction is called the numerator. It tells us how many parts of the whole the fraction contains.

obtuse angle [11.1] An obtuse angle is an angle whose measure is more than 90° and less than 180°.

obtuse triangle [11.1] An obtuse triangle is a triangle with one obtuse angle.

opposites [7.1] Two numbers that are the same distance from 0 on the number line but on opposite sides of 0 are called opposites.

parallel lines [11.1] Parallel lines are two lines in the same plane that do not intersect.

parallelogram [11.1] A parallelogram is a quadrilateral with both pairs of opposite sides parallel. Opposite sides are equal in length, and opposite angles have equal measures.

percent (or rate) [6.1] A percent is a ratio or fraction with denominator 100. A number written with the % sign means "divided by 100."

percent decrease [6.3] In a percent problem, if the quantity is decreasing, it is called a percent decrease.

percent increase [6.3] In a percent problem, if the quantity is increasing, it is called a percent increase.

perfect square [1.5, 11.6] A perfect square is a number that is the square of any whole number.

perimeter [1.2, 11.2] The perimeter of a polygon is the distance around it.

period [1.1] A period is a group of three digits, which are separated by commas, when writing a large whole number in standard form.

perpendicular lines [11.1] Perpendicular lines are two lines that intersect to form right angles.

pictograph [8.2] A pictograph is a kind of graph in which images such as people, books, coins, etc., are used to represent and to compare quantities.

place value [1.1] Each of the digits in a whole number in standard form has place value.

plane [11.1] A plane is a flat surface that extends endlessly in all directions.

point [11.1] A point is an exact location in space. A point has no dimension.

polygon [11.1] A polygon is a closed plane figure made up of line segments.

positive number [7.1] A positive number is a number greater than 0.

prime factorization [2.1] Prime factorization of a whole number is the number written as the product of its prime factors.

prime number [2.1] A prime number is a whole number that has exactly two different factors: itself and 1.

principal [6.3] The principal is the amount of money borrowed.

product [1.3] The result of a multiplication problem is called the product.

proper fraction [2.2] A proper fraction is a fraction less than 1, that is, a fraction whose numerator is smaller than its denominator.

proportion [5.2] A proportion is a statement that two ratios are equal.

Pythagorean theorem [11.6] The Pythagorean theorem states that for every right triangle, the sum of the squares of the lengths of the two legs equals the square of the length of the hypotenuse. That is, $a^2 + b^2 = c^2$, where a and b are the legs, and c is the hypotenuse.

quadrilateral [11.1] A quadrilateral is a polygon with four sides.

quotient [1.4] The result of a division problem is called the quotient.

radius [11.1] A radius is a line segment with one endpoint on the circle and the other at the center.

range [8.1] The range of a set of numbers is the difference between the largest and the smallest number in the set.

rate [5.1] A rate is a ratio of unlike quantities.

ratio [5.1] A ratio is a comparison of two quantities expressed as a quotient.

ray [11.1] A ray is a part of a line having only one endpoint.

reciprocal [2.4] The reciprocal of the fraction $\frac{a}{b}$ is $\frac{b}{a}$.

rectangle [11.1] A rectangle is a parallelogram with four right angles.

rectangular solid [11.4] A rectangular solid is a solid in which all six faces are rectangles.

right angle [11.1] A right angle is an angle whose measure is 90°.

right triangle [11.1] A right triangle is a triangle with one right angle.

GLOSSARY

rounding [1.1] Rounding is the process of approximating an exact answer by a number that ends in a given number of zeros.

scalene triangle [11.1] A scalene triangle is a triangle with no sides equal in length.

signed number [7.1] A signed number is a number with a sign that is either positive or negative.

similar triangles [11.5] Similar triangles are triangles that have the same shape but not necessarily the same size.

simplified form (or written in lowest terms) [2.2] A fraction is said to be in simplified form when the only common factor of its numerator and denominator is 1.

solution [9.1] A solution to an equation is a value of the variable that makes the equation a true statement.

solve [9.1] To solve an equation means to find all solutions of the equation.

sphere [11.4] A sphere is a three-dimensional figure made up of all points a given distance from the center.

square [11.1] A square is a rectangle with four sides equal in length.

square root [11.6] The (principal) square root of a number n, written $\sqrt{n}$, is the positive number whose square is n.

statistics [8.1] Statistics is the branch of mathematics that deals with ways of handling large quantities of information.

straight angle [11.1] A straight angle is an angle whose measure is 180°.

subtrahend [1.2] In a subtraction problem, the number that is being subtracted is called the subtrahend.

sum [1.2] The result of an addition problem is called the sum.

supplementary angles [11.1] Two angles are supplementary if the sum of their measures is 180°.

table [8.2] A table is a rectangular display of data.

tessellation [11.1] A tessellation is any repeating pattern of interlocking shapes.

trapezoid [11.1] A trapezoid is a quadrilateral with only one pair of opposite sides parallel.

triangle [11.1] A triangle is a polygon with three sides.

unit fraction [2.3] A fraction with 1 as the numerator is called a unit fraction.

unit price [5.1] The unit price is the price of one item, or one unit.

unit rate [5.1] A unit rate is a rate in which the number in the denominator is 1.

unlike fractions [2.2] Unlike fractions are fractions with different denominators.

unlike quantities [5.1] Unlike quantities are quantities that have different units.

unlike terms [9.2] Unlike terms are terms that do not have the same variables with the same exponents.

variable [4.1] A variable is a letter that represents an unknown number.

vertex [11.1] A vertex is the common endpoint of an angle.

vertical angles [11.1] Vertical angles are two opposite angles with equal measure formed by two intersecting lines.

volume [11.4] Volume is the number of cubic units required to fill a three-dimensional figure.

weighted average [8.1] A weighted average is a special kind of average (mean) used when some numbers in a set count more heavily than others.

written in lowest terms (or simplified) [2.2] A fraction is said to be written in lowest terms when the only common factor of its numerator and denominator is 1.

GLOSSARY

Index

INDEX

INDEX

U.S. Customary Units

Length	Weight	Capacity	Time
12 in. = 1 ft	16 oz = 1 lb	16 fl oz = 1 pt	60 sec = 1 min
3 ft = 1 yd	2,000 lb = 1 ton	2 pt = 1 qt	60 min = 1 hr
5,280 ft = 1 mi		4 qt = 1 gal	24 hr = 1 day
			7 days = 1 wk
			52 wk = 1 yr
			12 mo = 1 yr
			365 days = 1 yr

Metric Units

Length	Weight	Capacity
1,000 mm = 1 m	1,000 mg = 1 g	1,000 mL = 1 L
100 cm = 1 m	1,000 g = 1 kg	1,000 L = 1 kL
1,000 m = 1 km		

Key U.S./Metric Conversions

Length	Weight	Capacity
1 in. ≈ 2.5 cm	1 oz ≈ 28 g	1 pt ≈ 470 mL
1 ft ≈ 30 cm	1 lb ≈ 450 g	1.1 qt ≈ 1 L
39 in. ≈ 1 m	2.2 lb ≈ 1 kg	1 gal ≈ 3.8 L
3.3 ft ≈ 1 m	1 ton ≈ 910 kg	
1 mi ≈ 1,600 m		
1 mi ≈ 1.6 km		